Zhong Li, Wolfgang A. Halang, Guanrong Chen (Eds.)

Integration of Fuzzy Logic and Chaos Theory

T0205263

Studies in Fuzziness and Soft Computing, Volume 187

Editor-in-chief
Prof. Janusz Kacprzyk
Systems Research Institute
Polish Academy of Sciences
ul. Newelska 6
01-447 Warsaw
Poland
E-mail: kacprzyk@ibspan.waw.pl

Further volumes of this series
can be found on our homepage:
springer.com

Vol. 173. Bogdan Gabrys, Kauko Leiviskä,
Jens Strackeljan (Eds.)
Do Smart Adaptive Systems Exist?, 2005
ISBN 3-540-24077-2

Vol. 174. Mircea Negoita, Daniel Neagu,
Vasile Palade
*Computational Intelligence: Engineering of
Hybrid Systems*, 2005
ISBN 3-540-23219-2

Vol. 175. Anna Maria Gil-Lafuente
Fuzzy Logic in Financial Analysis, 2005
ISBN 3-540-23213-3

Vol. 176. Udo Seiffert, Lakhmi C. Jain,
Patric Schweizer (Eds.)
*Bioinformatics Using Computational
Intelligence Paradigms*, 2005
ISBN 3-540-22901-9

Vol. 177. Lipo Wang (Ed.)
*Support Vector Machines: Theory and
Applications*, 2005
ISBN 3-540-24388-7

Vol. 178. Claude Ghaoui, Mitu Jain,
Vivek Bannore, Lakhmi C. Jain (Eds.)
Knowledge-Based Virtual Education, 2005
ISBN 3-540-25045-X

Vol. 179. Mircea Negoita,
Bernd Reusch (Eds.)
*Real World Applications of Computational
Intelligence*, 2005
ISBN 3-540-25006-9

Vol. 180. Wesley Chu,
Tsau Young Lin (Eds.)
Foundations and Advances in Data Mining,
2005
ISBN 3-540-25057-3

Vol. 181. Nadia Nedjah,
Luiza de Macedo Mourelle
Fuzzy Systems Engineering, 2005
ISBN 3-540-25322-X

Vol. 182. John N. Mordeson,
Kiran R. Bhutani, Azriel Rosenfeld
Fuzzy Group Theory, 2005
ISBN 3-540-25072-7

Vol. 183. Larry Bull, Tim Kovacs (Eds.)
Foundations of Learning Classifier Systems,
2005
ISBN 3-540-25073-5

Vol. 184. Barry G. Silverman, Ashlesha Jain,
Ajita Ichalkaranje, Lakhmi C. Jain (Eds.)
*Intelligent Paradigms for Healthcare
Enterprises*, 2005
ISBN 3-540-22903-5

Vol. 185. Spiros Sirmakessis (Ed.)
Knowledge Mining, 2005
ISBN 3-540-25070-0

Vol. 186. Radim Bělohlávek, Vilém
Vychodil
Fuzzy Equational Logic, 2005
ISBN 3-540-26254-7

Vol. 187. Zhong Li, Wolfgang A. Halang,
Guanrong Chen
*Integration of Fuzzy Logic and Chaos
Theory*, 2006
ISBN 3-540-26899-5

Zhong Li
Wolfgang A. Halang
Guanrong Chen
(Eds.)

Integration of Fuzzy Logic and Chaos Theory

 Springer

Dr. Zhong Li
Professor Wolfgang A. Halang
FernUniversität in Hagen
FB Elektrotechnik
Postfach 940, 55084 Hagen
Germany
E-mail: zhong.li@fer_n_uni-hagen.de
        wolfgang.halang@fernuni-hagen.de

Professor Guanrong Chen
Department of Electronic Engineering
City University of Hong Kong
Tat Chee Avenue, Kowloon
Hong Kong/PR China
E-mail: gchen@ee.cityu.edu.hk

ISBN 978-3-642-06594-1   e-ISBN 978-3-540-32502-4

ISSN print edition: 1434-9922
ISSN electronic edition: 1860-0808

Springer is a part of Springer Science+Business Media
springer.com
© Springer-Verlag Berlin Heidelberg 2006
Softcover reprint of the hardcover 1st edition 2006

Printed on acid-free paper

Preface

The 1960s were perhaps a decade of confusion, when scientists faced difficulties in dealing with imprecise information and complex dynamics. A new set theory and then an infinite-valued logic of Lotfi A. Zadeh were so confusing that they were called fuzzy set theory and fuzzy logic; a deterministic system found by E. N. Lorenz to have random behaviours was so unusual that it was lately named a chaotic system. Just like irrational and imaginary numbers, negative energy, anti-matter, etc., fuzzy logic and chaos were gradually and eventually accepted by many, if not all, scientists and engineers as fundamental concepts, theories, as well as technologies.

In particular, fuzzy systems technology has achieved its maturity with widespread applications in many industrial, commercial, and technical fields, ranging from control, automation, and artificial intelligence to image/signal processing, pattern recognition, and electronic commerce. Chaos, on the other hand, was considered one of the three monumental discoveries of the twentieth century together with the theory of relativity and quantum mechanics. As a very special nonlinear dynamical phenomenon, chaos has reached its current outstanding status from being merely a scientific curiosity in the mid-1960s to an applicable technology in the late 1990s.

Finding the intrinsic relation between fuzzy logic and chaos theory is certainly of significant interest and of potential importance. The past 20 years have indeed witnessed some serious explorations of the interactions between fuzzy logic and chaos theory, leading to such research topics as fuzzy modeling of chaotic systems using Takagi–Sugeno models, linguistic descriptions of chaotic systems, fuzzy control of chaos, and a combination of fuzzy control technology and chaos theory for various engineering practices.

A deep-seated reason to study the interactions between fuzzy logic and chaos theory is that they are related at least within the context of human reasoning and information processing. In fact, fuzzy logic resembles human approximate reasoning using imprecise and incomplete information with inaccurate and even self-conflicting data to generate reasonable decisions under such uncertain environments, while chaotic dynamics play a key role in human brains for processing massive amounts of information instantly. It is believed that the capability of humans in controlling chaotic dynamics in their brains is more than just an accidental by-product of the brain's complexity, but rather, it could be the chief property that makes the human brain different from any artificial-intelligence machines. It is also believed that to understand the complex information processing within the human brain, fuzzy data and fuzzy logical inference are essential, since precise mathematical descriptions of such models and processes are clearly out of question with today's limited scientific knowledge.

With this book we attempt to present some current research progress and results on the interplay of fuzzy logic and chaos theory. More specifically, in this book we collect some state-of-the-art surveys, tutorials, and application examples written by some experts working in the interdisciplinary fields overlapping fuzzy logic and chaos theory. The content of the book covers fuzzy definition of chaos, fuzzy modeling and control of chaotic systems using both Mamdani and Takagi–Sugeno models, fuzzy model identification using genetic algorithms and neural network schemes, bifurcation phenomena and self-referencing in fuzzy systems, complex fuzzy systems and their collective behaviors, as well as some applications of combining fuzzy logic and chaotic dynamics, such as fuzzy–chaos hybrid controllers for nonlinear dynamic systems, and fuzzy model based chaotic cryptosystems.

It is our hope that this book can serve as a handy reference for researchers working in the interdisciplines related, among others, to both fuzzy logic and chaos theory.

We would like to thank all authors for their significant contributions, without which the publication of this book would have not been possible. We are very grateful to Prof. Janusz Kacprzyk for recommending this book to the Springer series, Studies in Fuzziness and Soft Computing, with appreciation going to the editorial and production staff of Springer-Verlag in Heidelberg for their fine work and kind cooperation.

May 2005   *Zhong Li*
Wolfgang A. Halang
Guanrong Chen

Contents

Beyond the Li–Yorke Definition of Chaos

Peter Kloeden and Zhong Li

Abstract. Extensions of the well-known definition of chaos due to Li and Yorke for difference equations in $\mathbb{R}^1$ are reviewed for difference equations in $\mathbb{R}^n$ with either a snap-back repeller or saddle point as well as for mappings in Banach spaces and complete metric spaces. A further extension applicable to mappings in a space of fuzzy sets, namely the metric space (ξ^n, D) of fuzzy sets on the base space $\mathbb{R}^n$, is then discussed and some illustrative examples are presented. The aim is to provide a theoretical foundation for further studies on the interaction between fuzzy logic and chaos theory.

1 Introduction

Chaos may well be considered together with relativity and quantum mechanics as one of the three monumental discoveries of the twentieth century. Over the past four decades chaos has matured as a science (though is still evolving) and has given us deep insights into previously intractable and inherently non-linear natural phenomena. The term *chaos* associated with an interval map was first formally introduced into mathematics by Li and Yorke in 1975 [1], where they established a simple criterion for chaos in one-dimensional difference equations, i.e., the well-known "period three implies chaos."

There is, however, still no unified, universally accepted, and rigorous mathematical definition of chaos in the scientific literature to provide a fundamental basis for studying such exotic phenomena. Various alternative, but closely related definitions of chaos have been proposed, among which those of Li–Yorke and Devaney seem to be the most popular.

Consider a one-dimensional discrete dynamical system [1, 2]:

$$x_{k+1} = f(x_k), \quad k = 0, 1, 2, \ldots, \tag{1}$$

where $x_k \in J$ (an interval) and $f : J \to J$ is a continuous mapping. For $x \in J$, $f^0(x)$ denotes x, while $f^{n+1}(x)$ denotes $f(f^n(x))$ for $n = 0, 1, 2, \ldots$. A point x^* is called a *period point with period n* (or an *n-period point*) if $x^* \in J$ and $x^* = f^n(x^*)$ but $x^* \neq f^k(x^*)$ for $1 \leq k < n$ and if $n = 1$, then $x^* = f(x^*)$ is called a *fixed point*. A point x^* is said to be *periodic* or

P. Kloeden and Z. Li: *Beyond the Li–Yorke Definition of Chaos*, StudFuzz **187**, 1–23 (2006)
www.springerlink.com

is called a *periodic point* if it is an n-periodic point for some $n \geq 1$. With this terminology, Li and Yorke introduced the first mathematical definition of chaos and established a very simple criterion, i.e., "period three implies chaos" for its existence. This criterion, which plays an key role in predicting and analyzing one-dimensional chaotic dynamic systems, was described by Li and Yorke as follows:

Theorem 1 (Li–Yorke Theorem) *Let J be an interval and $f : J \to J$ be continuous. Assume that there is one point $a \in J$, for which the points $b = f(a)$, $c = f^2(a)$, and $d = f^3(a)$ satisfy*

$$d \leq a < b < c \ (or \ d \geq a > b > c).$$

Then

(i) for every $k = 1, 2, \ldots$, there is a k-periodic point in J.
(ii) there is an uncountable set $S \subset J$, containing no periodic points, which
satisfies the following conditions:
(a) For every $p_s, q_s \in S$ with $p_s \neq q_s$,

$$\limsup_{n \to \infty} |f^n(p_s) - f^n(q_s)| > 0$$

and

$$\liminf_{n \to \infty} |f^n(p_s) - f^n(q_s)| = 0.$$

(b) For every $p_s \in S$ and periodic points $q_{per} \in J$, with $p_s \neq q_{per}$,

$$\limsup_{n \to \infty} |f^n(p_s) - f^n(q_{per})| > 0.$$

The set S in part (a) of conclusion (ii) was called a *a scrambled set* by Li and Yorke.

The first part of the Li–Yorke theorem is, in fact, a special case of Sharkovsky's theorem [3], which was proved by the Ukrainian mathematician A.N. Sharkovsky in 1964. It is, however, the second part of the Li–Yorke theorem that thoroughly unveils the nature and characteristics of chaos, specifically, the sensitive dependence on initial conditions and the resulting unpredictable nature of the long-term behavior of the dynamics.

In 1978 F.R. Marotto generalized the Li–Yorke theorem to higher dimensional discrete dynamical systems [4]. He proved that if a difference equation in $\mathbb{R}^n$ has a snap-back repeller, then it has a scrambled set similar to that defined in the Li–Yorke theorem and thus exhibits chaotic behavior.

Consider the following n-dimensional system:

$$x_{k+1} = f(x_k), \quad k = 0, 1, 2, \ldots, \tag{2}$$

where $x_k \in \mathbb{R}^n$ and $f : \mathbb{R}^n \to \mathbb{R}^n$ is a continuous mapping, which is usually nonlinear. Denote by $B_r(x)$ the closed ball in $\mathbb{R}^n$ of radius r centered at point x, and by $B_r^0(x)$ its interior. Also, let $\|x\|$ be the usual Euclidean norm of x in $\mathbb{R}^n$. Then, assuming f to be differentiable in $B_r(x)$, Marotto claimed that the logical relationship $A \Rightarrow B$ ($\Rightarrow$ means "implying") holds, where

(a) *all eigenvalues of the Jacobian $Df(z)$ of system (2) at the fixed point z*
    $= f(z)$ are greater than 1 in norm.
(b) *there exist some $s > 1$ and $r > 0$ such that $\|f(x) - f(y)\| > s\|x - y\|$*
    for all $x, y \in B_r(z)$.

In other words, if (a) is satisfied, then (b) also holds, i.e., f is expanding in $B_r(z)$. Then, Marotto introduced the following concepts.

Definition 1. (Marotto Definitions)

(1) *Expanding fixed point: Let f be differentiable in $B_r(z)$. The point $z \in \mathbb{R}^n$*
    is an expanding fixed point of f in $B_r(z)$ if $f(z) = z$ and all eigenvalues
    of $Df(x)$ exceed 1 in norm for all $x \in B_r(z)$.
(2) *Snap-back repeller: Assume that z is an expanding fixed point of f in*
    $B_r(z)$ for some $r > 0$. Then z is said to be a snap-back repeller of f
    if there exists a point $x_0 \in B_r(z)$ with $x_0 \neq z$, $f^M(x_0) = z$ and the
    determinant $|Df^M(x_0)| \neq 0$ for some positive integer M.

Marotto showed that the presence of a snap-back repeller is a sufficient criterion for the existence of chaos [4].

Theorem 2 (Marotto Theorem) *If f possesses a snap-back repeller, then system (2) is chaotic in the following generalized sense of Li–Yorke:*

(i) *There is a positive integer N such that for each integer $p \geq N$, f has a*
    point of period p.
(ii) *There is a "scrambled set" of f, i.e., an uncountable set S containing*
    no periodic points of f, such that
    (a) *$f(S) \subset S$.*
    (b) *for every $x_s, y_s \in S$ with $x_s \neq y_s$,*

$$\limsup_{k \to \infty} \|f^k(x_s) - f^k(y_s)\| > 0 .$$

(c) *for every $x_s \in S$ and any periodic point y_{per} of f,*

$$\limsup_{k \to \infty} \|f^k(x_s) - f^k(y_{per})\| > 0 .$$

(iii) *There is an uncountable subset S_0 of S such that for every $x_0, y_0 \in S_0$:*

$$\liminf_{k \to \infty} \|f^k(x_0) - f^k(y_0)\| = 0 .$$

It is apparent that the existence of a snap-back repeller for the one-dimensional mapping f is equivalent to the existence of a point of period-3 for the map f^n for some positive integer n, see [4].

Unfortunately, two counterexamples have been given in [2, 5] to show that $A \Rightarrow B$ is not necessarily true. Since the Marotto theorem is based on the concept of "snap-back repeller," which was introduced from the assertion of $A \Rightarrow B$, there exists an error in the proof given by Marotto. Recently, an improved and corrected version of Marotto's theorem was given by Li and Chen [2], where the essential meanings of the two concepts of an expanding fixed point and a snap-back repeller of continuously differentiable maps in $\mathbb{R}^n$ are clearly explained. For an earlier generalization of the Marroto theorem see [6] and for an extension to maps in metric spaces see [7] as well as below.

More generally, Devaney [8] calls a continuous map $f : X \to X$ in a metric space (X, d) *chaotic on X*, if

(i) f is transitive on X: for any pair of nonempty open sets $U, V \subset X$, there exists an integer $k > 0$ such that $f^k(U) \cap V$ is nonempty;
(ii) the periodic points of f are dense in X;
(iii) f has sensitive dependence on initial conditions: if there exists a $\delta > 0$ such that for any $x \in X$ and for any neighborhood D of x, there exists a $y \in D$ and an $k \geq 1$ such that $d(f^k(x), f^k(y)) > \delta$.

It has been observed that conditions (i) and (ii) in this definition imply condition (iii) if X is not a finite set [9] and that condition (i) implies conditions (ii) and (iii) if X is an interval [10]. Hence, condition (iii) is in fact redundant in the above definition.

For continuous time nonlinear autonomous systems it is much more difficult to give a mathematically rigorous proof to the existence of chaos. Even one of the classic icons of modern nonlinear dynamics, the Lorenz attractor, now known for 40 years, was not proved rigorously to be chaotic until 1999. Warwick Tucker of the University of Uppsala showed in his Ph.D. dissertation [11, 12], using normal form theory and careful computer simulations, that Lorenz equations do indeed possess a robust chaotic attractor. A commonly agreed analytic criterion for proving the existence of chaos in continuous time systems is based on the fundamental work of Shil'nikov, known as the Shil'nikov method or Shil'nikov criterion [13], whose role is in some sense equivalent to that of the Li–Yorke definition in the discrete setting. The Shil'nikov criterion guarantees that complex dynamics will occur near homoclinicity or heteroclinicity when an inequality (Shil'nikov inequality) is satisfied between the eigenvalues of the linearized flow around the saddle point(s), i.e., if the real eigenvalue is larger in modulus than the real part of the complex eigenvalue. Complex behavior always occurs when the saddle set is a limit cycle.

In this chapter we focus on discrete-time systems and discuss generalizations of Marotto's work, which are applicable to finite dimensional difference equations with saddle points as well as to those with repellers and to mappings in Banach spaces and in complete metric spaces including mappings from a metric space of fuzzy sets into itself.

2 Background

Consider the successive iterates $f^{k+1} = f^k \circ f$ of a mapping f from a topological space X into itself and sequences of points

$$x_{k+1} = f(x_k), \quad k = 0, 1, 2, \ldots \tag{3}$$

in X generated by such a mapping. Traditionally, research interest has focused on the regular asymptotical behavior of the sequences $x_0, x_1, x_2, \ldots, x_k = f^k(x_0), \ldots$, and in particular on conditions that ensure the existence of an asymptotically stable equilibrium point $\bar{x} = f(\bar{x})$ or of an asymptotically stable cycle $\bar{x}_2 = f(\bar{x}_1), \ldots, \bar{x}_p = f(\bar{x}_{p-1})$, $\bar{x}_1 = f(\bar{x}_p)$ for some period $p > 1$. (Such cyclic or periodic points $\bar{x}_j$ are fixed points of f^p.)

Over the years attention has turned to the investigation of the chaotic behavior of such iterated sequences [14–18]. This is readily seen in the simple logistic equation

$$x_{k+1} = 4x_k(1 - x_k) \quad \text{for } 0 \le x \le 1, \tag{4}$$

which describes the dynamics of a population with nonoverlapping generations, and in the Baker's equation

$$x_{k+1} = \begin{cases} 2x_k & \text{for } 0 \le x_k \le \frac{1}{2} \\ 2(1 - x_k) & \text{for } \frac{1}{2} < x_k \le 1, \end{cases} \tag{5}$$

which models the mixing of a dye spot on a strip of dough that is repeatedly stretched and folded over on itself. Both of the iterative schemes (4) and (5) involve mappings of the unit interval I into itself and display the highly irregular or chaotic behavior in the sense of Li and Yorke.

The logistic equation has the characteristic feature of all such chaotic difference equations in that the graph of f has a *hump* in it or *folds over* on itself. Actually, variations of the Li–Yorke result had appeared some years before Li and Yorke [1], namely, Barna [19] and Sharkovsky [20, 21]. The work of Sharkovsky is the most complete and far reaching in this regard. Of particular significance is his *cycle coexistence ordering*, which says that if a one-dimensional difference equation (3) has a cycle of period p then it also has a cycle of period p' when $p \prec p'$ in the following *Sharkovsky ordering:*

$$3 \prec 5 \prec 7 \cdots \prec 2 \cdot 3 \prec 2 \cdot 5 \prec 2 \cdot 7 \prec \cdots \tag{6}$$
$$\prec 2^k \cdot 3 \prec 2^k \cdot 5 \prec 2^k \cdot 7 \prec \cdots$$
$$\prec 2^n \prec 2^{n-1} \prec \cdots \prec 2 \prec 1.$$

Consequently, if f has only finitely many periodic points, then they must have all periods which are powers of two. Furthermore, if there is a periodic point of period 3, then there are periodic points of all other periods.

Sharkovsky's theorem does not state that there are stable cycles of those periods, just that there are cycles of those periods. For systems such as the logistic map, bifurcation diagrams show a range of parameter values for which apparently the only cycle has period 3. In fact, there must be cycles of all periods there, but they are not stable and therefore usually not visible on the computer-generated picture.

Interestingly, the above ordering of the positive integers also occurs in a slightly different manner in connection with the logistic map: the stable cycles appear in this order in the bifurcation diagram, starting with 1 and ending with 3, as the parameter is increased.

Furthermore, from this ordering and from other results of Sharkovsky or the Li and Yorke result it follows [22] that a scalar difference equation (3) behaves chaotically if it has a cycle of period $(2l + 1)2^k$ for some $l \geq 1$ and $k \geq 0$, for then the iterate $f^{2^{k+1}}$ has a cycle of period 3.

For the difference equation (2) defined in terms of a mapping $f : X \rightarrow X$, where X is a closed subset of $\mathbb{R}^n$ for $n \geq 2$, the properties of the iterated sequences and cycles are not as well understood as for their one-dimensional counterparts. However, it has been long known from the numerical calculations of Stein and Ulam [23] for 2-dimensional difference equations defined in terms of piecewise linear mappings that higher dimensional difference equations can display quite complicated, seemingly chaotic behavior. Moreover, on the theoretical level, difference equations defined in terms of Smale's horseshoe mapping [24] are known to have infinitely many cycles of different periods and something similar to the scrambled set above.

What is of research interest is that to what extent the one-dimensional results of Li and Yorke [1] and Sharkovsky [20, 21] carry over to higher dimensional difference equations. The following example shows that they will not without some modification or some restriction to the class of mappings f. To see this, consider the difference equation defined in terms of the rigid rotation mapping

$$f(x_1, x_2) = \begin{pmatrix} f_1(x_1, x_2) \\ f_2(x_1, x_2) \end{pmatrix} = \begin{pmatrix} -\frac{1}{2}x_1 - \frac{\sqrt{3}}{2}x_2 \\ \frac{\sqrt{3}}{2}x_1 - \frac{1}{2}x_2 \end{pmatrix} \qquad (7)$$

of the unit disc $X = \{(x_1, x_2) \in \mathbb{R}^2 \ x_1^2 + x_2^2 \leq 1\}$ into itself. Then each point $(x_1, x_2) \in X \setminus (0,0)$ belongs to a cycle of period 3, whereas $(0,0)$ belongs to a cycle of period one. This is obvious in terms of the complex variable $z = x_1 + ix_2$ in which case the mapping f can be written as $f(z) = az$, where $a = -\frac{1}{2} + i\frac{\sqrt{3}}{2}$. For this example neither Sharkovsky's cycle coexistence ordering (6) nor the "period three implies chaos" result of Li and Yorke is valid.

However, it is possible to generate the one-dimensional results subject to suitably restricting the mapping f. For example, Kloeden [25] has shown that the Sharkovsky cycle coexistence ordering (6) holds for triangular difference equations (2) where the continuous mapping f is defined on a compact n-dimensional rectangle $X = \prod_{i=1}^{n}[a_i, b_i]$ with the ith component of f depending only on the first i components of the vector $\mathbf{x} = (x_1, x_2, \ldots, x_n)$, i.e.,

$$f_i(\mathbf{x}) = f_i(x_1, x_2, \ldots, x_i), \tag{8}$$

for $i = 1, 2, \ldots, n$. These difference equations include the one-dimensional equations considered by Sharkovsky and also some important higher dimensional equations such as the *twisted horseshoe* difference equation of Guckenheimer et al. [14], for which $X = [0, 1]^2$, the unit square, and the mapping f has components

$$\left. \begin{array}{l} f_1(x_1) = \left\{ \begin{array}{ll} 2x_1 & \text{for } 0 \leq x_1 \leq \frac{1}{2} \\ 2 - 2x_1 & \text{for } \frac{1}{2} < x_1 \leq 1 \end{array} \right\} \\ f_2(x_1, x_2) = \frac{1}{2}x_1 + \frac{1}{10}x_2 + \frac{1}{4} \quad \text{for } 0 \leq x_1, x_2 \leq 1 \end{array} \right\}. \tag{9}$$

To determine a suitable class of mappings for which the results of the Li and Yorke type might hold in higher dimensions, it is noted that the one-dimensional mappings for which the difference equation (3) have cycles of period 3 all have graphs, which have a hump or fold over on themselves, namely, are not one to one mappings. Note also that the two-dimensional difference equation with the linear mapping (7), which is a one to one mapping, has cycles of period 3, but does not behave chaotically. This suggests that attention might profitably be restricted to mappings which are not one to one. (This is not the only possible approach as both the Smale horseshoe mapping [21] and the Hénnon mapping [26] are diffeomeorphisms, but have difference equations that behave chaotically). This was done by Marotto [4] who showed that difference equations on $\mathbb{R}^n$ defined in terms of continuously differentiable mappings with *snap-back repellers*, so consequently not one to one, behave chaotically in the sense of Li and Yorke. His proof used the inverse function theorem for one to one local restrictions of the mappings and the Brouwer fixed point theorem, but otherwise paralleled the proof of Li and Yorke for one-dimensional mappings.

3 Chaos of Difference Equations in $\mathbb{R}^n$ with a Saddle Point

The Marotto theorem says that a difference equations in $\mathbb{R}^n$ with a snap-back repeller behaves chaotically, and is thus a generalization of the Li–Yorke theorem for difference equations in $\mathbb{R}^1$. It differs in that the mappings defining the difference equations are required to be continuously differentiable rather than only continuous. Thus, the inverse mapping theorem and the Brouwer fixed point theorem can be used to prove the existence of continuous inverse functions and periodic points. Marotto's result is however applicable only to difference equations with repellers but not to those with saddle points. Therefore, it cannot be used for difference equations involving the horseshoe mappings of Smale [24] or the twisted-horseshoe mappings of Guckenheimer et al. [14].

In this section, sufficient conditions for the chaotic behavior of difference equations in $\mathbb{R}^n$ are given, which are applicable to difference equations with saddle points as well as to those with repellers. These conditions are valid for difference equations defined in terms of continuous mappings, and in the special case of a difference equation with a snap-back repeller, they are easier tested than those given by Marotto [4]. The proof is a modification of that used by Marotto, but there are two important differences. Firstly the mappings in the difference equations are assumed to be continuous rather than continuously differentiable. The existence of continuous inverse mapping follows from the fact that continuous one to one mappings have continuous inverses on compact sets. Using this result rather than the inverse mapping theorem considerably simplifies the proof. Secondly, the Brouwer fixed point theorem is used on a homeomorph of an l-ball for some $1 \leq l \leq n$ rather than on a homeomorph of an n-ball as in Marotto's proof. Thus, this allows saddle points to be considered as well as repellers.

3.1 Sufficient Conditions for Chaos in $\mathbb{R}^n$

In [15] a first-order difference equation

$$x_{k+1} = f(x_k) , \tag{10}$$

where $f : \mathbb{R}^n \to \mathbb{R}^n$ be a continuous mapping, was said to be *chaotic* if it is chaotic in the sense of the Marotto theorem, i.e., if there exist

(i) *a positive integer N such that (10) has a periodic point of period p for each $p \geq N$;*

(ii) *a scrambled set of (10) that is an uncountable set S containing no periodic points of (10) such that*
  (a) *$f(S) \subset S$,*
  (b) *for every $x_0, y_0 \in S$ with $x_0 \neq y_0$*

$$\limsup_{x \to \infty} \| f^k(x_0) - f^k(y_0) \| > 0 ,$$

  (c) *for every $x_0 \in S$ and any periodic point y_{per} of (10)*

$$\limsup_{x \to \infty} \| f^k(x_0) - f^k(y_{per}) \| > 0 ;$$

(iii) *an uncountable subset S_0 of S such that for every $x_0, y_0 \in S_0$:*

$$\liminf_{x \to \infty} \| f^k(x_0) - f^k(y_0) \| = 0 .$$

An l-ball is defined as a closed ball of finite radius in $\mathbb{R}^l$ in terms of the Euclidean distance on $\mathbb{R}^l$. Such a ball of radius r centred on a point $z_0 \in \mathbb{R}^l$ is denoted by $B^l(z_0; r)$. A mapping $f : \mathbb{R}^n \to \mathbb{R}^n$ is called *expanding* on a set $A \subset \mathbb{R}^n$ if there exists a constant $\lambda > 1$ such that

$$\lambda\|x - y\| \le \|f(x) - f(y)\| \tag{11}$$

for all $x, y \in A$. Note that such a mapping is one to one on A.

The following two lemmas will be used in the proof of the theorem below. The proof of the first one is straightforward and is thus omitted, while a proof of the second lemma can be found in [27].

Lemma 1. *Let $f : \mathbb{R}^n \to \mathbb{R}^n$ be a continuous mapping, which is one to one on a compact subset $K \subset \mathbb{R}^n$. Then there exists a continuous mapping $g : f(K) \to K$ such that $g(f(x)) = x$ for all $x \in K$.*

The mapping g in Lemma 1 is a continuous inverse of mapping f on the compact set K. It is denoted by f_K^{-1} in the sequel.

Lemma 2. *Let $f : \mathbb{R}^n \to \mathbb{R}^n$ be a continuous mapping and let $\{K_i\}_{i=0}^{\infty}$ be a sequence of compact sets in $\mathbb{R}^n$ such that $K_{i+1} \subseteq f(K_i)$ for $i = 0, 1, 2, \ldots$. Then there exists a nonempty compact set $K \subseteq K_0$ such that $f^i(x_0) \in K_i$ for all $x_0 \in K$ and all $i \ge 0$.*

The principal result in this section is the following generalization of the Marroto theorem due to Kloeden [15].

Theorem 3 (Kloeden) *Let $f : \mathbb{R}^n \to \mathbb{R}^n$ be a continuous mapping and suppose that there exist nonempty compact sets A and B, and integers $1 \le l \le n$ and $n_1, n_2 \ge 1$ such that*

(a) A is homeomorphic to an l-ball;
(b) $A \subseteq f(A)$;
(c) f is expanding on A;
(d) $B \subseteq A$;
(e) $f^{n_1}(B) \cap A = \emptyset$;
(f) $A \subseteq f^{n_1+n_2}(B)$;
(g) $f^{n_1+n_2}$ is one to one on B.

Then difference equation (10) defined in terms of the mapping f is chaotic in the sense of the Marotto theorem.

Proof. The proof is similar to that used by Marotto in [4], except that Lemma 1 is used instead of the inverse mapping theorem and the Brouwer fixed point theorem is used on homeomorphisms of l-balls rather than n-balls. The Brouwer fixed point theorem says that a smooth mapping from an n-dimensional closed ball into itself must have fixed point.

From the continuity of f and assumption (f) there exists a nonempty, compact subset $C \subseteq B$ such that $A = f^{n_1+n_2}(C)$. By (g) $f^{n_1+n_2}$ is one to one on C, and by Lemma 1 there exists a continuous function $g : A \to C$ such that $g(f^{n_1+n_2}(x)) = x$ for all $x \in C$. Note that $f^{n_1}(C) \cap A = \emptyset$ by (e).

Now f is one to one on A by (c), so by Lemma 1 f has a continuous inverse $f_A^{-1} : f(A) \to A$. By (b) $C \subset A \subseteq f(A)$, so $f_A^{-k}(C) \subset A$ holds for all $k \geq 0$.

For each $k \geq 0$, the mapping $f_A^{-k} \circ g : A \to A$ is a continuous mapping from a homeomorph of an l-ball into itself, so by the Brouwer fixed point theorem there exists a point $y_k \in A$ such that $f_A^{-k}(g(y_k)) = y_k$. In fact $y_k \in f^{-k}(C)$ and so $f^{n_1+k}(y_k) = f^{n_1+k}(f_A^{-k}(g(y_k))) = f^{n_1}(g(y_k)) \in f^{n_1}(C)$ as $g(y_k) \in C$. Hence $f^{n_1+n_k}(y_k) \notin A$ as $f^{n_1}(C) \cap A = \emptyset$. Also $f^{n_1+n_2+k}(y_k) = f^{n_1+n_2}(g(y_k)) = y_k$.

Now for $k \geq n_1 + n_2$ the point y_k is a periodic point of period $p = n_1 + n_2 + k$. To see this note that p cannot be less than or equal to k because $f^j(y_k) \in f_A^{-k+j}(C) \subset A$ for $1 \leq j \leq k$ and then the whole cycle would belong to A in contradiction to the fact that $f^{n_1+k}(y_k) \notin A$. Also p cannot lie between k and $n_1 + n_2 + k$ when $k \geq n_1 + n_2$ because $f^{n_1+n_2+k}(y_k) = y_k$ and so p would have to divide $n_1 + n_2 + k$ exactly, which is impossible when $k \geq n_1 + n_2$. Hence the difference equation (10) has a periodic point of period p for each $p \geq N = 2(n_1 + n_2)$.

Write $D = f^{n_1}(C)$ and $h = f^N$. Then $A \cap D = \emptyset$ and

$$h(D) = f^N(D) = f^{2n_1+n_2}(f^{n_2}(D)) = f^{2n_1+n_2}(A) \supseteq A \qquad (12)$$

in view of (b) and the definition of C. Also

$$h(A) = f^N(A) \supseteq A \qquad (13)$$

by (b) and

$$h(A) = f^N(A) \supseteq f^{2(n_1+n_2)}(f_A^{-n_1-2n_2}(C)) = f^{n_1}(C) = D \qquad (14)$$

as $f_A^{-n_1-2n_2}(C) \subset A$. Moreover as A and D are nonempty, disjoint compact sets it follows that

$$\inf\{\|x - y\|; x \in A, y \in D\} > 0 . \qquad (15)$$

The existence of a scrambled set S then follows exactly as in Marotto's proof [4] or in Li and Yorke [1]. It will be briefly outlined here for completeness.

Let $\mathcal{E}$ be the set of sequences $\xi = \{E_k\}_{k=1}^{\infty}$, where E_k is either A or D, and $E_{k+1} = E_{k+2} = A$ if $E_k = D$. Let $r(\xi, k)$ be the number of sets E_j equal to D for $1 \leq j \leq k$ and for each $\eta \in (0, 1)$ choose $\xi^\eta = \{E_k^\eta\}_{k=1}^{\infty}$ to be a sequence in $\mathcal{E}$ satisfying

$$\lim_{k \to \infty} \frac{r(\xi^\eta, k^2)}{k} = \eta .$$

Let $\mathcal{F} = \{\xi^\eta \; ; \; \eta \in (0,1)\} \subset \mathcal{E}$. Then $\mathcal{F}$ is uncountable. Also from (12)–(14), $h(E_k^\eta) \supseteq E_{k+1}^\eta$ and so by Lemma 2 for each $\xi^\eta \in \mathcal{F}$ there is a point $x_\eta \in A \cup D$ with $h^k(x_\eta) \in E_k^\eta$ for all $k \geq 1$. Let $S_h = \{h^k(x_\eta) \; ; \; k \geq 0$ and $\xi^\eta \in \mathcal{F}\}$. Then $h(S_h) \subset S_h$, S_h contains no periodic points of h, and there exists an infinite number of k's such that $h^k(x) \in A$ and $h^k(y) \in D$ for any $x, y \in S_h$ with $x \neq y$. Hence from (15) for any $x, y \in S_h$ with $x \neq y$

$$L_1 = \limsup_{k \to \infty} \|h^k(x) - h^k(y)\| > 0 \, .$$

Thus letting $S = \{f^k(x); x \in S_h$ and $k \geq 0\}$ it follows that $f(S) \subset S$, S contains no periodic points of f and for any $x, y \in S$ with $x \neq y$

$$\limsup_{k \to \infty} \|f^k(x) - f^k(y)\| \geq L_1 > 0 \, .$$

This proves that the set S has properties (iia) and (iib) of a scrambled set. The remaining property (iic) can be proven similarly. For further details see [1].

It remains now to establish the existence of an uncountable subset S_0 of the scrambled set S with the properties listed in part (iii) of the definition of chaotic behavior. In contrast with Marotto's proof this is the first place where assumption (c) that f is expanding on A is required. Until now all that has been required is that f is one to one on A. From this, (b) and Lemma 1 follows the existence of a continuous inverse $f_A^{-1} : A \to A$. Hence by the Brouwer fixed point theorem there exists a point $a \in A$ such that $f_A^{-1}(a) = a$, or equivalently $f(a) = a$.

Now because f is expanding on A it follows that f_A^{-1} is contracting A, i.e.,

$$\|f_A^{-1}(x) - f_A^{-1}(y)\| \leq \lambda^{-1}\|x - y\|$$

for all $x, y \in A$, where $\lambda > 1$ is the coefficient of expansion of f on A. Hence for any $k \geq 1$ and all $x, y \in A$

$$\|f_A^{-k}(x) - f_A^{-k}(y)\| \leq \lambda^{-k}\|x - y\| \, ,$$

and in particular for any $x \in C \subset A$ and for $y = a$

$$\|f_A^{-k}(x) - a\| \leq \lambda^{-k}\|x - a\| \, , \tag{16}$$

so $f_A^{-k}(x) \to a$ as $k \to \infty$ for all $x \in C$. Consequently for any $\varepsilon > 0$ there exists an integer $j = j(x, \varepsilon)$ such that $f_A^{-j}(x) \in A \cap B^n(a; \varepsilon)$. Then by continuity there exists a $\delta = \delta(x, \varepsilon) > 0$ such that $f_A^{-1}(A \cap \text{int} B^n(x; \delta)) \subset A \cap B^n(a; \varepsilon)$. Now the collection $\varsigma = \{\text{int } B^n(x; \delta); x \in C\}$ constitutes an open cover of the compact set C, so there exists a finite subcollection $\varsigma_0 = \{\text{int } B^n(x_i; \delta_i); i = 1, 2, \ldots, L\}$ which also covers C. Let $T = T(\varepsilon) = \max\{j(x_i; \varepsilon); i = 1, 2, \ldots, L\}$. Then $f_A^{-T}(x) \in B^n(a; \varepsilon) \cap A$ for all $x \in C$ and so by (16) $f_A^{-k}(C) \subset B^n(a; \varepsilon) \cap A$ for all $k \geq T(\varepsilon)$.

Let $H_k = h_A^{-k}(C)$ for all $k \geq 0$, where h_A^{-1} is a continuous inverse of $h = f^N$ on A. Then for any $\varepsilon > 0$ there exists a $J = J(\varepsilon)$ such that $\|x - a\| < \varepsilon/2$ for all $x \in H_k$ and all $k > J$.

The remainder of the proof parallels that in Marotto [4] and in Li and Yorke [1]. The sequences $\xi^n = \{E_k^n\}_{k=1}^\infty \in \mathcal{E}$ will be further restricted as follows: if $E_k^n = D$ then $k = m^2$ for some integer m, and if $E_k^n = D$ for both $k = m^2$ and $k = (m+1)^2$ then $E_k^\eta = H_{2m-j}$, for $k = m^2 + j$ and for $j = 1, 2, \ldots, 2m$. Finally for the remaining k's, $E_k^\eta = A$. Now these sequences still satisfy $h(E_k^\eta) \supset E_{k+1}^\eta$, so by Lemma 2 there exists a point x_η with $h^k(x_\eta) \in E_k^\eta$ for all $k \geq 0$. Let $S_0 = \{x_\eta : \eta \in (\frac{4}{5}, 1)\}$. Then S_0 is uncountable, $S_0 \subset S_h \subset S$ and for any $s, t \in (\frac{4}{5}, 1)$ there exist infinitely many m's such that $h^k(x_s) \in E_k^s = H_{2m-1}$ and $h^k(x_t) \in E_k^t = H_{2ml-1}$, where $k = m^2 + 1$. But from above, given any $\varepsilon > 0$, $\|x - a\| < \varepsilon/2$ for all $x \in H_{2m-1}$ provided m is sufficiently large. Hence for any $\varepsilon > 0$ there exists an integer m such that $\|h^k(x_s) - h^k(x_t)\| < \varepsilon$, where $k = m^2 + 1$. As $\varepsilon > 0$ is arbitrary it follows that

$$L_2 = \liminf_{k \to \infty} \|h^k(x_s) - h^k(x_t)\| = 0 \,.$$

Thus for any $x, y \in S_0$

$$\liminf_{k \to \infty} \|hk(x_s) - h^k(x_t)\| \leq L_2 = 0 \,.$$

This completes the proof of Theorem 3.

3.2 Examples

Two examples are given here to illustrate the application of Theorem 3. The first example is a one-dimensional difference equation with a snap-back repeller involving the tent or Baker's mapping. It forms one of the components of the second example, the two-dimensional twisted-horseshoe difference equation of Guckenheimer et al. [14], which has a saddle point.

Example 1 Consider the difference equation on the unit interval $I = [0, 1]$, which is defined in terms of the Baker's mapping

$$f(x) = \begin{cases} 2x & \text{for } 0 \leq x \leq \frac{1}{2} \,, \\ 2 - 2x & \text{for } \frac{1}{2} < x \leq 1 \,. \end{cases}$$

This mapping f maps I into itself and has two fixed points 0 and $\frac{2}{3}$, both of which are easily seen to be snap-back repellers.

The conditions of Theorem 3 are satisfied by $A = [\frac{9}{16}, \frac{7}{8}]$, $B = [\frac{3}{4}, \frac{7}{8}]$, $n = l = 1$, and, $n_1 = n_2 = 1$. To see this note that

$$f(A) = \left[\frac{1}{4}, \frac{7}{8}\right], \qquad f(b) = \left[\frac{1}{4}, \frac{1}{2}\right], \qquad f^2(B) = \left[\frac{1}{2}, 1\right] \,,$$

so

$$f(A) \supset A, \qquad f(B) \cap A = \emptyset, \qquad f^2(B) \supset A .$$

Also f is expanding on A because for $x, y \in A$

$$|f(x) - f(y)| = |(2 - 2x) - (2 - 2y)| = 2|x - y|$$

and f^2 is one to one on B because for all $x \in B$

$$f^2(x) = 2(2 - 2x) = 4 - 4x .$$

Hence this difference equation exhibits chaotic behavior.

Example 2 Consider the difference equation on the unit square I^2 in $\mathbb{R}^2$, which is defined in terms of the continuous mapping $f = (f_1, f_2)$, where

$$f_1(x, y) = \begin{cases} 2x & \text{for } 0 \le x \le \frac{1}{2}, \\ 2 - 2x & \text{for } \frac{1}{2} < x \le 1, \end{cases} \qquad f_2(x, y) = \frac{x}{2} + \frac{y}{10} + \frac{1}{4} .$$

This mapping describes a twisted horseshoe on I^2 and has been investigated in detail by Guckenheimer et al. [14]. It has a fixed point $(\bar{x}, \bar{y}) = (\frac{2}{3}, \frac{35}{54})$, which is a saddle point with eigenvalues -2 and $\frac{1}{10}$. Consequently Marotto's snap-back repeller theorem cannot be used here, but Theorem 3 can.

Let L_1 be the line $90x + 378y = 305$ and L_2 the line $90x - 378y = -125$. Then $(\bar{x}, \bar{y}) \in L_1$. Also let

$$A = \left\{ (x, y) \in L_1 ; \frac{9}{16} \le x \le \frac{7}{8} \right\}, \qquad B = \left\{ (x, y) \in L_1 ; \frac{3}{4} \le x \le \frac{7}{8} \right\} .$$

Then

$$f(A) = \left\{ (x, y) \in L_1 ; \frac{1}{4} \le x \le \frac{7}{8} \right\}, \qquad f(B) = \left\{ (x, y) \in L_1 ; \frac{1}{4} \le x \le \frac{1}{2} \right\},$$

$$f^2(B) = \left\{ (x, y) \in L_2 ; \frac{1}{2} \le x \le 1 \right\},$$

and $f^3(B) \supset L_1 \cap I^2$.

Hence $f(A) \supset A$, $f(B) \cap A = \emptyset$, and $f^3(B) \supset A$, so conditions (b), (d), (e), and (f) of Theorem 3 are satisfied with $n_1 = 1$ and $n_2 = 2$. Also A is homeomorphic to a 1-ball and f is expanding on A because for $(x, y) \in A$

$$f_1(x, y) = 2 - 2x, \qquad f_2(x, y) = \frac{35}{18} - 2y ,$$

so for any two points $(x', y'), (x'', y'') \in A$

$$\|f(x', y') - f(x'', y'')\| = 2\|(x', y') - (x'', y'')\| .$$

Finally for all $(x, y) \in B$

$$f_1^3(x, y) = 2 \cdot 2 \cdot (2 - 2x)) = 8 - 8x$$

and

$$f_2^3(x, y) = \frac{381}{200}x + \frac{1}{1000}y - \frac{249}{400} \,,$$

which gives the nonsingular Jacobian matrix

$$\begin{bmatrix} -8 & 0 \\ \frac{381}{200} & \frac{1}{1000} \end{bmatrix}.$$

Hence f^3 is one to one on B.

All the conditions of Theorem 3 are thus satisfied, so this twisted-horseshoe difference equation behaves chaotically.

4 Chaotic Mappings in Banach Spaces

The proof of Theorem 3 above can be easily modified by using the Schauder fixed point theorem, in which case X can be a Banach space, rather than the finite dimensional Euclidean space $\mathbb{R}^n$ [28]. The Schauder fixed point theorem states that a compact mapping f from a closed bounded convex set K in a Banach space X into itself has a fixed point. In this setting Theorem 3 becomes

Theorem 4 (Kloeden [16]) *Let $f : X \to X$ be a continuous mapping of a Banach space X into itself and suppose that there exist non-empty compact subsets A and B of X, and integers $n_1, n_2 \geq 1$ such that*

(i)   A is homeomorphic to a convex subset of X,
(ii) $A \subseteq f(A)$,
(iii) f is expanding on A, i.e., there exists a constant $\lambda > 1$ such that

$$\lambda \|x - y\| \leq \|f(x) - f(y)\|$$

*   for all $x, y \in A$,*
(iv) $B \subset A$,
(v) $f^{n_1}(B) \cap A = \varnothing$,
(vi) $A \subseteq f^{n_1+n_2}(B)$, and
(vii) $f^{n_1+n_2}$ is one to one on B.

Then the mapping f is chaotic in the generalized sense of Li and Yorke given in Theorem 3.

The proof is essentially a repetition of that given above for $X = \mathbb{R}^n$, but requires the Schauder fixed point theorem rather than the Brouwer fixed point theorem. For difference equations on $\mathbb{R}^1$ conditions (iii) and (vii) of the theorem are superfluous as the intermediate value theorem can be used instead of the Schauder fixed theorem to establish the existence of cyclic points. Without these conditions the theorem then contains the sufficient conditions for chaotic behavior of Barna [19] and Sharkovsky [20] as special cases.

This theorem applies to the Baker's mapping and to the twisted horseshoe mapping with the same sets A and B as in the previous section, noting that intervals and connected segments of straight lines are convex sets.

However, the above theorem is not applicable to diffeomorphisms such as the Hénnon mapping and the Smale horseshoe mapping, and hence in general not to the Poincaré mappings for ordinary differential equations.

5 Chaos of Discrete Systems in Complete Metric Spaces

Even more generally, some criteria for chaos of difference equations in general complete metric spaces will be given in this section. In contrast to the Euclidean spaces and Banach spaces these metric spaces may not have a linear structure which allows one to, say, take the difference of two points. Recall that the n-dimensional Euclidean space $\mathbb{R}^n$ is complete and any bounded and closed subset therein is compact. Furthermore, a compact subset of a general metric space is complete as a subspace. Therefore, difference equations defined in terms of continuous mappings in compact subsets of metric spaces and the corresponding criteria of chaos will be discussed. Thus, the existing relevant results of chaos in $\mathbb{R}^n$ and Banach spaces are extended and improved [7]. Here, we just list the main results of [7] without giving proofs. Readers interested in the details can refer to [7].

The criteria of chaos obtained in this section are related to Cantor sets in metric spaces and a symbolic dynamical system, which has rich dynamical structures.

Definition 2. *Let X be a topological space and Λ be a subset in X. Then Λ is called a Cantor set if it is compact, totally disconnected, and perfect. A set in X is totally disconnected if its each connected component is a single point; a set is perfect if it is closed and every point in it is an accumulation point or a limit point of other points in the set.*

Consider the space of sequences

$$\Sigma_2^+ := \{s = (s_0, s_1, s_2, \ldots) : s_j = 0 \text{ or } 1\}$$

and define a distance between two points $s = (s_0, s_1, s_2, \ldots)$ and $t = (t_0, t_1, t_2, \ldots)$ by

$$\rho(s,t) = \sum_{i=0}^{\infty} 2^i |s_i - t_i| \, .$$

For any $s, t \in \Sigma_2^+$, $\rho(s,t) \leq 2^{-n}$ if $s_i = t_i$ for $0 \leq i \leq n$. Conversely, if $\rho(s,t) < 2^{-n}$, then $s_i = t_i$ for $0 \leq i \leq n$.

Lemma 3. (Σ_2^+, ρ) *is a complete, compact, totally disconnected, and perfect metric space.*

Definition 3. *The shift map* $\sigma : \Sigma_2^+ \rightarrow \Sigma_2^+$ *defined by* $\sigma(s_0, s_1, s_2, \ldots) = (s_1, s_2, \ldots)$ *is continuous. The dynamical system governed by* σ *is called a symbolic dynamical system on* Σ_2^+.

The shift map σ has the following properties:

1. $Card \ Per_n(\sigma) = 2^n$,
2. $Per(\sigma)$ is dense in Σ_2^+, and
3. there exists a dense orbit of σ in Σ_2^+,

where $Card \ Per_n(\sigma)$ denotes the number of periodic points of period n for σ.

Theorem 5 *Let* (X, d) *be a complete metric space and* V_0, V_1 *be nonempty, closed, and bounded subsets of* X *with* $d(V_0, V_1) > 0$. *If a continuous map* $f : V_0 \cup V_1 \rightarrow X$ *satisfies*

1. $f(V_j) \supset V_0 \cup V_1$ *for* $j = 0, 1$;
2. f *is expanding in* V_0 *and* V_1, *respectively, i.e., there exists a constant* $\lambda_0 > 1$ *such that*

$$d(f(x), f(y)) \geq \lambda_0 d(x, y) \ \forall x, y \in V_0 \ and \ \forall x, y \in V_1;$$

3. *there exists a constant* $\mu_0 > 0$ *such that*

$$d(f(x), f(y)) \leq \mu_0 d(x, y) \ \forall x, y \in V_0 \ and \ \forall x, y \in V_1;$$

then there exist a Cantor set $\Lambda \subset V_0 \cup V_1$ *such that* $f : \Lambda \rightarrow \Lambda$ *is topologically conjugate to the symbolic dynamical system* $\sigma : \Sigma_2^+ \rightarrow \Sigma_2^+$. *Consequently,* f *is chaotic on* Λ *in the sense of Devaney.*

Recall from the fundamental theory of topology that a compact subset of a metric space is closed, bounded, and complete as a subspace; a closed subset of a compact space is compact; and the distance between two disjoint compact subsets of a metric space is positive. Therefore, if V_0 and V_1 are compact subsets of a metric space (X, d), assumption (3) in Theorem 5 can be dropped.

The following is the corresponding result for chaos of difference equations defined in terms of continuous mappings in two compact subsets of a metric space.

Theorem 6 *Let (X, d) be a metric space and V_0, V_1 be two disjoint compact subset of X. If the continuous map $f : V_0 \cup V_1 \to X$ satisfies*

1. $f(V_j) \supset V_0 \cup V_1$ *for* $j = 0, 1$ *and*
2. *there exists a constant* $\lambda_0 > 1$ *such that*

$$d(f(x), f(y)) \geq \lambda_0 d(x, y) \; \forall x, \quad y \in V_0 \text{ and } \forall x, y \in V_1 ,$$

then there exists a Cantor set $\Lambda \in V_0 \cup V_1$ such that $f : \Lambda \to \Lambda$ is topologically conjugate to the symbolic dynamical system $\sigma : \Sigma_2^+ \to \Sigma_2^+$. Consequently, f is chaotic on Λ in the sense of Devaney.

It should be noticed that by Theorems 5 and 6 the appearance of chaos of f is only relevant to the properties of f on V_0 and V_1, but has no relationship with the properties of f at any other points. The following example is used to illustrate the application of the Theorems.

Example 3 Consider the discrete dynamical system

$$x_{n+1} = \mu x_n (1 - x_n)$$

governed by the logistic mapping $f(x) = \mu x(1 - x)$, where $\mu > 0$ is the parameter.

This mapping has exactly two fixed points: $x_1^* = 0$ and $x_2^* = 1 - \mu^{-1}$. It is clear that f is continuously differentiable on $\mathbb{R}$ and if $\mu > 2 + \sqrt{5}$ then

$$|f'(x)| > 1 \quad \text{for} \ x \in [0, x_1] \cup [x_2, 1] ,$$

where $x_1 = 2^{-1} - \sqrt{4^{-1} - \mu^{-1}} > 0$ and $x_2 = 2^{-1} + \sqrt{4^{-1} - \mu^{-1}} < x_2^*$. This implies that

$$|f(x) - f(y)| \geq \lambda_0 |x \quad y| \; \forall x, y \in [0, x_1] \text{ and } \forall x, y \in [x_2, 1] ,$$

where $\lambda_0 = \sqrt{\mu^2 - 4\mu} > 1$. On the other hand, we have

$$f([0, x_1]) = [0, 1] \supset [0, x_1] \cup [x_2, 1] ,$$

$$f([x_2, 1]) = [0, 1] \supset [0, x_1] \cup [x_2, 1] ,$$

Clearly, $[0, x_1]$ and $[x_2, 1]$ are compact, so that all assumptions in Theorem 6 are satisfied, and for $\mu > 2 + \sqrt{5}$, there exists a Cantor set $\Lambda \in [0, x_1] \cup [x_2, 1]$ such that $f : \Lambda \to \Lambda$ is topologically conjugate to the symbolic dynamical system $\sigma : \Sigma_2^+ \to \Sigma_2^+$.

Now, consider the following mapping:

$$g(x) = \begin{cases} \mu x(1 - x), & x \in [0, x_1], \\ h(x), & x \in (x_1, x_2), \\ \mu x(1 - x), & x \in [x_2, 1] , \end{cases}$$

where $h(x)$ can be any function on (x_1, x_2). It is noted that g may not even be continuous on $[x_1, x_2]$. By the above discussion on the logistic mapping and by Theorem 6, one can conclude that for $\mu > 2 + \sqrt{5}$, there exists a Cantor set $\Lambda' \subset [0, x_1] \cup [x_2, 1]$ such that $g : \Lambda' \to \Lambda'$ is topologically conjugate to the symbolic dynamical system $\sigma : \Sigma_2^+ \to \Sigma_2^+$ and, consequently, g is chaotic on Λ'. We notice that the Cantor set Λ' may be taken to be the set Λ.

Furthermore, by means of snap-back repeller arguments, two criteria of chaos for difference equations defined in terms of continuous mappings in complete metric spaces and compact subsets of metric spaces will be established in the following.

Theorem 7 *Let (X, d) be a complete metric space and $f : X \to X$ be a mapping. Assume that*

1. *f has a regular nondegenerate snap-back repeller $z \in X$, i.e., there exist positive constants r_1 and $\lambda_1 > 1$ such that $f(B_{r_1}(z))$ is open and*

$$d(f(x), f(y)) \geq \lambda_1 d(x, y) \ \forall x, y \in \bar{B}_{r_1}(z) ,$$

and there exist a point $x_0 B_{r_1}(z)$, $x_0 \neq z$, a positive integer m, and positive constant δ_1 and γ, such that $f^m(x_0) = z$, $B_{\delta_1}(x_0) \subset B_{r_1}(z)$, z is an interior point of $f^m(B_{\delta_1}(x_0))$, and

$$d(f^m(x), f^m(y)) \geq \gamma d(x, y) \ \forall x, y \in \bar{B}_{\delta_1}(x_0) ; \tag{17}$$

2. *there exists a positive constant μ_1 such that*

$$d(f(x), f(y)) \leq \mu_1 d(x, y) \ \forall x, y \in \bar{B}_{r_1}(z) ; \tag{18}$$

3. *there exists a positive constant μ_2 such that*

$$d(f^m(x), f^m(y)) \leq \mu_2 d(x, y) \ \forall x, y \in \bar{B}_{\delta_1}(x_0) . \tag{19}$$

In addition, assume that f is continuous on $\bar{B}_{r_1}(z)$ and that f^m is continuous on $\bar{B}_{\delta_1}(x_0)$. Then, for each neighborhood U of z, there exist a positive integer $n > m$ and a Cantor set $\Lambda \subset U$ such that $f^n : \Lambda \to \Lambda$ is topologically conjugate to the symbolic dynamical system $\sigma : \Sigma_2^+ \to \Sigma_2^+$. Consequently, f^n is chaotic on Λ in the sense of Devaney.

By Theorem 6, the following result for metric spaces with a certain compactness property similar to that of finite dimensional Euclidean spaces can be established.

Theorem 8 *Let (X, d) be a metric space in which each bounded and closed subset is compact. Assume that $f : X \to X$ has a regular nondegenerate snap-back repeller z, associated with x_0, m, and r as specified in Marotto's definitions, f is continuous on $\bar{B}_r(z)$, and f^m is continuous in a neighborhood of x_0. Then, for each neighborhood U of z, there exist a positive integer n and a Cantor set $\Lambda \subset U$ such that $f^n : \Lambda \to \Lambda$ is topologically conjugate to the symbolic dynamical system $\sigma : \Sigma_2^+ \to \Sigma_2^+$. Consequently, f^n is chaotic on Λ in the sense of Devaney.*

6 Chaos of Difference Equations in Metric Spaces of Fuzzy Sets

In this section, the Li–Yorke and Marotto definitions are generalized to be applicable to mappings from a space of fuzzy sets into itself, namely the metric space $(\mathcal{E}^n, D)$ of fuzzy sets on the base space $\mathbb{R}^n$.

6.1 Chaotic Mappings on Fuzzy Sets

The following definitions and results are taken from [29], see also [30].

The set $\mathcal{E}^n$ consists of all functions, called fuzzy sets here, $u : \mathbb{R}^n \to [0, 1]$ for which

(i) u is normal, i.e., there exists an $x_0 \in \mathbb{R}^n$ such that $u(x_0) - 1$,
(ii) u is fuzzy convex, i.e., for any $x, y \in \mathbb{R}^n$ and $0 \leq \lambda \leq 1$

$$u(\lambda x + (1 - \lambda)y) \geq \min\{u(x), u(y)\} ,$$

(iii) u is uppersemicontinuous, and
(iv) the closure of $\{x \in \mathbb{R}^n; u(x) > 0\}$, denoted by $[u]^0$, is compact.

Let $u \in \mathcal{E}^n$. Then for each $0 < \alpha \leq 1$ the α-level set $[u]^\alpha$ of u, defined by

$$[u]^\alpha = \{x \in \mathbb{R}^n; u(x) \geq \alpha\} ,$$

is a nonempty compact convex subset of $\mathbb{R}^n$, as is the support $[u]^0$ of u. Let d be the Hausdorff metric for nonempty compact subsets of $\mathbb{R}^n$. Then

$$D(u, v) = \sup_{0 \leq \alpha \leq 1} d([u]^\alpha, [v]^\alpha) ,$$

where $u, v \in \mathcal{E}^n$, is a metric on $\mathcal{E}^n$. Moreover, $(\mathcal{E}^n, D)$ is a complete metric space.

Let $u, v \in \mathcal{E}^n$ and let c be a positive number. Then addition $u + v$ and (positive) scalar multiplication cu in $\mathcal{E}^n$ are defined in terms of the α-level sets by

$$[u + v]^\alpha = [u]^\alpha + [v]^\alpha ,$$

and

$$[cu]^\alpha = c[u]^\alpha ,$$

for each $0 \leq \alpha \leq 1$, where

$$A + B = \{x + y; x \in A, y \in B\} \quad \text{and} \quad cA = \{cx; x \in A\}$$

for nonempty subsets A and B of $\mathbb{R}^n$. This defines a linear structure (but without subtraction) on $\mathcal{E}^n$, such that

$$D(u + w, v + w) = D(u, v)$$

and

$$D(cu, cv) = cD(u, v) \, ,$$

for all $u, v, w \in \mathcal{E}^n$ and $c > 0$. There is however no norm on $\mathcal{E}^n$ equivalent to the metric D, which makes $\mathcal{E}^n$ into a normed linear space with the above natural linear structure. Nevertheless by an embedding theorem of Rådström, $\mathcal{E}^n$ can be embedded isometrically isomorphically as a convex cone in some Banach space. Consequently many well-known results for Banach spaces can be adapted to the metric space $(\mathcal{E}^n, D)$ of fuzzy sets. An example is the following fixed point theorem of Kaleva [29].

Theorem 9 (Kaleva) *Let* $f : \mathcal{E}^n \to \mathcal{E}^n$ *be continuous and let* $\mathcal{X}$ *be a nonempty compact convex subset of* $\mathcal{E}^n$ *such that* $f(\mathcal{X}) \subseteq \mathcal{X}$. *Then* f *has a fixed point* $\bar{u} = f(\bar{u}) \in \mathcal{X}$.

Consider now an iterative scheme of fuzzy sets

$$u_{k+1} = f(u_k), \quad k = 1, 2, \dots \, , \tag{20}$$

where f is a continuous mapping from the space of fuzzy sets $\mathcal{E}^n$ into itself. Such an iterative scheme or mapping f will be called to be chaotic if conditions analogous to those of Theorem 1 are satisfied. Using the Kaleva fixed point theorem, the following analogue of Theorem 3 can be shown to hold. It provides sufficient conditions for a mapping on fuzzy sets to be chaotic. (Alternatively, one could apply the results of the previous section here).

Theorem 10 (Kloeden [31]) *Let* $f : \mathcal{E}^n \to \mathcal{E}^n$ *be continuous and suppose that there exist nonempty compact subsets* $\mathcal{A}$ *and* $\mathcal{B}$ *of* $\mathcal{E}^n$, *and integers* $n_1, n_2 \geq 1$ *such that*

(i) $\mathcal{A}$ *is homeomorphic to a convex subset of* $\mathcal{E}^n$,
(ii) $\mathcal{A} \subseteq f(\mathcal{A})$,
(iii) f *is expanding on* $\mathcal{A}$, *that there exists a constant* $\lambda > 1$ *such that*

$$\lambda D(u, v) \leq D(f(u), f(v))$$

for all $u, v \in \mathcal{A}$,
(iv) $\mathcal{B} \subset \mathcal{A}$,
(v) $f^{n_1}(\mathcal{B}) \cap \mathcal{A} = \varnothing$,
(vi) $\mathcal{A} \subseteq f^{n_1 + n_2}(\mathcal{B})$, *and*
(vii) $f^{n_1 + n_2}$ *is one to one on* $\mathcal{B}$.

Then the mapping f *is chaotic.*

The proof of Theorem 10 essentially mimics the proof of Theorem 3, using the Kaleva fixed point theorem instead of the Brouwer fixed point theorem. Thus, it is omitted. A useful characterization of compact subsets of $\mathcal{E}^n$ has been presented by Diamond and Kloeden in [30] and [32].

6.2 An Example of a Chaotic Mapping on Fuzzy Sets

To illustrate the application of Theorem 10 a simple example of a chaotic mapping $f : \mathcal{E}^1 \to \mathcal{E}^1$, which satisfies the hypotheses of the theorem, will be constructed. For this purpose observe that for each $u \in \mathcal{E}^1$ there exist functions (their dependence on u is not explicitly stated) $a, b : [0,1] \to \mathbb{R}$ such that the α-level sets of u are the intervals $[u]^\alpha = [a(\alpha), b(\alpha)]$. Moreover for any $0 \le \alpha \le \alpha' \le 1$ the following inequalities hold:

$$a(0) \le a(\alpha) \le a(\alpha') \le a(1) \le b(1) \le b(\alpha') \le b(\alpha) \le b(0) .$$

Consider the following subsets of $\mathcal{E}^1$:

(a) $\mathcal{E}_0^1 = \{u \in \mathcal{E}^1 \; ; \; a(0) = 0\}$,
(b) $\Im_0^1 = \{u \in \mathcal{E}_0^1 \; ; \; a(\alpha) = \frac{1}{2}\alpha(b(0) - L)$ and $b(\alpha) = b(0) - \frac{1}{2}\alpha(b(0) - L)$ for some $0 \le L \le b(0)\}$,
(c) $\Delta_0^1 = \{u \in \Im_0^1 \; ; \; L = 0\}$.

For any $u \in \mathcal{E}_0^1$, the support $[u]^0$ is a nonnegative interval anchored on $x = 0$. The endograph of any $u \in \Im_0^1$ is a symmetric trapezium centred on $x = \frac{1}{2}b(0)$, with base length $b(0)$ and top length L. For any $u \in \Delta_0^1$ the endograph is an isosceles triangle.

Now define the following mappings on fuzzy sets:

1. $f_1 : \mathcal{E}^1 \to \mathcal{E}_0^1$ by $[f_1(u)]^\alpha = [a(\alpha) - a(0), b(\alpha) - b(0)]$,
2. $f_2 : \mathcal{E}_0^1 \to \Im_0^1$ by $[f_2(u)]^\alpha = [\alpha M, b(0) - \alpha M]$, where $M = \frac{1}{2}b(0) - \frac{1}{8}(b(1) - a(1))$,
3. $f_3 : \Im_0^1 \to \Im_0^1$ by $[f_3(u)]^\alpha = g(b(0))[u]^\alpha = [g(b(0))a(\alpha), g(b(0))b(\alpha)]$,

where $g : \mathbb{R}^+ \to \mathbb{R}^+$ is the function

$$g(x) = \begin{cases} 2 & \text{if } 0 \le x \le \frac{1}{2} , \\ -2 + 2/x & \text{if } \frac{1}{2} \le x \le 1 , \\ 0 & \text{if } 1 \le x. \end{cases}$$

Also define $h : \mathbb{R}^+ \to \mathbb{R}^+$ by

$$h(x) = xg(x) = \begin{cases} 2x & \text{if } 0 \le x \le \frac{1}{2} , \\ 2 - 2x & \text{if } \frac{1}{2} \le x \le 1 , \\ 0 & \text{if } 1 \le x. \end{cases}$$

Finally, define $f : \mathcal{E}^1 \to \mathcal{E}^1$ by $f = f_3 \circ f_2 \circ f_1$. This mapping f is clearly continuous with respect to the D-metric and maps Δ_0^1 into itself. Now any $u \in \Delta_0^1$ is determined uniquely by its value of $b(0)$, written b henceforth, and will be denoted by u_b. Then $f(u_b) = u_{h(b)}$.

Theorem 10 applies for the mapping f with $n_1 = n_2 = 1$ and with the compact convex subsets of $\mathcal{E}^1$

$$\mathcal{A} = \left\{ u_b \in \Delta_0^1 \; ; \; \frac{9}{16} \le b \le \frac{7}{8} \right\}, \qquad \mathcal{B} = \left\{ u_b \in \Delta_0^1 \; ; \; \frac{3}{4} \le b \le \frac{7}{8} \right\},$$

with

$$f(\mathcal{A}) = \left\{ u_b \in \Delta_0^1 \; ; \; \frac{1}{4} \le b \le \frac{7}{8} \right\}, \qquad f(\mathcal{B}) = \left\{ u_b \in \Delta_0^1 \; \frac{1}{4} \le b \le \frac{1}{2} \right\}$$

and

$$f^2(\mathcal{B}) = \left\{ u_b \in \Delta_0^1 \; \frac{1}{2} \le b \le 1 \right\},$$

so

$$\mathcal{A} \subseteq f(\mathcal{A}), \qquad f(\mathcal{B}) \cap \mathcal{A} = \emptyset, \qquad \mathcal{A} \subseteq f^2(\mathcal{B}).$$

Moreover, f is expanding on $\mathcal{A}$ because h is expanding on $[\frac{9}{16}, \frac{7}{8}]$ with

$$|h(x) - h(y)| = |(2 - 2x) - (2 - 2y)| = 2|x - y| \, ,$$

so

$$D(f(u_x), f(u_y)) = 2D(u_x, u_y)$$

for any $u_x, u_y \in \mathcal{A}$. Finally, h^2 is one to one on $[\frac{3}{4}, \frac{7}{8}]$ because

$$h^2(x) = 2(2 - 2x) = 4 - 4x \, ,$$

which implies that f^2 is one to one on $\mathcal{B}$.

The mapping f is thus chaotic. Its chaotic action is most apparent in the compact subset $\{u_b \in \Delta_0^1 ; 0 \le b \le 1\}$ of $\mathcal{E}^1$. It is not hard to show that for any $u \in \mathcal{E}^1$, the successive iterates $f^k(u)$ asymptote toward this set. Their endographs become more and more triangular in shape, unless $b(0) - a(0) > 1$, in which case they collapse onto the singleton fuzzy set χ_0.

7 Conclusions

Studies thus far on criteria of chaos for difference equations defined in terms of continuous mappings in various spaces ranging from the simplest one dimensional space $\mathbb{R}$, to $\mathbb{R}^n$ through Banach spaces and complete metric spaces, and finally to metric spaces of fuzzy sets have been reviewed. In practice, however, to establish the existence of chaos in a particular dynamical system often still depends mainly on numerical calculations to estimate quantities such as the maximum Lyapunov exponent and topological entropy. A unified, well accepted, easy-to-test, and rigorous mathematical definition of chaos is still in the process of being revealed, and this work is far from complete.

References

1. T.Y. Li, J.A. Yorke: Amer. Math. Monthly **82**, 481–485 (1975)
2. C.P. Li, G. Chen: Chaos Solitons Fractals **18**, 69–77 (2003)
3. R. Devaney: *A First Course in Chaotic Dynamical Systems: Theory and Experiment* (Perseus Books, Cambridge, MA, 1992)
4. F.R. Marotto: J. Math. Anal. Appl. **63**, 199–223 (1978)
5. G. Chen, S. Hsu, J. Zhou: J. Math. Phys **39**(12), 6459–6489 (1998)
6. K. Shiraiwa, M. Kurata: Nagoya Math J. **82**, 83–97 (1981)
7. Y.M. Shi, G. Chen: Chaos Solitons Fractals **22**, 555–571 (2004)
8. R. Devaney: *An Introduction to Chaotic Dynamical Systems* (Addison-Wesley, New York, 1989)
9. J. Banks, J. Brooks, G. Cairns, G. Davis, P. Stacey: Amer. Math. Monthly **99**(4), 332–334 (1992)
10. M. Vellekoop, R. Berglund: Amer. Math. Monthly **101**(4), 353–355 (1994)
11. I. Stewart: Nature **406**, 948–949 (2000)
12. W. Tucker, C.R. Acad. Sci. Paris **328**, 1197–1202 (1999)
13. C.P. Silva: IEEE Trans. Circuits Syst.-I **40**, 675–682 (1993)
14. J. Guckenheimer, G. Oster, A. Ipaktachi: J. Math. Biol. **4**, 101–147 (1977)
15. P.E. Kloeden: J. Austral. Math. Soc. (Ser A) **31**, 217–225 (1981)
16. P.E. Kloeden: Cycles and chaos in higher dimensional difference equations. In: *Proc. 9th Int. Conf. Nonlinear Oscillations*, vol 2, ed by Yu. A. Mitropolsky (Naukova Dumka, Kiev, 1984), 184–187
17. P.E. Kloeden, A.I. Mees: Bull. Math. Biol. **47**, 697–738 (1985)
18. R.M. May: Nature, **261**, 459–467 (1976)
19. B. Barna: Publ. Math. Debrecen **22**, 269–278 (1975)
20. A.N. Sharkovsky: Ukrain. Math. Zh., **16**(1), (1964)
21. A.N. Sharkovsky: Ukrain. Math. Zh., **17**(3), 104–111 (1965)
22. P. E. Kloeden, M. Deakin, A. Tirkel: Nature **264**, 295 (1976)
23. P.R. Stein, S.M. Ulam: Rozprawy Matematyczne **XXXIX**, 1–65 (1964)
24. S. Smale: Bull. Amer. Math. Soc. **73**, 747–817 (1967)
25. P.E. Kloeden: Bull. Austral. Math. Soc. **20**, 171–177 (1979)
26. M. Hénnon: Commun. Math. Phys. **50**, 69–77 (1976)
27. P. Diamond: Int. J. Systems Sci. **7**, 953–956 (1976)
28. Y.M. Shi, G. Chen: Sci. China Ser. A: Math. **48**(2), 222–238 (2005)
29. O. Kaleva, Fuzzy Sets Syst. **17**, 53–65 (1985)
30. P. Diamond, P.E. Kloeden: *Metric Spaces of Fuzzy Sets: Theory and Applications* (World Scientific, Singapore, 1994)
31. P.E. Kloeden: Fuzzy Sets Syst. **42**(1), 37–42 (1991)
32. P. Diamond, P.E. Kloeden: Fuzzy Sets Syst. **35**, 241–249 (1990)

Chaotic Dynamics with Fuzzy Systems

Domenico M. Porto

Abstract. In this chapter a new approach for modeling chaotic dynamics is proposed. It is based on a linguistic description of chaotic phenomena, which can be easily related to a fuzzy system design. This approach allows building chaotic generators by means of few fuzzy sets and using a small number of fuzzy rules. It is also possible to create chaotic signals with assigned characteristics (e.g., Lyapunov exponents). Fuzzy descriptions of well-known discrete chaotic maps are therefore introduced, denoting an improved robustness to parameter changes.

1 Introduction

The possible interactions between fuzzy logic and chaos theory have been explored since the 1980s, but these explorations have been carried on mainly in three directions: the fuzzy control of chaotic systems [1, 2], the definition of an adaptive fuzzy system by data from a chaotic time series [3], and the study of the theoretical relations between fuzzy logic and chaos [4]. We shall follow none of these approaches, but take as a starting point the work in [5], to design fuzzy systems, which exhibit a chaotic behavior via a linguistic description of chaotic dynamics.

By giving a linguistic description of a chaotic system and by translating this description in a fuzzy model, we aim first to obtain fuzzy chaotic systems with desired characteristics, denoting an improved robustness to parameter changes, and second to show that a simple fuzzy system with few fuzzy sets and few rules is capable of being a good model of a complex and cryptic chaotic system.

2 A Brief Review of Chaos

During the last decade, the study of chaos has become increasingly important among physicists and engineers [6] due to the large number of its possible applications. But what is "chaos"? Firstly, all those behaviors in some sense unpredictable due to the inadequate feature of measurement methodology (e.g., weather evolution) were considered "chaotic". Nevertheless, technological improvements demonstrated that long-term prediction of certain phenomena fails because of their intrinsic complexity (highly nonlinear behavior)

D.M. Porto: *Chaotic Dynamics with Fuzzy Systems*, StudFuzz **187**, 25–44 (2006)
www.springerlink.com   © Springer-Verlag Berlin Heidelberg 2006

and not because of computational limitation. This quasirandom behavior has been observed even in simple nonlinear systems, which demonstrated to be very sensitive to changes in parameters. What was initially considered only as a "curious phenomenon" nowadays is found everywhere in nature, showing a chaotic feature of our physical world. But the random behavior of a deterministic system may also have some useful and surprising application in cryptography [7], signal processing [8], and, more generally, in most fields of industrial process control.

The peculiarities of a chaotic system can be listed as follows:

1. Strong dependence of the behavior on initial conditions
2. The sensitivity to the changes of system parameters
3. Presence of strong harmonics in the signals
4. Fractional dimension of space state trajectories
5. Presence of a stretch direction, represented by a positive Lyapunov exponent [9]

The last can be considered as an "index" that quantifies a chaotic behavior.

Famous artificial chaotic systems are Chua's circuit [10], the Duffing oscillator [11], and the Roessler system, which can be represented as third-order nonlinear autonomous systems. However, chaotic dynamics can also be generated by simple discrete maps, like the logistic map:

$$x(k+1) = ax(k)(1 - x(k)) \tag{1}$$

or the Henon map:

$$\begin{cases} x(k+1) = y(k) + 1 - ax^2(k) \\ y(k+1) = bx(k) \end{cases} \tag{2}$$

The case of the one-dimensional map $x(k+1) = f(x(k))$ is quite interesting because there exists a simple expression for Lyapunov exponents:

$$\lambda = \lim_{n \to \infty} \frac{1}{n} \sum_{k=1}^{n} \ln |f'(x(k))| \tag{3}$$

It can be shown that even in this case slight changes of the parameters may lead to very different behaviors (see Figs.1–3) to this end it has become necessary to find an alternative description of chaos in order to design more robust chaotic generators.

3 Fuzzy Modeling of Chaotic Behaviors

To model the evolution of a chaotic signal $x(.)$, two variables need to be considered as inputs: the "center" value $x(k)$, which is the nominal value of

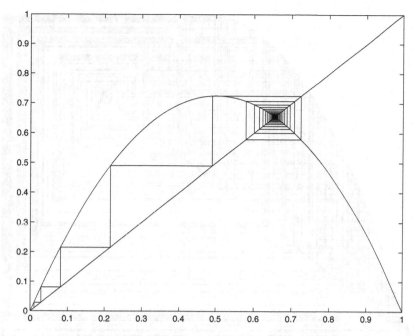

Fig. 1. Logistic map: $a = 2.9$ (fixed point motion)

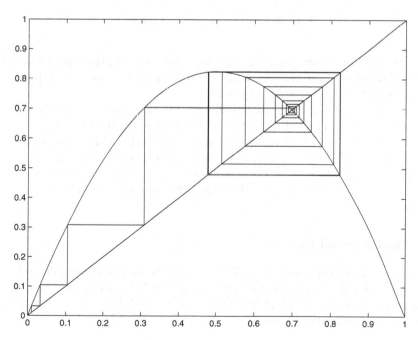

Fig. 2. Logistic map: $a = 3.3$ (stable circle motion)

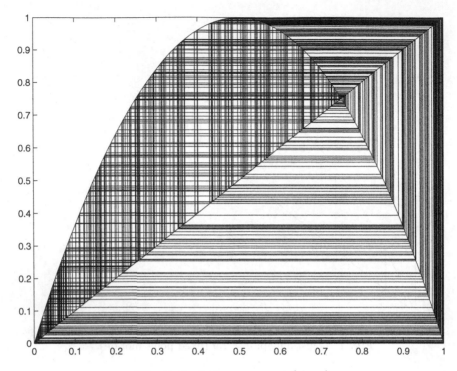

Fig. 3. Logistic map: $a = 4$ (chaos)

the state $x(k)$ at the step k, and the uncertainty $d(k)$ on the center value. In terms of fuzzy description, this means that the model contains four linguistic variables [12]: $x(k+1)$, $x(k)$ $d(k)$, and $d(k+1)$.

The fuzzy rules we use to model the iteration must assert the values $x(k+1)$ and $d(k+1)$ from the values $x(k)$ and $d(k)$. In the following we shall consider two types of chaotic signals, single scroll and double scroll, according to the cases in which it is possible to distinguish two different zones of chaos in the space state. In both cases we shall use a fuzzy model with the Mamdani implication, the center-of-sums defuzzification method, and the product as t-norm [13].

3.1 Double Scroll System

In the fuzzy model of a double scroll system [5] the linguistic variables $x(k+1)$ and $x(k)$ can take four linguistic values: large left (LL), small left (SL), small right (SR), and large right (LR). The linguistic variables $d(k+1)$ and $d(k)$ can take five linguistic values:

zero (Z), small (S), medium (M), large (L), and very large (VL).

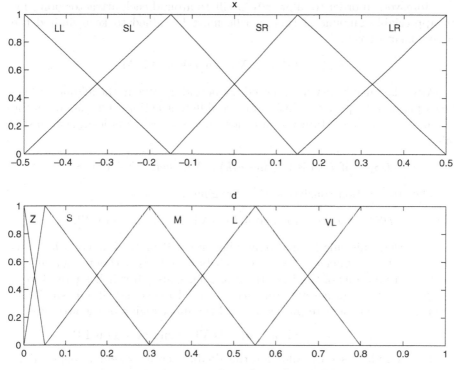

Fig. 4. The fuzzy sets for x (upper) and d (lower)

The fuzzy sets associated with these linguistic values are shown in Fig. 4. A double scroll system is characterized by two attractor points, one in the positive axis and one in the negative axis, that make the state oscillate around each of these points. Thus, considering the positive attractor, it happens that if the value of $x(k)$ is LR then the value of $x(k+1)$ is SR and vice versa. The scheme of this rule is

$$DS_1 \quad \text{if } x(k) \text{ is LR then } x(k+1) \text{ is SR}$$

$$DS_2 \quad \text{if } x(k) \text{ is SR then } x(k+1) \text{ is LR}$$

Obviously, the same behavior is observed when $x(k)$ is negative (i.e., when *x(k) is* LL or SL).

Once the region where the attractor is embedded is defined, the uncertainty d must enlarge until the boundary region is reached.

$$DS_3 \quad \text{if } d(k) \text{ is S then } x(k+1) \text{ is M}$$

$$DS_4 \quad \text{if } d(k) \text{ is L then } x(k+1) \text{ is VL}$$

Moreover, in order to allow x to oscillate around each attractor point for an appreciable amount of time, a rule must be added to keep limited the enlargement of $x(k)$.

$$DS_5 \quad \text{if } d(k) \text{ is M then } x(k+1) \text{ is M}$$

After the uncertainty has reached the boundary region, the folding process takes place. This process consists of two different actions. One action is the shrinking of the uncertainty d in order to avoid behaviors leading to instability.

$$DS_6 \quad \text{if } x(k) \text{ is LR and } d(k) \text{ is VL then } d(k+1) \text{ is S}$$

The other action consists in the changing of the lobe.

$$DS_7 \quad \text{if } x(k) \text{ is LR and } d(k) \text{ is VL then } x(k+1) \text{ is SL}$$

Note that this rule is fired only when x is LR or LL but not when x is SR or LR; this happens so that it cannot bounce back without beginning to oscillate around one of the two attractor points just after it has passed from the positive to negative axis (or vice versa). To this end, if x is small the diverging trajectories are pushed toward the inner region of the attractor.

$$DS_8 \quad \text{if } x(k) \text{ is SR and } d(k) \text{ is VL then } x(k+1) \text{ is LR}$$

The complete set of rules for the double scroll system is depicted in Table 1.

Table 1. The sets of rules implementing a double scroll chaotic system

$x(k)/d(k)$	Z	S	M	L	VL
LL	SL/Z	SL/M	SL/M	SL/VL	SR/L
SL	LL/Z	LL/M	LL/M	LL/VL	LL/S
SR	LR/Z	LR/M	LR/M	LR/VL	LR/S
LR	SR/Z	SR/M	SR/M	SR/VL	SL/L

Figure 5 represents the evolution of the $x(k)$ time series generated by the fuzzy system for $x(0) = 0.01$ and $d(0) = 0.01$.

As expected, the state x oscillates around one of the attractor points, then jumps and begins oscillating around the other attractor point, then jumps again, and so on. Thus the behavior of the state reproduces that of Chua's circuit.

The system we have modeled is a function $F : [-0.5, 0.5] \times [0, 1] \mapsto [-0.5, 0.5] \times [0, 1]$, such that $F(x(k), d(k)) = (x(k+1), d(k+1))$. Figure 6 is obtained projecting the F point to a point on the plane $x(k) \times x(k+1)$ showing the nonlinear map implemented. From this figure, it is quite evident

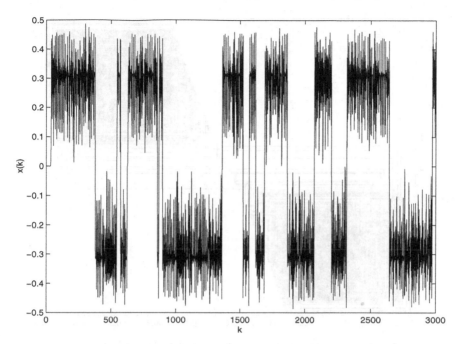

Fig. 5. The chaotic time series generated by the fuzzy system

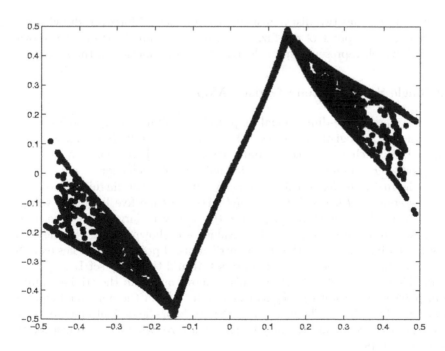

Fig. 6. The center value $x(k+1)$ on $x(k)$

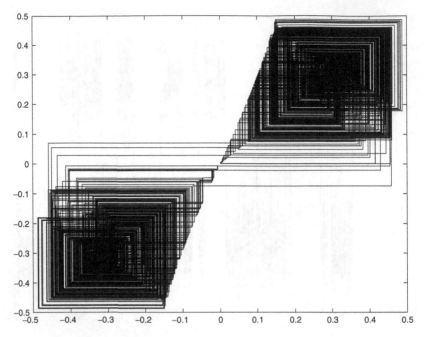

Fig. 7. The map generated by the fuzzy model of the two-lobe system

that the system has two distinguished zones of chaos. Moreover, drawing the evolution of each point of the fuzzy system, we obtain the trajectory shown in Fig. 7, which represents more clearly the chaotic motion of the system.

3.2 Single Scroll System: Logistic Map

The most known one-dimensional map, which exhibits a single scroll chaotic behavior, is the logistic map $x(k+1) = 4(1 - x(k))x(k)$ (see Fig. 3). This map, which is characterized by only one zone of chaos, has two unstable fixed points, $x = 0$ and $x = 3/4$, which influence its behavior.

In the fuzzy model of this chaotic dynamics, all the linguistic variables of the system ($x(k)$, $d(k)$, $x(k+1)$, $d(k+1)$) can take five linguistic values: zero (Z), small (S), medium (M), large (L), and very large (VL). The fuzzy sets associated with these linguistic values are shown in Fig. 8; they are constructed in such a way that the nontrivial fixed point $x = 3/4$ lies exactly in the middle of the peaks of the fuzzy set M and the fuzzy set L.

In this single scroll system x tends to move out from the trivial equilibrium point ($x = 0$) until it begins to oscillate around the nontrivial equilibrium point ($x = 3/4$). The increasing amplitude of the oscillations forces the trajectory to reach again the neighborhood of zero, where, due to instability, the process repeats.

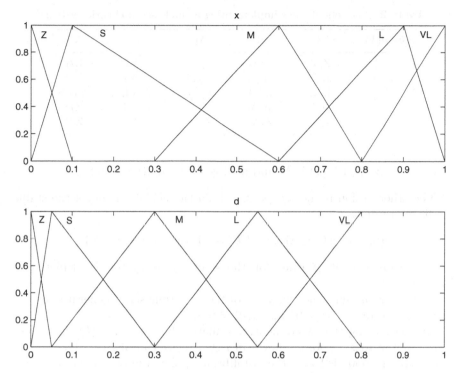

Fig. 8. The fuzzy sets for x (upper) and d (lower): logistic map

In other words, when x is smaller than the nontrivial equilibrium point $(x = 3/4)$, it tends to increase, and when it is very large, it tends to decrease very fast.

$$SS_1 \text{ if } x(k) \text{ is S then } x(k+1) \text{ is M}$$

$$SS_2 \text{ if } x(k) \text{ is VL then } x(k+1) \text{ is Z}$$

On the contrary, when x is close enough to the nontrivial equilibrium point $(x = 3/4)$, it tends to oscillate around that point. Thus it happens that if the value of $x(k)$ is medium then the value of $x(k+1)$ is large, and if the value of $x(k)$ is large then the value of $x(k+1)$ is medium.

$$SS_3 \text{ if } x(k) \text{ is M then } x(k+1) \text{ is L}$$

$$SS_4 \text{ if } x(k) \text{ is L then } x(k+1) \text{ is M}$$

Moreover, the asymmetric shape of these fuzzy sets (L and M) is necessary to obtain a motion like that in Fig. 3.

As in the case of the double scroll system, after the uncertainty has reached the boundary region, two different actions take place. One action is the shrinking of the uncertainty d.

Table 2. The sets of rules implementing a single scroll chaotic system

$x(k)/d(k)$	Z	S	M	L	VL
Z	Z/Z	Z/M	Z/M	S/VL	L/L
S	M/Z	M/M	M/M	M/VL	L/S
M	L/Z	L/M	L/M	L/VL	VL/S
L	M/Z	M/M	M/M	M/VL	Z/S
VL	Z/Z	Z/M	Z/M	Z/VL	Z/L

SS_5 if $x(k)$ is M and $d(k)$ is VL then $d(k+1)$ is S

The other action is the escape of x from the neighborhood of the stable equilibrium point.

SS_6 if $x(k)$ is M and $d(k)$ is VL then $x(k+1)$ is VL

The complete set of rules for the single scroll system is depicted in Table 2.

Figure 9 represents the evolution of the $x(k)$ time series generated by the fuzzy system for $x(0) = 0.01$ and $d(0) = 0.01$.

By considering the two-dimensional map $x(k + 1) = F(x(k), d(k))$ generated by the fuzzy system (shown in Fig. 10), we can observe the range in which $x(k)$ is bounded and the equilibrium point around which the fuzzy map evolves.

Figure 11 is obtained by drawing the trajectory of the fuzzy system in the state space. By comparing this figure with Fig. 3, we can note the similarity between the two. Moreover, it is worth observing that the approximation is better where the fuzzy sets are dense and worst where they are sparse; thus, if we want to have a better approximation in a fixed region, we only need to increase the number of fuzzy sets describing $x(k)$.

3.3 Lyapunov Exponents

For the one-dimensional map $x(k + 1) = f(x(k))$ a simple expression for Lyapunov exponents has already been given in Sect. 2:

$$\lambda = \lim_{n \to \infty} \frac{1}{n} \sum_{k=1}^{n} \ln |f'(x(k))|$$

Moreover, in general, this expression can be written only by means of the values of uncertainty

$$\lambda = \lim_{n \to \infty} \frac{1}{n} \sum_{k=1}^{n} \ln \left| \frac{E_{k+1}}{E_k} \right| ,$$

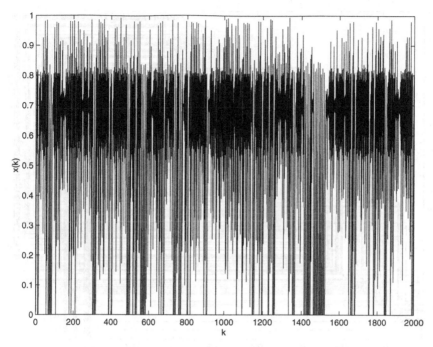

Fig. 9. The chaotic time series generated by the fuzzy system: logistic map

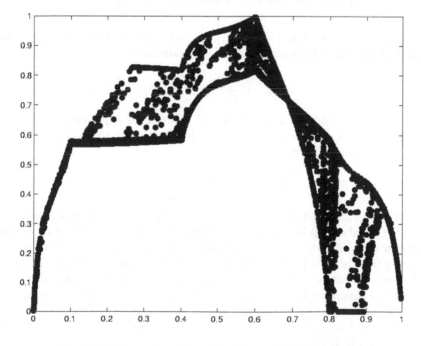

Fig. 10. The center value $x(k+1)$ on $x(k)$: logistic map

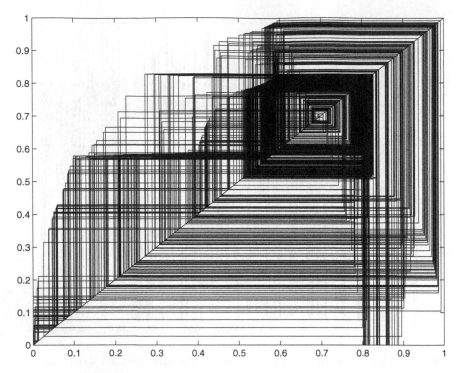

Fig. 11. The logistic map generated by the fuzzy model

where $E_{k+1} = f(x(k) + E_k) - f(x(k))$ only when the exact expression of the mapping function is known. It can also be written as

$$\left| \frac{E_{k+1}}{E_k} \right| = e^{\bar{\lambda}}$$

where $\bar{\lambda}$ can be considered as the "desired" value of the exponent. Considering our fuzzy models, whose uncertainty evolves following the rules previously explained, it can be written that

$$\left| \frac{C_M}{C_S} \right| = e^{\bar{\lambda}}, \qquad \left| \frac{C_L}{C_M} \right| = e^{\bar{\lambda}}$$

where C_S, C_M, and C_L denote, respectively, the centers of small, medium, and large membership functions of uncertainty. This means that it is possible to design a chaotic time series and "drive" its Lyapunov exponent by only changing the distance between the centers of the fuzzy sets. Figure 12 shows that there is a quite good accordance among computed values, obtained using the method described in [14], and desired values of Lyapunov exponents, at least for values between 0 and 0.25. Even if this range of values seems to be a small one, the Lyapunov exponents of a very large number of chaotic systems are contained in it [15].

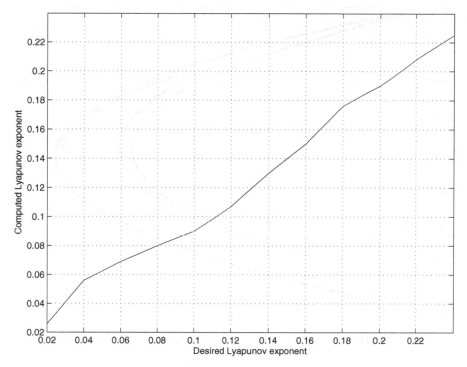

Fig. 12. The desired Lyapunov exponents against calculated ones

4 Two-Dimensional Maps

The Henon map, introduced in Sect. 2, can be considered in some sense the two-dimensional extension of the logistic map. Here we rewrite the state equations

$$\begin{cases} x(k+1) = y(k) + 1 - ax^2(k) \\ k(k+1) = bx(k) \end{cases} \tag{2}$$

For some values of parameters a and b, the discrete state space plot (or *Poincarè* map) denotes a fractal-like limit set typical of a chaotic system (see Fig. 13). The time evolution of the state variable x is depicted in Fig. 14. The same behavior can be observed for y, which is proportional to a one-sample delayed sequence of x, as can be seen from (2).

An approach similar to that described in Sect. 3.1 and in [16] could be attempted. In this case a function $F(x(k), d_x(k), y(k), d_y(k)) = (x(k + 1), d_x(k + 1), y(k + 1), d_y(k + 1))$ should be modeled, taking into account uncertainties d_x and d_y for each one of the two variables. However, this fact may lead to a quite complex definition of the qualitative fuzzy model, if compared with the analytical description of the system. In order to avoid these complications, we assume to have only the uncertainty d_x, considering

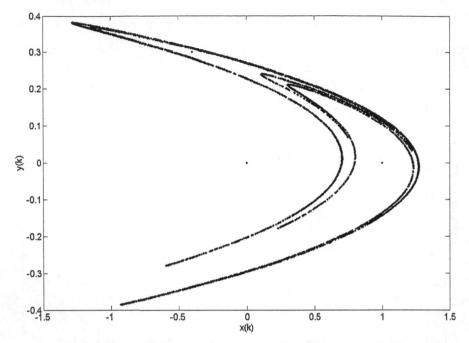

Fig. 13. Henon map state space plot (parameters $a. = 1.4, b = 0.3$)

that it can influence also y through the second equation of (2) in a linear way. $y(k+1)$ being proportional to $x(k)$, the membership function of x and y could be chosen to be identical. However, the slight influence of $y(k)$ on $x(k+1)$ with given parameters ($a = 1.4$ and $b = 0.3$) suggests us to use fewer fuzzy sets for $y(k)$, which acts only if its absolute value is high. The choice for $x(k)$ is therefore similar to that adopted in the logistic map, because of a similar parabolic behavior (but different range). Even in this case there is a stable equilibrium point ($X = 0.63$, $Y = 0.19$) asymmetric with respect to the range of x (and of y).

The fuzzy sets associated with the linguistic values are shown in Fig. 15; they have been constructed in such a way that the nontrivial equilibrium point is between the fuzzy set M and the fuzzy set L. The case of y is simpler: only three fuzzy sets are adopted, only small or large values of y (medium values, close to zero, are in this case neglected) being remarkable for the evolution of x (first equation of (2)). Fuzzy sets of the uncertainty d_x are exactly the same as considered in the previous section, but with different ranges.

Qualitatively x evolves similarly to the logistic map: x tends to move toward the nontrivial equilibrium point, until it begins to oscillate around that point. This happens until the influence of the unstable equilibrium point perturbs its trajectory, moving it out from the neighborhood of the nontrivial

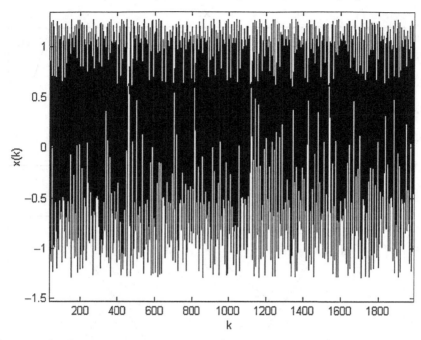

Fig. 14. The chaotic time series generated by the Henon map (parameters $a = 1.4$, $b = 0.3$)

equilibrium point. To this end, when the influence of $y(k)$ on $x(k+1)$ is slight (approximately when $y(k)$ is M), it is possible to keep the same rule of the previous section (see Table 3).

The influence becomes relevant when $y(k)$ is L or S, even if limited to the transition L-VL and S-Z of x. Therefore, when $y(k)$ is S, the new rules that have to be added are the following:

HS_1 if $x(k)$ is Z and $d(k)$ is L and $y(k)$ is S then $x(k+1)$ is Z (instead of S)

HS_2 if $x(k)$ is M and $d(k)$ is VL and $y(k)$ is S then $x(k+1)$ is L (instead of VL)

Table 3. The set of rules for the evaluation of $x(k+1)$ and $d(k+1)$ when $y(k)$ is M. (It is the same as in Table 2)

$x(k)/d(k)$	Z	S	M	L	VL
Z	Z/Z	Z/M	Z/M	S/VL	L/L
S	M/Z	M/M	M/M	M/VL	L/S
M	L/Z	L/M	L/M	L/VL	VL/S
L	M/Z	M/M	M/M	M/VL	Z/S
VL	Z/Z	Z/M	Z/M	Z/VL	Z/L

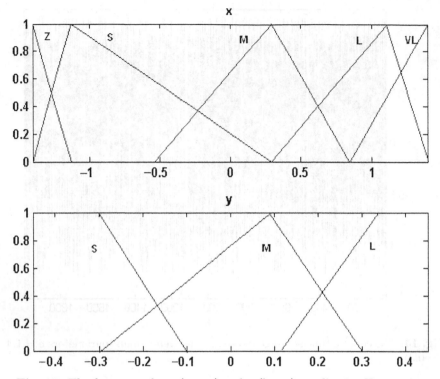

Fig. 15. The fuzzy sets for x (upper) and y (lower): qualitative Henon map

The complete set of rules for this case is given in Table 4. The differences with respect to Table 3 are underlined. The action of y decreases $x(k+1)$ from S to Z or from VL to L.

On the other side, when $y(k)$ is L, only increasing transitions (from L to VL or from S to Z) take place. Therefore, the new rules to be added are the following:

HS_3 if $x(k)$ is Z and $d(k)$ is Z and $y(k)$ is L then $x(k+1)$ is S (instead of Z)

HS_4 if $x(k)$ is Z and $d(k)$ is S and $y(k)$ is L then $x(k+1)$ is S (instead of Z)

HS_5 if $x(k)$ is Z and $d(k)$ is M and $y(k)$ is L then $x(k+1)$ is S (instead of Z)

HS_6 if $x(k)$ is Z and $d(k)$ is VL and $y(k)$ is L then $x(k+1)$ is VL (instead of L)

HS_7 if $x(k)$ is M and $y(k)$ is L then $x(k+1)$ is VL (instead of L)

HS_8 if $x(k)$ is L and $d(k)$ is VL and $y(k)$ is L then $x(k+1)$ is S (instead of Z).

The complete set of rules for this case is shown in Table 5. The changes made with respect to Table 3 are underlined.

In order to complete the whole set of rules for this fuzzy system, the dynamic of $y(k)$ has to be considered. Due to the second equation of (2), these simple rules can be added:

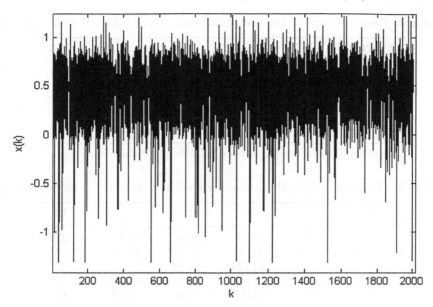

Fig. 16. Time evolution of state variable $x(k)$ generated by the fuzzy system

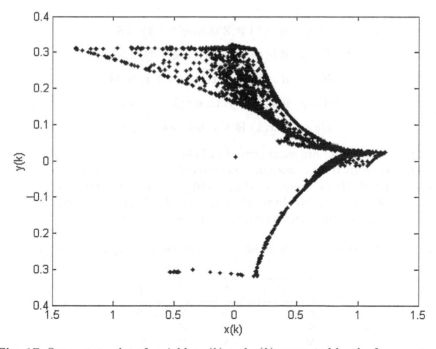

Fig. 17. Space state plot of variables $x(k)$ and $y(k)$ generated by the fuzzy system

Table 4. The set of rules for the evaluation of $x(k+1)$ and $d(k+1)$ when $y(k)$ is S

$x(k)/d(k)$	Z	S	M	L	VL
Z	Z/Z	Z/M	Z/M	Z/VL	L/L
S	M/Z	M/M	M/M	M/VL	L/S
M	L/Z	L/M	L/M	L/VL	L/S
L	M/Z	M/M	M/M	M/VL	Z/S
VL	Z/Z	Z/M	Z/M	Z/VL	Z/L

Note: The differences with respect to Table 3 are underlined

Table 5. The set of rules for the evaluation of $x(k+1)$ and $d(k+1)$ when $y(k)$ is L

$x(k)/d(k)$	Z	S	M	L	VL
Z	S/Z	S/M	S/M	S/VL	VL/L
S	M/Z	M/M	M/M	M/VL	L/S
M	VL/Z	VL/M	VL/M	VL/VL	VL/S
L	M/Z	M/M	M/M	M/VL	S/S
VL	Z/Z	Z/M	Z/M	Z/VL	Z/L

Note: The differences with respect to Table 3 are underlined

HS_9 if $x(k)$ is Z then $y(k+1)$ is S

HS_{10} if $x(k)$ is S then $y(k+1)$ is S

HS_{11} if $x(k)$ is M then $y(k+1)$ is M

HS_{12} if $x(k)$ is L then $y(k+1)$ is L

HS_{13} if $x(k)$ is VL then $y(k+1)$ is L

These statements are summarized in Table 6.

The so-designed fuzzy system is now completed and its dynamic evolution can be derived. By choosing $[x(0)\ y(0)\ d(0)] = [0.01\ 0.01\ 0.01]$ as an initial condition, it is possible to obtain the behavior of $x(k)$ (Fig. 16) and the space state plot representing both variables x and y (Fig. 17).

Table 6. The set of rules which allows to evaluate $y(k+1)$ (it depends only on $x(k)$)

$x(k)/y(k)$	S	M	L
Z	S	S	S
S	S	S	S
M	M	M	M
L	L	L	L
VL	L	L	L

5 Conclusions

In this chapter a qualitative approach for fuzzy modeling of chaotic dynamics has been discussed. This analysis has pointed out several facts regarding both fuzzy logic and chaos theory:

1. Simple fuzzy systems are able to generate complex dynamics.
2. The precision in the approximation of the time series depends only on the number and the shape of the fuzzy sets for x.
3. The "chaoticity" of the system depends only on the shape of the fuzzy sets for d.
4. The analysis of a chaotic system via a linguistic description allows a better understanding of the system itself.
5. Accurate generators of chaos with desired characteristics can be built using the fuzzy model.
6. Multidimensional chaotic maps in some cases do not need a large number of rules in order to be represented.

Future researches on fuzzy modeling of chaotic systems may be developed in several directions. Qualitative analysis should evolve into the automatic design of the fuzzy system, improving the precision of the resulting model. These results could be used for a great number of applications, above all the generation of chaotic signals for cryptographic purposes, which could require well-defined statistic properties of the signals.

References

1. H.O. Wang, K. Tanaka, T. Ikeda: Fuzzy modeling and control of chaotic systems. In: *Proc. ISCAS'96*, pp 209–212
2. O. Calvo, J.L. Cartwright: Int. J. Bifurc. Chaos **8**, 1743–1747 (1998)
3. H.L. Hiew: Inf. Sci. **81**, 193–212 (1994)
4. P.E. Kloeden: Fuzzy Sets Syst. **42**, 37–42 (1991)
5. S. Baglio, L. Fortuna, G. Manganaro: Uncertainty analysis to generate chaotic time series with assigned Lyapunov exponent. In: *Proc. NOLTA'95*, pp 861–864
6. T.S. Parker, L.O. Chua: Proc. IEEE **75**(8), (1987)
7. L.J. Kocarev, K.S. Halle, K. Eckert, L.O. Chua, U. Parlitz: Int. J. Bifurc. Chaos **2**, 709–713 (1992)
8. A.V. Oppenheim, G.W. Wornell, S.H. Isabelle, K.M. Cuomo: Signal processing on the context of chaotic signals. In: *Proc. IEEE ICCASP'92*, pp IV-117–IV-120
9. H. Peitgen, H. Wornell, S.H. Isabelle, K.M. Cuomo: *Chaos and Fractals* (Springer, Berlin Heidelberg New York, 1992)
10. R. Madan: *Chua's Circuit: A Paradigm for Chaos* (World Scientific, Singapore, 1993)
11. P. Arena, R. Caponetto, L. Fortuna, D. Porto: Chaos in fractional order Duffing system. In: *Proc. ECCTD '97*, pp 1259–1272
12. L.A. Zadeh: Inf. Sci. **8**, 199–249 (1975)

13. D. Driankov, H. Hellendoorn, M. Reinfrank: *An Introduction to Fuzzy Control* (Springer, Berlin Heidelberg New York, 1993)
14. S. Baglio, L. Fortuna: A singular value decomposition approach to detect chaos in nonlinear circuits and dynamical systems. In: *IEEE Trans. Circuits Systems-I*, Dec. 1994
15. J.P. Eckmann, S. Oliffson Khamphorst, D. Ruelle, S. Ciliberto: Lyapunov exponents from time series. In: *Coping with Chaos*, ed by E. Ott, T. Saver, J.A. Yourke (Wiley-Interscience, New York, 1994)
16. D.M. Porto, P. Amato: A fuzzy approach for modeling chaotic dynamics with assigned properties. In: *9th IEEE Int. Conf. Fuzzy Syst.* San Antonio, TX, May 2000, pp 435–440

Fuzzy Modeling and Control
of Chaotic Systems

Hua O. Wang and Kazuo Tanaka

Abstract. In this chapter, fuzzy modeling techniques based on Takagi-Sugeno
(TS) fuzzy model are first proposed to model chaotic systems; then, a unified approach is presented for stabilization, synchronization, and chaotic model following
control for the chaotic TS fuzzy systems using linear matrix inequality (LMI) technique; finally, illustrative examples are presented.

1 Introduction

Chaotic behavior is a seemingly random behavior of a deterministic system
that is characterized by sensitive dependence on initial conditions. There are
several distinct definitions of chaotic behavior of dynamical systems. From
a practical point of view, a chaotic motion can be defined as a bounded
invariant motion of a deterministic system that is not an equilibrium solution
or a periodic solution or a quasiperiodic solution [1]. Chaotic behavior of a
physical system can either be desirable or undesirable, depending on the
application. It can be beneficial in many circumstances, such as enhanced
mixing of chemical reactants. Chaos can, on the other hand, entails large
amplitude motions and oscillations that might lead to system failure. The
Ott, Grebogi and Yorke (OGY) method [2, 3] for controlling chaos sparked
significant interest and activity in control of chaos (see, e.g., [4–11] for more
recent developments).

In this chapter we explore the interplay between fuzzy control systems
and chaos. First, we show that fuzzy modeling techniques can be used to
model chaotic dynamical systems, which also implies that fuzzy system can
be chaotic. This is not surprising given the fact that fuzzy systems are essentially nonlinear. This chapter presents a unified approach on the subject of
controlling chaos, [12–17], addressing a number of nonstandard control problems via a linear matrix inequality (LMI) based fuzzy control system design
scheme.

In the literature, most of the work on chaos control have as their goal the
replacement of chaotic behavior by a nonchaotic steady-state behavior. This is
the same as the regulation problem in conventional control engineering. Regulation is no doubt one of the most important problems in control engineering.
For chaotic systems, however, there are a number of interesting nonstandard

H.O. Wang and K. Tanaka: *Fuzzy Modeling and Control of Chaotic Systems*,
StudFuzz **187**, 45–80 (2006)
www.springerlink.com

46   H.O. Wang and K. Tanaka

control problems besides stabilization. In this chapter, we develop a unified approach to address some of these problems including stabilization, synchronization, and chaotic model following control (CMFC) for chaotic systems. The unified approach is based on the Takagi–Sugeno (TS) fuzzy modeling and the associated parallel distributed compensation (PDC) control design methodology [17]. In this framework, a nonlinear dynamical system is first approximated by the TS fuzzy model. In this type of fuzzy model, local dynamics in different state space regions are represented by linear models. The overall model of the system is achieved by fuzzy "blending" of these linear models. The control design is carried out based on the fuzzy model. For each local linear model, a linear feedback control is designed. The resulting overall controller, which is nonlinear in general, is again a fuzzy blending of each individual linear controller. This control design scheme is referred to as the PDC technique in the literature [17]. More importantly, it has been shown in [17] that the associated stability analysis and control design can be aided by convex programming techniques for LMIs.

In this chapter, for chaos control, a cancellation technique (CT) is presented as a main result for stabilization of chaotic systems. The CT also plays an important role in the synchronization and the CMFC. Two cases are considered in the synchronization. The first one deals with the feasible case of the cancellation problem. The other one addresses the infeasible case of the cancellation problem. Furthermore, the CMFC problem, which is more difficult than the synchronization problem, is discussed using the CT method. One of the most important aspects is that the approach described here can be applied not only to stabilization and synchronization but also to the CMFC in the same control framework. That is, it is a rather unified approach to a class of chaos control problems. In fact, the stabilization and the synchronization discussed here can be regarded as a special case of the CMFC. Simulation results demonstrate the utility of the unified design approach.

2 Fuzzy Modeling of Chaotic Systems

To utilize the LMI-based fuzzy system design techniques, we start with representing chaotic systems using TS fuzzy models. In this regard, the techniques described in [17] are employed to construct fuzzy models for chaotic systems. In the following, a number of typical chaotic systems with the control input term added are represented in the TS modeling framework.

Lorenz's equation with input term

$$\dot{x}_1(t) = -ax_1(t) + ax_2(t) + u(t) ,$$
$$\dot{x}_2(t) = cx_1(t) - x_2(t) - x_1(t)x_3(t) ,$$
$$\dot{x}_3(t) = x_1(t)x_2(t) - bx_3(t) ,$$

where a, b, and c are constants and $u(t)$ is the input term. Assume that $x_1(t) \in [-d \quad d]$ and $d > 0$. Then, we can have the following fuzzy model which exactly represents the nonlinear equation under $x_1(t) \in [-d \quad d]$:

Rule 1: IF $x_1(t)$ is M_1 THEN $\dot{x}(t) = A_1 x(t) + B u(t)$,

Rule 2: IF $x_1(t)$ is M_2 THEN $\dot{x}(t) = A_2 x(t) + B u(t)$,

where $x(t) = [x_1(t) \quad x_2(t) \quad x_3(t)]^T$,

$$
A_1 = \begin{bmatrix} -a & a & 0 \\ c & -1 & -d \\ 0 & d & -b \end{bmatrix}, \quad A_2 = \begin{bmatrix} -a & a & 0 \\ c & -1 & d \\ 0 & -d & -b \end{bmatrix},
$$

$$
B = \begin{bmatrix} 1 \\ 0 \\ 0 \end{bmatrix},
$$

$$
M_1(x_1(t)) = \frac{1}{2}\left(1 + \frac{x_1(t)}{d}\right), \quad M_2(x_1(t)) = \frac{1}{2}\left(1 - \frac{x_1(t)}{d}\right).
$$

Here $a = 10$, $b = 8/3$, $c = 28$, and $d = 30$.

Rossler's equation with input term

$$
\dot{x}_1(t) = -x_2(t) - x_3(t) ,
$$
$$
\dot{x}_2(t) = x_1(t) + a x_2(t) ,
$$
$$
\dot{x}_3(t) = b x_1(t) - \{c - x_1(t)\} x_3(t) + u(t) ,
$$

where a, b and c are constants. Assume that $x_1(t) \in [c - d \quad c + d]$ and $d > 0$. Then, we obtain the following fuzzy model which exactly represents the nonlinear equation under $x_1(t) \in [c - d \quad c + d]$:

Rule 1: IF $x_1(t)$ is M_1 THEN $\dot{x}(t) = A_1 x(t) + B u(t)$,

Rule 2: IF $x_1(t)$ is M_2 THEN $\dot{x}(t) = A_2 x(t) + B u(t)$,

where $x(t) = [x_1(t) \quad x_2(t) \quad x_3(t)]^T$,

$$
A_1 = \begin{bmatrix} 0 & -1 & -1 \\ 1 & a & 0 \\ b & 0 & -d \end{bmatrix}, \quad A_2 = \begin{bmatrix} 0 & -1 & -1 \\ 1 & a & 0 \\ b & 0 & d \end{bmatrix},
$$

$$
B = \begin{bmatrix} 0 \\ 0 \\ 1 \end{bmatrix},
$$

$$
M_1(x_1(t)) = \frac{1}{2}\left(1 + \frac{c - x_1(t)}{d}\right), \quad M_2(x_1(t)) = \frac{1}{2}\left(1 - \frac{c - x_1(t)}{d}\right).
$$

Here $a = 0.34$, $b = 0.4$, $c = 4.5$, and $d = 10$.

Duffing forced-oscillation model

$$\dot{x}_1(t) = x_2(t)$$
$$\dot{x}_2(t) = -x_1^3(t) - 0.1x_2(t) + 12\cos(t) + u(t)$$

Assume that $x_1(t) \in [-d \quad d]$ and $d > 0$. Then, we can have the following fuzzy model as well:

Rule 1:   IF $x_1(t)$ is M_1 THEN $\dot{x}(t) = A_1 x(t) + Bu^*(t)$,
Rule 2:   IF $x_1(t)$ is M_2 THEN $\dot{x}(t) = A_2 x(t) + Bu^*(t)$,

where $x(t) = [x_1(t) \quad x_2(t)]^T$ and $u^*(t) = u(t) + 12\cos(t)$,

$$A_1 = \begin{bmatrix} 0 & 1 \\ 0 & -0.1 \end{bmatrix}, \quad A_2 = \begin{bmatrix} 0 & 1 \\ -d^2 & -0.1 \end{bmatrix},$$

$$B = \begin{bmatrix} 0 \\ 1 \end{bmatrix},$$

$$M_1(x_1(t)) = 1 - \frac{x_1^2(t)}{d^2}, \quad M_2(x_1(t)) = \frac{x_1^2(t)}{d^2} .$$

Here $d = 50$.

Henon mapping model

$$x_1(t+1) = -x_1^2(t) + 0.3x_2(t) + 1.4 + u(t) ,$$
$$x_2(t+1) = x_1(t) .$$

Assume that $x_1(t) \in [-d \quad d]$ and $d > 0$. The following equivalent fuzzy model can be constructed as well:

Rule 1:   IF $x_1(t)$ is M_1 THEN $x(t+1) = A_1 x(t) + Bu^*(t)$,
Rule 2:   IF $x_1(t)$ is M_2 THEN $x(t+1) = A_2 x(t) + Bu^*(t)$,

where $x(t) = [x_1(t) \quad x_2(t)]^T$ and $u^*(t) = u(t) + 1.4$,

$$A_1 = \begin{bmatrix} d & 0.3 \\ 1 & 0 \end{bmatrix}, \quad A_2 = \begin{bmatrix} -d & 0.3 \\ 1 & 0 \end{bmatrix},$$

$$B = \begin{bmatrix} 1 \\ 0 \end{bmatrix},$$

$$M_1(x_1(t)) = \frac{1}{2}\left(1 - \frac{x_1(t)}{d}\right), \quad M_2(x_1(t)) = \frac{1}{2}\left(1 + \frac{x_1(t)}{d}\right) .$$

Here $d = 30$.

In all cases above, the fuzzy models exactly represent the original systems. As shown in [17], the TS fuzzy model is a universal approximator for nonlinear dynamical systems. Other chaotic systems can be approximated by the TS fuzzy models.

The fuzzy models above have the common B matrix in the consequent parts and $x_1(t)$ in the premise parts. In this chapter, all the fuzzy models are assumed to be the common B matrix case, i.e., the fuzzy model (1) is considered.

Plant Rule i:

$$\text{If } z_1(t) \text{ is } M_{i1} \text{ and } \cdots \text{ and } z_p(t) \text{ is } M_{ip} \text{ ,}$$

$$\text{then } s\boldsymbol{x}(t) = \boldsymbol{A}_i\boldsymbol{x}(t) + \boldsymbol{B}u(t), \quad i = 1, 2, \ldots, r \text{ ,} \tag{1}$$

where $p = 1$ and $z_1(t) = x_1(t)$. Equation (1) is represented by the defuzzification form

$$s\boldsymbol{x}(t) = \frac{\sum_{i=1}^{r} w_i(\boldsymbol{z}(t)) \{\boldsymbol{A}_i\boldsymbol{x}(t) + \boldsymbol{B}u(t)\}}{\sum_{i=1}^{r} w_i(\boldsymbol{z}(t))}$$

$$= \sum_{i=1}^{r} h_i(\boldsymbol{z}(t)) \{\boldsymbol{A}_i\boldsymbol{x}(t) + \boldsymbol{B}u(t)\} \text{ ,} \tag{2}$$

where $s\boldsymbol{x}(t)$ denotes $\dot{\boldsymbol{x}}(t)$ and $\boldsymbol{x}(t+1)$ for continuous-time fuzzy systems (CFS) and discrete-time fuzzy systems (DFS), respectively. In the fuzzy models above for chaotic systems, $\boldsymbol{z}(t) = z_1(t) = x_1(t)$.

Remark 1. The fuzzy models above have a single input. We can also consider multi-inputs case. For instance, we may consider Lorenz's equation with multi-inputs:

$$\dot{x}_1(t) = -ax_1(t) + ax_2(t) + u_1(t) \text{ ,}$$
$$\dot{x}_2(t) = cx_1(t) - x_2(t) - x_1(t)x_3(t) + u_2(t) \text{ ,}$$
$$\dot{x}_3(t) = x_1(t)x_2(t) - bx_3(t) + u_3(t) \text{ .}$$

Same as before, we can derive the the following fuzzy model to exactly represents the nonlinear equation under $x_1(t) \in [-d \quad d]$:

$$\begin{array}{ll} \text{Rule 1:} & \text{IF } x_1(t) \text{ is } M_1 \text{ THEN } \dot{\boldsymbol{x}}(t) = \boldsymbol{A}_1\boldsymbol{x}(t) + \boldsymbol{B}u(t) \text{ ,} \\ \text{Rule 2:} & \text{IF } x_1(t) \text{ is } M_2 \text{ THEN } \dot{\boldsymbol{x}}(t) = \boldsymbol{A}_2\boldsymbol{x}(t) + \boldsymbol{B}u(t) \text{ ,} \end{array} \tag{3}$$

where $\boldsymbol{u}(t) = [u_1(t) \ u_2(t) \ u_3(t)]^{\mathrm{T}}$ and $\boldsymbol{x}(t) = [x_1(t) \ x_2(t) \ x_3(t)]^{\mathrm{T}}$,

$$\boldsymbol{A}_1 = \begin{bmatrix} -a & a & 0 \\ c & -1 & -d \\ 0 & d & -b \end{bmatrix}, \quad \boldsymbol{A}_2 = \begin{bmatrix} -a & a & 0 \\ c & -1 & d \\ 0 & -d & -b \end{bmatrix},$$

$$\boldsymbol{B} = \begin{bmatrix} 1 & 0 & 0 \\ 0 & 1 & 0 \\ 0 & 0 & 1 \end{bmatrix},$$

$$M_1(x_1(t)) = \frac{1}{2}\left(1 + \frac{x_1(t)}{d}\right), \quad M_2(x_1(t)) = \frac{1}{2}\left(1 - \frac{x_1(t)}{d}\right).$$

This fuzzy model with three inputs is used as a design example later in this chapter.

3 Stabilization

Two techniques for the stabilization of chaotic systems (or nonlinear systems) are presented in this section. We first consider the common stabilization problem followed by a so-called cancellation technique (CT). In particular, the CT plays an important role in synchronization and CMFC, which are discussed in Sects. 4 and 5, respectively.

3.1 Stabilization via Parallel Distributed Compensation

Equation (4) shows the PDC controller for the fuzzy models given in Sect. 2.

$$\begin{array}{lll} \text{Rule 1:} & \text{IF } x_1(t) \text{ is } M_1 \text{ THEN } \boldsymbol{u}(t) = -\boldsymbol{F}_1\boldsymbol{x}(t) , \\ \text{Rule 2:} & \text{IF } x_1(t) \text{ is } M_2 \text{ THEN } \boldsymbol{u}(t) = -\boldsymbol{F}_2\boldsymbol{x}(t) . \end{array} \tag{4}$$

Please note that the chaotic systems under consideration in the previous section are represented (coincidentally) by simple TS fuzzy models with two rules. Therefore the following PDC fuzzy controller also has only two rules:

$$\boldsymbol{u}(t) = -\frac{\sum_{i=1}^{2} w_i(\boldsymbol{z}(t))\boldsymbol{F}_i\boldsymbol{x}(t)}{\sum_{i=1}^{2} w_i(\boldsymbol{z}(t))} = -\sum_{i=1}^{2} h_i(\boldsymbol{z}(t))\boldsymbol{F}_i\boldsymbol{x}(t) . \tag{5}$$

By substituting (5) into (2), we have

$$s\boldsymbol{x}(t) = \sum_{i=1}^{r} h_i(\boldsymbol{z}(t))\big(\boldsymbol{A}_i - \boldsymbol{BF}_i\big)\boldsymbol{x}(t) , \tag{6}$$

where $r = 2$. We recall stable and decay rate fuzzy controller designs for CFS and DFS cases, where the following conditions are simplified due to the common $\boldsymbol{B}$ matrix case. These design conditions are all given for the general TS model with r number of rules.

Stable fuzzy controller design: CFS

Find $\boldsymbol{X} > \boldsymbol{0}$ and $\boldsymbol{M}_i (i = 1 \sim r)$ satisfying

$$-\boldsymbol{X}\boldsymbol{A}_i^{\mathrm{T}} - \boldsymbol{A}_i\boldsymbol{X} + \boldsymbol{M}_i^{\mathrm{T}}\boldsymbol{B}^{\mathrm{T}} + \boldsymbol{B}\boldsymbol{M}_i > \boldsymbol{0} \,,$$

where $\boldsymbol{X} = \boldsymbol{P}^{-1}$ and $\boldsymbol{M}_i = \boldsymbol{F}_i\boldsymbol{X}$.

Stable fuzzy controller design: DFS

Find $\boldsymbol{X} > \boldsymbol{0}$ and $\boldsymbol{M}_i (i = 1 \sim r)$ satisfying

$$\begin{bmatrix} \boldsymbol{X} & \boldsymbol{X}\boldsymbol{A}_i^{\mathrm{T}} - \boldsymbol{M}_i^{\mathrm{T}}\boldsymbol{B}^{\mathrm{T}} \\ \boldsymbol{A}_i\boldsymbol{X} - \boldsymbol{B}\boldsymbol{M}_i & \boldsymbol{X} \end{bmatrix} > \boldsymbol{0} \,,$$

where $\boldsymbol{X} = \boldsymbol{P}^{-1}$ and $\boldsymbol{M}_i = \boldsymbol{F}_i\boldsymbol{X}$.

Decay rate fuzzy controller design: CFS

$$\underset{\boldsymbol{X}, \boldsymbol{M}_1, \dots, \boldsymbol{M}_r}{\text{maximize}} \quad \alpha$$

$$\text{subject to} \quad \boldsymbol{X} > \boldsymbol{0}$$

$$-\boldsymbol{X}\boldsymbol{A}_i^{\mathrm{T}} - \boldsymbol{A}_i\boldsymbol{X} + \boldsymbol{M}_i^{\mathrm{T}}\boldsymbol{B}^{\mathrm{T}} + \boldsymbol{B}\boldsymbol{M}_i - 2\alpha\boldsymbol{X} > \boldsymbol{0} \,,$$

where $\alpha > 0$, $\boldsymbol{X} = \boldsymbol{P}^{-1}$, and $\boldsymbol{M}_i = \boldsymbol{F}_i\boldsymbol{X}$.

Decay rate fuzzy controller design: DFS

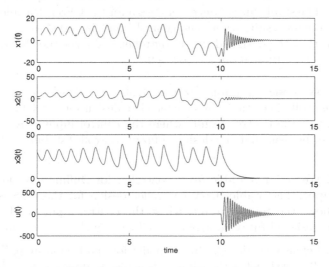

Fig. 1. Control result (Example 1)

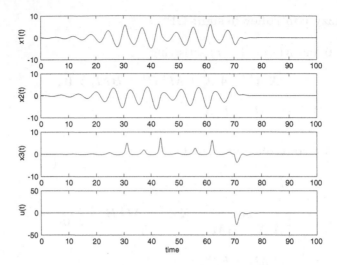

Fig. 2. Control result (Example 2)

$$\begin{array}{cc} \text{minimize} & \beta \\ X, M_1, \ldots, M_r & \end{array}$$

$$\text{subject to} \quad X > 0$$

$$\begin{bmatrix} \beta X & X A_i^{\mathrm{T}} - M_i^{\mathrm{T}} B^{\mathrm{T}} \\ A_i X - B M_i & X \end{bmatrix} > 0 ,$$

where $X = P^{-1}$ and $M_i = F_i X$. It should be noted that $0 \leq \beta < 1$.

Example 1. Let us consider the fuzzy model for Lorenz's equation with the input term. The stable fuzzy controller design for the CFS is feasible. Figure 1 shows the control result, where the control input is added at $t > 10$ (s). It can be seen that the designed fuzzy controller stabilizes the chaotic system, i.e., $x_1(0) \to 0$, $x_2(0) \to 0$, and $x_3(0) \to 0$.

Example 2. We design a stable fuzzy controller for Rossler's equation with the input as well. The stable fuzzy controller design for the CFS is feasible. Figure 2 shows the control result, where the control input is added at $t > 70$ (s). It can be seen that the designed fuzzy controller stabilizes the chaotic system.

Example 3. We design a stable fuzzy controller for Duffing forced-oscillation with the input. The stable fuzzy controller design for the CFS is feasible. Figure 3 shows the control result, where the control input is added at $t > 30$ (s). The designed fuzzy controller stabilizes the chaotic system.

Example 4. Let us consider the fuzzy model for the Henon map. The stable fuzzy controller design for the DFS is feasible. Figure 4 shows the control result, where the control input is added at $t > 20$ (s).

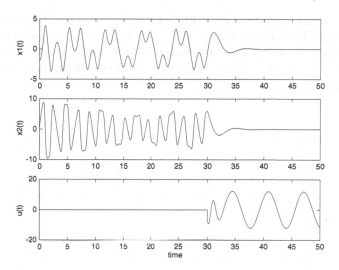

Fig. 3. Control result (Example 3)

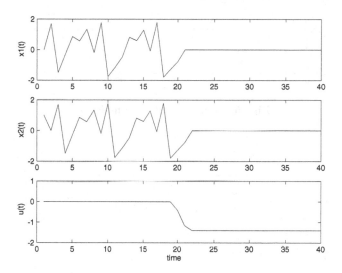

Fig. 4. Control result (Example 4)

Example 5. Consider the fuzzy model for Lorenz's equation with the input term. The decay rate fuzzy controller design for the CFS is feasible. Figure 5 shows the control result, where the control input is added at $t > 10$ (s). Note that the speed of response of the decay rate fuzzy controller is better than that of the stable fuzzy controller in Example 1.

Example 6. Consider the fuzzy model for Lorenz's equation with the input term. The fuzzy controller design satisfying the stability conditions and the constraint on the output for the CFS is feasible, where $\lambda = 9$ and $C = C_1 = C_2 = [1\ 0\ 0]$. This means that $x_1(t)$ is selected as the output, i.e., $y(t) = x_1(t) = Cx(t)$. Figure 6 shows the control result, where the control input is added at $t > 10$ (s). Note that the fuzzy controller satisfies $\max\limits_{t}\|x_1(t)\| \leq \lambda$, but the control effort is very large.

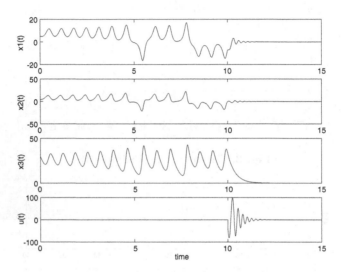

Fig. 5. Control result (Example 5)

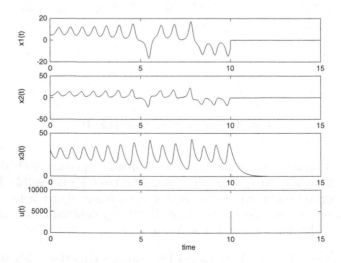

Fig. 6. Control result (Example 6)

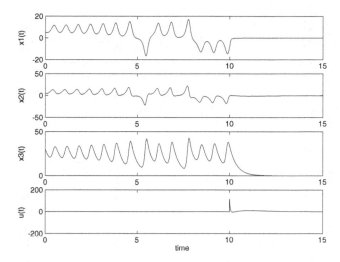

Fig. 7. Control result (Example 7)

Example 7. To solve the excessive control effort problem, the constraint on the control input is added to the design of Example 6. The fuzzy controller design satisfying the stability conditions and the constraints on the output and the control input for the CFS is feasible, where $\lambda = 9$, $\mu = 500$ and $C = C_1 = C_2 = [1\ 0\ 0]$. Figure 7 shows the control result, where the control input is added at $t > 10$ (s). The designed fuzzy controller stabilizes the chaotic system. It should be emphasized that the control input and output satisfy the constraints, i.e., $\max_t \|u(t)\|_2 \leq \mu$ and $\max_t \|x_1(t)\|_2 \leq \lambda$.

Example 8. Consider the Lorenz's equation with three inputs described in Remark 1. The fuzzy controller design satisfying the stability condition and the constraints on the output and the control input for the CFS is feasible, where $\lambda = 9$, $\mu = 500$, and $C = C_1 = C_2 = [1\ 0\ 0]$. Figure 8 shows the control result, where the control input is added at $t > 10$ (s). Note that the control input and output also satisfy the constraints, i.e., $\max_t \|\boldsymbol{u}(t)\|_2 \leq \mu$ and $\max_t \|x_1(t)\|_2 \leq \lambda$.

3.2 Cancellation Technique

This subsection discusses a CT. This approach attempts to cancel the non-linearity of a chaotic system via a PDC controller. If this problem is feasible, the resulting controller can be considered as a solution to the so-called global linearization and the feedback linearization problems. The conditions for re-alizing the cancellation via the PDC are given in the following theorem.

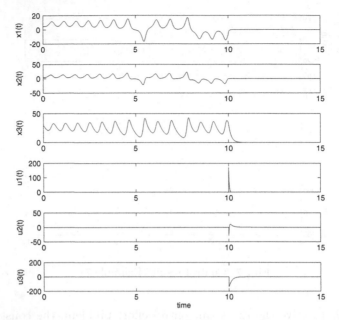

Fig. 8. Control result (Example 8)

Theorem 1. *Chaotic system represented by the fuzzy system* (2) *is exactly linearized via the fuzzy controller* (5) *if there exist the feedback gains F_i such that*

$$\left\{(A_1 - BF_1) - (A_i - BF_i)\right\}^{\mathrm{T}}$$
$$\times \left\{(A_1 - BF_1) - (A_i - BF_i)\right\} = 0, \quad i = 2, 3, \dots, r. \quad (7)$$

Then, the overall control system is linearized as $s\boldsymbol{x}(t) = G\boldsymbol{x}(t)$, where $G = A_1 - BF_1 = A_i - BF_i$.

Proof. It is obvious that $G = A_1 - BF_1 = A_i - BF_i$ if condition (7) holds.

The conditions are applicable to both the CFS and the DFS. If B is a nonsingular matrix, the system is exactly linearized using $F_i = B^{-1}(G - A_i)$. However, the assumption that B is a nonsingular matrix is very strict. If B is not a nonsingular matrix, the conditions of Theorem 1 can still be utilized by the following approximation technique. That is, the equality conditions of Theorem 1 are approximately by the following inequality conditions:

$$X\left\{(A_1 - BF_1) - (A_i - BF_i)\right\}^{\mathrm{T}}$$
$$\times \left\{(A_1 - BF_1) - (A_i - BF_i)\right\}X < \beta S, \quad i = 2, 3, \dots, r,$$

where X is a positive definite matrix and S is a positive definite matrix such that $S^{\mathrm{T}}S < I$. The conditions (7) are likely to be satisfied if the elements in βS are near zero, i.e., $\beta S \approx 0$, in the above inequality. Using Schur complement, we obtain

$$\begin{bmatrix} \beta S \\ \{(A_1 - BF_1) - (A_i - BF_i)\}X \\ X\{(A_1 - BF_1) - (A_i - BF_i)\}^{\mathrm{T}} \\ I \end{bmatrix} > 0, \quad i = 2, 3, \ldots, r.$$

Define $M_i = F_i X$ so that for $X > 0$ we have $F_i = M_i X^{-1}$. Substituting into the inequalities above yields

$$\begin{bmatrix} \beta S \\ \{(A_1 X - BM_1) - (A_i X - BM_i)\} \\ \{(A_1 X - BM_1) - (A_i X - BM_i)\}^{\mathrm{T}} \\ I \end{bmatrix} > 0, \quad i = 2, 3, \ldots, r.$$

Note that G is not always a stable matrix even if the condition of Theorem 1 holds.

From the discussion above as well as the stability conditions described in this section, we define the following design problems using the CT:

Stable fuzzy controller design using the CT: CFS

$$\underset{X, S, M_1, M_2, \ldots, M_r}{\text{minimize}} \quad \beta$$

$$\text{subject to} \quad X > 0, \beta > 0, S > 0$$

$$\begin{bmatrix} I & S \\ S & I \end{bmatrix} > 0,$$

$$-A_i X + BM_i - XA_i^{\mathrm{T}} + M_i^{\mathrm{T}} B^{\mathrm{T}} > 0, \quad i = 1, 2, \ldots, r,$$

$$\begin{bmatrix} \beta S \\ \{(A_1 X - BM_1) - (A_i X - BM_i)\} \\ \{(A_1 X - BM_1) - (A_i X - BM_i)\}^{\mathrm{T}} \\ I \end{bmatrix} > 0, \quad i = 2, 3, \ldots, r,$$

where $X = P^{-1}$ and $M_i = F_i X$.

Stable fuzzy controller design using the CT: DFS

$$\underset{X, S, M_1, M_2, \ldots, M_r}{\text{minimize}} \quad \beta$$

$$\text{subject to} \quad X > 0, \beta > 0, S > 0$$

$$\begin{bmatrix} I & S \\ S & I \end{bmatrix} > 0 \,,$$

$$\begin{bmatrix} X & XA_i - M_i^{\mathrm{T}}B^{\mathrm{T}} \\ A_iX - BM_i & X \end{bmatrix} > 0, \quad i = 1, 2, \ldots, r \,,$$

$$\begin{bmatrix} \beta S & \\ \{(A_1X - BM_1) - (A_iX - BM_i)\} & \end{bmatrix}$$

$$\left.\begin{matrix} \{(A_1X - BM_1) - (A_iX - BM_i)\}^{\mathrm{T}} \\ I \end{matrix}\right] > 0, \quad i = 2, 3, \ldots, r \,,$$

where $X = P^{-1}$ and $M_i = F_iX$.

Decay rate fuzzy controller design using the CT: CFS

$$\underset{X,S,M_1,M_2,\ldots,M_r}{\text{maximize}} \quad \alpha$$

$$\underset{X,S,M_1,M_2,\ldots,M_r}{\text{minimize}} \quad \beta$$

$$\text{subject to} \quad X > 0, \beta > 0, \alpha > 0, S > 0$$

$$\begin{bmatrix} I & S \\ S & I \end{bmatrix} > 0 \,,$$

$$-A_iX + BM_i - XA_i^{\mathrm{T}} + M_i^{\mathrm{T}}B^{\mathrm{T}} - 2\alpha X > 0, \quad i = 1, 2, \ldots, r \,,$$

$$\begin{bmatrix} \beta S & \\ \{(A_1X - BM_1) - (A_iX - BM_i)\} & \end{bmatrix}$$

$$\left.\begin{matrix} \{(A_1X - BM_1) - (A_iX - BM_i)\}^{\mathrm{T}} \\ I \end{matrix}\right] > 0, \quad i = 2, 3, \cdots, r \,,$$

where $X = P^{-1}$ and $M_i = F_iX$.

Decay rate fuzzy controller design using the CT: DFS

$$\underset{X,S,M_1,M_2,\ldots,M_r}{\text{minimize}} \quad \alpha$$

$$\underset{X,S,M_1,M_2,\ldots,M_r}{\text{minimize}} \quad \beta$$

$$\text{subject to} \quad X > 0, \beta > 0, 0 \le \alpha < 1, S > 0$$

$$\begin{bmatrix} I & S \\ S & I \end{bmatrix} > 0 \,,$$

$$\begin{bmatrix} \alpha X & XA_i - M_i^{\mathrm{T}}B^{\mathrm{T}} \\ A_iX - BM_i & X \end{bmatrix} > 0, \quad i = 1, 2, \ldots, r \,,$$

$$
\left[\begin{matrix} \beta S \\ \{(A_1 X - BM_1) - (A_i X - BM_i)\} \end{matrix} \right.
$$

$$
\left. \begin{matrix} \{(A_1 X - BM_1) - (A_i X - BM_i)\}^{\mathrm{T}} \\ I \end{matrix} \right] > 0, \quad i = 2, 3, \ldots, r,
$$

where $X = P^{-1}$ and $M_i = F_i X$.

Remark 2. In the LMIs above, if the elements in $\beta \cdot S$ are near zero, i.e., $\beta \cdot S \approx 0$, the CT problems are feasible. In this case, $G = A_i - BF_i$ for all i and G is a stable matrix.

Remark 3. The decay rate design problems have two parameters α and β to be maximized or minimized. These problems can be solved as follows: For instance, first minimize β, where $\alpha = 0$. After β is fixed, α can be minimized or maximized. This procedure may be repeated to obtain a tighter solution.

Of course, other LMI conditions, e.g., the constraints on control input and output can be added to the design problem. Thus, by combining a variety of control performances represented by LMIs, we can realize multiobjective control.

Example 9. The stable fuzzy controller design to realize the CT for Lorenz's equation with three inputs term is feasible. Figure 9 shows the control result, where the control input is added at $t > 10$ (s). The designed fuzzy controller linearizes and stabilizes the chaotic system.

Example 10. Let us consider the fuzzy model for Rossler's equation with the input term. The stable fuzzy controller design using the CT is feasible. Figure 10 shows the control result, where the control input is added at $t > 70$ (s). It can be seen that the designed fuzzy controller linearizes and stabilizes the chaotic system.

4 Synchronization

In addition to the stabilization of chaotic systems (Sect. 3), chaos synchronization and model following are perhaps more stimulating problems in that chaotic behavior is exploited for potential applications such as secure communications.

In this section, we consider the following synchronization problem: design the control input so that the controlled system achieves asymptotic synchronization with the reference system given the two systems start from different initial conditions. Here the reference system and controlled system are taken

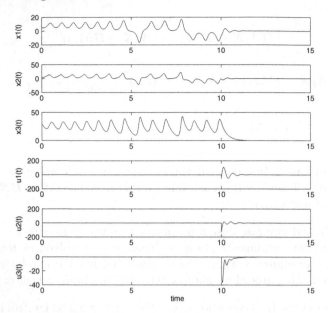

Fig. 9. Control result (Example 9)

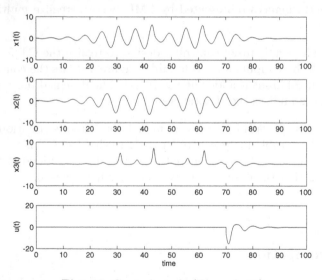

Fig. 10. Control result (Example 10)

to be the same chaotic oscillator except that the controlled system has control input(s) (the controlled system can be viewed as an observer of the reference system). In this section, only the special case of full state feedback based on the CT is considered. Two cases of the cancellation problem are discussed:

Case 1: The cancellation problem is feasible, i.e., all the elements in $\beta \cdot S$ are near zero.

Case 2: The cancellation problem is infeasible, i.e., all the elements in $\beta \cdot S$ are not near zero.

4.1 Case 1

Consider a reference fuzzy model which represents a reference chaotic system.

Reference Rule i:

If $z_{R1}(t)$ is M_{i1} and $\cdots$ and $z_{Rp}(t)$ is M_{ip} ,

$$\text{then } s\boldsymbol{x}_R(t) = \boldsymbol{A}_i \boldsymbol{x}_R(t), \quad i = 1, 2, \ldots, r , \tag{8}$$

where $\boldsymbol{z}_R(t) = [\boldsymbol{z}_{R1}(t) \; \boldsymbol{z}_{R2}(t) \; \cdots \; \boldsymbol{z}_{Rp}(t)]^{\mathrm{T}}$. The defuzzification process is given as

$$s\boldsymbol{x}_R(t) = \sum_{i=1}^{r} h_i(\boldsymbol{z}_R(t)) \boldsymbol{A}_i \boldsymbol{x}_R(t) . \tag{9}$$

Assume that $\boldsymbol{e}(t) = \boldsymbol{x}(t) - \boldsymbol{x}_R(t)$. Then, from (2) and (9), we have

$$s\boldsymbol{e}(t) = \sum_{i=1}^{r} h_i(\boldsymbol{z}(t)) \boldsymbol{A}_i \boldsymbol{x}(t) - \sum_{i=1}^{r} h_i(\boldsymbol{z}_R(t)) \boldsymbol{A}_i \boldsymbol{x}_R(t) + \boldsymbol{B}\boldsymbol{u}(t) . \tag{10}$$

We design two fuzzy subcontrollers to realize the synchronization:

Sub-controller A

Control Rule i:

If $z_1(t)$ is M_{i1} and $\cdots$ and $z_p(t)$ is M_{ip} ,

$$\text{then } \boldsymbol{u}_A(t) = -\boldsymbol{F}_i \boldsymbol{x}(t), \quad i = 1, 2, \ldots, r . \tag{11}$$

Subcontroller B

Control Rule i :

If $z_{R1}(t)$ is M_{i1} and $\cdots$ and $z_{Rp}(t)$ is M_{ip} ,

$$\text{then } \boldsymbol{u}_B(t) = \boldsymbol{F}_i \boldsymbol{x}_R(t), \quad i = 1, 2, \ldots, r . \tag{12}$$

The overall fuzzy controller is constructed by combining the two subcontrollers, i.e.,

$$u(t) = u_A(t) + u_B(t)$$

$$= -\sum_{i=1}^{r} h_i(z(t)) F_i x(t) + \sum_{i=1}^{r} h_i(z_R(t)) F_i x_R(t) \,. \qquad (13)$$

The design is to determine the feedback gains F_i. By substituting (13) into (10), we obtain

$$se(t) = \sum_{i=1}^{r} h_i(z(t))(A_i - BF_i)x(t)$$

$$- \sum_{i=1}^{r} h_i(z_R(t))(A_i - BF_i)x_R(t) \,. \qquad (14)$$

Applying Theorem 1 to the error system (14), we attempt to linearize the error system using the fuzzy control law (13). If the conditions of Theorem 1 hold, the linearized error system becomes $se(t) = Ge(t)$, where $G = A_i - BF_i$. As mentioned before the G is not always a stable matrix even if the conditions of Theorem 1 hold. If we can find feedback gains F_i such that G is a stable matrix, the fuzzy controller linearizes and stabilizes the error system. The linearizable and stable fuzzy controllers with the feedback gains F_i can be designed by solving the LMI-based design problems using the approximate CT algorithm described in Sect. 3.

Example 11. The decay rate fuzzy controller design to realize the synchronization for Lorenz's equation with three input terms is feasible. Figures 11 and 12 show the control result, where the control input is added at $t > 20$ (s) and the initial values of $x(0)$ are slightly different from those of $x_R(0)$. It can be seen that the designed fuzzy controller linearizes and stabilizes the error system, *i.e.*, $e_1(t) \to 0$, $e_2(t) \to 0$, and $e_3(t) \to 0$.

Example 12. Consider the Lorenz's equation with three inputs. The fuzzy controller design satisfying the stability conditions and the constraints on the output and the control input for the CFS is feasible, where $\lambda = 100, \mu = 500$, and $C = C_1 = C_2 = I_3$. This means that $e_1(t), e_2(t)$, and $e_3(t)$ are selected as the outputs, *i.e.*, $e(t) = [e_1(t) \; e_2(t) \; e_3(t)] = Cx(t)$. Figures 13 and 14 show the control result. The designed fuzzy controller linearizes and stabilizes the error system. It should be emphasized that the control input and output satisfy the constraints, *i.e.*, $\max_t \|u(t)\|_2 \le \mu$ and $\max_t \|e(t)\|_2 \le \lambda$.

Example 13. Consider the Rossler's equation with the input term. The fuzzy controller design satisfying the stability conditions and the constraints on the output and the control input for the CFS is feasible, where $\lambda = 10, \mu = 30$, and $C = C_1 = C_2 = I_3$. Figures 15 and 16 show the control result, where the control input is added at $t > 30$ (s). It can be seen that the designed

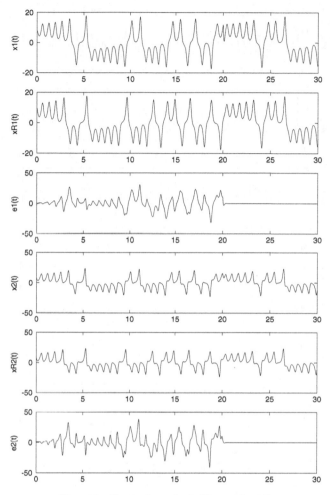

Fig. 11. Control result 1 (Example 11)

fuzzy controller linearizes and stabilizes the error system. Note that the control input and the output satisfy the constraints, i.e., $\max_{t}\|u(t)\|_2 \leq \mu$ and $\max_{t}\|e(t)\|_2 \leq \lambda$.

Example 14. Consider the Rossler's equation with the input term. The fuzzy controller design satisfying the stability conditions and the constraints on the output and the control input for the CFS is feasible, where $\lambda = 10, \mu = 30$, and $C = C_1 = C_2 = I_3$. Figure 17 and 18 show the control result. It can be seen that the designed fuzzy controller linearizes and stabilizes the error system. It should be emphasized that the control input and the output satisfy the constraints, i.e., $\max_{t}\|u(t)\|_2 \leq \mu$ and $\max_{t}\|e(t)\|_2 \leq \lambda$. In addition, note

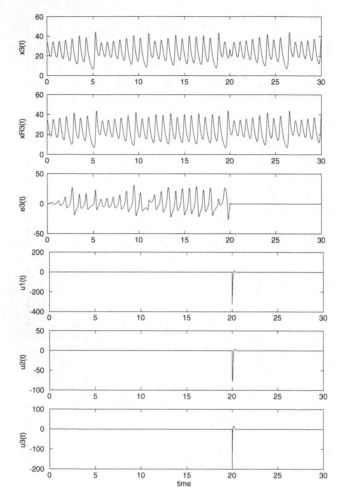

Fig. 12. Control result 2 (Example 11)

that this control result is better than that of Example 13 since the decay rate is considered in the design.

4.2 Case 2

If the cancellation problem is infeasible, i.e., all the elements in $\beta \cdot S$ are not near zero, the error system cannot be linearized. Then, we have

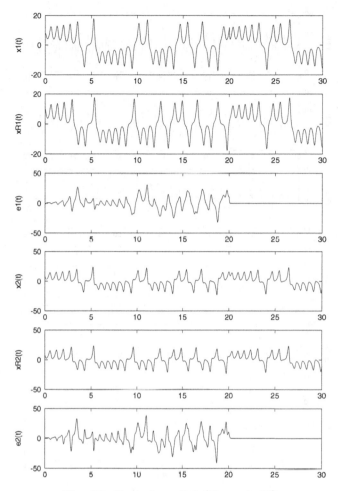

Fig. 13. Control result 1 (Example 12)

$$se(t) = \sum_{i=1}^{r} h_i(\boldsymbol{z}(t))\boldsymbol{A}_i\boldsymbol{x}(t) - \sum_{i=1}^{r} h_i(\boldsymbol{z}_R(t))\boldsymbol{A}_i\boldsymbol{x}_R(t) + \boldsymbol{B}\boldsymbol{u}(t)$$

$$= \sum_{i=1}^{r} h_i(\boldsymbol{z}(t))\boldsymbol{A}_i\boldsymbol{e}(t)$$

$$+ \sum_{i=1}^{r} \{h_i(\boldsymbol{z}(t)) - h_i(\boldsymbol{z}_R(t))\}\boldsymbol{A}_i\boldsymbol{x}_R(t) + \boldsymbol{B}\boldsymbol{u}(t) \,. \tag{15}$$

Assume that $\boldsymbol{z}(t) = \boldsymbol{x}(t)$ and $\boldsymbol{z}_R(t) = \boldsymbol{x}_R(t)$. Then, the second term is almost zero, i.e.,

$$\sum_{i=1}^{r} \{h_i(\boldsymbol{z}(t)) - h_i(\boldsymbol{z}_R(t))\}\boldsymbol{A}_i\boldsymbol{x}_R(t) \approx 0$$

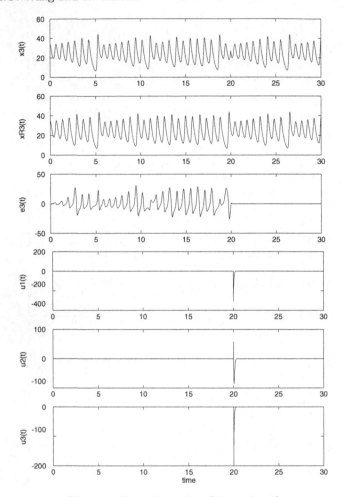

Fig. 14. Control result 2 (Example 12)

if $\|e(t)\| \leq \delta$, where δ is a small value. As a result, the overall system is approximated as

$$\dot{e}(t) = \sum_{i=1}^{r} h_i(z(t)) A_i e(t) + B u(t) \ .$$

Consider the following fuzzy feedback law for the error system:

$$u(t) = \begin{cases} -\sum_{i=1}^{r} h_i(z(t)) F_i e(t) & \|e(t)\| \leq \delta \\ \mathbf{0} & \text{o.w.} \end{cases}$$

Then, if there exist the feedback gains F_i satisfying the stability conditions described in Chap. 3 of [17], the stability of the error system is guaranteed

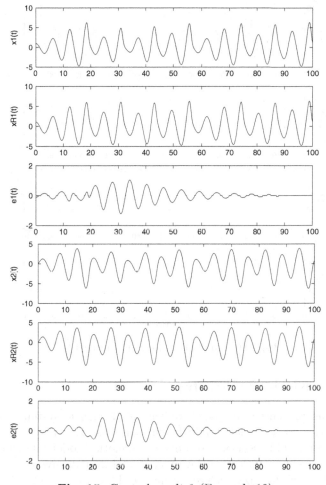

Fig. 15. Control result 1 (Example 13)

near the equilibrium points, i.e., $\|e(t)\| \leq \delta$. The feedback gains $\boldsymbol{F}_i$ can be found by solving the design problems in Sect. 3. It should be noted that this approach guarantees only the local stability. This is the same idea as the OGY method [2]. Therefore, the converging time to an equilibrium point is very long in general, but the control effort is small.

Example 15. We design a stable fuzzy controller for Rossler's equation with the input using the "case 2" design technique. The design problem is feasible. Figures 19 and 20 show the control result, where the control starts at $t = 40$ (s). However, the control input is added around 83 s and stabilizes the error system and the synchronization is realized.

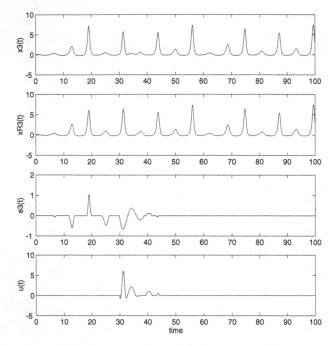

Fig. 16. Control result 2 (Example 13)

5 Chaotic Model Following Control

Section 4 has presented the synchronization of chaotic systems, where A_i matrices of the fuzzy model should be the same as A_i matrices of the fuzzy reference model. This section presents the chaotic model following control (CMFC), where A_i matrices of the fuzzy model do not have to be the same as A_i matrices of the fuzzy reference model. Therefore, the CMFC is more difficult than the synchronization. In this section, the controlled objects are assumed to be chaotic systems. However, note that the CMFC can be designed for general nonlinear systems represented by TS fuzzy models.

Consider a reference fuzzy model which represents a reference chaotic system.

Reference Rule i:

If $z_{R1}(t)$ is N_{i1} and $\cdots$ and $z_{Rp}(t)$ is N_{ip} ,

$$\text{then } s x_R(t) = D_i x_R(t), \quad i = 1, 2, \ldots, r_R . \tag{16}$$

Assume that $x_R(t) \in R^n$ and $A_i \neq D_i$. The defuzzification process is given as

$$s x_R(t) = \sum_{i=1}^{r_R} v_i(z_R(t)) D_i x_R(t) . \tag{17}$$

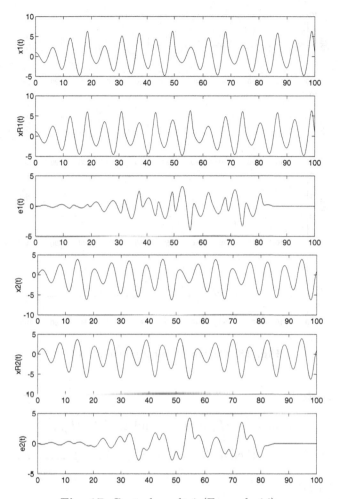

Fig. 17. Control result 1 (Example 14)

The CMFC can be regarded as nonlinear model following control for the reference fuzzy model (17). Assume that $e(t) = x(t) - x_R(t)$. Then, from (2) and (17), we have

$$se(t) = \sum_{i=1}^{r} h_i(z(t)) A_i x(t)$$
$$- \sum_{i=1}^{r_R} v_i(z_R(t)) D_i x_R(t) + B u(t) . \qquad (18)$$

Consider two sub-fuzzy controllers to realize the CMFC:

Subcontroller A

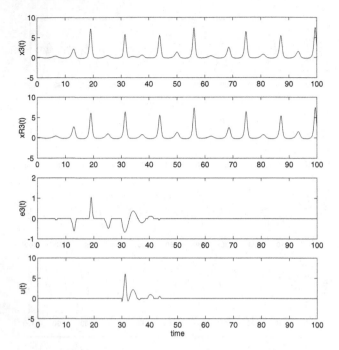

Fig. 18. Control result 2 (Example 14)

Control Rule i:

$$\text{If } z_1(t) \text{ is } M_{i1} \text{ and } \cdots \text{ and } z_p(t) \text{ is } M_{ip} ,$$

$$\text{then } \boldsymbol{u}_A(t) = -\boldsymbol{F}_i\boldsymbol{x}(t), \quad i = 1, 2, \ldots, r . \tag{19}$$

Subcontroller B

Control Rule i:

$$\text{If } z_{R1}(t) \text{ is } N_{i1} \text{ and } \cdots \text{ and } z_{Rp}(t) \text{ is } N_{ip} ,$$

$$\text{then } \boldsymbol{u}_B(t) = \boldsymbol{K}_i\boldsymbol{x}_R(t), \quad i = 1, 2, \ldots, r_R . \tag{20}$$

The combination of the subcontroller A and the subcontroller B is represented as

$$\begin{aligned}
\boldsymbol{u}(t) &= \boldsymbol{u}_A(t) + \boldsymbol{u}_B(t) \\
&= -\sum_{i=1}^{r} h_i(\boldsymbol{z}(t))\boldsymbol{F}_i\boldsymbol{x}(t) + \sum_{i=1}^{r_R} v_i(\boldsymbol{z}_R(t))\boldsymbol{K}_i\boldsymbol{x}_R(t) .
\end{aligned} \tag{21}$$

By substituting (21) into (18), the overall control system is represented as

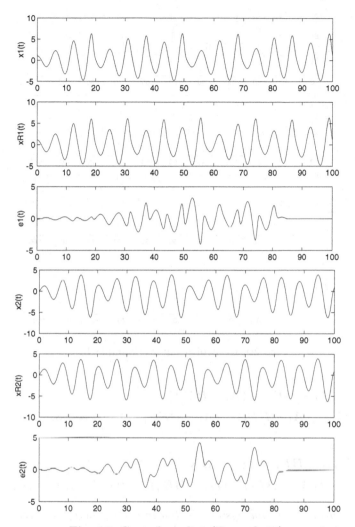

Fig. 19. Control result 1 (Example 15)

$$se(t) = \sum_{i=1}^{r} h_i(\boldsymbol{z}(t))(\boldsymbol{A}_i - \boldsymbol{BF}_i)\boldsymbol{x}(t)$$

$$- \sum_{i=1}^{r_R} v_i(\boldsymbol{z}_R(t))(\boldsymbol{D}_i - \boldsymbol{BK}_i)\boldsymbol{x}_R(t) \,. \tag{22}$$

Theorem 2. *The chaotic system represented by the fuzzy system* (2) *is exactly linearized via the fuzzy controller* (21) *if there exist the feedback gains* $\boldsymbol{F}_i$ *and* $\boldsymbol{K}_j$ *such that*

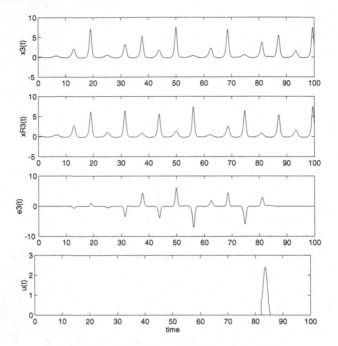

Fig. 20. Control result 2 (Example 15)

$$\left\{(A_1 - BF_1) - (A_i - BF_i)\right\}^{\mathrm{T}} \\ \times \left\{(A_1 - BF_1) - (A_i - BF_i)\right\} = 0, \quad i = 2, 3, \ldots, r \,, \qquad (23)$$

$$\left\{(A_1 - BF_1) - (D_j - BK_j)\right\}^{\mathrm{T}} \\ \times \left\{(A_1 - BF_1) - (D_j - BK_j)\right\} = 0, \quad i = 1, 2, \ldots, r_R \,. \qquad (24)$$

Then, the overall control system is linearized as $s\boldsymbol{x}(t) = \boldsymbol{G}\boldsymbol{x}(t)$, *where* $\boldsymbol{G} = \boldsymbol{A}_1 - \boldsymbol{B}\boldsymbol{F}_1 = \boldsymbol{A}_i - \boldsymbol{B}\boldsymbol{F}_i = \boldsymbol{D}_j - \boldsymbol{B}\boldsymbol{K}_j$.

Proof. It is obvious that $\boldsymbol{G} = \boldsymbol{A}_1 - \boldsymbol{B}\boldsymbol{F}_1 = \boldsymbol{A}_i - \boldsymbol{B}\boldsymbol{F}_i = \boldsymbol{D}_j - \boldsymbol{B}\boldsymbol{K}_j$ if the conditions (23) and (24) hold.

An important remark is in order here.

Remark 4. The CMFC reduces to the synchronization problem when $r = r_R$ and $\boldsymbol{A}_i = \boldsymbol{D}_j$ for $i = 1 \sim r$ and $j = 1 \sim r_R$. The CMFC reduces to the stabilization problem when $\boldsymbol{D}_i = \boldsymbol{0}$ and $\boldsymbol{x}_R(0) = \boldsymbol{0}$ for $i = 1 \sim r_R$. Therefore, as mentioned above, the CMFC problem is more general and difficult than the stabilization and synchronization problems. In addition, the controller design described here can be applied not only to stabilization and synchronization but also to the CMFC in the same control framework. Therefore the LMI-based methodology represents a unified approach to the problem of controlling chaos.

If B is a nonsingular matrix, the error system is exactly linearized and stabilized using $F_i = B^{-1}(G - A_i)$ and $K_i = B^{-1}(G - D_i)$. However, the assumption that B is a nonsingular matrix is very strict. On the other hand, if B is not a nonsingular matrix, Theorem 2 can be utilized by the approximation CT technique. LMI conditions can be derived from Theorem 2 in the same way as described in Sect. 3.

Note that G is not always a stable matrix even if the conditions of Theorem 2 hold. From Theorem 2 and the stability conditions, we define the following design problems:

Stable fuzzy controller design using the CT: CFS

$$\underset{X,S,M_1,M_2,\cdots,M_r}{\text{minimize}} \quad \beta$$

$$\text{subject to} \quad X > 0, \beta > 0, S > 0$$

$$\begin{bmatrix} I & S \\ S & I \end{bmatrix} > 0,$$

$$-A_i X + BM_i - XA_i^{\mathrm{T}} + M_i^{\mathrm{T}} B^{\mathrm{T}} > 0, \quad i = 1, 2, \ldots, r,$$

$$\begin{bmatrix} \beta S & \\ \{(A_1 X - BM_1) - (A_i X - BM_i)\} & \end{bmatrix}$$

$$\left. \{(A_1 X - BM_1) - (A_i X - BM_i)\}^{\mathrm{T}} \\ I \right] > 0, \quad i = 2, 3, \ldots, r,$$

$$\begin{bmatrix} \beta S & \\ \{(A_1 X - BM_1) - (D_j X - BN_j)\} & \end{bmatrix}$$

$$\left. \{(A_1 X - BM_1) - (D_j X - BN_j)\}^{\mathrm{T}} \\ I \right] > 0, \quad j = 1, 2, \ldots, r_R,$$

where $X = P^{-1}$, $M_1 = F_1 X$, $M_i = F_i X$, and $N_j = K_j X$.

Stable fuzzy controller design using the CT: DFS

$$\underset{X,S,M_1,M_2,\ldots,M_r}{\text{minimize}} \quad \beta$$

$$\text{subject to} \quad X > 0, \beta > 0, S > 0$$

$$\begin{bmatrix} I & S \\ S & I \end{bmatrix} > 0,$$

$$\begin{bmatrix} X & XA_i - M_i^{\mathrm{T}}B^{\mathrm{T}} \\ A_iX - BM_i & X \end{bmatrix} > 0, \quad i = 1, 2, \ldots, r,$$

$$\begin{bmatrix} \beta S \\ \{(A_1X - BM_1) - (A_iX - BM_i)\} \end{bmatrix}$$

$$\left. \begin{matrix} \{(A_1X - BM_1) - (A_iX - BM_i)\}^{\mathrm{T}} \\ I \end{matrix} \right] > 0, \quad i = 2, 3, \ldots, r,$$

$$\begin{bmatrix} \beta S \\ \{(A_1X - BM_1) - (D_jX - BN_j)\} \end{bmatrix}$$

$$\left. \begin{matrix} \{(A_1X - BM_1) - (D_jX - BN_j)\}^{\mathrm{T}} \\ I \end{matrix} \right] > 0, \quad j = 1, 2, \ldots, r_R,$$

where $X = P^{-1}$, $M_1 = F_1X$, $M_i = F_iX$, and $N_j = K_jX$.

Decay rate fuzzy controller design using the CT: CFS

$$\underset{X,S,M_1,M_2,\ldots,M_r}{\text{maximize}} \quad \alpha$$

$$\underset{X,S,M_1,M_2,\ldots,M_r}{\text{minimize}} \quad \beta$$

$$\text{subject to} \quad X > 0, \beta > 0, \alpha > 0, S > 0$$

$$\begin{bmatrix} I & S \\ S & I \end{bmatrix} > 0,$$

$$-A_iX + BM_i - XA_i^{\mathrm{T}} + M_i^{\mathrm{T}}B^{\mathrm{T}} - 2\alpha X > 0, \quad i = 1, 2, \cdots, r,$$

$$\begin{bmatrix} \beta S \\ \{(A_1X - BM_1) - (A_iX - BM_i)\} \end{bmatrix}$$

$$\left. \begin{matrix} \{(A_1X - BM_1) - (A_iX - BM_i)\}^{\mathrm{T}} \\ I \end{matrix} \right] > 0, \quad i = 2, 3, \ldots, r,$$

$$\begin{bmatrix} \beta S \\ \{(A_1X - BM_1) - (D_jX - BN_j)\} \end{bmatrix}$$

$$\left. \begin{matrix} \{(A_1X - BM_1) - (D_jX - BN_j)\}^{\mathrm{T}} \\ I \end{matrix} \right] > 0, \quad j = 1, 2, \ldots, r_R,$$

where $X = P^{-1}$, $M_1 = F_1X$, $M_i = F_iX$, and $N_j = K_jX$.

Decay rate fuzzy controller design using the CT: DFS

$$\begin{array}{c} \text{minimize} \\ \boldsymbol{X},\boldsymbol{S},\boldsymbol{M}_1,\boldsymbol{M}_2,...,\boldsymbol{M}_r \end{array} \quad \alpha$$

$$\begin{array}{c} \text{minimize} \\ \boldsymbol{X},\boldsymbol{S},\boldsymbol{M}_1,\boldsymbol{M}_2,\cdots,\boldsymbol{M}_r \end{array} \quad \beta$$

subject to $\quad \boldsymbol{X} > 0, \beta > 0, 0 \le \alpha < 1, \boldsymbol{S} > 0$

$$\begin{bmatrix} \boldsymbol{I} & \boldsymbol{S} \\ \boldsymbol{S} & \boldsymbol{I} \end{bmatrix} > \boldsymbol{0},$$

$$\begin{bmatrix} \alpha\boldsymbol{X} & \boldsymbol{X}\boldsymbol{A}_i - \boldsymbol{M}_i^{\mathrm{T}}\boldsymbol{B}^{\mathrm{T}} \\ \boldsymbol{A}_i\boldsymbol{X} - \boldsymbol{B}\boldsymbol{M}_i & \boldsymbol{X} \end{bmatrix} < \boldsymbol{0}, \quad i = 1, 2, \ldots, r,$$

$$\begin{bmatrix} \beta\boldsymbol{S} \\ \{(\boldsymbol{A}_1\boldsymbol{X} - \boldsymbol{B}\boldsymbol{M}_1) - (\boldsymbol{A}_i\boldsymbol{X} - \boldsymbol{B}\boldsymbol{M}_i)\} \end{bmatrix}$$

$$\left. \begin{array}{c} \{(\boldsymbol{A}_1\boldsymbol{X} - \boldsymbol{B}\boldsymbol{M}_1) - (\boldsymbol{A}_i\boldsymbol{X} - \boldsymbol{B}\boldsymbol{M}_i)\}^{\mathrm{T}} \\ \boldsymbol{I} \end{array} \right] > \boldsymbol{0}, \quad i = 2, 3, \ldots, r$$

$$\begin{bmatrix} \beta\boldsymbol{S} \\ \{(\boldsymbol{A}_1\boldsymbol{X} - \boldsymbol{B}\boldsymbol{M}_1) - (\boldsymbol{D}_j\boldsymbol{X} - \boldsymbol{B}\boldsymbol{N}_j)\} \end{bmatrix}$$

$$\left. \begin{array}{c} \{(\boldsymbol{A}_1\boldsymbol{X} - \boldsymbol{B}\boldsymbol{M}_1) - (\boldsymbol{D}_j\boldsymbol{X} - \boldsymbol{B}\boldsymbol{N}_j)\}^{\mathrm{T}} \\ \boldsymbol{I} \end{array} \right] > \boldsymbol{0}, \quad j = 1, 2, \ldots, r_R,$$

where $\boldsymbol{X} = \boldsymbol{P}^{-1}$, $\boldsymbol{M}_1 = \boldsymbol{F}_1\boldsymbol{X}$, $\boldsymbol{M}_i = \boldsymbol{F}_i\boldsymbol{X}$, and $\boldsymbol{N}_j = \boldsymbol{K}_j\boldsymbol{X}$.

Remark 5. In the LMIs, if all elements in $\beta \cdot \boldsymbol{S}$ are near zero, i.e., $\beta \cdot \boldsymbol{S} \approx 0$, the cancellation problems for decay rate fuzzy controller designs are feasible. In this case, $\boldsymbol{G} = \boldsymbol{A}_1 - \boldsymbol{B}\boldsymbol{F}_1 = \boldsymbol{A}_i - \boldsymbol{B}\boldsymbol{F}_i = \boldsymbol{D}_j - \boldsymbol{B}\boldsymbol{K}_j \ \forall i, j$ and $\boldsymbol{G}$ is a stable matrix.

Example 16. Let us consider the fuzzy model for Lorenz's equation with three inputs term. The parameters are set as follows:

Rule 1:　　IF $x_1(t)$ is M_1 THEN $\dot{\boldsymbol{x}}(t) = \boldsymbol{A}_1\boldsymbol{x}(t) + \boldsymbol{B}\boldsymbol{u}(t)$,
Rule 2 :　　IF $x_1(t)$ is M_2 THEN $\dot{\boldsymbol{x}}(t) = \boldsymbol{A}_2\boldsymbol{x}(t) + \boldsymbol{B}\boldsymbol{u}(t)$,

where $\boldsymbol{x}(t) = [x_1(t) \quad x_2(t) \quad x_3(t)]^{\mathrm{T}}$,

$$\boldsymbol{A}_1 = \begin{bmatrix} -0.5 \cdot a & 0.5 \cdot a & 0 \\ 2 \cdot c & -1 & -d \\ 0 & d & -0.5 \cdot b \end{bmatrix},$$

$$\boldsymbol{A}_2 = \begin{bmatrix} -0.5 \cdot a & 0.5 \cdot a & 0 \\ 2 \cdot c & -1 & d \\ 0 & -d & -0.5 \cdot b \end{bmatrix},$$

$$\boldsymbol{B} = \begin{bmatrix} 1 & 0 & 0 \\ 0 & 1 & 0 \\ 0 & 0 & 1 \end{bmatrix},$$

$$M_1(x_1(t)) = \frac{1}{2}\left(1 + \frac{x_1(t)}{d}\right), \qquad M_2(x_1(t)) = \frac{1}{2}\left(1 - \frac{x_1(t)}{d}\right).$$

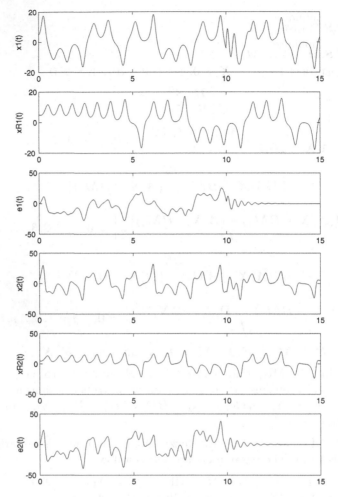

Fig. 21. Control result 1 (Example 16)

Consider the following reference fuzzy model:

Reference Rule 1 :   IF $x_{1R}(t)$ is N_1 THEN $\dot{\boldsymbol{x}}_R(t) = \boldsymbol{D}_1 \boldsymbol{x}_R(t)$,
Reference Rule 2 :   IF $x_{1R}(t)$ is N_2 THEN $\dot{\boldsymbol{x}}_R(t) = \boldsymbol{D}_2 \boldsymbol{x}_R(t)$,

where $\boldsymbol{x}_R(t) = [x_{R1}(t) \quad x_{R2}(t) \quad x_{R3}(t)]^{\mathrm{T}}$,

$$
\boldsymbol{D}_1 = \begin{bmatrix} -a & a & 0 \\ c & -1 & -d \\ 0 & d & -b \end{bmatrix}, \qquad
\boldsymbol{D}_2 = \begin{bmatrix} -a & a & 0 \\ c & -1 & d \\ 0 & -d & -b \end{bmatrix},
$$

$$
N_1(x_{R1}(t)) = \frac{1}{2}\left(1 + \frac{x_{R1}(t)}{d}\right), \qquad
N_2(x_{R1}(t)) = \frac{1}{2}\left(1 - \frac{x_{R1}(t)}{d}\right) ,
$$

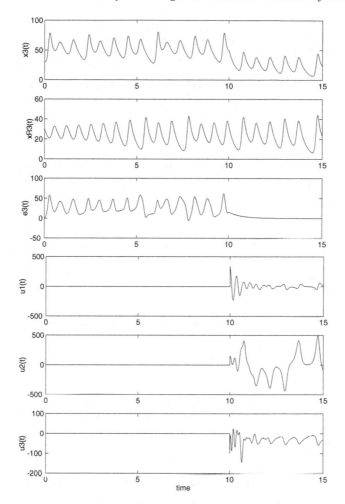

Fig. 22. Control result 2 (Example 16)

where $x_{R1}(t) \in [-d \quad d]$. The stable fuzzy controller design using the CT is feasible. Figures 21 and 22 show the control result, where the control input is added at $t > 10$ (s). It can be seen that the designed fuzzy controller realizes the CMFC i.e., $e_1(t) \rightarrow 0$, $e_2(t) \rightarrow 0$, and $e_3(t) \rightarrow 0$.

Example 17. Let us consider the fuzzy model for Rossler's equation with the input term. The parameters are set as follows:

Rule 1:   IF $x_1(t)$ is M_1 THEN $\dot{\boldsymbol{x}}(t) = \boldsymbol{A}_1\boldsymbol{x}(t) + \boldsymbol{B}\mathrm{u}(t)$,
Rule 2 :   IF $x_1(t)$ is M_2 THEN $\dot{\boldsymbol{x}}(t) = \boldsymbol{A}_2\boldsymbol{x}(t) + \boldsymbol{B}\mathrm{u}(t)$,

where $\boldsymbol{x}(t) = [x_1(t) \quad x_2(t) \quad x_3(t)]^{\mathrm{T}}$,

78   H.O. Wang and K. Tanaka

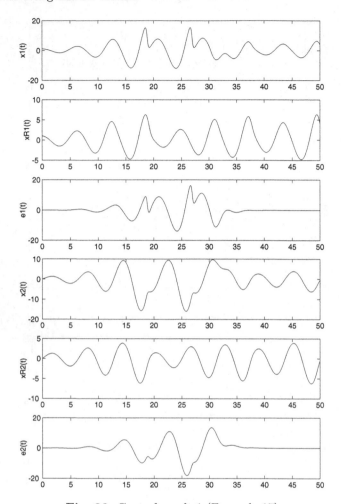

Fig. 23. Control result 1 (Example 17)

$$A_1 = \begin{bmatrix} 0 & -1 & -1 \\ 1 & a & 0 \\ 0.5 \cdot b & 0 & -d \end{bmatrix}, \qquad A_2 = \begin{bmatrix} 0 & -1 & -1 \\ 1 & a & 0 \\ 0.5 \cdot b & 0 & d \end{bmatrix},$$

$$B = \begin{bmatrix} 0 \\ 0 \\ 1 \end{bmatrix},$$

$$M_1(x_1(t)) = \frac{1}{2}\left(1 + \frac{2 \cdot c - x_1(t)}{d}\right), \quad M_2(x_1(t)) = \frac{1}{2}\left(1 - \frac{2 \cdot c - x_1(t)}{d}\right).$$

Consider the following reference fuzzy model:

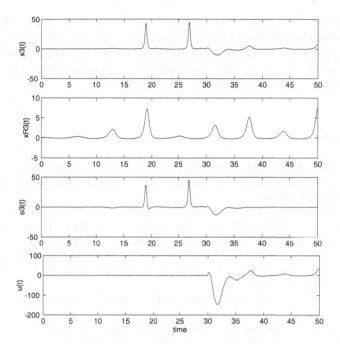

Fig. 24. Control result 2 (Example 17)

Reference Rule 1:   IF $x_{1R}(t)$ is N_1 THEN $\dot{\boldsymbol{x}}_R(t) = \boldsymbol{D}_1 \boldsymbol{x}_R(t)$,
Reference Rule 2:   IF $x_{1R}(t)$ is N_2 THEN $\dot{\boldsymbol{x}}_R(t) = \boldsymbol{D}_2 \boldsymbol{x}_R(t)$,

where $\boldsymbol{x}_R(t) = [x_{R1}(t) \quad x_{R2}(t) \quad x_{R3}(t)]^{\mathrm{T}}$,

$$\boldsymbol{D}_1 = \begin{bmatrix} 0 & -1 & -1 \\ 1 & a & 0 \\ b & 0 & -d \end{bmatrix}, \qquad \boldsymbol{D}_2 = \begin{bmatrix} 0 & -1 & -1 \\ 1 & a & 0 \\ b & 0 & d \end{bmatrix},$$

$$N_1(x_{R1}(t)) = \frac{1}{2}\left(1 + \frac{c - x_{R1}(t)}{d}\right), \quad N_2(x_{R1}(t)) = \frac{1}{2}\left(1 - \frac{c - x_{R1}(t)}{d}\right),$$

where $x_{R1}(t) \in [c-d \quad c+d]$. The stable fuzzy controller design using the CT is feasible. Figures 23 and 24 show the control result, where the control input is added at $t > 30$ (s). The designed fuzzy controller realizes the CMFC.

6 Concluding Remarks

In this chapter we have explored the interplay between fuzzy modeling and control systems and chaos. A comprehensive framework based on TS fuzzy models and PDC control methodology has been proposed for the modeling and control of chaotic systems. We have shown that fuzzy modeling techniques can be used to model chaotic dynamical systems. We have presented a unified approach on controlling chaos, addressing a number of control problems such as stabilization, synchronization and chaotic model following via the LMI-based fuzzy control system design scheme.

References

1. A.H. Nayfeh, B. Balachandran: *Applied Nonlinear Dynamics: Analytical, Computational, and Experimental Methods*, Wiley Series in Nonlinear Science (Wiley, New York, 1995)
2. E. Ott, C. Grebogi, J.A. Yorke: Phys. Rev. Lett. **64**, 1196–1199 (1990)
3. T. Shinbort, C. Grebogi, E. Ott, J.A. Yorke: Nature **363**, 411–417 (1993)
4. G. Chen, X. Dong: Int. J. Bifurcation Chaos **3**(6), 1363–1409 (1993)
5. E.H. Abed, H.O. Wang, A. Tesi: Control of bifurcations and chaos. In: *The Control Handbook*, ed by W.S. Levine (CRC Press & IEEE Press, Boca Raton, FL, 1995) pp. 951–966
6. F.J. Romeiras, C. Grebogi, E. Ott, W.P. Dayawansa: Physica D **58**, 165–192 (1992)
7. E. Ott: *Chaos in Dynamical Systems* (Cambridge, New York, 1993)
8. H.O. Wang, E.H. Abed: Automatica **31**(9), 1213–1226 (1995)
9. L.O. Chua, M. Komuro, T. Matsumoto: IEEE Trans. Circuits Syst. **33**, 1072–1118 (1996)
10. K. Pyragas: Phys. Lett. A **170**, 421–428 (1992)
11. G. Chen, D.J. Hill, X. Yu (ed): *Chaos and Bifurcation Control: Theory and Applications*, Lecture Notes in Control and Information Sciences (Springer, Berlin Heidelberg New York, 2003)
12. H.O. Wang, K. Tanaka, T. Ikeda: Fuzzy modeling and control of chaotic systems. In: *1996 IEEE Int. Symp. on Circuits and Systems*, Atlanta, vol 3, pp 209–212
13. H.O. Wang, K. Tanaka: An LMI-based stable fuzzy control of nonlinear systems and its applications to control of chaos. In: *5th IEEE Int. Conf. on Fuzzy Systems*, New Orleans, 1996, vol 2, pp 1433–1438
14. K. Tanaka, T. Ikeda, H.O. Wang: Controlling chaos via an LMI-based fuzzy control system Design. In: *36th IEEE Conf. on Decision and Control*, San Diego, 1997, vol 2, pp 1488–1493
15. K. Tanaka, T. Ikeda, H.O. Wang: Fuzzy control of chaotic systems using LMIs: Regulation, synchronization and chaos model following. In: *7th Int. IEEE Conf. on Fuzzy Systems*, Alaska, 1998, pp 434–439
16. K. Tanaka, T. Ikeda, H.O. Wang: IEEE Trans. Circuits Syst. **45**(10), 1021–1040 (1998)
17. K. Tanaka, H.O. Wang: *Stability Analysis and Design of Fuzzy Control Systems—A Linear Matrix Inequality Approach*, a research monograph (Wiley, New York, 2001)

Fuzzy Model Identification Using a Hybrid mGA Scheme with Application to Chaotic System Modeling

Ho Jae Lee, Jin Bae Park, and Young Hoon Joo

Abstract. In constructing a successful fuzzy model for a complex chaotic system, identification of its constituent parameters is an important yet difficult problem, which is traditionally tackled by a time-consuming trial-and-error process. In this chapter, we develop an automatic fuzzy-rule-based learning method for approximating the concerned system from a set of input–output data. The approach consists of two stages: (1) Using the hybrid messy genetic algorithm (mGA) together with a new coding technique, both structure and parameters of the zero-order Takagi–Sugeno fuzzy model are coarsely optimized. The mGA is well suited to this task because of its flexible representability of fuzzy inference systems: (2) The identified fuzzy inference system is then fine-tuned by the gradient descent method. In order to demonstrate the usefulness of the proposed scheme, we finally apply the method to approximating the chaotic Mackey–Glass equation.

1 Introduction

Identification of chaotic dynamical systems has recently attracted increasing attention from engineering, physics, mathematics, and biomedical communities [1]. Various difficulties encountered in tackling this subject have posed a real need for using some kind of intelligent approaches. To date, much endeavor have been devoted; to name a few, see [2–7] and references therein. Among them, fuzzy logic theory has been remarkably successful in a wide spectrum of applications such as control [8, 9] as well as identification [2, 6, 7, 10–12] of complex chaotic systems.

Since fuzzy logic controllers and identifiers are implemented by fuzzy inference systems, it is crucial to design such inference systems with minimum numbers of rules and optimized parameters. In general, a fuzzy modeling procedure is composed of two stages: structure identification and parameter identification. There have been many studies on automatic identification technologies of fuzzy inference systems. For example, Horikawa proposed three types of fuzzy neural networks with back-propagation learning and studied an automatic identification method of fuzzy models for nonlinear function approximation problems [5]. These neural network-based methods, however, suffer from convergence to local minima, and their learning performance seriously depends on initial parameter settings.

H.J. Lee et al.: *Fuzzy Model Identification Using a Hybrid mGA Scheme with Application to Chaotic System Modeling*, StudFuzz **187**, 81–97 (2006)
www.springerlink.com

On the contrary, genetic algorithms (GA), which are based on the mechanism of biological genetics and natural selection, are studied as an alternative method for the identification of fuzzy inference systems [2, 6, 7, 12, 13]. However, these works are based on the standard genetic algorithm (SGA) algorithm proposed by Goldberg [14]. To assure convergence to a global optima, strings in GAs must be coded so that short, highly fit allele combination can well combine to form the optima. If the linkage between necessary allele combinations is too weak, GAs will converge to suboptimal solutions [2]. In 1989, Goldberg proposed the messy genetic algorithm (mGA) to tackle this problem and has lately been further developed in [15]. Chowdhury successfully applied the mGA to the design of a fuzzy neural network-based controller for an inverted pendulum [16].

In this chapter, we propose an automatic scheme for identification of fuzzy inference systems with a hybrid algorithm using the mGA and the gradient descent method. The proposed identification method consists of two stages: (1) The first step is to determine the structure and parameters of the fuzzy inference system for the given complex nonlinear, even chaotic, system coarsely by the mGA, which processes the variable-length strings in contrast to conventional GAs that work with fixed-length strings. Since the coding scheme of the mGA is much more flexible than that of the SGA, we use a two-dimensional string representation of the fuzzy inference system:(2) The second step is the procedure that fine-tunes the parameters of the fuzzy model, obtained from the mGA in the first step, by the gradient descent method.

This chapter is organized as follows: Section 2 describes the Takagi–Sugeno fuzzy inference system adopted for identification. Section 3, briefly introduces, the mGA and shows how the fuzzy rule base is represented using the mGA. In Sect. 4, the proposed method is verified through the fuzzy model identification problem of chaotic time series generated by the Mackey–Glass delay differential equation. Conclusions are drawn in Sect. 5.

2 Takagi–Sugeno Fuzzy Systems

A fuzzy model can provide a linguistic description for a complex, ill-defined, and even chaotic systems, in which conventional mathematical model may fail to give a satisfactory result. Fuzzy modeling is an approach to constructing fuzzy inference systems based on given input–output data or knowledge of experienced human experts [10].

In this chapter, we use the zero-order Takagi–Sugeno (TS) fuzzy system, in which the consequent parts are represented as crisp numbers [10, 17]. The ith rule is described by

$$R^i : \text{IF } x_1 \text{ is about } \Gamma_1^i \text{ and } \cdots \text{ and } x_p \text{ is about } \Gamma_p^i$$
$$\text{THEN } y^i = w^i \tag{1}$$

where R^i, $i \in \mathcal{I}_R = \{1, 2, \ldots, r\}$, denotes the ith fuzzy inference rule, Γ_j^i, $j \in \mathcal{I}_P = \{1, 2, \ldots, p\}$, is fuzzy set, $x_j \in \mathbb{R}$ is input variable, $y^i \in \mathbb{R}$ is an output variable, and $w^i \in \mathbb{R}$ takes fuzzy singleton. To measure the membership value of x_j in Γ_j^i, we use the triangular membership function $\Gamma_j^i : U_{x_j} \subset \mathbb{R}_{[a_j^i, c_i^j]} \rightarrow \mathbb{R}_{[0,1]}$ mathematically modeled by

$$\Gamma_j^i(x_j) = \max \left\{ \min \left\{ \frac{x_j - (b_j^i - a_j^i)}{a_j^i}, \frac{b_j^i + c_j^i - x_j}{c_j^i} \right\}, 0 \right\} \qquad (2)$$

where U_{x_j} is the universe of discourse of x_j and the triplet $\{a_j^i, b_j^i, c_j^i\}$, $(i, j) \in \mathcal{I}_R \times \mathcal{I}_P$, are left width, center, and right width for (2). By using singleton fuzzifier, product inference, and center average defuzzifier, (1) is globally inferred as

$$\widehat{y} = \frac{\sum_i^r \theta_i(x) w^i}{\sum_{i=1}^r \theta_i(x)}$$

where $\theta_i(x) = \prod_{j=1}^p \Gamma_j^i(x_j)$.

2.1 Identification Objective

The identification objective is to construct (1) that minimizes the modeling error, which will be defined in the sequel together with the size of fuzzy inference system. This includes the following tasks:

- *Structure identification:* Find a minimal structure of the fuzzy inference system with a minimal number of rules, r.
- *Parameter identification:* Optimize the membership function parameters, a_j^i, b_j^i, c_j^i, and w^i.

Structure identification is related to the selection of the input variables and partition of the input space into some fuzzy subspaces. The determination of the partition of the input space leads to the determination of the number of rules of the TS fuzzy model in our case. Parameter identification is related to describing an input–output relation in each subspace.

3 Fuzzy Model Identification by Using mGA Hybrid Scheme

3.1 GA Preliminaries

GAs are optimization methods in which a stochastic search algorithm is performed based on the basic biological principles of selection, crossover, and mutation. A GA scheme encodes each point in a solution space into a string composing of binary, integer, or real values, called a chromosome. Each point is assigned a fitness value from zero to one, which is usually taken to be the same as the objective function to be maximized, although they can be different.

Unlike other optimization methods, such as the familiar gradient descent method, a GA scheme keeps a set of points as a population, which is evolved repeatedly toward a better and then even better fitness value. In each generation, the GA generates a new population using genetic operators such as crossover and mutation. Through these operations, individuals with higher fitness values are more likely to survive and to participate in the next genetic operations. After a number of generations, individuals with higher fitness values are kept in the population while the others are eliminated. GAs, therefore, can ensure a gradual increasing of improving solutions, till a desired optimal or suboptimal solution is obtained.

A pseudocode outline of a usual GA is shown in Algorithm 1, where the population at time t is represented by a time-dependent variable, $P = P(t)$, with an initial population $P(0)$, which can be randomly estimated when the algorithm is run for an application.

Algorithm 1 GA elements

1: **set** $t = 0$; **initialize** $P(t)$.
2: **evaluate** $P(t)$.
3: **while** not finished **do**
4:     $t \Leftarrow t + 1$.
5:     **reproduce** $P(t)$ from $P(t-1)$.
6:     **crossover** individuals in $P(t)$.
7:     **mutate** individuals in $P(t)$.
8:     **evaluate** $P(t)$.
9: **end while**

GA was initiated by Holland [18] and has been applied to fuzzy modeling and fuzzy control since the late 1980s. Recently, the GA has been successfully applied to a wide variety of problems such as search, optimization, and machine learning in science, commerce [7], and engineering fields [12, 13, 16]. The major reason for this wide range of application is that GA searches for optima within the entire complex solution space, which grants the robustness against localization of an optimal solution.

3.2 The mGA

Despite empirical success of GAs, there have been some objections to their use. The most crucial objection is the so-called *linkage problem* [15]. The linkage problem arises because of the coding of the problem parameters. To guarantee convergence to the global optima, strings in GA must be coded so that highly fit allele combinations can well combine to form the optima. Unfortunately, the necessary linkages associated with a given problem are usually not known. mGA works by searching for tight building blocks and then combining them together to form the optima in a way that respects a version of the schema theorem [15, 19]. Unlike the conventional GAs, mGAs use variable-length strings that may be overspecified or underspecified with respect to the problem being solved. mGAs use simple *cut-and-splice* operators rather than fixed-length crossover operations. As shown in Fig. 1, an mGA divides the evolutionary into two phases: a primordial phase and a juxtapositional phase.

In the primordial phase, individual that just evolves out of many candidate strings of a population is selected; that is, there is no evolution, but only the selection operation is used to increasingly sample the highly fit strings and the size of the population is reduced during the selection process. As shown in Fig. 1, the size is fixed after certain generations. In the juxtapositional phase, instead of crossover and mutation used in conventional GAs, the cut-and-splice operator is used to evolve individuals.

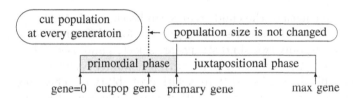

Fig. 1. Typical population reduction schedule

3.3 mGA Operators

In a crossover operation of the conventional GA, the position of the crossover point for two parent individuals has a common locus. But the standard crossover operator is no longer suitable to handle strings of variable length. Therefore, mGAs use a cut-and-splice operator instead. A schematic diagram of the cut-and-splice operation is shown as Fig. 2. The cut operator simply cuts the string in two parts at a randomly chosen position. The splice operator concatenates two strings, which could have been previously cut, in a randomly chosen order. When the cut-and-splice operators are applied simultaneously to two parent strings they alter in a similar way to the ordinary crossover operator.

The cut-and-splice operation is similar to the crossover operation of the conventional GA, but there are many differences between them. While the parent strings have the same crossover point in the crossover operation, the cut-and-splice operation strings do not need to have the same crossover point.

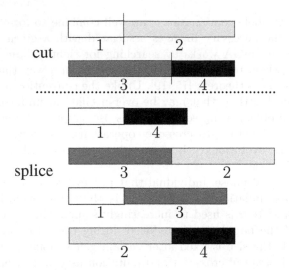

Fig. 2. The cut-and-splice operation

Therefore, the length of the child strings can be changed. As shown in Fig. 2, it is possible that two parent strings generate four different child strings. In real applications, only two child strings are selected and used. The cut operator cuts a string at an arbitrary point with probability p_c. The splice operator splices two arbitrarily selected strings with probability p_s. These probabilities are applied as mating rate and mutation rate in the conventional GA.

3.4 Coding Method for Fuzzy Modeling Using mGA

In the conventional mGA, genes are composed of the index of a gene and the value corresponds to it. For example, a gene $(1, 3)$ corresponds to the first gene in the string whose allele value is 3. Unlike the conventional GA, the order of genes in the string is not important in mGA, i.e., the strings $\{(1,3)(3,1)(2,1)\}$ and $\{(2,1)(1,3)(3,1)\}$ are considered identical. Notice that we have not required all genes to be present, nor have we precluded the possibility of multiple, possibly contradictory, genes. For example, string $\{(1,3)(2,1)\}$ and string $\{(1,3)(2,1)(3,2)(1,1)\}$ are both valid. The former is said to be underspecified because there is no gene that corresponds to the third gene $(3, \bullet)$, and the latter is said to be overspecified. However, in most problems, the string needs to be fully complemented with genes. In mGA, this underspecification problem is easily tackled by using *templates*. Templates are used to fill unspecified genes in the string with locally optimal solutions. Since a local optimal solution is not known a priori, the levelwise mGA is adopted in this chapter, which is said to be overspecified. One way of solving this problem is to select a gene from among conflicted genes based on the *first-come-first-serve* rule.

		premise part					consequent part	weight
	1	2	3	4	5	6	7	8
rule $\{$ 1	0.0694	0.3046	0.0076	0.4659	0.6822	0.3361	0.3406	1
number $\{$ 2	0.1014	0.1897	0.3734	0.2330	0.3028	0.4191	0.1897	0

Fig. 3. An example of parameter matrix of fuzzy inference system

The next step is to design a proper structure of string that best represents a given fuzzy inference system. Herein, we use the zero-order TS fuzzy model in a scatter partition style. This string may contain one or more substring(s). Each string, therefore, contains a possible solution to the problem. Fitness function is used to evaluate how well a string solves the problem. In the conventional mGA coding of a fuzzy inference system, integer number coding is adopted to represent fuzzy inference systems [20]. In this chapter, real number coding and integer number coding are used. The parameters and the structure of the fuzzy model are encoded into one or more substring(s) in the string. Parameters and structure of the fuzzy inference system can be represented in a two-dimensional matrix form as shown in Fig. 3.

Figure 3 is an example of a parameter matrix of a fuzzy inference system and of the raw structure of string not represented by the mGA coding. The fuzzy inference system in Fig. 3 has two inputs and one output. Here, we slightly modify the coding of the conventional mGA string to effectively represent the fuzzy inference system.

In the proposed method, one gene is composed of three elements, i.e., the gene $\{(i, j, h)\}$ corresponds to the (i, j)th element in the parameter matrix with value h. Figure 4 shows an example of string and decoding of the string, while the obtained fuzzy inference system is shown in Fig. 5.

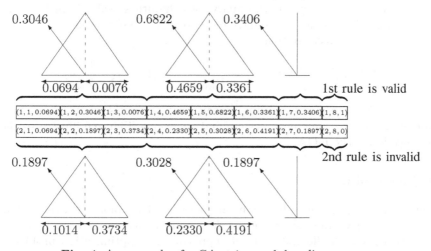

Fig. 4. An example of mGA string and decoding process

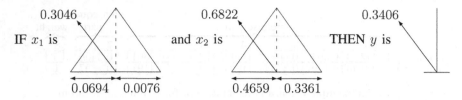

Fig. 5. The fuzzy inference system obtained by the proposed decoding process

As can be seen above, the proposed mGA coding scheme has a more flexible structure than those of the conventional mGA coding schemes. It is more suitable to fuzzy modeling of chaotic nonlinear systems. By using the above structure, the new string can efficiently describe the accuracy and the size of the fuzzy inference system.

3.5 Multiobjective Fitness Function

Since the GA is guided by the fitness function, we also have to consider more delicate fitness functions to determine the accuracy and the size of the fuzzy inference system. The performance measures of accuracy and size of the fuzzy inference system are

$$\mathcal{P}_{\text{accuracy}} = \frac{1}{n} \sum_{k=1}^{n} (\widehat{y}_k - y_k)^2$$

$$\mathcal{P}_{\text{size}} = r$$

where y_k and $\widehat{y}_k$ are training and inferred outputs matched with the training input data $\{\{x_j\}_{j \in \mathcal{I}_P}^h\}_{h \in \mathcal{I}_N}$. The purpose of identification is to reduce $\mathcal{P}_{\text{accuracy}}$ and $\mathcal{P}_{\text{size}}$. However, the mGA uses the fitness value, which has to be maximized in general. So we have to determine the performance index for fitness function transformation:

$$f(\mathcal{P}_{\text{accuracy}}, \mathcal{P}_{\text{size}}) = \lambda \frac{1}{1 + \mathcal{P}_{\text{accuracy}}} + (1 - \lambda) \frac{1}{1 + \mathcal{P}_{\text{size}}} \tag{3}$$

where $\lambda \in \mathbb{R}_{[0,1]}$ is the weighting factor; large λ values would result in a more accurate fuzzy inference system and vice versa. Figure 6 illustrates the pseudocode for the mGA.

3.6 Fine-tuning

As stated in the previous subsection, the fact that GA exploits only the coding and the objective function value to determine plausible trials in the next generation gives the flexibility for its application to optimization problems. Since GA is a parallel searching algorithm through the solution space, the local convergence problem does not appear. However, GA is a blind search algorithm; hence it has disadvantages when compared to other methods that make use of problem-specific information. Although GA can effectively find a near global optimal solution, it will take a long time to converge to the optimal solution. When problem-specific information exists, it may be advantageous to combine this information with GA to improve the ultimate genetic search performance and to guarantee the convergence to a global optimum.

In the subsequent procedure, the gradient descent method carries out the identification of parameters, which fine-tunes the parameters of the membership functions in premise part and the fuzzy singletones in consequence, after achieving a simultaneous parameter/structure identification of fuzzy model by the mGA. Using this hybrid scheme, we are able to identify the best fuzzy model in the view of global optimization.

In this hybrid scheme, we set an objective function to be minimized as

$$\mathcal{P}_{\mathrm{FT}} = \frac{1}{n} \sum_{k=1}^{n} (\widehat{y}_k - y_k)^2$$

The parameter update rules for a fuzzy inference system can be easily derived by using the chain rule. The following are the obtained learning rules:

$$a_j^i(k+1) = a_j^i(k) - \kappa_a \frac{\partial \mathcal{P}_{\mathrm{FT}}}{\partial a_j^i}$$

$$b_j^i(k+1) = b_j^i(k) - \kappa_b \frac{\partial \mathcal{P}_{\mathrm{FT}}}{\partial b_j^i}$$

$$c_j^i(k+1) = c_j^i(k) - \kappa_c \frac{\partial \mathcal{P}_{\mathrm{FT}}}{\partial c_j^i}$$

$$w^i(k+1) = w^i(k) - \kappa_w \frac{\partial \mathcal{P}_{\mathrm{FT}}}{\partial w^i}$$

where κ_a, κ_b, κ_c, and κ_w are learning rates and the partial derivative are further computed by

$$\frac{\partial \mathcal{P}_{\mathrm{FT}}}{\partial a_j^i} = \frac{1}{\sum_{i=1}^{r} \theta_i(x)} (\widehat{y} - y)(w^i - \widehat{y}) \frac{\partial \theta_i(x)}{\partial a_j^i}$$

$$\frac{\partial \mathcal{P}_{\mathrm{FT}}}{\partial b_j^i} = \frac{1}{\sum_{i=1}^{r} \theta_i(x)} (\widehat{y} - y)(w^i - \widehat{y}) \frac{\partial \theta_i(x)}{\partial b_j^i}$$

$$\frac{\partial \mathcal{P}_{\mathrm{FT}}}{\partial c_j^i} = \frac{1}{\sum_{i=1}^{r} \theta_i(x)} (\widehat{y} - y)(w^i - \widehat{y}) \frac{\partial \theta_i(x)}{\partial c_j^i}$$

$$\frac{\partial \mathcal{P}_{\mathrm{FT}}}{\partial w^i} = \frac{\theta_i(x)}{\sum_{i=1}^{r} \theta_i(x)} (\widehat{y} - y) \ .$$

```
 1  void mGA(maxera, popsize, maxgen, pcut, psplice, pmut) {
 2      template = make_initial_template();
 3      for (era = 1; era <= maxera; era++) {pop = init_pop(era);
 4          fitness = evalpopu(pop);
 5          // Check if there are better individual in the initialized pop
 6          // than TEMPLATE
 7          if (check_init_pop(pop)) {
 8              // Primordial phase
 9              while(primoridal_phase) {
10                  pop = binary_tournament_selection(pop);
11              }
12              // Juxtapositionl phase
13              while(juxtapositional_phase) {
14                  fitness = evalpopu(pop);
15                  // Generate next population via cut, splice,
16  // and mutation
17                  pop = nextpopu(pop, fitness, pcut, psplice, pmut);
18              }
19              // Update template
20              best = find_best(pop);
21              template = string2template(best);
22          }
23  // No individual in the initial population is better than the template
24          else {
25              // Do nothing
26  } } }
```

Fig. 6. Pseudocode of mGA

3.7 Summary: Algorithm Description

Step 1: Obtain a sample input–output data set for identification. The initial parameters such as population size, maximum number of rules of the fuzzy inference system, maximum number of generations, or the tolerance, cut, splice, and mutation rate are initially provided by the designer.

Step 2: Perform the mGA identification routine shown in Fig. 6.

Step 3: Choose the best individual and using this individual construct initial parameters of the fuzzy inference system. Apply the gradient descent algorithm to fine-tune the parameters of the fuzzy inference system.

4 An Example: The Chaotic Mackey–Glass Time Series

We seek to find a TS fuzzy model of the Mackey–Glass chaotic equation. The Mackey–Glass equation is a time delay differential equation first proposed as a model of white blood cell production in a human body [21]. Because rates of stem cell proliferation entail a time delay, periodic dynamics and chaos can be obtained. Indeed, Mackey and Glass have suggested that long-term fluctuations in cell counts observed in certain forms of leukemia are evidence for these behaviors in vivo:

$$\frac{dx}{dt} = \frac{ax(t-\tau)}{1 + x(t-\tau)^c} - bx(t) \qquad (4)$$

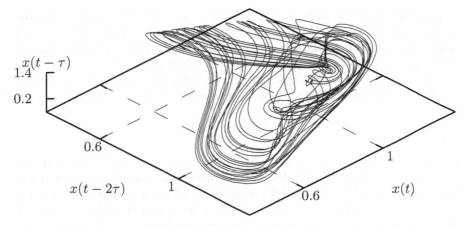

Fig. 7. Mackey–Glass chaotic attractor

where $x(t) \in \mathbb{R}$ is a change of blood cell concentration; the constants are commonly chosen as $a = 0.2$, $b = 0.1$, and $c = 10$. The delay parameter τ determines the behavior of (4): for $\tau < 4.53$, there is a stable fixed point attractor; for $4.53 \leqslant \tau < 13.3$, there is a stable limit cycle attractor; period doubling begins at $\tau = 13.3$ and continues until $\tau = 16.8$; for $\tau > 16.8$, the chaos emerges from (4). Figure 7 shows a spatial orbit of $(x(t), x(t-\tau), x(t-2\tau))$, where $\tau = 17$ and $x(t) = 1.2$, $t \in \mathbb{R}_{\leqslant 0}$.

Our goal is to find some TS fuzzy inference system to produce a prediction of $x(k + D)$, $\widehat{y}$ by using a subset of the time series data up to k. Mathematically, the task is formulated to identify some chaotic mapping F_D with several parameters as

$$\widehat{y} = F_D(x(k), x(k - \Delta), \dots, x(k - (p - 1)\Delta)$$

in terms of zero-order TS fuzzy inference system, where Δ is a lag time and p is an embedding dimension. Mapping F_D implies that prediction of x, $\widehat{x}$ at the time ahead D can be obtained through a proper combinations of p points of the Mackey–Glass series space Δ apart. We choose the parameters as $p = 4$ and $\Delta = D = 6$, i.e., $\{x_i\}_{i \in \mathcal{I}_4} = \{x(k-18), x(k-12), x(k-6), x(k)\}$. Numerical pointwise solutions are generated based on *dde23.m* function in MATLAB with an initial condition $x(0) = 1.2$, a history function $x(t) = 1.2$ for $t \in \mathbb{R}_{\leqslant 0}$, and time lag $\tau = 17$, from which 1000 input–output data pairs, $\{x_i; y\}_{i \in \mathcal{I}_4} = \{x(k - 18), x(k - 12), x(k - 6), x(k); x(k + 6)\}$, $k \in \mathcal{I}_{1023} = \{24, 25, \dots, 1023\}$, are extracted. For convenience, all training data pair are normalized to be between 0 and 1. In order to use levelwise processing of the mGA, we first randomly select an initial template. The standard mGA provides all possible building blocks as an initial population in each level. However, it is impossible for the standard mGA to provide all possible building blocks as an initial population since real number coding is adopted in the

proposed method. We solve this problem by simply constructing a large number of strings in the initial population. The mean-square-error (MSE) is used
as the cost function. Although we generate initial population and template
at random, we apply some problem-specific information in the initialization
process. First, the initial widths of the membership functions are random real
numbers of the interval $[0, W]$, which divides input space properly:

$$W = \frac{2(x_{\max} - x_{\min})}{R}$$

where $x_{\min} = \min_{k \in \mathcal{I}_{[24,1023]}} x(k)$, $x_{\max} = \max_{k \in \mathcal{I}_{[24,1023]}} x(k)$, and R is the
initial number of the fuzzy rules. Second, we extend the initial range of the
input space, in which the initial center values of membership functions are
randomly set to improve the accuracy of the fuzzy model in the extremes of
the input space [22]. In this example, we use the extend input space $[0.2, 1.4]$.

Initial parameters for the mGA hybrid identification scheme are as follows:
Maximum number of era is 10, maximum size of initial population in each
era is 500, population size in juxtapositional phase is 400, splice probability
is $p_s = 1.0$, cut probability is $p_c = 0.2$, mutation probability is 0.2, and λ is
0.95. Large λ value means that we would have a more accurate fuzzy model
while the size of the fuzzy model is not reduced. The fitness function (3) is
adopted in the simulation.

Figure 8 shows the change of $f(\mathcal{P}_{\text{accuracy}}, \mathcal{P}_{\text{size}})$ during the mGA optimization process. The change of $f(\mathcal{P}_{\text{accuracy}}, \mathcal{P}_{\text{size}})$ from era 4 to era 10 is

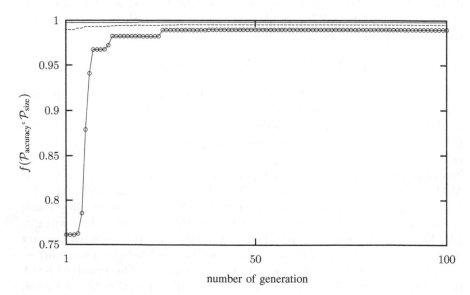

Fig. 8. Change of $f(\mathcal{P}_{\text{accuracy}}, \mathcal{P}_{\text{size}})$ in the mGA tuning state: first era (line with
circle), second era (dashed line), third era (solid line)

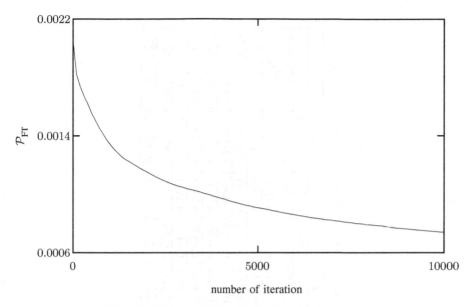

Fig. 9. Change of $\mathcal{P}_{\mathrm{FT}}$ in the fine–tuning state

omitted since there are no better individual in the initial population than the template. The value of $f(\mathcal{P}_{\mathrm{accuracy}}, \mathcal{P}_{\mathrm{size}})$ from the obtained fuzzy model is 0.9979. In the fine-tuning stage, the learning rates κ_a, κ_b, κ_c, and κ_w are 1.0^{-5}, 1.0^{-5}, 1.0^{-5}, and 1.0^{-4}, respectively. The number of iteration is 10 000. Figure 9 shows the change of $\mathcal{P}_{\mathrm{FT}}$ in the fine-tuning stage. Since we focus on the accuracy of the fuzzy model, the MSE of the fuzzy model obtained by the proposed method is superior to other methods. However, the number of fuzzy rules are not reduced very much.

The parameters of the fuzzy model identified by the proposed method are listed in Table 1 and its membership function are illustrated in Fig. 10 for visual understanding. In Table 2, we compare the performance of our fuzzy model with other models, in which our model outperforms the method in [7] in both performance and number of rules. Figure 11 compares the actual output of the Mackey–Glass and the output from the identified fuzzy model.

Table 1. Parameters of identified fuzzy model of the Mackey–Glass chaotic time series by the proposed method

Rule Number	x_1			x_2			x_3			x_4			y^i
	a_1^i	b_1^i	c_1^i	a_2^i	b_2^i	c_2^i	a_3^i	b_3^i	c_3^i	a_4^i	b_4^i	c_4^i	w_i
1	0.4516	0.3785	0.7870	0.3235	0.3698	0.7533	0.6086	0.4974	1.1744	0.2725	0.5168	0.7524	1.0539
2	0.9263	0.9894	0.7755	0.7043	0.3078	0.3865	0.7163	0.6217	0.1928	0.1276	-0.0076	0.3353	0.1684
3	0.00971	0.3045	0.4799	0.5097	0.1680	0.7809	0.6503	0.8340	0.3695	0.7638	0.4231	0.4147	0.9438
4	0.6157	0.1456	0.9164	0.9912	0.1601	0.7949	0.7845	0.4209	0.9854	0.5430	0.2054	0.6313	0.4223
5	0.7817	0.6271	0.2313	0.7771	0.5276	0.6778	0.4822	0.8355	0.5953	0.0093	0.0028	0.3266	0.6453
6	0.0416	0.4273	0.2930	0.0349	0.9512	0.5547	0.0953	0.9606	0.2696	0.5808	0.5861	0.5329	0.0195
7	0.3202	0.3681	0.5738	0.8170	0.0360	0.8782	0.5705	0.3110	0.7333	0.7449	0.1229	0.4278	0.7165
8	0.0388	0.6213	0.8916	0.6156	0.8733	0.8541	0.3086	0.4618	0.8414	0.5423	0.4516	0.5704	0
9	0.9862	0.5828	0.0351	0.3916	0.9468	0.0422	0.7358	0.9631	0.2558	0.9276	0.2731	0.8347	0.1641
10	0.9657	0.6525	0.311	0.9584	0.2530	0.9522	0.5282	0.3199	0.7365	0.9393	0.7675	0.6325	0.7474

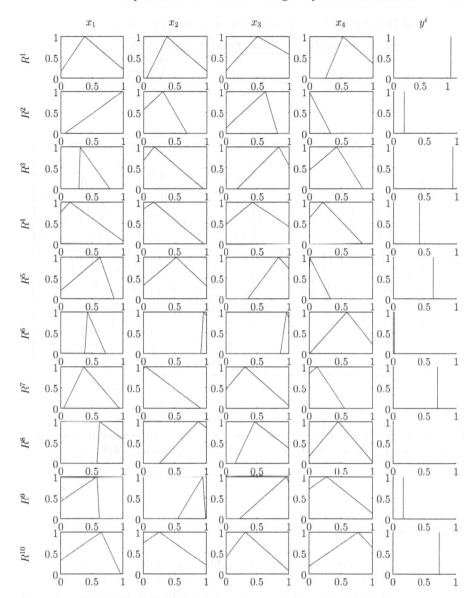

Fig. 10. Identified fuzzy rules by the proposed method

Table 2. Comparison of the proposed method with other method

Reference	Number of Rules	RMSE
[7]	164	0.0378
Ours	10	0.0245

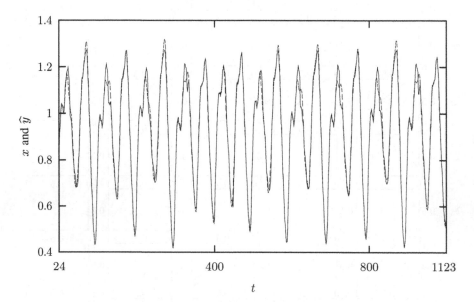

Fig. 11. Comparison of the actual output of Mackey–Glass chaotic time series (dashed line) and the output from the identified fuzzy model (solid line)

5 Conclusions

In predicting the chaotic time series via fuzzy inference systems, the most difficult obstacle is the identification of an optimal fuzzy model. In order to resolve this problem, this study presented an approach to constructing an optimal fuzzy model from a given complex chaotic input–output data by hybridly combining the merits of the mGA with our new coding scheme for coarsely optimizing structure and parameters of the concerned fuzzy model, together with the gradient descent method for fine-tuning. The simulation result on the chaotic Mackey–Glass time series data convincingly demonstrated the advantage of the developed identification method. It implies the potential of the proposed method for reliable chaos applications.

References

1. L. Chen, G. Chen: IEEE Trans. Circuits Syst. I **47**(10), 1527–1531 (2000)
2. Y.H. Joo, H.S. Hwang, K.B. Kim, K.B. Woo: Fuzzy Sets Syst. **86**, 279–288 (1997)
3. W. Pedrycz: Fuzzy Sets Syst. **13**, 153–167 (1984)
4. J.J.R. Jang, C.T. Sun: Predicting chaotic time series with fuzzy if-then rules. In: *Proc. IEEE Int. Conf. Fuzzy Syst.*, 1993, 1079–1082
5. S. Horikawa, T. Furuhashi, Y. Uchikawa: IEEE Trans. Neural Network **3**(5), 801–806 (1992)
6. Y.H. Joo, H.S. Hwang, K.B. Kim, K.B. Woo: Electron. Lett. **31**(4), 330–331 (1995)
7. D. Kim, C. Kim: IEEE Trans. Fuzzy Syst. **5**(4), 523–535 (1997)
8. Y.H. Joo, L.S. Shieh, G. Chen, IEEE Trans. Fuzzy Syst. **7**(4), 394–408 (1999)
9. H.J. Lee, J.B. Park, G. Chen: IEEE Trans. Fuzzy Syst. **9**(2), 369–379 (2001)
10. M. Sugeno, G.T. Kang: Fuzzy Sets Syst. **18**, 329–346 (1986)
11. T. Takagi, M. Sugeno: IEEE Trans. Syst. Man, Cybern. **15**(1), 116–132 (1985)
12. H.S. Hwang, Y.H. Joo, H.K. Kim, K.B. Woo: Control Eng. Practice **1**(1), 249–254 (1993)
13. B. Wu, X. Yu: Fuzzy Sets Syst. **113**, 351–365 (2000)
14. D.E. Goldberg: *Genetic Algorithms in Search, Optimization and Machine Learning* (Addison-Wesley, Reading, MA, 1990)
15. D.E. Goldberg, B. Korb, K. Deb: Complex Syst. **3**(5), 493–530 (1989)
16. M. Chowdhury, Y. Li: Messy genetic algorithm based new learning method for structurally optimized neurofuzzy controllers. In: *Proc. IEEE Int. Conf. Ind. Tech.*, 1996, pp 274–278
17. H. Takagi, M. Sugeno: IEEE Trans. Syst. Man Cybern. **15**, 116–132 (1985)
18. J.H. Holland: *Adaptation in Natural and Artificial Systems* (MIT Press, Cambridge, MA, 1975)
19. K. Deb, D.E. Goldberg: "mGA in C: A messy genetic algorithm in C," *IlliGAL Rep.*, 91008 (1991)
20. F. Hoffmann, G. Pfister: A new learning method for the design of hierarchical fuzzy controllers using messy genetic algorithms. In: *Proc. Sixth Int. Fuzzy Syst. Association World Cong.*, 1995, vol. 1, pp. 249–252
21. M.C. Mackey, L. Glass: *Science* 197–287 (1977)
22. B. Carse, T.C. Fofarty, A. Munro: Fuzzy Sets Syst. **80**, 273–293 (1996)

Fuzzy Control of Chaos

Oscar Calvo

Abstract. In this chapter a Mamdani fuzzy model based fuzzy control technique is proposed to control chaotic systems, whose dynamics is complex and unknown, to the unstable periodic orbits (UPO). Some empirical tricks are introduced for building up a proper fuzzy rule base and designing a fuzzy controller. Finally, an example of fuzzy control of the Chua's circuit is presented to illustrate the effectiveness of the proposed approach.

1 Control of Chaos

1.1 Introduction

Chaos is present in many aspects of life. Initially regarded as a curiosity that interested only the mathematics community, it was later revamped when observed in meteorology, physics, chemistry, and biology. The observations of chaos in Nature, in many forms, caught the attention and span a huge number of publications in that area. In particular there has been a recent interest in the field of synchronization of neurons in human brain. Further developments on chaos quickly triggered practical applications in the fields of mechanical and electrical engineering, communications, laser dynamics, chemistry, biology, oceanography, and economics to name a few. Furthermore, beneficial aspect of chaos fueled the study of synchronization and control. Examples are found in practically all the areas mentioned above.

A recent review by Andrievskii [1] summarizes the main uses of chaos control. For instance, in mechanical engineering where the applications started as experimental demos for educational purposes, such as the control of pendulum [2], beams, and plates, these were quickly developed in more realistic applications such as the control of vibroformers [3], stabilization of crane oscillations [4], spacecrafts [5], satellite, and others. Physics is probably the field where chaos control became a paradigm and a discipline itself. Important contribution has been made in control of turbulence [6], dynamics of lasers [7], plasma [8], and many others. In chemistry, it has been used in the control of chemical reactions, such as the stabilization of the Belousov–Zhabotinsky [9] reaction, electrodissociation of nickel-based electrodes in sulfuric acid [10], and many more. In biology, chaos suppression has been considered in ecological systems, control of algae growth [11], beetle populations, and many

O. Calvo: *Fuzzy Control of Chaos*, StudFuzz **187**, 99–125 (2006)
www.springerlink.com   © Springer-Verlag Berlin Heidelberg 2006

applications in medicine. In particular, very interesting results have been obtained in the treatment of cardiac arrhythmias [12], reduction of seasonal epidemics, and brain synchronization. In electrical and electronic engineering, applications initially started for the simulation of the classical sets of nonlinear differential equations (Rossler, Lorenz, Chua) with analog circuits but further developments have been made in the control of chaos in DC–DC converters [13], DC motors [14], transducer and power systems [15], magnetic suspension [16], etc. In the last decade, chaos synchronization has proven to be useful to encrypt communications [17], as presented in the special issue of *IEEE Transactions on Circuits and Systems* [18]. In chemical industries, chaos control has been used for chaotic stirring of liquids, materials [19], and gases. There are many more applications of chaos control; those mentioned in this chapter are only a few samples of the numerous applications of chaos control existing in the literature.

1.2 Different Techniques of Chaos Control

Chaos is usually associated with randomness and intermittency [20], since chaotic systems show random conduct with bursts of synchronization and almost periodic behavior. According to Ditto [20], a definition of chaos would consist of a superposition of (unstable) periodic motions. Another important observation of the chaotic dynamics is its sensitivity to small changes in the parameters or to the initial conditions (butterfly effect). A distinction of the techniques of chaos control could be made based on the use of feedback [21]. Closed-loop techniques monitor some variable in the phase plane and by perturbating temporarily a parameter or variable bring the dynamics to the desired orbit. On the other hand, open-loop system produces the same effect by changing slightly some parameter or property of the system, permanently, without the feedback.

One of the techniques based on the closed-loop approach consists of the following: while observing carefully the motion on the phase plane, we can detect the presence of unstable periodic orbits (UPOs). These orbits are present in an infinite number and are embedded in a compact area of the phase plane, a chaotic attractor. A small perturbation in the parameters of the system kicks the dynamics to jump from one orbit to another, very close to it. This property was utilized by Ott et al. [22] to propose method for the control of chaos that became a benchmark in the field.

1.3 OGY Method

A key aspect in the application of the Ott–Grebogi–York (OGY) methodology is the observation of the dynamics of the particle in the phase plane. The trajectory in a simple dynamical system such as a pendulum would be a circle. If the pendulum is composed of two masses joined by a link (double pendulum), the dynamics gets much more complicated. There will be trajectories of double period, observed as two circles in the phase plane. If we intersect the trajectory with a plane (Poincaré surface) and plot the results, we would visualize a single point for the first case and two points for the second (periodicity two). If a system shows a random behavior, the Poincaré surface would be populated by dots. In chaotic systems, since they perform as a superposition of infinite periodic motions, there will be infinite points confined in a geometric structure called chaotic attractor. A slight perturbation to the system will change the position of the attractor on the surface and hence the location of the unstable fixed points. The idea underneath the method is to perturb the attractor slightly, so that future intersections of the chaotic orbits with the plane will bring them back to the stable manifolds of the attractor, confining the dynamics always in the vicinity of the UPOs.

In order to apply the OGY method, the following assumptions are usually made [21, 22]:

a) The dynamics of the system can be described by an n-dimensional map $\xi_{n+1} = F(\xi_n, p)$, where in the continuous case the map is obtained using a Poincaré surface.
b) p is a parameter that can be changed slightly around its nominal value.
c) For this value of p_{nominal}, there is a periodic orbit embedded in the attractor where the system will stay when stabilized.
d) The orbit changes smoothly with changes in p and the global dynamics of the system is slightly affected.

Let us assume that we have a three-dimensional continuous time differential equation

$$\frac{\partial x}{\partial t} = F(x, p) \text{ with } x \in \Re \text{ and } p \in \Re \tag{1}$$

where p is a parameter that can be changed within a range $p \pm \Delta p_{\text{max}}$.

We choose a Poincaré surface Σ which defines a map P. Given ξ_0 (point belonging to the map P), we identify it as the point where the trajectory intersects the surface Σ. Since F depends on p, the Poincaré map also depends on $p : P(\xi, p) \in \Re^2$. We can choose one of the UPOs of the attractors as the target of our control. For simplicity, let us consider a period-one orbit (fixed point on map P). Let us call ξ_F an unstable point of P. For $p = p_0, P(\xi_F, p_0) = \xi_F$. Now, consider a neighborhood of (ξ_F, p_0). By a linear approximation procedure, we can obtain

$$P(\xi, p) \approx P(\xi_F, p_0) + A(\xi - \xi_F) + \omega(p - p_0) \tag{2}$$

where A is the Jacobian matrix of $P(\xi_F, p_0)$ and $\omega = \frac{\partial P(\xi_F, p)}{\partial p}$ is the derivative of P with respect to p. Stabilization is achieved by introducing feedback $P(\xi) = p_0 + c^T(\xi - \xi_F)$, where $c^T = \frac{\lambda_\mu}{f_\mu^T} f_\mu^T$ and λ_u is the unstable eigenvalue and f_μ is the eigenvector of A.

1.4 Occasional Proportional Feedback Method

A derivation of the OGY method for one dimensions is the occasional proportional feedback (OPF) method. Hunt proposed this method to stabilize the orbits of a chaotic circuit [23]. In the OPF method, the control signal is computed using only one variable:

$$P(\xi) = P_0 = c(\xi - \xi_F) \tag{3}$$

We are interested in the values of c for which ξ is a fixed point, that is, $\xi = \xi_F$. An application of the OPF to a double scroll dynamical system was presented by Ogorzalek [24]. He applied the OPF to stabilize the orbits of a Chua's circuit by using a shock absorber concept.

Satisfying the conditions enumerated above to the one-dimensional case: Let ξ_F be the fixed point of the map P, for the parameter value p^*. In the close vicinity of ξ_F we assume that the dynamics is linear and given by $\xi_{n+1} - \xi_F = M(\xi_n - \xi_F)$. The elements of the matrix M can be computed using the time series in the neighbourhood of ξ_F. We can obtain the eigenvalues λ_s and λ_u (stable and unstable) and the corresponding eigenvectors e_s and e_u that provide the stable and unstable directions around the fixed point. Defining f_s and f_u the contravariant eigenvectors $f_s e_s = f_u e_u = 1$ and $f_s e_u = f_u e_u$ we find a linear approximation for small $|p_n - p^*|$:

$$\xi_{n+1} = sp_n + [\lambda_u e_u f_u + \lambda_s e_s f_s][\xi_n - sp_n] \tag{4}$$

where $s = \frac{\partial \xi_F(p)}{\partial p}$.

1.4.1 Implementation Details

In the implementation made by Hunt, the OPF algorithm used to stabilize the amplitude of the limit cycles was based on measuring the local maximum of the output. The control law suggested that

$$u_k \begin{cases} k y_k & \text{if } |\overline{y_k}| < \Delta \\ 0 & \text{otherwise} \end{cases} \tag{5}$$

$$\overline{y_k} = y_k - y_* \quad \text{and} \quad y_* = h(x_0) \tag{6}$$

where y_* represents the desired upper level (target) of the oscillation. A problem of this method resides in estimation of the Poincaré map because of the uncertainty of the linearized plant model.

2 Fuzzy Control of Chaos

2.1 Introduction

As mentioned above, chaos control exploits the sensitivity to initial conditions and to perturbations that is inherent in chaos as a means to stabilize

UPOs within a chaotic attractor. The control can operate by altering system variables or system parameters, and either by discrete corrections or by continuous feedback. Many methods of chaos control have been derived and tested [19, 22, 25]. Why then consider fuzzy control of chaos?

A fuzzy controller works by controlling a conventional control method. We propose that fuzzy control can become useful together with one of these other methods—as an extra layer of control—in order to improve the electiveness of the control in terms of the size of the region over which control is possible, the robustness to noise, and the ability to control long period orbits. We will put forward the idea of fuzzy control of chaos, and we provide an example showing how a fuzzy controller applying OPF to one of the system parameters can control chaos in Chua's circuit.

2.2 Fuzzy Control

Fuzzy control [26, 27] is based on the theory of fuzzy sets and fuzzy logic [28, 29]. The principle behind the technique is that imprecise data can be classified into sets having fuzzy rather than sharp boundaries, which can be manipulated to provide a framework for approximate reasoning in the face of imprecise and uncertain information. Given a datum, x, a fuzzy set A is said to contain x with a degree of membership $\mu_A(x)$, where $\mu_A(x)$ can take any value in the domain [0,1]. Fuzzy sets are often given descriptive names (called linguistic variables) such as $FAST$; the membership function $\mu_{FAST}(x)$ is then used to reject the similarity between values of x and the contextual meaning of $FAST$. For example, if x represents the speed of a car in kilometers per hour and $FAST$ is to be used to classify cars traveling fast, then $FAST$ might have a membership function equal to zero for speeds below 90 km/h and equal to one for speeds above 130 km/h, with a curve joining these two extremes for speeds between these values. The degree of truth of the statement *the car is travelling fast* is then evaluated by reading off the value of the membership function corresponding to the car's speed.

Logical operations on fuzzy sets require an extension of the rules of classical logic. The three fundamental Boolean logic operations intersection, union, and complement have fuzzy counterparts defined by extension of the rules of Boolean logic. A fuzzy expert system uses a set of membership functions and fuzzy logic rules to reason about data. The rules are of the form "if x is $FAST$ and y is $SLOW$ then z is $MEDIUM$," where x and y are input variables, z is an output variable, and $SLOW, MEDIUM,$ and $FAST$ are linguistic variables. The set of rules in a fuzzy expert system is known as the rule base, and together with the database of input and output membership functions it comprises the knowledge base of the system.

A fuzzy expert system functions in four steps. The first is *fuzzification*, during which the membership functions defined on the input variables are applied to their actual values, to determine the degree of truth for each rule premise. Next is *inference*, during which the truth-value for the premise of each rule is computed and applied to the conclusion part of each rule. This results in one fuzzy set to be assigned to each output variable for each rule. Third is *composition* in which all of the fuzzy sets assigned to each output variable are combined together to form a single fuzzy set for each output variable. Finally comes *defuzzification*, which converts the fuzzy output set to a crisp (nonfuzzy) number.

A fuzzy controller may then be designed using a fuzzy expert system to perform fuzzy logic operations on fuzzy sets representing linguistic variables in a qualitative set of control rules (see Fig. 1).

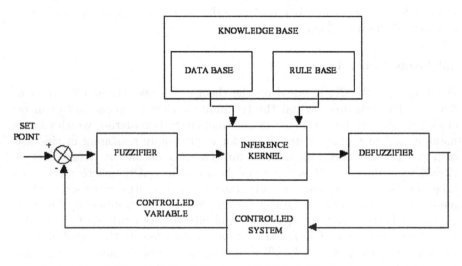

Fig. 1. Fuzzy logic controller block diagram

As a simple metaphor of fuzzy control in practice, consider the experience of balancing a stick vertically on the palm of ones hand. The equations of motion for the stick (a pendulum at its unstable fixed point) are well known, but we do not integrate these equations in order to balance the stick. Rather, we stare at the top of the stick and carry out a type of fuzzy control to keep the stick in the air: we move our hand slowly when the stick leans by a small angle and fast when it leans by a larger angle. Our ability to balance the stick despite the imprecision of our knowledge of the system is at the heart of fuzzy control.

3 Fuzzy Logic Controller

A fuzzy logic controller (FLC) is a special controller that is used to modify the dynamics of a closed-loop system based on heuristic rules. It elaborates a control law from a set of rules that mimic the reactions of a human expert to various situations, mainly, when the system to be controlled is vaguely defined, is very complex and nonlinear, or when its dynamics is unknown and the sensors provide noisy and incomplete data.

The diagram shown in Fig. 1 corresponds to a single-input–single-output (SISO) controller. The input variable, usually an analog signal, must be sampled and converted to a discrete signal for its further processing. An FLC can be seen as a special case of a digital controller with a nonlinear behavior as we will show later. The main elements of the FLC are a fuzzification unit, an inference engine, a knowledge base, and the defuzzification unit. We will analyze each of these units separately.

Knowledge Base: The fuzzy knowledge base (KB) contains the knowledge necessary to solve a given problem along with the proper control objectives. The main components of the KB are the rules, mapping the fuzzy inputs to fuzzy values of the outputs, and the memberships functions of the variables to the fuzzy sets. The rules contain all the knowledge expressing the control relations and are expressed in a format like

$$\text{IF } E \text{ IS PB AND } \Delta E \text{ is PM THEN } u \text{ IS NB} \qquad (7)$$

where E and ΔE are either input variables coming from the sensors or state variables and u is an output variable that will influence the actuators. PB, PM, and NB are the linguistic labels (fuzzy sets) defined for the variables E, ΔE, and u in their universes of discourse.

In general, if X is a collection of objects denoted generically by x, then a fuzzy set A in X is defined as a set of ordered pairs: $\boldsymbol{A} = \{(\boldsymbol{x}, \mu_A(x)) \in X\}$, where $\mu_A(x)\}$ is called the membership function for the fuzzy set A. The membership function maps each element of X (the universe of discourse) to a membership grade between 0 and 1. In human terms (natural language) a rule could state "If the temperature error is positive big and its derivative is positive medium the fuel injected should be decreased."

Inference Engine: The inference engine is responsible for the implications, using the information stored in the database and applying recursively the rules by a recursive composition inference mechanism. It performs a fuzzy inference produce control actions, evaluating the KB to the fuzzified inputs.

Fuzzifier: This component associates a membership value to the inputs, for a given partitioning of the universe of discourse. It transforms a crisp data in a fuzzy variable by a verbalization process. The fuzzification mechanism involves the following operations:

a) Measure the input value.
b) Scale the inputs, normalizing them to the universe of discourse.
c) Determine the degree of match of every input with and the defined fuzzy sets, using the membership functions

Defuzzifier: The fuzzy consequences of the fired rules must be combined to provide a unique control action. Defuzzification implies a mapping from a fuzzy space into a crisp value. No single optimal strategy exists. There are several methods, and the matter is still a subject of research. Nevertheless, only a few methods cover most of the practical cases. They are the center of area (COA), center of largest area (COLA), Maximum (MAX), and mean of maximum (MOM).

3.1 Fuzzy Logic Controller Design

Unfortunately there is no standard procedure for the design of an FLC, but it is possible to define the sequence of steps involved to achieve a good design. The methodology employed follows the general guidelines described clearly in the literature [30, 31]. As it will be demonstrated later, the solution is no unique and the design is based on heuristic knowledge of the process to be controlled and a trial-and-error process until an adequate response is obtained (covering our specifications) in a iterative tuning procedure. Many methods exist nowadays to automate the tuning procedure. These can be based on genetic algorithms, neural networks [32], or other optimization technique. When the adjustment is performed on-line, we have the self-organized fuzzy logic controller (SOFLC), though stability considerations must be taken.

3.1.1 Methodology

The methodology proposed to build a static FLC can be summarized by the following steps:

1. Variables and universe of discourse selection.
2. Adoption of a proper fuzzification strategy, including
   2.1 input and output spaces partitioning,
   2.2 selection of membership functions of primary fuzzy sets,
   2.3 discretization and normalization of the universes of discourse, and
   2.4 completeness of the spaces.
3. Rule base construction:
   3.1 input and control variable selection.
   3.2 choice of the control rules.
4. Decision making logic selection:
   4.1 implication definition,
   4.2 interpretation of AND and OR connectives,
   4.3 definition of composition operator,
   4.4 inference mechanism, and
   4.5 defuzzifier strategy.

3.1.2 Variables and Universe of Discourse Selection

The first step in any control system is to identify the input and output variables in the controller. It is also necessary to assert the ranges of these variables. Two different scenarios could be observed. If there is previous expertise, the FLC develops from human knowledge, in which case this stage is completed to mimic the human expert. But if there is no previous knowledge about the behavior of the systems we have to select the control action (output variable of the controller) and monitor relevant variables, usually error signal and its first derivative. These signals come directly from the sensors. It is necessary to identify the ranges and the standby condition of the different signals involved.

3.1.3 Fuzzification Strategy

This is related with the vagueness and imprecision and transforms a measure in a subjective value. It is a mapping between crisp measured data into a membership value for a given fuzzy set of the input universe of discourse. It is necessary to define the fuzzy sets and their support and the partitioning of the universes. By increasing the number of fuzzy sets we obtain a more powerful and flexible control. If we use the *error* and its *first derivative* as input variables, we can adjust the parameters to obtain an adequate dynamic response. Typical sets are Positive Small (PM), Negative Medium (NM), etc. The increase in power on describing the problem must be balanced by the growth in the number of rules involved, since they grow as the product of the number of fuzzy sets for each input. Typical number of linguistic labels ranges from two to nine for each input and output variable, but usually a trial-and-error procedure will give us a logical number.

There are many articles discussing the shape of the membership functions, but no big difference exists on the results of the control action obtained for the different possible shapes (triangular, trapezoidal, gaussian, sigmoid, etc.). Most of the commercial software tools prefer trapezoidal (and triangular as a special case) because of its simple implementation.

The fuzzification operator converts a crisp value into a fuzzy singleton for a given universe of discourse. A singleton is not a real fuzzy variable, in the sense that it represents an input x_0 as a fuzzy set A with a membership $\mu_a(x)$ that is equal to zero, except at x_0, where it equals one. Now, if the input is noisy, like in most practical cases and we can characterize the noise by its standard deviation, the support of the fuzzy set for the given input signal must be at least twice the standard deviation of the noise to make sense. In some cases designers take the probability distribution of the noise as the membership function, but it should be clear that the approximation to fuzzy sets is purely a deterministic process. It must also be observed that the partitioning process explained for the inputs is also valid for the outputs,

Table 1. Discretization and partitioning example

Level	Range	NB	NM	NS	ZE	PS	PM	PB
−6	$x_0 < -3.2$	1.0	0.3	0.0	0.0	0.0	0.0	0.0
−5	$-3.2 < x_0 < -1.6$	0.7	0.7	0.0	0.0	0.0	0.0	0.0
−4	$-1.6 < x_0 < -08$	0.3	1.0	0.3	0.0	0.0	0.0	0.0
−3	$-0.8 < x_0 < -0.4$	0.0	0.7	0.7	0.0	0.0	0.0	0.0
−2	$-0.4 < x_0 < -0.2$	0.0	0.3	1.0	0.3	0.0	0.0	0.0
−1	$-0.2 < x_0 < -0.1$	0.0	0.0	0.7	0.7	0.0	0.0	0.0
0	$-0.1 < x_0 < 0.1$	0.0	0.0	0.3	1.0	0.3	0.0	0.0
1	$0.1 < x_0 < 0.2$	0.0	0.0	0.0	0.7	0.7	0.0	0.0
2	$0.2 < x_0 < 0.4$	0.0	0.0	0.0	0.3	1.0	0.3	0.0
3	$0.4 < x_0 < 0.8$	0.0	0.0	0.0	0.0	0.7	0.7	0.0
4	$0.8 < x_0 < 1.6$	0.0	0.0	0.0	0.0	0.3	1.0	0.3
5	$1.6 < x_0 < 3.2$	0.0	0.0	0.0	0.0	0.0	0.7	0.7
6	$3.2 < x_0$	0.0	0.0	0.0	0.0	0.0	0.3	1.0

keeping in mind that the output sent to the perform a control action (valve, current, etc.) must be a real number.

3.1.4 Discretization and Normalization of the Universes of Discourse

When a digital computer is used to process the information on a fuzzy system, such information must be quantified first. If the universe of discourse is continuous, it must also be discretized. Besides, to simplify the processing, the universe of discourse may also be normalized.

Discretization is quantification into segments (quantification levels). An example of discretization that includes a non-linear mapping between the physical inputs and the segments of the universe of discourse is shown in Table 1. The number of levels influence directly on the smoothness of the control action (fine control) but commensurable with the word-length memory saving constrains. Also the time is discrete through the sampling process, turning a continuous $x(t)$ signal into a discrete time variable $x(k)$. Furthermore, since it is expected that the controller provides a control action for any possible state of the system under control, it is necessary that every value of inputs and outputs variables of the FLC should be assigned to a fuzzy set (completeness of the sets). The union of the supports should cover all the ranges of input and output variables. There is even an intrinsic overlapping of sets and the crossover points of consecutive fuzzy sets are usually in 0.5. This means that when there is no dominant rule between two conflicting ones, they contribute with equal weight to the output.

The normalization implies that the physical values of both the inputs and the outputs are mapped on a predetermined normalized domain. This

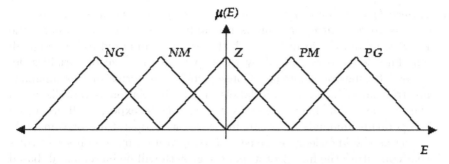

Fig. 2. Symmetric partitioning of universe of discourse in fuzzy sets

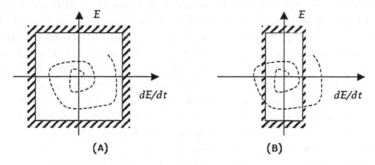

Fig. 3. Scales adjustment. (a) Correct; (b) incorrect

process involves a scaling, or multiplication, of the physical input by a normalization factor. For example, to obtain a symmetric partitioning as shown in Fig. 2, where all the membership functions are symmetrical and have the same support length, some (or all) of the inputs and output variables should be multiplied by certain gain. By inspecting the evolution of the variables in the phase plane we can decide the proper values of the gains. Figure 3 shows and example of correct (a) and incorrect (b) gain adjustments. The advantages obtained is that fuzzification, rule firing, and defuzzification can be designed independently of the physical domains of inputs and output variables. This facilitates the implementation of an FLC in a microprocessor. In most of the cases the fuzzy sets are symmetrical with equal support length.

3.1.5 Rule Base Design

A fuzzy system is defined by a set of linguistic statements that embed the knowledge of the expert. This is usually in the form of if–then–else rules, which are implemented in fuzzy logic. The rules are linguistic relations between linguistic variables. The choice of variables and memberships has a strong influence on the behavior of the FLC. There are essentially four modes of obtaining the fuzzy control rules as reported in [31].

a) *Knowledge and experience of a control engineer knowledgeable on the process to be controlled*: This is probably the easiest way to elicit the knowledge and the form by which the KB was supposed to be designed. The knowledge is organized by the expert in the form of variables described by natural language. Consequently, the terms of this language are translated into linguistic labels and their relation is established in the rule base. The knowledge may came from the expert itself, as stated in operation manuals or by an interview with operators, or somebody acting as the knowledge engineer. Starting from a minimum set of rules that constitute the first prototype, the growth will be incremental, based mainly on trial and error until the desired response is achieved.

b) *Operator's control actions*: In many situations it is difficult to find a model that describes mathematically the process to be controlled, or the model is too complex or with too many variables involved. Nevertheless the operator can bring the system to a successful behavior. A simple example of this is parking a car backward—a simple chore for a human that can be put as a set of rules that mimic the human actions.

c) *Fuzzy model of the process*: It is possible to make a linguistic description of the dynamical characteristics of the process under control. In general terms, this is a fuzzy model of the process. Given the model, fuzzy rules can be obtained to control it, even optimizing some figure of merit. This strategy has given very good results. A well-known model is the parametric model, proposed by Sugeno [33, 34] and used widely nowadays. The rules of the TS model have the following format:

$$R_i : \text{IF } s_1 \text{ is } S_1^i \text{ AND } s_2 \text{ is } S_2^i \text{ AND } s_3 \text{ is } S_3^i \text{ AND } \cdots s_p \text{ is } S_p^i$$
$$\text{THEN } v_i = a_0^i + a_1^i s_1 + a_2^i s_2 + \cdots + a_p^i s_p$$

d) *Learning*: This technique implies the automatic modification of the rulebase based on some optimization criteria. The precursor of this method was Mamdani [35, 36] who built a self organized controller (SOC) composed two sets of rules. The first one is the standard FLC that takes care of the control actions. The second rulebase acts as a supervisor exhibiting a human like behavior, tuning the control rules. In this hierarchy of two sets of rules, the supervisory rules, known as "meta rules", modify the rulebase based on performance index. This technique developed into the adaptive FLC, becoming very popular nowadays. The popular one is the neuro-fuzzy controllers, like ANFIS [32] based on Genetic algorithms.

e) *Phase plane*: When little experience is available on the process to be controlled, this methodology is useful since we can learn about the process dynamics and obtain control actions based on intuitive ideas. We will develop this approach in detail and use it later in this chapter [31].

f) Looking at the transient waveform of the controlled variable it is possible to divide it in time segments where the variable and its derivative keep

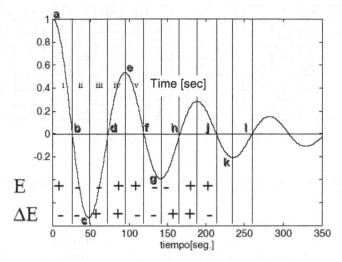

Fig. 4. Transient response to step inputs and time partitioning for rule derivation

the same sign as shown in Fig. 4. It is useful to define as critical points, the time instants where any of the two variables becomes zero (a, b, c, d, etc., in Fig. 4). For each of the zones just defined (i, ii, iii, iv, v, etc.), a control action will be applied with a double purpose: *to reduce overshoot and risetime.* To achieve this goal simple rules must be followed. For instance, in zone i, *when the error is big and positive and the derivative is negative, to decrease the risetime, the control action should be large and positive.* In region ii, the error is negative and the slope is still negative, so the best action would be to try decrease the overshoot, slowing down the response with a negative control action (like pressing the brake pedal).

With these "common sense" rules Table 2 is obtained as a preliminary rulebase with three linguistic variables for the two inputs and the control action. Rules are interpreted as follows: *"If the error is Positive (P) and the derivative is Zero (Z) then the control action is Positive (P)."* This rule (Rule 1) is applicable to points a, e, and i and to the rules that follow the same pattern.

From these initial rules we can build a control table of two inputs, also known as inference matrix. Plotting the matrix in three dimensions would yield the standard control surface. In effect, entering the surface with the error and its derivative, the height of the surface is directly the control action. The matrix just developed, with 19 rules, is shown in Fig. 5a.

The same zones and critical values identified in the transient response can be seen in a phase plot, as shown in Fig. 6. Using this plot and with the same objectives set before, reducing overshoot and risetime, we can infer the control actions to be applied. After dividing the phase plane in five areas ($A1, A2, A3, A4$, and ZE), and considering the critical points defined before

	NB	NM	NS	ZE	PS	PM	PB
PB			PM	PB			
PM				PM			
PS	NM		ZE	PS			
ZE	NB	NM	NS	ZE	PS	PM	PB
NS				NS	ZE		
NM				NM			
NB				NB	NM		

	NB	NM	NS	ZE	PS	PM	PB
PB	ZE	PS	PM	PB	PB	PB	PB
PM	NS	ZE	PS	PM	PB	PB	PB
PS	NM	NS	ZE	PS	PM	PB	PB
ZE	NB	NM	NS	ZE	PS	PM	PB
NS	NB	NB	NM	NS	ZE	PS	PM
NM	NB	NB	NB	NM	NS	ZE	PS
NB	NB	NB	NB	NB	NM	NS	ZE

Fig. 5. Inference matrix: a) incomplete; b) complete

$(a, b, c, \ldots$ etc.$)$ we can deduct some of the general rules that govern the FLC (metarules):

a) If $E = \Delta E$, keep the output of the FLC $\Delta u = 0$.
b) If E goes naturally to zero, at a right speed, do nothing.
c) If E does not corrects itself, ΔU will be different from zero. Its sign and magnitude will depend on E and ΔE according to the following rules:
   1. At the critical points, when the signal crosses the zero axis $(E = 0)$ $(b, d, f, \ldots) \operatorname{sgn}(\Delta u) = \operatorname{sgn}(\Delta E)$.
   2. At peaks and valleys $(\Delta E = 0)$ $(c, e, g, \ldots)$: $\operatorname{sgn}(\Delta u) = \operatorname{sgn}(\Delta E)$.
   3. In the area denoted as $A1$ we want to shorten the risetime when E is big and in $A2$ we want to prevent overshoot when E is small, then $\Delta u > 0$ when we are far from zero.

Table 2. Initial rule base

Rule	E	ΔE	Control	Reference Value
1	PB	ZE	PB	a
2	PM	ZE	PM	e
3	PS	ZE	PS	i
4	ZE	NB	NB	b
5	ZE	NN	NM	f
6	ZE	NS	NS	j
7	NB	ZE	NB	c
8	NM	ZE	NM	j
9	NS	ZE	NS	k
10	ZE	PB	PB	d
11	ZE	PM	PM	h
12	ZE	PS	PS	i
13	ZE	ZE	ZE	Desired Value
14	PB	NS	PM	i (rise time)
15	PS	NB	NM	i (overshoot)
16	NB	PS	NM	iii
17	NS	PB	PM	iii
18	PS	NS	ZE	ix
19	NS	PS	ZE	xi

Table 3.

E	NB	NM	NS	ZE	PS	PM	PB
μ_0	NB	NM	NS	ZE	PS	PM	PB

Δu lower or equal to zero when we are getting closer to the way point

4. In the area $A2$ we want to prevent the overshoot by making $\Delta u < 0$.
5. In the area $A3$ (similar reasoning as in A1)
   Δu greater or equal zero when we are converging to the setpoint
   $\Delta u < 0$ when $|E|$ is far from zero.
6. In the Area $A4$ we want to reduce the overshoot on the valley, then
   $\Delta u > 0$.

The meta rules allow us to determine the sign of Δu. For its magnitude we should consider that when $\Delta E \approx 0$, then $u = u_0$, where u_0 takes the values shown in Table 3. When ΔE is different from 0, then $u = (u_0 + \Delta E) + C$, where C is a compensation factor, usually zero.

The sum, can be considered as a *linguistic sum*

$$PM + PS = PL, PS + PL = PL, PS + NM = NS, \text{etc.}$$

When E is big (*NM, NL, PM, PL*), ΔE has little influence, and we choose C to speed up the response. When $|E|$ is small (*ZS, PS, NS*), ΔE has a big influence and we can choose C to have a small $|u|$ and prevent overshoot.

The relation between the phase plane and the inference matrix is depicted in Fig. 7.

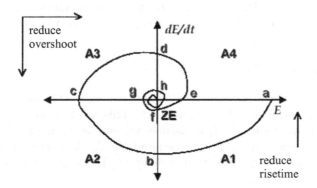

Fig. 6. Rule determination by phase plane partitionnig

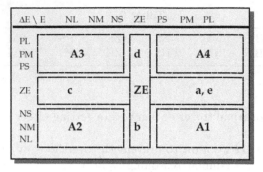

Fig. 7. Phase plane and inference matrix relation

Completeness

The controller must generate outputs for any fuzzy input state. Assuming these states are represented by x, the previous condition can be stated as follows:

$$\forall x \in X \underset{1 \le i \le n}{\exists} X_t(x) > \varepsilon \quad \varepsilon \in [0.1]$$

It is desirable that E exceeds 0.5 to be sure that at least a dominant rule fires (situation contemplated at the KB level). This feature guarantees that linguistic labels cover all the universes of discourse. In effect, if for every value of the input and the output variables there exists a fuzzy set to which they belong, and a rule that is related with those sets, the controller will provide an output for any combination of the inputs. If, instead we detect that this is not the case, a rule should be added to cover the unforeseen situation and rule out any possible hole on the rulebase. For instance, by simple inspection of Fig. 5 we could check if there is a case not contemplated. There will be also a violation of this condition when any of the linguistic labels is not present; in other words, there are holes in the description of part of the support of the universe of discourse.

Control Rules Interaction

In an FLC, there is a strong interaction between rules. The presence of any rule affects the behavior of the whole set. This set is responsible for the shape of the control surface. Lets assume we add an i-Rule to the base. If the surface is represented by R and the inputs by X, the composition of both will produce a new fuzzy set Y that is different. This can be stated formally as

$$\underset{1 \le j < i \le n}{\exists} X_i \cdot R \ne Y_j$$

It has been shown [37] that the interaction between the rules depends on the type of implication chosen.

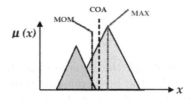

Fig. 8. Comparison of defuzzification strategies

Rulebase Consistency

Contradictory rules proposing opposed control actions must be eliminated or replaced in the KB. In an FLC with few rules, the consistency can be determined by simple inspection, but in cases with a large number of rules, it is necessary to make the test more systematic. There are methods based on the possibility (PI) and necessity (ND) definitions that can provide an *inconsistency index* for each rule in the set [38].

3.1.6 Defuzzification Strategy

Once we are finished with the above steps, the result of the rules aggregation (the inference process) will be a fuzzy set. Nevertheless, a physical *actuator* must be driven by a nonfuzzy signal. For instance, if the actuator is a valve, the controller's output acting over a servomotor controlling the valve does not support commands such as *apply a small positive voltage*. It is therefore necessary to have a block capable of converting the fuzzy signal into a *crisp* value, being the best possible representation of the distribution of probabilities of the control action inferred. This is the purpose of the *deffuzzifier*. The most popular strategies employed for this purpose are the criteria of Maximum (MAX), the Mean of Maximum (MOM), and the Center of Area (COA). The different results obtained with each method are shown in Fig. 8.

MAX Criteria

This method forces an output at the support value of the fuzzy set where the possibility distribution of the control output has a maximum value. It may present the problem of multiple maximum, mainly when the implication is done with Mamdani minimum function. In that case the crisp value is not unequivocally defined.

Mean of Maximum Criteria

The MOM generates a control action u_0 that represents the average value of the control actions with a maximum value in their membership functions:

$$u_0 = \sum_{j=1}^{l} \frac{w_j}{l}$$

where w_j is the support at which the membership function reaches a maximum. This criterion provides the best transient response. Unfortunately it does not take into account the shape of the membership functions of the output variables. The information lost in the process takes into account the elicited knowledge of each rule and could affect the shape of the responses.

Center of Area Method

This is by far the most widely used criterion. It computes the center of gravity of the output fuzzy set and uses its support as the crisp output of the controller. In the case of a discrete universe, with n quantification levels, the method provides

$$u_0 = \frac{\sum_{j=1}^{n} \mu(w_j) \cdot w_j}{\sum_{j=1}^{n} \mu(w_j)} \tag{8}$$

The behavior of the FLC for this criterion is similar to the one obtained with a conventional PI controller with variable gains. This method yields the best static behavior since the quadratic error is the smallest of all.

3.1.7 Inference Mechanisms

So far we have shown how the data coming from the sensor is fuzzified and depending on the similarity degree that present with the antecedents of the rules, they fired with that level of activation (minimum of the AND for Mamdani). Later, the rules combine their consequents (max for Mamdani OR) on a unique fuzzy set through a procedure known as aggregation. Finally, a defuzzification operation is applied over this set, yielding to a crisp output, as explained before, that is sent to the actuators. The whole procedure is known as inference mechanism and is shown in Fig. 9. So far we have utilized the

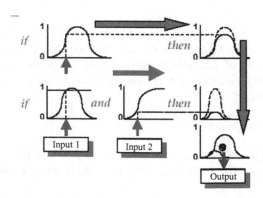

Fig. 9. Inference mechanisms for two inputs and one output

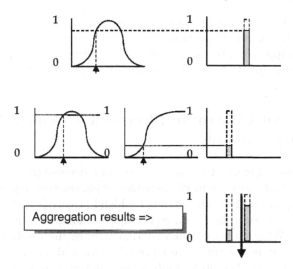

Fig. 10. Sugeno inference method

inference mechanism proposed by Ebrahim Mamdani and his colleagues from the Queen Mary College [35].

Nevertheless, it is possible to imagine that instead of having fuzzy sets at the consequents of the rules, we could have single values (singletons) with amplitude proportional to the degree of firing of the rule (see Fig. 10). This mechanism of inference was proposed by Sugeno [33] and it is widely used nowadays in multiple applications. It presents some advantages that can be easily appreciated. For instance, the defuzzification process is much simpler and easier to implement in a computer since the activation value of each rule is considered as the defuzzified output. For this reason it is usually the choice for hardware implementations. The Sugeno systems are classified by their *order*. For instance, a typical rule of order zero would be given as

$$\text{if } x \text{ is } A \text{ and } y \text{ is } B \quad \text{then } z = k$$

where A and B are fuzzy variables and k is a constant. Likewise, a Sugeno rule of order one would be written as

$$\text{if } x \text{ is } A \text{ and } y \text{ is } B \quad \text{then } z = p^*x + q^*y + r$$

where A and B are fuzzy variables and p, q, and r are constants.

We can look at these outputs as if they shift along the support depending on the values of the inputs. Higher orders of the Sugeno inference do not show a commensurable advantage on the results obtained, but they increase the complexity of the computation. And that is why they are rarely employed.

Another attractive feature of this type of inference is the possibility of guaranteeing continuity in the control surface simplifying its mathematical treatment. Besides, it adapts easily to adaptive control and identification since it facilitates the incorporation of lineal techniques. It can be used easily as a non-linear approximator or as a supervisor, interpolating between several linear controllers adjusted for different operating points [32].

4 Fuzzy Chaos Control in Electronic Circuits: An Introductory Example

As we mentioned before, to control a system necessitates perturbing it. Whether to perturb the system via variables or parameters depends on which are more readily accessible to be changed, which in turn depends on what type of system is to be controlled, i.e., electronic, mechanical, optical, chemical, biological, etc. Whether to perturb continuously or discretely is a question of intrusiveness; it is less intrusive to the system, and less expensive to the controller, to perturb discretely. Only when discrete control is not effective might continuous control be considered.

As discussed in Sect. 1.3, Ott et al. [21] invented a method of applying small feedback perturbations to an accessible system parameter in order to control chaos. The OGY method uses the dynamics of the linearized map around the orbit one wishes to control. Using the OGY method, one can pick any UPO that exists within the attractor and stabilize it. The control is imposed when the orbit crosses a Poincaré section constructed close to the desired UPO. Since the perturbation applied is small, it is supposed that the UPO is unaffected by the control.

As discussed in Sect. 1.4, OPF [19, 38] is a variant of the original OGY chaos control method. Instead of using the unstable manifold of the attractor to compute corrections, it uses one of the dynamical variables in a type of one-dimensional OGY method.

This feedback could be applied continuously or discretely in time; in OPF it is applied discretely. An OPF method exploits the strongly dissipative nature of the flows often encountered, enabling one to control them with a one-dimensional map. The method is easy to implement, and in many cases one can stabilize high period unstable orbits by using multiple corrections per period. It is a suitable method to base a fuzzy logic technique for the control of chaos, since it requires no knowledge of a system model, but merely an accessible system parameter.

4.1 Implementation Details and Experimental Results

Chua's circuit [39, 40] exhibits chaotic behavior, which has been extensively studied and whose dynamics is well known [41]. Recently, the OPF methods has been used to control the circuit [23]. The control used an electronic circuit to sample the peaks of the voltage across the negative resistance and if it fell

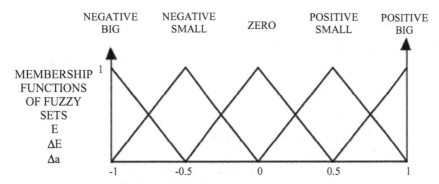

Fig. 11. Membership functions of the input and output variables E, ΔE, and Δa

within a window, centered about a by a set-point value, modified the slope of the negative resistance by an amount proportional to the difference between the set point and the peak value. The non-linear nature of this system and the heuristic approach used to find the best set of parameters to take the system to a given periodic orbit suggest that a fuzzy controller that can include knowledge rules to achieve periodic orbits may provide significant gains over the OPF alone.

We have implemented a fuzzy controller to control the nonlinearity of the nonlinear element (a three segment nonlinear resistance) within Chua's circuit. We have followed the steps described in Sect. 3. The block diagram of the controller is like the one already shown in Fig. 3. As we described earlier, it consists of four blocks: knowledge base (KB), fuzzification, inference, and defuzzification. The KB is composed of a data base and a rule base. The data base consists of the input and output membership functions (Fig. 11). Table 4 shows the quantification levels for these memberships functions and the membership values for all five linguistic

Table 4. Ouantification levels and membership functions

Error, E	−1	−0.75	−0.5	−0.25	0	0.25	0.5	0.75	1
Change in error, ΔE	−1	−0.75	−0.5	−0.25	0	0.25	0.5	0.75	1
Control, Δa	−1	−0.75	−0.5	−0.25	0	0.25	0.5	0.75	1
Quantification level	−4	−3	−2	−1	0	1	2	3	4
Linguistic Labels			Membership	Functions					
Positive Big, PB	0	0	0	0	0	0	0	0.5	1
Positive Small, PS	0	0	0	0	0	0.5	1	0.5	0
Approximately Zero, AZ	0	0	0	0.5	1	0.5	0	0	0
Negative Small, NS	0	0.5	1	0.5	0	0	0	0	0
Negative Big, NB	1	0.5	0	0	0	0	0	0	0

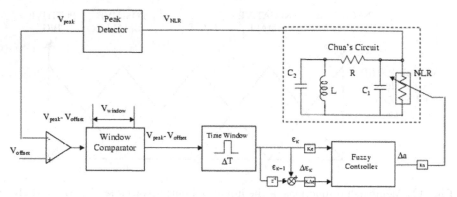

Fig. 12. The whole controller and control systme in the form of a block diagram, including the fuzzy controller, the peak detector, the window comparator, and the chua's circuit system being controlled

Table 5. Rule table for the linguistic variables defined in Table 3

	E				
	NB	NS	AZ	PS	PB
De	NB	NS	NS	AZ	AZ
NB	NB	NS	AZ	AZ	PS
NS	NS	NS	AZ	PS	PB
AZ	NS	AZ	AZ	PS	PB
PS	NS	AZ	AZ	PS	PB
PB	AZ	AZ	PS	PS	PB

sets. It provides the basis for the fuzzification, defuzzification, and inference mechanisms. The rule base is made up of a set of linguistic rules mapping inputs to control actions. Fuzzification converts the input signals E and ΔE into fuzzified signals with membership values assigned to linguistic sets. The inference mechanisms operate on each rule, applying fuzzy operations on the antecedents and by compositional inference methods derives the consequents. Finally, defuzzification converts the fuzzy outputs to control signals, which in our case control is the slope of the negative resistance Δa in Chua's circuit (Fig. 12). The fuzzification maps the error E and the change in the error ΔE to the labels of the fuzzy sets. Scaling and quantification operations are applied to the inputs. The knowledge rules (Table 5) are represented as control statements such as

"if E is *NEGATIVE BIG* and ΔE is *NEGATIVE SMALL* then Δa is *NEGATIVE BIG*"

The normalized equations representing the circuit are

$$\Delta a = \text{Fuzzy Controller Output} \times \text{Gain} \times a$$

where $f(x)$ represents the nonlinear element of the circuit. Changes in the negative resistance were made by changing a by an amount

$$\frac{\partial x}{\partial t} = \alpha(y - x - f(x))$$

$$\frac{\partial y}{\partial t} = x - y + z$$

$$\frac{\partial z}{\partial t} = -\beta y$$

and

$$f(x) = bx + \frac{1}{2}(a - b)((x + 1)(x - 1))$$

We have performed numerical simulations, both in C and in Simulink, of Chua's circuit controlled by the FLC. Figure 12 shows the whole control system in the form of a block diagram, including Chua's circuit, the fuzzy controller, the peak detector, and the window comparator. Figure 13 gives a sample output of the fuzzy controller stabilizing an unstable period-one orbit by applying a single correction pulse per cycle of oscillation. By changing the control parameters we can stabilize orbits of different periods. In Fig. 14 we illustrate more complex higher period orbits stabilized by the controller.

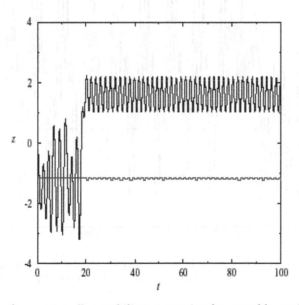

Fig. 13. The fuzzy controller stabilizes a previously unstable period-one orbit. The control is switched on at time 20. The lower trace shows the correction pulses applied by the controller

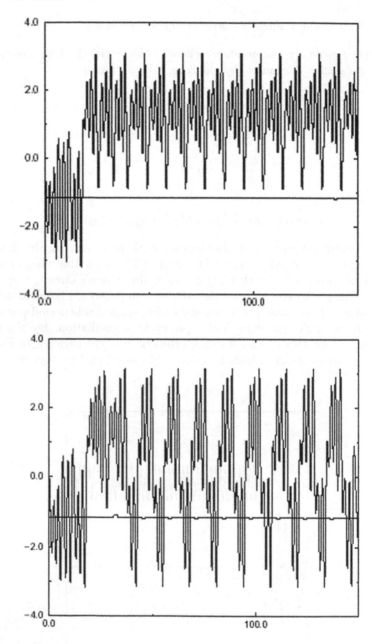

Fig. 14. Trajectory traces show higher period orbits stabilized by the controller. As before, the lower trace shows the correction pulses applied by the fuzzy control

One can tune the fuzzy control over the circuit to achieve the type of response required in a given situation by modifying some or all of the rules in the KB of the system.

Of course, in the case of Chua's circuit the system equations are available and fuzzy logic is thus not necessary for control, but this simple example permits us to see the possibilities that fuzzy control provides, by allowing a nonlinear gain implemented in the form of knowledge based rules.

5 Conclusions

We have introduced the idea of using fuzzy logic for the control of chaos. The FLCs are commonly used to control systems whose dynamics is complex and unknown, but for expositional clarity here we have given an example of its use with a well-studied chaotic system. We have shown that it is possible to control chaos in Chua's circuit using fuzzy control. Further work is necessary to quantify the effectiveness of fuzzy control of chaos compared to alternative methods, to identify ways in which to systematically build the knowledge base for fuzzy control of a particular chaotic system, and to apply the fuzzy controller to further chaotic systems.

Acknowledgments

I greatly appreciate the help and material received from and the fruitful discussions with Dr. Gerardo Gabriel Acosta, Department of Engineering of the Universidad Nacional del Centro, Olavarria, Argentina. Dr. Acosta is actually a visiting researcher of the University of Balearic Islands under the Marie Curie Incoming Fellowship Program.

References

1. B.R. Andrievskii, A.L. Fradkov: Appl. Autom. Remote Control **65**(4), 505–533 (2004)
2. B.R. Andrievsky, K.B. Boykov, A.L. Fradkov: Numerical and experimental excitability analysis of multi-pendulum mechatronics system. In: *15th IFAC World Congress on Automatic Control*, Barcelona, July 2002
3. M. Paskota: Chaos Solitons Fractals, **9**, 323–335 (1998)
4. B. Kimiaghalam, A. Homaifar, M. Bikdash: Pendulation suppression of a shipboard crane using fuzzy controller. In: *Proc. Am. Control Conf. (ACC'99)*, San Diego, CA, June 2–4, 1999, pp 586–590
5. P.A. Meehan, S.F. Asokanthan: J. Guid. Control Dyn. **25**(2), 209–214 (2002)
6. M. Gad-el-Hak: *Flow Control: Passive, Active, and Reactive Flow Management* (Cambridge University Press, London, 2000)

7. R. Roy, T.W. Murphy, T.D. Maier, Z. Gills, E.R. Phys. Rev. Lett. **68**, 1259–1262 (1992)
8. R.A. Chowdhury, P. Saha, S. Banerjee: Chaos Solitons Fractals **12**, 2421–2426 (2002)
9. V. Petrov, V. Gaspar, J. Masere, K. Showalter: Nature **361**, 240–243 (1993)
10. I.Z. Kiss, V. Gáspár: J. Phys. Chem. A **104**, 8033–8037 (2000)
11. V.V. Alekseev, A.Yu. Loskutov: On the possibility of controlling a system with strage attractor. In: *Problemy ekologicheskogo monitoringa i modelirovaniya ekosistem,* vol VIII [Problems of Ecological Monitoring and Modeling of Ecosystems]: (Gidrometeoizdat, Leningrad, 1985)
12. A. Garnkel, M.L. Spano, W.L. Ditto, J.N. Weiss: Science **257**, 1230–1235 (1992)
13. C. Batlle, E. Fossas, G. Olivar: Int. J. Circuit Theory Appl. **27**, 617–631 (1999)
14. J.H. Chen, K.T. Chau, S.M. Siu, C.C. Chan: IEEE Trans. Circuits Syst. I No. 47, 1093–1095 (2000)
15. K.N. Srivastava, S.C. Srivastava: Eur. Trans. Electr. Power **9**, 241–245 (1999)
16. B.Z. Kaplan, Y. Horen, G. Cohen, Y. Hellerman: IEEE Trans. Magn. **38**(5), 3475–3481 (2002)
17. A.S. Dmitriev, A.I. Panas, S.O. Starkov: Zarub. Radioelektron. **10**, 4–26 (1997)
18. J.M. Ottino: *The Kinematics of Mixing: Stretching, Chaos, and Transport* (Cambridge University Press, New York, 1989)
19. W. Ditto, M. Spano, J. Linder: Physica D **86**, 198–211 (1995)
20. T. Kapitaniak: *Controlling Chaos* (Academic Press, New York, 1996)
21. E. Ott, C. Grebogi, J.A. Yorke: Phys. Rev. Lett. **64**, 1196–1199 (1990)
22. G. Chen, X. Dong: Int. J. Bifurcation Chaos **3**, 1363–1409 (1993)
23. G.A. Johnson, T.E. Tigner, E.R. Hunt: J. Circuits Syst. Comput. **3**, 109–117 (1993)
24. M.J. Ogorzalek: IEEE Trans. Circuits Syst. **40**, 700–706 (1993)
25. W.L. Ditto, S.N. Rauseo, M.L. Spano: Phys. Rev. Lett. **65**, 3211 (1990)
26. T. Terano, K. Asai, M. Sugeno (ed): *Applied Fuzzy Systems* (Academic Press, New York, 1994)
27. D. Driankov, H. Hellendoorn, M. Reinfrank: *An Introduction to Fuzzy Control* (Springer, Berlin, Heidelberg, New York, 1993)
28. J.C. Bezdek: IEEE Trans. Fuzzy Syst. **1**, 1–6 (1993)
29. R.R. Yager, L.A. Zadeh: *An Introduction to Fuzzy Logic Applications in Intelligent Systems* (Kluwer, Norwell, MA, 1991)
30. G. Acosta: Fuzzy logic controller design, Course notes, Faculty of Engineering, UNCPBA, Argentina
31. Ch. Ch. Lee: IEEE Trans. on Syst. Man. Cybern. **20**(2), 404–435 (1990)
32. R. Jang, N. Gulley: *Fuzzy Logic Toolbox for Use with Matlab* (The MathWorks Inc., Natick, MA, 1996)
33. M. Sugeno: *Industrial Applications of Fuzzy Control* (Elsevier, New York, 1985)
34. T. Takagi, M. Sugeno: IEEE Trans. Syst. Man. Cybern. **15**, 116–132 (1985)
35. E. Mamdani, S. Assilian: Int. J. Man-Machine Stud. **7**(1), 1–13, (1975)
36. T.J. Procyk, E. Mamdani: Automatica **15**, 15–30 (1979)
37. W. Pedrycz: *Fuzzy Control and Fuzzy Systems* (Wiley, New York, 1988)
38. E.R. Hunt: Phys. Rev. Lett. **67**, 1953–1955 (1991)
39. T. Matsumoto: IEEE Trans. Circuits Syst. **31**, 1055–1058 (1984)
40. M.P. Kennedy: IEEE Trans. Circuits Syst. **40**, 657–674 (1993)

41. R.N. Madan (ed): *Chua's Circuit: A Paradigm for Chaos* (World Scientific, Singapore, 1993)
42. R.A. Desharnais, R.F. Costantino, J.M.Cushing et al.: Ecology Lett. **4**, 229–235 (2001)
43. J.F. Lindner, W.L. Ditto: Appl. Mech. Rev. **48**, 795–808 (1995)
44. G.G. Acosta, M.A. Mayosky, J.M. Catalfo: Controlador PID Autoadaptable. In: *Actas del XIII Simposio Nacional de Control Automático AADECA '92*, Buenos Aires, Argentina, Sept. 14–18, 1992, pp 28–32
45. G.G. Acosta, M.A. Mayosky, J.M. Catalfo: Int. J. Appl. Intell. **4**(1), 64–78 (1994)

Chaos Control Using Fuzzy Controllers (Mamdani Model)

Ahmad M. Harb and Issam Al-Smadi

Abstract. Controlling a strange attractor, or say, a chaotic attractor, is introduced in this chapter. Because of the importance to control the undesirable behavior in systems, researchers are investigating the use of linear and nonlinear controllers either to get rid of such oscillations (in power systems) or to match two chaotic systems (in secure communications). The idea of using the fuzzy logic concept for controlling chaotic behavior is presented. There are two good reasons for using the fuzzy control: first, mathematical model is not required for the process, and second, the nonlinear controller can be developed empirically, without complicated mathematics. The two systems are well-known models, so the first reason is not a big deal, but we can take advantage from the second reason.

1 Introduction

Modern nonlinear theories, such as bifurcation and chaos, have been widely used in many fields. Many researchers have used such theories to investigate and analyze the stability problem. Abed and Varaiya [1], Dobson et al. [2], and Harb et al. [3] used the bifurcation theory to analyze the stability of voltage collapse and SSR phenomena in electrical power systems. Endo and Chua [4] and Harb and Harb [5] analyzed the stability of phase locked loop (PLL) in communication systems. Nayfeh and Balachandran [6] and Harb et al. [7] analyzed the stability of Duffing oscillator in mechanical systems.

Recently, research has been devoted toword the bifurcation and chaos control of such mentioned systems. The main goal of bifurcation and chaos control is stabilizing bifurcation branches, changing the type of bifurcation from subcritical to supercritical Hopf bifurcation, and delaying the bifurcations. Abed et al. [8–10] used state feedback nonlinear controllers to change the type of the Hopf bifurcation and to suppress the amplitude of the limit cycles at the vicinity of the Hopf bifurcation points. Ikhouane and Krstic [11], Harb et al. [12, 13], and Zaher et al. [14–17] used recursive backstepping algorithms to design nonlinear controllers to stabilize systems of chaotic behavior.

Fuzzy set theory has been used successfully in virtually all technical fields, including modeling, control, and signal/image processing. Fuzzy control is a rule-base system that is based on fuzzy logic. Since fuzzy is described as computing with words rather than numbers, then fuzzy control can be described as control with sentences rather than equations. In 1974, Professor Mamdani

A.M. Harb and I. Al-Smadi: *Chaos Control Using Fuzzy Controllers (Mamdani Model)*,
StudFuzz **187**, 127–155 (2006)
www.springerlink.com

was the first to develop the concept of the fuzzy controller. Driankov et al. [18] and Calvo and Cartwright [19] introduced the idea of fuzzy in chaos control. Tang et al. [20], Mann et al. [21], Hu et al. [22], and Gradjevac [23] used the PID fuzzy controller, while Hsu and Cheng [24] and Toliyat et al. [25] designed a fuzzy controller to enhance power system stability. In this chapter we are going to discuss how to design control signals based on fuzzy theory to stabilize two chaotic systems.

The structure of this chapter is as follows: The next section introduces basic definitions of fuzzy theory and discusses fuzzy logic controllers. In Sect. 3 the mathematical models of Lorenz equation and Chua's circuit are introduced. Section 4 discusses the numerical simulations in order to test the performance of fuzzy controllers. Finally, some conclusions and comments are given.

2 Fuzzy Logic Control Preliminaries and Background

2.1 Fuzzy Logic Control

After being mostly viewed as a controversial technology for more than two decades, fuzzy logic has finally been accepted as an emerging technology since the late 1980s. This is largely due to the wide array of successful applications ranging from image processing to industrial control [26]. Compared with conventional control approaches, fuzzy logic control (FLC) utilizes information from domain experts and relies less on mathematical modeling of physical systems. The term fuzzy logic has been used in two different senses. In a narrow sense, fuzzy logic refers to a logic system that generalizes classical two-valued logic for reasoning under uncertainty. In a broad sense, fuzzy logic refers to all of the theories and technologies that employ fuzzy sets, which are classes with nonsharp boundaries [27]. Fuzzy logic control and modeling use only a small portion of fuzzy mathematics that is available [26]. Fuzzy set theory has been used successfully in virtually all technical fields, including modeling, control, signal/image processing, except systems.

In the next section, the necessary background and preliminaries of the theory of fuzzy logic, especially of FLC, will be discussed.

2.2 Basic Concepts of Fuzzy Logic

Fuzzy set theory was developed by Lotfi A. Zadeh, professor of computer science at the University of California at Berkeley, to provide a mathematical tool for dealing with the concepts used in natural language (linguistic variables). Fuzzy logic is basically multivalue logic that allows intermediate values to be defined between conventional evaluations. In this section, we give an introduction to the concept of fuzzy sets and membership functions (MFs).

2.2.1 Fuzzy Set

A fuzzy set is a set with smooth boundaries [26–28]. Fuzzy set theory generalizes the classical set theory to allow partial membership. A set in classical set
theory "Boolean or conventional logic" has a sharp boundary, which uses the
black-and-white concept to represent the MF. An object either completely
belongs to the set or does not belong to the set at all [28]. It makes us draw
lines in the sand. A definition of fuzzy sets can be given as follows:

A fuzzy set is a set with smooth boundaries, which consists of a universe
of discourse (X) and an $\mu(x)$, that maps every element in the discourse to a
membership value between 0 and 1. Mathematically this can be represented
as follows: Let X be a collection of objects (X is the universel of discourse set),
then a fuzzy set A in X can be defined as a set of ordered pairs [26, 29, 30]:

$$A = \{(x, \mu_A(x)) \,|\, x \in X\} \tag{1}$$

where $\mu_A(x)$ is called the MF of x in A, and normally its values are limited
between 0 and 1. A value of $\mu_A(x)$ which is close to 1 implies that it is very
likely for x to be in A; on the other hand, a value of $\mu_A(x)$ near 0 denotes
nonmembership. Equivalently, $\mu_A(x)$ is the degree to which $x \in A$. When the
membership space contains only two points 0 and 1, A is a nonfuzzy (crisp)
set, and $\mu_A(x)$ is identical to the characteristic function of a nonfuzzy set [5].
Elements with zero degree of membership are not usually listed. When A is
a discrete (finite) set, the fuzzy sets may be expressed as in [29, 30].

$$A = \mu_A(x_1)/x_1 + \mu_A(x_2)/x_2 + \cdots + \mu_A(x_n)/x_n = \sum_{i=1}^{n} \frac{\mu_A(x_i)}{x_i} \tag{2}$$

where "+" denotes the set theory union operator rather than the arithmetic
sum. The oblique line "/" does not denote division. Instead it denotes a
particular MF for a value on the universe of discourse. Fuzzy sets provide a
systematic means for dealing with uncertain and imprecise notation.

2.2.2 Membership Function

It was shown that a fuzzy set is completely characterized by its membership
functions (MFs), which can be described by a mathematical formula. In this
subsection, we will describe some of the most commonly used MF's with
respect to their input and parameter, and describe in detail the MF types
utilized in this thesis.

Triangular membership function: A triangular MF is specified by three parameters a, b, c, with $a < b < c$, as follows [27, 29]:

$$trimf(x; a, b, c) = \begin{cases} 0, & x < a \\ \frac{x-a}{b-a}, & a \leq x \leq b \\ \frac{c-x}{c-b}, & b \leq x \leq c \\ 0, & x > c \end{cases} \tag{3}$$

Trapezoidal membership function: A trapezoidal MF is specified by four parameters a, b, c, d, with $a < b < c \leq d$, as follows [27, 29]:

$$trapmf(x; a, b, c, d) = \begin{cases} 0, & x \leq a \\ \frac{x-a}{b-a}, & a \leq x \leq b \\ 1, & b \leq x \leq c \\ \frac{d-x}{d-c}, & c \leq x \leq d \\ 0, & d \leq x \end{cases} \qquad (4)$$

The triangular MF is a special case of the trapezoidal MF. Due to their simple formulas and computational efficiency, both triangular and trapezoidal MFs have been used widely in fuzzy logic control and modeling [26, 27, 30].

Gaussian membership function: A Gaussian MF is specified by two parameters $m\sigma$, where m represents the center and σ determines the width as follows:

$$gaussmf(x; m, \sigma) = \exp\left(-\left(\frac{x - m}{\sigma}\right)^2\right) \qquad (5)$$

We can control the shape of the function by adjusting the parameter σ. A small σ will generate a "thin" MF, while a big σ will lead to a "flat" MF.

Bell-shaped membership function: A bell-shaped MF is specified by three parameters a, b, c as follows:

$$bellmf(x; a, b, c) = \frac{1}{1 + \left|\frac{x-c}{a}\right|^{2b}} \qquad (6)$$

where the parameter b is usually positive (if b is negative, the shape becomes an upside-down bell); a and c are used to vary the center, while b is used to control the slopes at the crossover.

Figure 1 shows the difference between fuzzy and crisp sets for the linguistically variable Tall.

Figure 2 shows triangular, trapezoidal, Gaussian, and bell-shaped MFs, respectively. Since the MF essentially embodies all fuzziness for all particular fuzzy sets, one of the key issues in the theory and practice of fuzzy sets is how to define the proper MF of a fuzzy set. The following approaches can be used to define the proper MFs for a fuzzy set [26, 30]:

Intuition: Asking the experts to define them. This can lead to different fuzzy sets to the same vague concept.

Inference: Using data from the controlled/modeled system to generate them.

Neural network: Here a neural network is first used to learn the relationship between the system input and output by using collected data from the controlled/modeled system, and then it is used to describe the MF.

Trial and error: From a practical point of view this method works effectively and efficiently in many real-time applications.

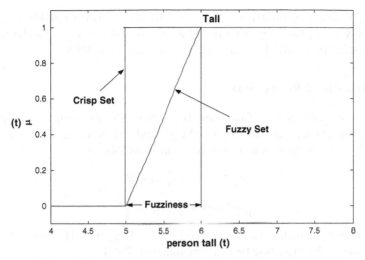

Fig. 1. Fuzzy and crisp sets for linguistic variable Tall

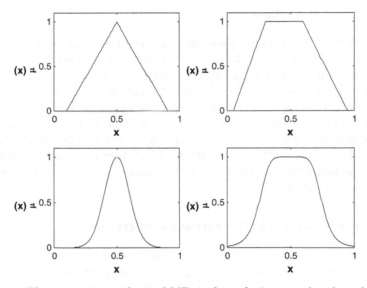

Fig. 2. The most commonly used MFs in fuzzy logic control and modeling

From the stability viewpoint, the MF of input variable fuzzy sets should be designed in a way such that the following conditions are satisfied:

$$1.\ A_i \cap A_j = \phi \quad \forall\, j \neq i,\, i+1,\, i-1 \qquad (7)$$

$$2.\ \sum_i \mu_{Ai}(x) \cong 1 \qquad (8)$$

In other words, each MF overlaps only with the closest neighboring MFs, and for any time, the membership values in all active fuzzy sets should sum to (or nearly to) 1, which means the use of symmetry MFs.

2.2.3 Hedges of Fuzzy Sets

A hedge is a modifier to a fuzzy set. It modifies the meaning of the original set to create a compound fuzzy set. "Very" and "More or Less" are the most commonly used hedges, which can be defined as follows:

$$\mu_A^{\text{very}} = [\mu_A(x)]^2 \tag{9}$$

$$\mu_A^{\text{more or less}} = \sqrt{\mu_A(x)} \tag{10}$$

In principle, a hedge can be applied to any fuzzy set. However, it should be used when the modified term is meaningful [28, 30].

2.2.4 Linguistic Variable

A linguistic term is used to express concepts and knowledge in human communication. A linguistic variable enables its value to be described both qualitatively by a linguistic term (symbol serving as the name of a fuzzy set) and quantitatively by a corresponding MF which expresses the meaning of a fuzzy set [30]. Linguistic terms represent the process states and control variables in a fuzzy logic controller. Their values are defined in linguistic terms and they can be words or sentences in a natural or artificial language [26, 27]. For example, the sentence "The temperature is high" uses the fuzzy set "high" to describe the linguistic variable temperature. More formally, this is expressed as "Temperature is high."

2.3 Basic Set Theoretic Operation for Fuzzy Sets

As a fuzzy set is defined in terms of its MF, a set operation of fuzzy sets will be defined via their MFs. Fuzzy logic AND (intersection or conjunction), OR (union or disjunction), and NOT (complement) operations are used in fuzzy controller and modeling. Unlike the binary AND, OR, and NOT operators whose operations are uniquely defined, their fuzzy counterparts are not unique [27, 28]. A variety of consistent definitions can be given for the fuzzy operations. The following are the most widely accepted for FLC. Let A, B, and C be three fuzzy sets in X, with MFs $\mu_A(x)$, $\mu_B(x)$, and $\mu_c(x)$ respectively. The basic connective operations in conventional set theory are those of intersection, union, complement, and subsethood.

2.3.1 Intersection (Conjunction or AND)

The definition of the intersection operator $\cap$ (or the logical AND connective) of fuzzy sets is defined as follows: The MF $\mu_{A\cap B}(x)$ of the *intersection* (conjunction) $A \cap B$ is a pointwise function defined by

$$\mu_{A\cap B}(x) = \mu(x)\; t\; \mu_B(x) \leq \min\{\mu_A(x), \mu_B(x)\} \quad \text{for} \quad x \in X \qquad (11)$$

where "t" is the t- or triangular norm defined as a two-place function mapping from $[0,1] \times [0,1]$ to $[0,1]$ which is nondecreasing (monotonically) in each element, $(x\,t\,w \leq y\,t\,z$ for $x \leq y,\, w \leq z)$; it is commutative, associative, and satisfies boundary conditions $x\,t\,0 = 0$ and $x\,t\,1 = x$, for $x, y, z, w \in [0,1]$ (corresponds to the properties of intersection).

The t-norms include intersection, algebraic product, logarithmic product, inverse product, bounded product, and drastic product. The most common t-norms are the intersection and the algebraic product. Some t-norms for $x, y \in [0,1]$ are

Intersection: $x\,t\,y = x \wedge y = min(x,y)$
Algebraic product: $x\,t\,y = x * y = xy$

Logarithmic product: $x\,t\,y = x \otimes_l y = \log_w\left[1 + \dfrac{(w^x - 1)(w^y - 1)}{(w - 1)}\right],$
$$0 < w < \infty, w \neq 1 \qquad (12)$$

Bounded product: $x\,t\,y = x \otimes_b$
$$y = \max[0, (\lambda + 1)(x + y - 1 - \lambda xy)], \quad \lambda \geq -1$$

Drastic product: $x\,t\,y = x \otimes_d y = \begin{cases} x, y = 1 \\ y, x = 1 \\ 0, x, y < 1 \end{cases}$

The usual t-norms are intersection and algebraic product with

$$x \wedge y \geq x * y \qquad (13)$$

t-norms are employed for defining conjunction in approximate reasoning.

2.3.2 Union (Disjunction or OR)

The definition of the union operator (the logical OR connective) of fuzzy sets is defined as follows: The MF $\mu_{A\cup B}(x)$ of the union (disjunction) $A \cup B$ is a pointwise function defined by

$$\mu_{A\cap B}(x) = \mu_A(x)s\mu_B(x) \geq \max\{\mu_A(x), \mu_B(x)\} \quad \text{for } x \in X \qquad (14)$$

where "s" is the s- or triangular co-norm defined as a two-place function mapping from $[0, 1] \times [0, 1]$ to $[0,1]$ which is nondecreasing (monotonically) in each argument; it is commutative, associative, and satisfies boundary conditions $x s 0 = x$, and $x s 1 = 1$, for $x \in [0, 1]$ (corresponds to the properties of union). The relation between s- and t-norms is given by the equivalent of the De-Morgan law in set theory

$$x \, s \, y = (1 - x)t(1 - y) \quad \text{with} \, x, y \in [0, 1] \tag{15}$$

The triangular co-norms or s-norms include union, algebraic sum, bounded sum, logarithmic sum, ration sum, drastic sum, and disjoint sum operations. The most common s-norms are the union and algebraic sum. Some s-norms for $x, y \in [0, 1]$ are

Union: $\qquad\qquad x \, s \, y = x \vee y = \max(x, y)$

Algebraic sum: $\qquad x \, s \, y = x \oplus y = x + y = xy$

Logarithmic sum: $\;\; x \, s \, y = x \oplus_1 y = 1 - \log_w \left[\dfrac{(w^{1-x} - 1)(w^{1-y} - 1)}{(w - 1)} \right]$,

$$0 < w < \infty, \, w \neq 1$$

Drastic sum: $\qquad\quad x \, s \, y = x \oplus_d y = \begin{cases} x, y = 0 \\ y, x = 0 \\ 1, x, y > 0 \end{cases}$

Disjoint sum: $\qquad\;\; x \, s \, y = x \Delta y = \max\{\min(1, \, 1 - y), \min(1 - x, y)\}$

$$\tag{16}$$

The most used s-norms are union and algebraic sum, for which

$$A \oplus B \geq A \wedge B \tag{17}$$

Clearly the t- and s-norms provide a wide range of models for connective fuzzy sets, each selected as appropriate to the problem domain. For the set operation of complement, we can define the negation of the fuzzy set by the following definition.

2.3.3 Fuzzy Set Complement

The MF $\mu_{\bar{A}}(x)$ of the *complement* of a fuzzy set A is defined for all $x \in X$ as

$$\mu_{\bar{A}}(x) = 1 - \mu_A(x) \tag{18}$$

This corresponds to the logical NOT operation; the idea of the complement reflects the negation.

2.3.4 Subsethood

A fuzzy set A in the universe X is a subset of another fuzzy set B if for every element x in X its membership degree in A is less or equal to its membership degree in B. Mathematically, this can formulated as

$$A \subseteq B \Leftrightarrow \forall x \in X \quad \mu_A(x) \leq \mu_B(x) \tag{19}$$

The notation of subset itself can also be extending to a matter of degree, which is called fuzzy subsethood. Figure 3 shows some operations on fuzzy sets.

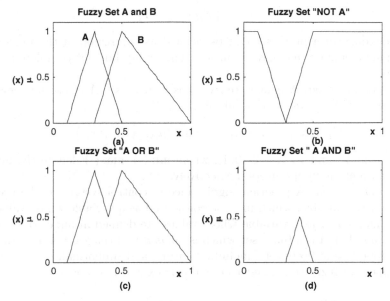

Fig. 3. Operation on fuzzy sets: (**a**) Two fuzzy sets A and B; (**b**) complement of fuzzy set A; (**c**) union of A and B fuzzy sets; (**d**) intersection of A and B

2.4 Properties of Fuzzy Set

We now introduce some definitions of fuzzy sets, which are needed to describe fuzzy logic controllers.

Continuous fuzzy sets: A fuzzy set is said to be continuous if its MF is continuous.

Support, fuzzy singleton, and α cut: For a fuzzy set whose universe of discourse is X, all elements in X that have nonzero membership value form the support of the fuzzy set $S(A)$. Mathematically this can be set as

$$S(A) = \{x \in X | \mu_A(x) > 0\} \tag{20}$$

In practice, the fuzzy set whose support is a single point in X with $\mu_A(x) = 1$ is referred to as a *fuzzy singleton*. A fuzzy set A has a *compact set* if its support is finite. If (20) is modified so that we have the definitions of α-cut for $\mu_A(x) > 0$ and strong α-cut for $\mu_A(x) \geq 0$, respectively.

Height, normal, and subnormal fuzzy sets: The largest membership value of a fuzzy set is called the height $H(A)$ of the fuzzy sets, mathematically,

$$H(A) = \max \{\mu_A(x)\} \tag{21}$$

If the height of a fuzzy set is less than unity, the fuzzy set is called subnormal. While if it is unity the fuzzy set is called a normal fuzzy set.

Core and boundary of fuzzy set: The core of a fuzzy set $A, C(A)$, is the set of all points $x \in X$ such that

$$C(A) = \{x|\mu_A(x) = 1\} \tag{22}$$

The boundary of a fuzzy set can be defined as the region of the universe-containing elements that have nonzero membership but not complete membership.

Convex fuzzy set: A convex fuzzy set is a fuzzy set A whose universe of discourse $[a, b]$ is convex if and only if

$$\mu_A(\lambda x_1 + (1 - \lambda)x_2) \geq \min[\mu_A(x_1), \mu_A(x_2)] \tag{23}$$

$\forall x_1, x_2 \in [a, b]$, and $\forall \lambda \in [0, 1]$, where min( ), max( ) denote the minimum and maximum operators, respectively.

A fuzzy number: A practical significance in control systems is the use of the linguistic variable, which may be considered as a variable whose value is a fuzzy number or as a variable whose values are defined in linguistic terms. A fuzzy number A is a fuzzy set which is *normal* and *convex*. Figure 4 shows some properties of fuzzy sets, including normal and subnormal, the core and the support of fuzzy sets, singleton, and convex and nonconvex fuzzy sets.

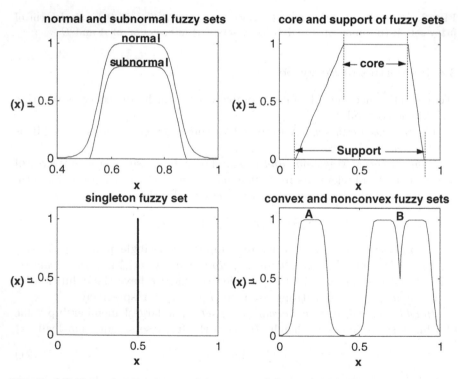

Fig. 4. (a) Normal and subnormal fuzzy sets; (b) core and support of fuzzy sets; (c) singleton fuzzy set; (d) convex (A) and nonconvex (B) fuzzy sets

2.5 Fuzzy Relation

A fuzzy relation generalizes the notation of the classical black-and-white relation into one that allows partial membership [27]. The classical notion of relation is described as the relationship that holds between two or more objects. In control systems relationships are defined between system inputs and outputs. In fuzzy systems, these relationships or mappings are between fuzzy variables defined on different universes of discourse through fuzzy conditional statements or linguistic implications.

2.5.1 Fuzzy IF–THEN Rules

Fuzzy control is a rule-base system that is based on fuzzy logic. Among all the techniques developed using fuzzy sets, the fuzzy IF–THEN is by far the most visible one due to its wide range of successful applications. Fuzzy IF–THEN rules play a critical role in industrial applications ranging from robotics manufacturing, process control, medical imaging to financial trading. A fuzzy logic controller (FLC) uses fuzzy rules, which are, linguistically, IF–THEN statements involving fuzzy sets, fuzzy logic, and fuzzy inference. Fuzzy rules play a key role in representing expert control knowledge and experience and in linking the input variables of FLC to output variables [26, 30]. Fuzzy logic controller (FLC) is a rule-based controller in which the control action depends on a set of rules and on selected input variables. In fuzzy logic, the truth of any statement becomes a matter of degree. The main feature of fuzzy rule-based inference is its capability to perform inference under partial matching. The general jth IF–THEN rule in fuzzy rule-based system takes the following form:

$$\text{Rule } j\colon \text{IF} < antecedent_j > \text{THEN} < consequent_j >$$

where $<antecedent_j>$ is the premise (condition) of the rule j with respect to a certain input variable, and $<consequent_j>$ is the control action (*conclusion*) of the rule j. The FLC rules may use several variables both in the condition (*antecedent*) and the conclusion (*consequent*) of the rules. The controllers can therefore be applied to both multi-input–multi-output (MIMO) problems and single-input–single-output (SISO) problems. Mamdani and Takagi–Sugeno (TS) fuzzy rules are the major types of fuzzy rules used. Next, we will give a brief discussion of only the Mamdani type of fuzzy rule.

2.6 Mamdani Fuzzy Rules

One of the most widely used fuzzy rule based systems is the Mamdani fuzzy rule, which consists of the following linguistic rules that describe a mapping from $X_1 \times X_2 \ldots X_n$ to U. The general jth IF–THEN rule in fuzzy rule based system takes the form

Rule j: IF x_1 is X_{j1} AND x_2 is X_{j2} AND ... AND x_n is X_{jn} THEN u_j is A

where $X_{jk}(k = 1, 2, \ldots, n)$ are the fuzzy sets (levels), $x_1, x_2, \ldots, x_n$ are the prime (input) variables, and u is A is a fuzzy consequent which is the control action of the rule j. This rule is an example of MISO Mamdani fuzzy rule, which can also be modified to perform SISO and MIMO. A MIMO Mamdani fuzzy rule in general form can be formulated as follows:

$$\text{Rule } j: \text{ IF } x_1 \text{ is } X_{j1} \text{ AND } x_2 \text{ is } X_{j2} \text{ AND} \ldots \text{AND } x_n \text{ is } X_{jn}$$
$$\text{THEN } u_1 \text{ is } A, \ u_2 \text{ is } B, \ldots, u_m \text{ is } Z$$

As mentioned before, for most FLCs, the input fuzzy sets are continuous, normal, convex and usually of the four common types of MFs. If the output fuzzy sets are of the singleton type then the general Mamdani fuzzy rule can be reduced to

$$\text{Rule } j: \text{ IF } x_1 \text{ is } X_{j1} \text{ AND } x_2 \text{ is } X_{j2} \text{ AND} \ldots \text{AND } x_n \text{ is } X_{jn}$$
$$\text{THEN } u_j \text{ is } \alpha_j$$

where α_j represents a singleton fuzzy set.

2.7 Fuzzy Rule Based Inference

Fuzzy inference system (FIS) is the process of formulating the mapping from a given input to an output using fuzzy logic that uses the theory of fuzzy set which will give the decisions to be made. Fuzzy inference is sometimes called reasoning or approximate reasoning. It is used in a fuzzy rule to determine the rule outcome from the input information and known facts [26, 30].

The algorithm of fuzzy rule based inference consists of three steps and an additional optional step, which can be summarized as follows:

- *Fuzzy Matching*: Calcualate the degree to which the input data match the condition of the fuzzy rule. In this first step, fuzzify the input variables, which means use the fuzzification method, which is a mathematical procedure for converting an element in the universe of discourse to the membership value of the fuzzy set. Then apply fuzzy operators if the antecedent (IF part) of the rules has more than one input.
- *Inference*: Calculate the rules conclusion based on its matching degree. There are two major methods, namely the clipping method and the scaling method. Both methods generate an inferred conclusion by suppressing the MF of the consequent, depending on the degree to which the rule is matched.
- *Combination*: Combine the conclusions inferred by all the fuzzy rules into a final conclusion. Combining fuzzy conclusions through superposition is based on applying the max fuzzy disjunction operator to multiple possibility distribution of the output variables.

- *Defuzzification*: For an application that needs a crisp output (e.g., control systems), an additional step called defuzzification is used to convert a fuzzy conclusion into a crisp one. Defuzzification is a mathematical process used to convert a fuzzy set or sets to a real number (crisp value).

Next we will discuss the theoretical foundation of the fuzzy inference and defuzzification steps.

2.7.1 Fuzzy Inference

Fuzzy inference is an inference procedure that drives conclusion from a set of fuzzy IF–THEN rules and known facts. There are a number of fuzzy inference methods that can be used for Mamdani fuzzy rule based system. The most common used are shown in Table 1. They are the Mamdani minimum inference, the Larsen product inference, and the bounded product inference methods [26, 30].

Remark 1: If Mamdani FLC employs singleton output set as the rule consequent, then all the inference methods produce the same inference result.

2.7.2 Defuzzification

Defuzzification is a mathematical operation with the aim to produce a non-fuzzy control action from fuzzy sets. It is a necessary step because fuzzy sets generated by fuzzy inference in fuzzy rules must be somehow mathematically combined to come up with a crisp value such as an FLC output. There are many proposed methods in the literature. We will present some of the most widely used:

- *Max-membership principle*: Also known as the height method, this scheme is limited to peak output function. This principle can be formulated algebraically as

$$\mu_A(x^*) \geq \mu_A(x), \quad \text{for all } x \in X \tag{24}$$

  where x^* corresponds to the maximum value of the output fuzzy set.
- *Center of area (COA) method*: This method is sometimes called center of gravity, weighted average defuzzification method. COA calculates the weighted average of fuzzy sets. The result of applying COA defuzzification to a fuzzy set conclusion can be expressed by the formula

Table 1. Definition of different types of inference methods

Fuzzy Inference Method	Definition
Mamdani minimum inference	$\min(\mu, \mu_A(x))$, for all x
Larsen product inference	$\mu \times \mu_A(x)$, for all x
Bounded product inference	$\max(\mu + \mu_A(x)\text{-}1, 0)$, for all x

$$U = \frac{\sum_{i=1}^{R} \mu_A(x)\mu_i}{\sum_{i=1}^{R} \mu_i(x)} \tag{25}$$

where R is the total number of rules.

- *The mean of maximum (MOM) method*: The MOM method generates a crisp control action by averaging the support values when their membership values reach the maximum

$$u = \sum_{i=1}^{l} \frac{m_i}{l} \tag{26}$$

where l is the number of the quantized m values, which reach their maximum membership values.

2.8 General Viewpoint to Design Fuzzy Logic Controllers

In this section, we will give a basic example on how to design and implement an FLC Mamdani type for a SISO system, using a basic knowledge and linguistic IF–THEN rules. Thus the aim is to develop an FLC that maps the error $(e = r(t) - y(t))$ into the control action (u as shown in Fig. 5).

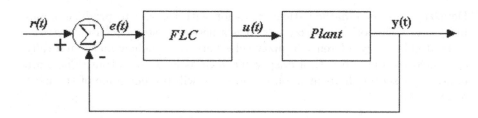

Fig. 5. Closed-loop fuzzy logic controller

Figure 6 shows system response and four rules which can be used to establishe an FLC. The FLC will depend on the error and with some logic, we can derive the following rules.

2.8.1 Mamdani FLC Type

For the Mamdani FLC type one may construct the following rules:

Rule 1: IF the error is Pos AND the change of error is Pos THEN u is Pos

Rule 2: IF the error is Pos AND the change of error is Neg THEN u is Zero

Rule 3: IF the error is Neg AND the change of error is Pos THEN u is Zero

Rule 4: If the error is Neg AND the change of error is Neg THEN u is Neg

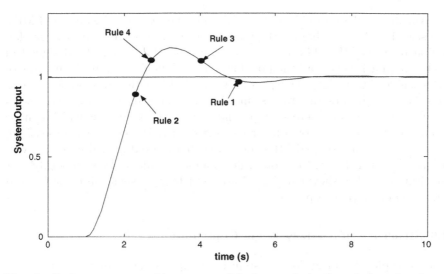

Fig. 6. System response with some possible fuzzy rule in the rule-based system

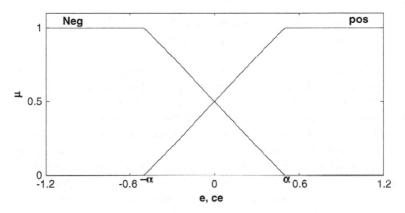

Fig. 7. The MFs for the error (e) and change of error (ce)

The input variables in the antecedent part of the rules are error and change of error, and the output variable in the consequent part is u (the control action). Pos, Zero, and Neg are fuzzy levels (sets) representing "Positive," "Zero," and "Negative," respectively. The rules are set heuristically. In rule 1, the error is positive and increasing (change of error is positive), so control action should be Positive. In rules 2 and 3 there is an error but it is decreasing, so it is desirable to let the FLC output be at the same level; hopping systems output will land on the set-point smoothly on its own. Finally, in rule 4, the error is negative and increasing, which is the opposite circumstance of rule 1, so control action should be Negative.

Among the design issues in the FLC Mamdani type is the design of MFs, the selection of their shapes, the universe of discourse MFs, and the overlap region between MFs. The universe of discourse of MFs can be designed in such way that it covers all the possible values of the FLC input. The shape of MFs is indeed a challenging aspect in designing FLC. However, some remarks may help in the design process: in FLC use the parameterized MFs with small number of parameters. A parameterized MF is an MF which can be defined mathematically by a few parameters, such as triangular, trapezoidal, Gaussian, and bell-shaped MFs. The MFs should be designed in such a way so that the translation from one rule to another rule is done smoothly. For the FLC the MFs shown in Fig. 7 are used to represent the error (e) and change of error (ce) respectively.

2.9 Structure of a Fuzzy Logic Controller

The principle of a fuzzy logic system is to express the human knowledge in the form of linguistic IF–THEN rules. These rules take the form

Rule 1:   IF x is Neg and z is Neg   THEN u is NB

Rule 2:   IF x is Neg and z is ZE   THEN u is NM

and so on, as shown in Table 2. Every rule has two parts: (i) the antecedent part, which is the IF part of the rule, and (ii) the consequent part, which is the THEN part of the rule. The input value "Neg" is the linguistic term used for the word *negative*; the output value "NB" stands for *negative big* and "NM" for *negative medium*. The collection of such rules is called rule base. A fuzzy expert system functions in four steps as shown in Fig. 8. The first is fuzzification, in which the crisp inputs are measured and translated into corresponding universes of discourse by scaling. Next under inference of the truth, value for the antecedent of each rule is computed, and applied on the consequent part of each rule. This results in one fuzzy set to be assigned to each output variable. By applying the implication method (use the degree of support for the entire rule to shape the output fuzzy set) all of the fuzzy sets assigned to each output variable combine together to form a single fuzzy set for each output. Finally, defuzzification is used to translate the processed fuzzy data into the crisp (nonfuzzy) data suited to real-world applications.

Table 2. Rule table for the linguistic variable (state) x, z

x/z	Neg	ZE	Pos
Neg	NB	NM	ZE
ZE	NM	ZE	PM
Pos	ZE	PM	PB

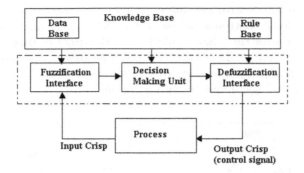

Fig. 8. General structure of fuzzy interference system

3 Mathematical Models

To validate the proposed Mamdani fuzzy controller, two chaotic systems are studied in this section. First is the Lorenz system, which discusses the thermally induced fluid convection in the atmosphere [6]. The second is the well-known Chua's circuit [19].

First Model: The mathematical model of the lorenz system is given by

$$
\begin{aligned}
x' &= \sigma(y - x) \\
y' &= \rho x - y - xz \\
z' &= -\beta z + xy
\end{aligned}
\tag{27}
$$

where σ, ρ, and β are system parameters.

Second Model: The mathematical model (normalized equation) of the Chua's circuit is given by

$$
\begin{aligned}
x' &= \alpha(y - x - f(x)) \\
y' &= x - y + z \\
z' &= -\beta y
\end{aligned}
\tag{28}
$$

where

$$
f(x) = bx + \frac{1}{2}(a - b)(|x + 1| - |x - 1|)
$$

which represents the nonlinear element of the circuit, and α and β are system parameters.

4 Numerical Simulations

Fuzzy logic controller (Mamdani model) is used to control the chaotic behavior of Lorenz system and Chua's circuit to be a stable constant or periodic solution, where the control parameter in Lorenz system is chosen to be ρ, and the fuzzy logic controller adjust ρ. On the other hand, a is used as a control parameter for the Chua's circuit, and the fuzzy controller adjust the same control parameter. In both cases, the used fuzzy logic controller is of the Mamdani type, which consists of two inputs and one output with nine rules, as shown in Figs. 9 and 10, and in Table 2.

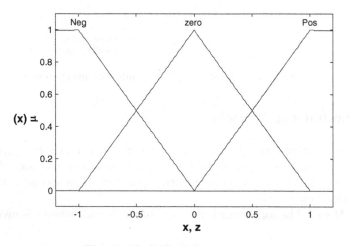

Fig. 9. The MF of the input x, z

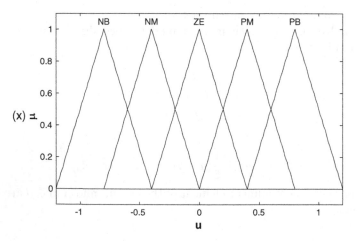

Fig. 10. The MF of the output u

The chaotic behavior, in both systems, is controlled to be constant or periodic solution through adjusting the control parameter. For the Lorenz system: Adjust ρ by $\Delta\rho$, where $\Delta\rho = \rho \times$ Fuzzy Controller Output $\times$ Gain. For the Chua's circuit: Adjust a by Δa, where; $\Delta a = a \times$ Fuzzy Controller Output $\times$ Gain. After we formulated the designed fuzzy logic controller as shown in Fig. 11, we used *Simulink* in Matlab Software.

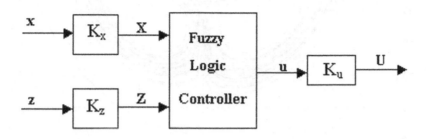

Fig. 11. Fuzzy logic controller in the form of block diagrams x, z is the linguistic variable and X, Z is the scaled variable through K_x and K_z respectively

4.1 Numerical Simulation for Lorenz Equation

Figure 12 shows the uncontrolled results, time history, as well as state-plane. It shows the chaotic behavior when $\sigma = 10$, $\rho = 20$, and $\beta = 1$. The control signal u is a fuzzy controller output and U is the adjusted fuzzy controller output by K_u. For the same parameters of the uncontrolled case and for the following parameters of the fuzzy control signal, $K_x = 0.5, K_z = 0.5$, and $K_u = 0.75$, Fig. 13 shows the time history and the state-plane. It can be clearly seen that the control signal brings the chaotic behavior to a stable periodic solution. While for $\sigma = 10, \rho_0 = 20, \beta = 1, K_x = 0.15, K_z = 0.15, K_u = 0.32$, Fig. 14 shows how the fuzzy control signal brings the system to the equilibrium solution (constant solution).

4.2 Numerical Simulation for Chua's Circuit

In this case, Fig. 15 shows the uncontrolled chaotic behavior of Chua's circuit, for $a = -1.139, b = -0.711, \alpha = 40$, and $\beta = 93.333$. Again, same as we did in the previous example, and for the same uncontrolled case ($a = -1.139, b = -0.711, \alpha = 40, \beta = 93.333$), we used the fuzzy control concept; the control parameters adjusted were $K_x = 0.05, K_z = 0.05$, and $K_u = 2$. Figure 16 shows the stable periodic solution. On the other hand, Fig. 17 shows the constant solution for $a_0 = -1.139, b = -0.711, \alpha = 40, \beta = 93.333$, but $K_x = 0.45, K_z = 0.45$, and $K_u = 2$.

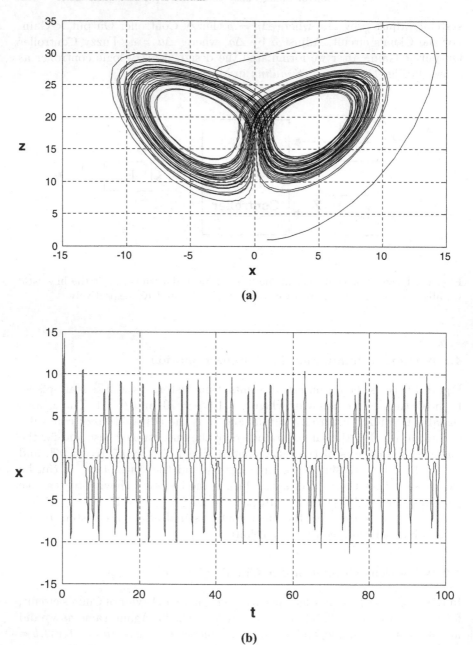

(a)

(b)

Fig. 12. Uncontrolled simulations of Lorenz equation

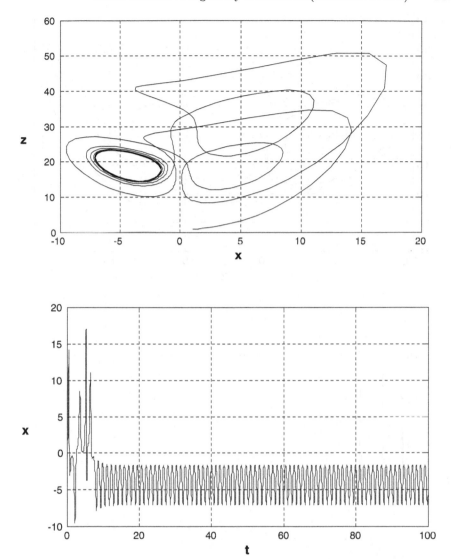

Fig. 13. Controlled system to get on a periodic solution using the fuzzy controller (Lorenz equation)

4.3 Remark

The fuzzy logic controller can be used when there is no mathematical model available for the process, this gives the robustness behavior for the proposed fuzzy logic controller design. To prove so, a systematic study of the control error versus the parameters of the nonlinear dynamical system is shown in Figs. 18–20 for the Lorenz attractor and the same thing can also be done

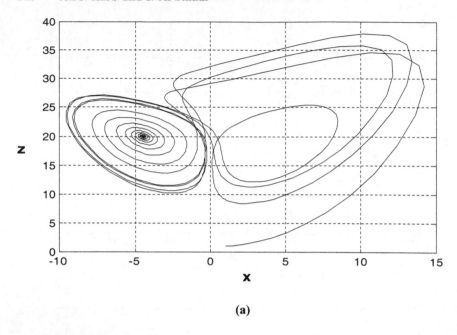

(a)

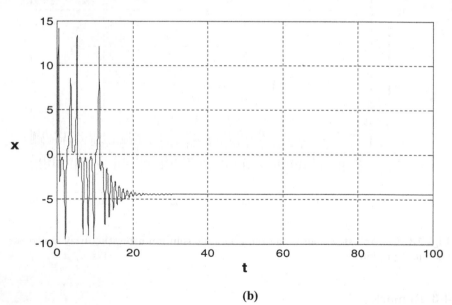

(b)

Fig. 14. Controlled system to get on a constant solution using the fuzzy controller (Lorenz equation)

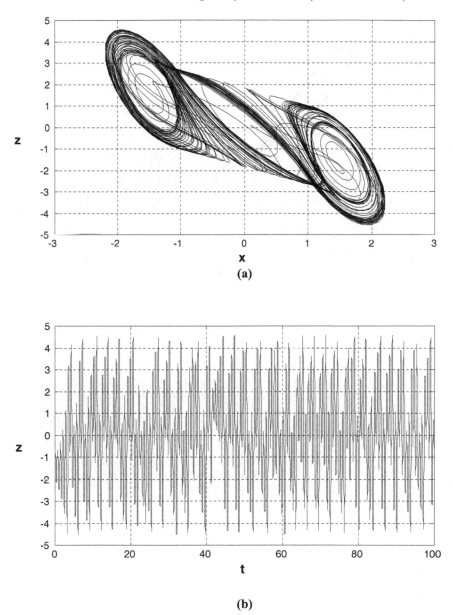

(a)

(b)

Fig. 15. Uncontrolled simulations of Chua's circuit

for Chua's circuit. Figure 18 shows the control error when σ changes to 10 (curve a), 15 (curve b), 20 (curve c), and 5 (curve d). On the other hand, Fig. 19 shows the control error when β changes; curve a for $\beta = 1$, curve b for $\beta = 0.5$, and curve c for $\beta = 1.5$. In addition, Fig. 20 shows the control

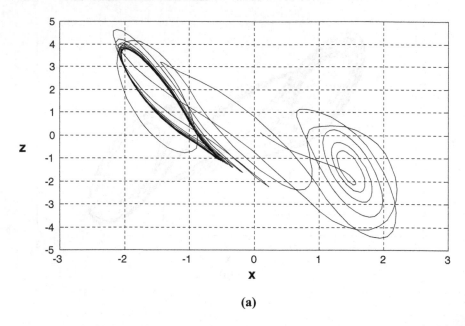

(a)

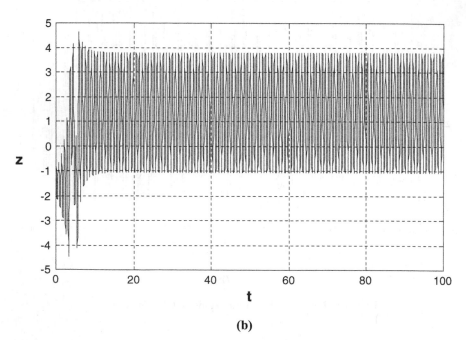

(b)

Fig. 16. Controlled system to get on a periodic solution using the fuzzy controller (Chua's circuit)

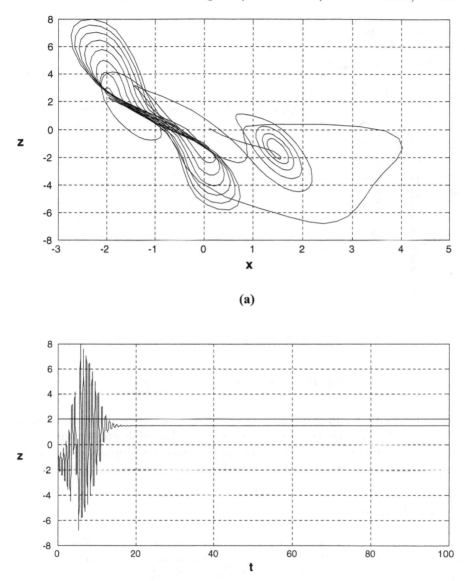

(a)

(b)

Fig. 17. Controlled system to get on a constant solution using the fuzzy controller (Chua's circuit)

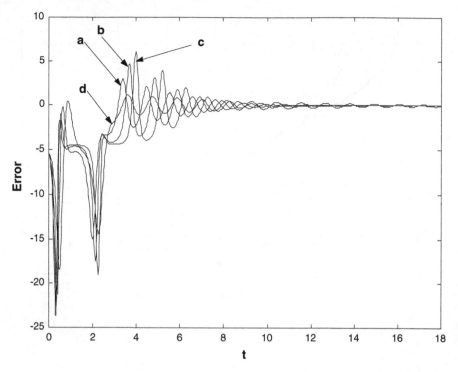

Fig. 18. Control error when σ changed

error when both σ and β change; curve a represents the control error when $\sigma = 15$ and $\beta = 1.5$, curve b for $\sigma = 12$ and $\beta = 1.3$, while curve c for $\sigma = 8$ and $\beta = 0.5$. From these results one can say that the proposed fuzzy logic controller is robust under parameter variations.

5 Conclusion

In this chapter, the idea of using fuzzy logic concept for controlling chaotic behavior is presented. There are two good reasons for using the fuzzy control: first, there is no mathematical model available for the process, and second, to satisfy nonlinear control that can be developed empirically, without complicated mathematics. The two systems are well-known models, so the first reason is not a big deal, but we can take advantage from the second reason. The results show the effectiveness of using fuzzy theory to control chaotic systems.

References

1. E.H. Abed, P.P. Varaiya: Int. J. Electr. Power Energy Syst. **6**, 37–43 (1989)

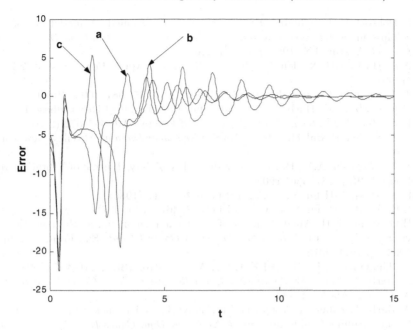

Fig. 19. Control error when β changed

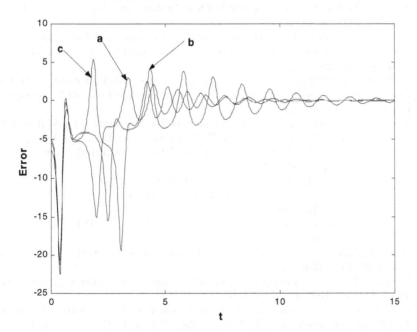

Fig. 20. Control error when σ and β changed

2. I. Dobson, H.-D. Chiang, J.S. Thorp, L. Fekih-Ahmed: A model of voltage collapse in electric power systems. In: *IEEE Proc. 27th Conf. on Decision and Control*, Austin, TX, 1988, pp 2104–2109
3. A.M. Harb, A.H. Nayfeh, C. Chin, L. Mili: Electr. Mach. Power Syst. J. **28**(11) (1999)
4. T. Endo, L.O. Chua: IEEE Trans. Circuits Syst. **35**, 987–1003 (1988)
5. B.A. Harb, A.M. Harb: Chaos and Bifurcation in Third-Order Phase Locked Loop, *Int. J. Electron.* Chaos, Solitons and Fractals **19**, 667–672 (2004)
6. A.H. Nayfeh, B. Balachandran: *Applied Nonlinear Dynamics* (Wiley, New York, 1994)
7. A.A. Al-Qaisia, A.M. Harb, A.A. Zaher, M.A. Zohdy, Journal of Vibration and Control, **9**(6), 537–556 (2003)
8. E.H. Abed, J.-H Fu: Syst. Control Lett. **7**, 11–17 (1986)
9. E.H. Abed, J.H. Fu: Syst. Control Lett. **8**, 467–473 (1987)
10. H.O. Wang, E.H. Abed: Control of nonlinear phenomena at the inception of voltage collapse. In: *Proc. 1993 American Control Conf.*, San Francisco, June 1993, pp 2071–2075
11. F. Ikhouane, M. Krstic: IEEE Trans. Aut. Control **43**(3), 431–437 (1998)
12. A. Harb, A. Zaher, M. Zohdy: *Nonlinear Recursive Chaos Control*, accepted in ACC 2002, Alaska, USA.
13. A. Harb, A. Zaher, A. Al-Qaisia, M. Zohdy: Recursive Backstepping Control of Chaotic Duffing Oscillators. *Int. J. Nonlinear Dyn. Chaos Eng. Appl.* Accepted in the Int. Journal of Modeling and Simulation, 2004.
14. A. Zaher, M. Zohdy: Robust control of biped robots. In: *Proc. of ACC*, Chicago IL, June 2000, pp 1473–1478
15. A. Zaher, M. Zohdy, F. Areed, K. Soliman: Real-time model-reference control of non-linear processes. In: *2nd Int. Conf. Computers in Industry*, Manama, Bahrain, Nov. 2000
16. A. Zaher, M. Zohdy, F. Areed, K. Soliman: Robust model-reference control for a class of non-linear and piece-wise linear systems. In: *Proc. ACC*, Arlington, VA, June 2001, pp 4514–4519
17. A. Zaher, M. Zohdy, F. Areed, K. Soliman: Robust estimation-based design for uncertain plants. To appear in AMSE Conference, Dearborn, MI, Oct. 2001
18. D. Drinkov, H. Hellendoorn, M. Reinfrank: (1993),"An Introduction to Fuzzy Control", Springer-Verlag, Berlin
19. O. Calvo, J. Cartwright: *Int. J. Bifurc. Chaos* **8**(8) (1998)
20. K.S. Tang, Kim Fung Man, Guanrong Chen, Sam Kwong: IEEE Trans. Ind. Electron. **48**(4) (Aug. 2001)
21. K.I. Mann George, Bao-Gang Hu, G. Gosine Raymond: IEEE Trans. Syst. Cybern. **29**(3) (June 1999)
22. Bao-Gang Hu, K.I. Mann George, G. Gosine Raymond: IEEE Trans. Fuzzy Syst. **7**(5) (Oct. 1999)
23. Jelena Godjevac," Comparison between PID and Fuzzy Control", Internal Report LAMI no. R93.361, June 1993
24. Y.-Y. Hsu, C.H. Cheng: Design of fuzzy power system stabilizer for multi-machine power system. In: IEEE Proc, May 1990, vol 137, pt. C, no. 3
25. H.A. Toliyat, J. Sadeh, R. Ghazi: *IEEE Trans. Energy Conversion* **11**(1) (March 1996)
26. H. Ying: Fuzzy control and modeling analytical foundation and applications, IEEE Press series on biomedical engineering, 2000

27. Yen, R. Langari: *Fuzzy Logic Intelligence, Control, and Information* (Prentice Hall, Englewood Cliffs, NJ, 1999)
28. M. Negnevitsky: *Artificial Intelligence: A Guide to Intelligent Systems* (Pearson Education, Harlow, 2002)
29. J.R. Jang, C. Sun, E. Mizutani: *Neuro-Fuzzy and Soft Computing* (Prentice Hall, Upper Saddle River, NJ, 1997)
30. J. Ross Timothy: *Fuzzy Logic with Engineering Applications* (McGraw-Hill, New York, 1995)

Digital Fuzzy Set-Point Regulating Chaotic Systems: Intelligent Digital Redesign Approach

Ho Jae Lee, Jin Bae Park, and Young Hoon Joo

Abstract. This chapter concerns digital control of chaotic systems represented by Takagi–Sugeno fuzzy systems, using intelligent digital redesign (IDR) technique. The term IDR involves converting an existing analog fuzzy set-point regulator into an equivalent digital counterpart in the sense of state-matching. The IDR problem is viewed as a minimization problem of norm distances between nonlinearly interpolated linear operators to be matched. The main features of the present method are that its constructive condition with global rather than local state-matching, for concerned chaotic systems, is formulated in terms of linear matrix inequalities; the stability property is preserved by the proposed IDR algorithm. A few set-point regulation examples of chaotic systems are demonstrated to visualize the feasibility of the developed methodology, which implies the safe digital implementation of chaos control systems.

1 Introduction

The research on controlling chaos has received increasing attention in the last few years [1–6]. There are many practical reasons for controlling chaos. For example, when chaotic mechanical vibrations occur, chaos is expected to be suppressed, where the most interesting mechanical systems are mathematically built in terms of continuous-time differential equations. It is, therefore, common practice and, in fact, advantageous to design a control in the continuous-time framework, which we call an analog control design.

It is now known that controlling chaos in continuous-time setting can be implemented by analog circuits. However, in order to take advantage of the modern high-speed computers and microelectronics, it is more preferable to digitally implement an analog control by use of digital devices [5]. Unfortunately, such an attempt often evolves an analog plant controlled by feeding sampled outputs back with analog-to-digital and digital-to-analog devices for interfacing in which continuous-time and discrete-time signals coexist. It makes traditional analysis tools for a homogeneous signal system unable to be directly used.

To fully enjoy the advancement of the digital technology in control engineering as well as surmount the theoretical obstacles, various digital control techniques have been consistently pursued with tremendous effort by many

H.J. Lee et al.: *Digital Fuzzy Set-Point Regulating Chaotic Systems: Intelligent Digital Redesign Approach*, StudFuzz **187**, 157–183 (2006)
www.springerlink.com

researchers. Among these, yet another efficient approach is the so-called digital redesign (DR) [3–5, 7–15], which is used to convert a well-designed analog control into an equivalent digital one maintaining the property of the original analog control system in the sense of state-matching, by which the benefits of both the analog control and the advanced digital technology can be achieved.

DR techniques were first considered in [16], and then developed by many others [3–5, 7–15]. It has been noticed that although the DR techniques are attractive for digital implementation of advanced analog control, these schemes basically work only for linear systems, but are not applicable to chaotic nonlinear systems.

For that reason, there has been high demand to develop some intelligent digital redesign (IDR) methodology for complex chaotic systems, and the first attempt in this direction was made by Joo et al. [2]. They synergistically merged both the Takagi–Sugeno (TS) fuzzy model based control and the DR technique for a class of complex chaotic systems. Chang et al. extended it to uncertain chaotic TS fuzzy systems [6].

Although the previous IDR techniques allowed the control engineer to enjoy the classical DR techniques for nonlinear systems represented by TS fuzzy systems, there were some issues to be addressed. The previous IDRs were acquired under the local state-matching criterion of each sub-closed-loop system—the control loop closed with the the ith plant and the ith control rule, not the global one. Moving to stability, preserving the stability has been assumed to be implicitly retained by closely matching the digitally controlled state with the analogously controlled stable state. Involving an explicit stability preserving condition into the IDR algorithm has not been treated. Thus far, the above-mentioned still remain theoretically challenging issues and thereby must be fully tackled.

In this chapter, we further develop a systematic method for the IDR of a fuzzy set-point regulator for complex chaotic systems. To alleviate the problems discussed above, we propose an alternative way: numerical optimization-based IDR involving linear matrix inequalities (LMIs). Casting the IDR problem into such a format is highly desirable since it allows to simultaneously reflect a variety of specifications in the design algorithm. The stabilizability of the hybrid digital chaos control system is also rigorously proved. Two numerical examples are provided to show how reliable our method is in the digital chaos set-point regulation.

2 Preliminaries

2.1 TS Fuzzy Systems

Most chaotic systems are quite complex in practice and have strong nonlinearities so that it is difficult, if not impossible, to build rigorous mathematical models [17]. Fortunately, a certain class of chaotic dynamical systems can be expressed in some forms of either a linear mathematical model locally or an aggregation of a set of linear mathematical models.

Consider a nonlinear dynamical system of the following form:

$$\begin{cases} \dot{x}_c(t) = f(x_c(t)) \\ y_c(t) = Cx_c(t) , \end{cases} \tag{1}$$

where $x_c(t) \in \mathbb{R}^n$ is the state and $y_c(t) \in \mathbb{R}^p$ is the output. The subscript c means that an analog control signal is applied to (1), for ordering the chaotic phenomenon evolved by (1). While the subscript d denotes that we inject the digital control signal into (1) in the sequel. The vector field $f : U_x \subset \mathbb{R}^n \to V_x \subset \mathbb{R}^n$ on some compact set U_x containing its equilibrium points is assumed to belong to C^q, $q \geqslant 1$, and $C : U_x \subset \mathbb{R}^n \to V_y \subset \mathbb{R}^{p \times n}$ is a linear output mapping.

One way to view a TS fuzzy system is that it performs a nonlinearly interpolated linear mapping $\chi(x_c(t)) : U_x \subset \mathbb{R}^n \to V_x \subset \mathbb{R}^n$ so as to satisfy

$$\sup_{x_c(t) \in U_x} \|f(x_c(t)) - \chi(x_c(t))\| \leqslant \delta$$

with arbitrary small scalar $\delta \in \mathbb{R}_{>0}$, which is known as the universal approximation in the literature.

Suppose there exist r singleton of $v_i = A_i$ which represent the local dynamic behavior of (1), such that the matrix polytope

$$\mathcal{F} = \mathrm{co}\{A_1, \dots, A_r\}$$

contains the domain U_x, where co $\{\cdot\}$ denotes a convex hull of the set $V = \{v_1, \dots, v_r\}$ and $A_i \in \mathbb{R}^{n \times n}$. Therefore, one can find an adequate mapping at time instant t with δ of the form

$$\chi(x_c(t)) = A(\theta)x_c(t),$$

where $A(\theta)$ ranges over a matrix polytope

$$A(\theta) \in \mathrm{co}\{A_1, \dots, A_r\}$$

with $\sum_{i=1}^r \theta_i = 1, \theta_i \in \mathbb{R}_{[0,1]}$, $i = \mathcal{I}_R = \{1, 2, \dots, r\}$. The key idea of the TS fuzzy inference system is to determine the coefficients θ_i in the convex combination of the given vertices V by virtue of the available qualitative knowledge from domain experts, which are quantified by "IF–THEN" rule base. More precisely, the ith rule of the TS fuzzy system is formulated in the following form [1, 2, 6, 7, 17–27]:

$$R^i : \text{IF } z_1(t) \text{ is about } \Gamma_1^i \text{ and } \cdots \text{ and } z_p(t) \text{ is about } \Gamma_p^i$$
$$\text{THEN } \dot{x}_c(t) = A_i x_c(t), \tag{2}$$

where R^i denotes the ith fuzzy inference rule; $z_j(t)$, $j \in \mathcal{I}_P = \{1,\ldots,p\}$, is the premise variable injectively mappled from $x_c(t)$; and Γ_j^i, $(i,j) \in \mathcal{I}_R \times \mathcal{I}_P$, is the fuzzy set of the jth premise variable in the ith fuzzy inference rule.

Using the center-average defuzzification, product inference, and singleton fuzzifier, the global dynamics of (2) is inferred as

$$\begin{cases} \dot{x}_c(t) = \sum_{i=1}^{r} \theta_i(z(t))A_i x_c(t) \\ y_c(t) = C x_c(t) \end{cases} \tag{3}$$

in which

$$\theta_i(z(t)) = \frac{\omega_i(z(t))}{\sum_{i=1}^{r} \omega_i(z(t))}, \quad \omega_i(z(t)) = \prod_{j=1}^{p} \Gamma_j^i(z_j(t))$$

and $\Gamma_j^i : U_{z_j} \subset \mathbb{R} \to \mathbb{R}_{[0,1]}$ is the membership value of $z_j(t)$ in Γ_j^i, where U_{z_j} the universe of discourse of $z_j(t)$ is a compact set. Some basic properties of $\theta_i(t)$ are

$$\theta_i(z(t)) \geqslant 0, \quad \sum_{i=1}^{r} \theta_i(z(t)) = 1 .$$

This TS fuzzy system is quite suitable to exactly represent a large number of chaotic systems [28]. We provide two examples.

Example 1. (Chen's Chaotic Attractor) Chen's attractor is a newly found chaotic attractor that was derived from the Lorenz system, with a similar but topologically not-equivalent structure [29]. This system has a sophisticated yet elegant and symmetric attractor, and is given by the following simple closed form.

The dynamics of the Chen's chaotic is as follows [29]:

$$\begin{cases} \dot{x} = a(y - x) \\ \dot{y} = (c - a)x - xz + cy \\ \dot{z} = xy - bz. \end{cases} \tag{4}$$

The nominal values of (a, b, c) are $(35, 3, 28)$ for chaos to emerge. System (4) has two nonlinear terms: xz and xy.

To construct a TS fuzzy system for Chen's attractor, TS fuzzy modeling of these nonlinear terms is first discussed. Consider a nonlinear function

$$f(x, y) = xy \tag{5}$$

with assumption $x \in U_x \subset \mathbb{R}_{[M_1, M_2]}$. Then (5) can be represented by a nonlinear weighted sum of linear functions of the form

$$\chi(x, y) = M(\theta)y , \tag{6}$$

where $M(\theta) \in \mathrm{co}\{M_1, M_2\}$ so that the following is satisfied:

$$\begin{cases} x = \theta_1 M_1 + \theta_2 M_2 \\ \theta_1 + \theta_2 = 1 \,. \end{cases}$$

Solving the above yields

$$\theta_1(x) = \frac{-x + M_2}{M_2 - M_1}, \qquad \theta_2(x) = \frac{x - M_1}{M_2 - M_1} \,. \tag{7}$$

Importing (7) as membership functions of fuzzy sets Γ_1^1 and Γ_1^2, (6) can be viewed as a defuzzified output of a TS fuzzy system for (5) with $\sup_{x \in U_x} \|f(x, y) - \chi(x, y)\| \leqslant \delta = 0$. Then, one can further construct a TS fuzzy system for (4). The result is

$$R^1: \quad \text{IF } x \text{ is about } \Gamma_1^1 \text{ THEN } \dot{x}_{\mathrm{c}}(t) = A_1 x_{\mathrm{c}}(t) \,,$$

$$R^2: \quad \text{IF } x \text{ is about } \Gamma_1^2 \text{ THEN } \dot{x}_{\mathrm{c}}(t) = A_2 x_{\mathrm{c}}(t) \,,$$

where

$$A_1 = \begin{bmatrix} -a & a & 0 \\ c - a & c & -M_1 \\ 0 & M_1 & -b \end{bmatrix}, \qquad A_2 = \begin{bmatrix} -a & a & 0 \\ c - a & c & -M_2 \\ 0 & M_2 & -b \end{bmatrix} \,.$$

In order to construct a TS fuzzy system for this Chen's attractor, we set $(M_1, M_2) = (-30, 30)$. This boundary is found through a numerical simulation of (4). Figure 1 shows very typical chaotic phenomena generated by the discussed TS fuzzy system with initial data $[x(0), y(0), z(0)] = [1, 1, 1]$, from an initial time $t_0 = 0$ (s) to $t_f = 30$ (s).

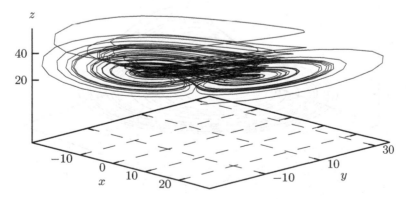

Fig. 1. Trajectory of TS fuzzy system for Chen's chaotic attractor

Example 2. Another famous chaotic attractor is by Otto Rössler, who set out to find the simplest set of differential equations capable of generating chaotic motion. Rössler's equations can be viewed as a metaphor for chemical chaos, in which regard it is worth noting that dynamics of the Rössler variety were subsequently discovered experimentally in the Belousov–Zhabotinsky and peroxidase–oxidase reactions.

$$\begin{cases} \dot{x} = -y - z \\ \dot{y} = x + by \\ \dot{z} = b + z(x - a)\,. \end{cases} \tag{8}$$

Rössler discovered that the system (8) has the spiral type chaotic behavior when $b = 0.2$ and $a = 5.7$ [30].

We can identify a TS fuzzy system of (8) as follows: since we are concerned in an autonomous form, let the state be $x_c(t) = [x, y, z, b]^T$. Next, along a similar line to Example 1, an autonomous TS fuzzy system for (8) is obtained as

$$R^1: \quad \text{IF } x \text{ is about } \Gamma_1^1 \text{ THEN } \dot{x}_c(t) = A_1 x_c(t)\,,$$
$$R^2: \quad \text{IF } x \text{ is about } \Gamma_1^2 \text{ THEN } \dot{x}_c(t) = A_2 x_c(t)\,,$$

where

$$A_1 = \begin{bmatrix} 0 & -1 & -1 & 0 \\ 1 & b & 0 & 0 \\ 0 & 0 & M_1 - a & 1 \\ 0 & 0 & 0 & 0 \end{bmatrix}, \quad A_2 = \begin{bmatrix} 0 & -1 & -1 & 0 \\ 1 & b & 0 & 0 \\ 0 & 0 & M_2 - a & 1 \\ 0 & 0 & 0 & 0 \end{bmatrix},$$

and $(M_1, M_2) = (-3, 3)$. The chaos generated by the TS fuzzy system is shown in Fig. 2 and is of the so-called spiral variety.

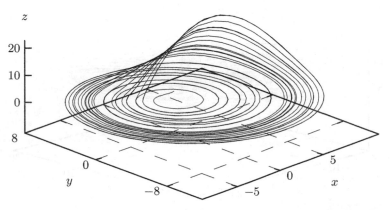

Fig. 2. Trajectory of TS fuzzy system for Rössler chaotic attractor

2.2 Analog Fuzzy Control Systems

For control purpose of (3), we consider a general analog TS fuzzy system equipped with input term

$$\begin{cases} \dot{x}_c(t) = \sum_{i=1}^{r} \theta_i(z(t))(A_i x_c(t) + B u_c(t)) \\ y_c(t) = C x_c(t) \,, \end{cases} \tag{9}$$

where $u_c(t) \in \mathbb{R}^{m \times 1}$ and $B \in \mathbb{R}^{n \times m}$ are arbitrarily chosen such that the local controllability of (9) is secured [6].

Throughout our discussion, a well-constructed analog fuzzy control guaranteeing global exponential stability (GES) with zero-reference is assumed to be predesigned, which will be used in intelligently redesigning a digital fuzzy control. The predesigned analog control rule takes the following form:

$$R^i: \text{IF } z_1(t) \text{ is about } \Gamma_1^i \text{ and } \cdots \text{ and } z_p(t) \text{ is about } \Gamma_p^i$$

$$\text{THEN } u_c(t) = K_c^i x_c(t) + E_c^i r(t) \,,$$

where K_c^i and E_c^i, $i \in \mathcal{I}_R$, are given analog control gain matrices. $r(t) \in \mathbb{R}^{p \times 1}$ is a reference to be tracked by $y_c(t)$, which is assumed to be piecewisely constant in any time interval $[kT, kT+T]$, $k \in \mathbb{Z}_{\geq 0}$, $\sup_{t \in \mathbb{R}_{\geq 0}} \|r(t)\| = r_M < \infty$.

Similar to (3), its defuzzified output is given by

$$u_c(t) = \sum_{i=1}^{r} \theta_i(z(t))(K_c^i x_c(t) + E_c^i r(t)) \,. \tag{10}$$

Then the overall analog closed-loop TS fuzzy system (9) cascaded by (10) is written as

$$\dot{x}_c(t) = \sum_{i=1}^{r} \theta_i(z(t))((A_i + B K_c^i) x_c(t) + B E_c^i r(t)) \,. \tag{11}$$

2.3 Digital Fuzzy Systems and Its Time Discretization

This subsection discusses a desired digital fuzzy control system and its time-discretization. By sharing the premise parts of (3), and interfacing an ideal sampler and a zero-order holder between a concerned system and a control, a desired digital fuzzy control system is represented by

$$\begin{cases} \dot{x}_d(t) = \sum_{i=1}^{r} \theta_i(z(t))(A_i x_d(t) + B u_d(t)) \\ y_d(t) = C x_d(t) \,, \end{cases} \tag{12}$$

where $u_d(t) = u_d(kT) \in \mathbb{R}^m$ is a piecewise constant control to be determined during a time interval $[kT, kT+T)$, $k \in \mathbb{Z}_{\geq 0}$, where $T \in \mathbb{R}_{>0}$ is a nonpathological sampling period.

The IDR problem is usually set up in discrete-time setting so as to find some relevant digital control satisfying the state-matching criterion at each sampling time instant. Thus, it is more convenient to timely discretize (12) for derivation of the IDR condition.

There are several methods in discretizing an LTI system. Unfortunately, these approaches are not directly applicable to the discretization of the TS fuzzy system, since it is not LTI but implicitly time-varying and nonlinear [18]. Moreover, it is further strongly desired to maintain the polytopic structure of the discretized TS fuzzy system for the construction of a digital fuzzy model based control. Thus we need a mathematical foundation for the discretization of TS fuzzy systems.

Assumption 1. Assume that the firing strength of the ith rule $\theta_i(z(t))$: $U_{z_i} \to \mathbb{R}_{[0,1]}$ is approximated by its value at time $t = kT$, that is, $\theta_i(z(t)) \approx \theta_i(z(kT))$ for $t \in [kT, kT + T)$. Consequently, the nonlinear matrix function $\sum_{i=1}^{r} \theta_i(z(t)) A_i : U_z \to \mathrm{co}\,\{A_1, \ldots, A_r\}$ can be approximated by constant matrices $\sum_{i=1}^{r} \theta_i(z(kT)) A_i$ over any interval $[kT, kT+T)$, $k \in \mathbb{Z}_{\geqslant 0}$ [1, 18, 31].

If a suitable small sampling period $T \in \mathbb{R}_{>0}$ is chosen, Assumption 1 is reasonable and we have the following result.

Proposition 1. *The pointwise dynamical behavior of* (12) *can be efficiently approximated by*

$$
\begin{cases}
x_d(kT + T) = \sum_{i=1}^{r} \theta_i(z(kT))(G_i x_d(kT) + H_i u_d(kT)) \\
y_d(kT) = C x_d(kT) ,
\end{cases}
\tag{13}
$$

where $G_i = \mathrm{e}^{A_i T}$ *and* $H_i = (G_i - I) A_i^{-1} B$.

Proof. The general solution $x_d(t; t_0, x_0)$ to (12) with any initial data $(t_0, x_0) \in \mathbb{R}_{\geqslant 0} \times U_x$ is

$$
x_d(t; t_0, x_0) = \phi(t, t_0) x_d(t_0) + \int_{t_0}^{t} \phi(t, \tau) B u_d(kT) \, \mathrm{d}\tau ,
$$

where the state transition map $\phi : \mathbb{R}_{>0} \times \mathbb{R}_{>0} \to \mathbb{R}^{n \times n} \in C^3$ satisfies $\phi(t, t_0) = \phi(t, t_1)\phi(t_1, t_0)$ and $\frac{\partial}{\partial t}\phi(t, t_0) = (\sum_{i=1}^{r} \theta_i(z(t)) A_i)\phi(t, t_0)$ with the initial condition $\phi(t_0, t_0) = I$. Let the initial time t_0 be kT, then the exact solution to (12) evaluated at $t = kT + T$ can be written as

$$
x_d(kT + T) = \phi(kT) x_d(kT) + \psi(kT) u_d(kT)
$$

where $\phi(kT) = \phi(kT + T, kT)$ and $\psi(kT) = \int_{kT}^{kT+T} \phi(kT + T, \tau) B \, \mathrm{d}\tau$.

The exact evaluation of $\phi(\cdot, \cdot)$ is very difficult, if not impossible, since (12) exhibits nonlinear dynamical behavior. To get around with this difficulty, suppose that Assumption 1 is satisfied, then $\phi(kT)$ and $\psi(kT)$ are approximately evaluated as follows:

$$\phi(kT) = e^{\sum_{i=1}^{r} \theta_i(z(kT))A_iT}$$

$$= I + \sum_{i=1}^{r} \theta_i(z(kT))A_iT + \mathcal{O}(T^2)$$

$$\approx \sum_{i=1}^{r} \theta_i(z(kT))(I + A_iT)$$

$$\approx \sum_{i=1}^{r} \theta_i(z(kT))\, e^{A_iT}$$

and

$$\psi(kT) = \int_{kT}^{kT+T} e^{\sum_{i=1}^{r} \theta_i(z(kT))A_i(kT+T-\tau)}\, B\, d\tau$$

$$= \left(e^{\sum_{i=1}^{r}\theta_i(z(kT))A_i\mathrm{T}} - I\right)\left(\sum_{i=1}^{r}\theta_i(z(kT))A_i\right)^{-1} B$$

$$= \left(\sum_{i=1}^{r}\theta_i(z(kT))A_iT + \mathcal{O}(T^2)\right)\left(\sum_{i=1}^{r}\theta_i(z(kT))A_i\right)^{-1} B$$

$$\approx \sum_{i=1}^{r} \theta_i(z(kT))BT$$

$$\approx \sum_{i=1}^{r} \theta_i(z(kT))H_i\,,$$

where $G_i = e^{A_iT}$ and $H_i = \int_{kT}^{kT+T} e^{A_i(kT+T-\tau)}\, B\, d\tau = (G_i - I)A_i^{-1}B$.

Remark 1. If A_i is singular, H_i can be computed by the following formula [20]:

$$H_i = \sum_{j=1}^{\infty} \frac{1}{j!}(A_iT)^{j-1}BT.$$

Remark 2. The discretized TS fuzzy system (13) contains the discretization error with order of $\mathcal{O}(T^2)$.

For digitally ordering (12), the intelligently redesigned digital fuzzy control takes the following form:

$$R^i\colon \text{IF } z_1(kT) \text{ is about } \Gamma_1^i \text{ and } \cdots \text{ and } z_p(kT) \text{ is about } \Gamma_p^i,$$
$$\text{THEN } u_d(t) = K_d^i x_d(kT) + E_d^i r(kT)$$

for $t \in [kT, kT + T)$, and the overall control is given by

$$u_d(t) = \sum_{i=1}^{r} \theta_i(z(kT))(K_d^i x_d(kT) + E_d^i r(kT)) \tag{14}$$

during the sampling time interval.

Then, the closed-loop system with (13) and (14) is constructed to yield

$$\begin{cases} x_d(kT + T) = \sum_{i=1}^{r} \sum_{j=1}^{r} \theta_i(z(kT))\theta_j(z(kT))((G_i + H_i K_d^j)x_d(kT) \\ \qquad + H_i E_d^j r(kT)) \\ y_d(kT) = C x_d(kT) \, . \end{cases}$$
$$\tag{15}$$

Corollary 1. *The pointwise dynamical behavior of* (11) *can also be approximately discretized as*

$$\begin{cases} x_c(kT + T) = \sum_{i=1}^{r} \sum_{j=1}^{r} \theta_i(z(kT))\theta_j(z(kT))(\phi_i x_c(kT) + \psi_{ij} r(kT)) \\ y_c(kT) = C x_d(kT) \, , \end{cases}$$
$$\tag{16}$$

where $\phi_i = e^{(A_i + BK_c^i)T}$ *and* $\psi_{ij} = \int_{kT}^{kT+T} e^{(A_i + BK_c^i)(kT+T-\tau)} BE_c^j \, d\tau$.

Proof. It can be straightforwardly proven by Proposition 1.

3 Intelligent Digital Redesign of Fuzzy Set-Point Regulator

For practical digital implementation of the predeveloped analog fuzzy ordering of chaotic phenomenon, one may desire to convert it into a digital control equivalent in some performance measure sense.

To attain this, one may try to simply insert sample-and-hold devices into the original analog control loop. However, such a digital implementation of analog control will heavily degrade the ordering performance, and what is worse is that it can be destabilized even in a small sampling period, as is shown in [1], because this method does not consider the state behavior nor preserves the stabilizability of the digitalized analog control.

Our goal is to develop an IDR technique for TS fuzzy systems revealing chaotic phenomenon so that the global dynamical behavior of (12) with the intelligently redesigned digital fuzzy control may retain that of the closed-loop TS fuzzy system with the existing analog fuzzy control, and the stability of the digitally controlled TS fuzzy system is preserved.

To this end, we formulate the following IDR problem:

Problem 1. (IDR Problem for a Fuzzy Set-Point Regulator) Given the well-designed analog control gain matrices K_c^i and E_c^i for (10) that ensures GES of (11), find the digital control gain matrices K_d^i and E_d^i for (14) such that the followings are satisfied:

1. The state $x_d(kT)$ of the discrete-time representation (15) of the digitally controlled system (12) with (14) matches the state $x_c(kT)$ of the discrete-time representation (16) of the analogously controlled system (11) at $t = kT$, $k \in \mathbb{Z}_{>0}$, as closely as possible.
2. The hybrid system (12) controlled by (14) with $r(t) = [0]_{p \times 1}$ is GES.

Consider the first objective of Problem 1. In comparison of (15) with (16), to realize $x_c(kT+T) = x_d(kT+T)$ under the assumption of $x_c(kT) = x_d(kT)$, it is necessary to determine K_d^i and E_d^i such that the following matrix equality constraints

$$\phi_i = G_i + H_i K_d^j \tag{17}$$

$$\psi_{ij} = H_i E_d^j \tag{18}$$

hold for all pair $(i,j) \in \mathcal{I}_R \times \mathcal{I}_R$.

The second objective in Problem 1 can be handled in a discrete manner by virtue of the following proposition:

Proposition 2. *Suppose* (15) *with* $r(kT) = [0]_{p \times 1}$ *is GES; then the zero equilibrium points* $x_{d_{eq}} = [0]_{n \times 1}$ *of the hybrid digital control system* (12) *and* (14) *are also GES.*

Proof. It follows from (12), (14), and Assumption 1 for $t \in [kT, kT+T)$ that

$$\|x_d(t)\| \leqslant \left\| e^{(\sum_{i=1}^r \theta_i(z(kT))A_i)(t-kT)} x_d(kT) \right\|$$

$$+ \left\| \int_{kT}^t e^{(\sum_{i=1}^r \theta_i(z(kT))A_i)(t-\tau)} B\left(\sum_{i=1}^r \theta_i(z(kT))K_d^i\right) x_d(kT)\, d\tau \right\|$$

$$\leqslant \sup_{z(kT)\in U_z} \left\{ e^{\left\|\sum_{i=1}^r \theta_i(z(kT))A_i\right\| T} \|x_d(kT)\| \right.$$

$$\left. + T\, e^{\left\|\sum_{i=1}^r \theta_i(z(kT))A_i\right\| T} \left\|\sum_{j=1}^r \theta_i(z(kT))BK_d^j\right\| \|x_d(kT)\| \right\}$$

$$\leqslant \sup_{(i,j)\in\mathcal{I}_R\times\mathcal{I}_R} \left\{ e^{\|A_i\| T} \|x_d(kT)\| + T\, e^{\|A_i\| T} \left\|BK_d^j\right\| \|x_d(kT)\| \right\}$$

$$= \mu \|x_d(kT)\| ,$$

where $\mu = \sup_{(i,j)\in\mathcal{I}_R\times\mathcal{I}_R} e^{\|A_i\| T}(1 + T\left\|BK_d^j\right\|)$ is independent of k. By the GES definition of (15): $\|x_d(k, k_0, x_0)\| \leqslant \varsigma_1 e^{-\varsigma_2(k-k_0)T} \|x_0\|$, for any initial data $(k_0, x_0) \in \mathbb{Z}_{[0,k)} \times \mathbb{R}^n$, we further proceed over the interval $[kT, kT+T)$

$$\|x_d(t; t_0, x_0)\| \leqslant \mu_{\varsigma_1} e^{-\varsigma_2(k-k_0)T} \|x_0\|$$
$$= \mu_{\varsigma_1} e^{\varsigma_2(t-kT)} e^{-\varsigma_2(t-k_0T)} \|x_0\|$$
$$\leqslant \mu_{\varsigma_1} e^{\varsigma_2 T} e^{-\varsigma_2(t-t_0)} \|x_0\|$$
$$= \varsigma_3 e^{-\varsigma_2(t-t_0)} \|x_0\| ,$$

where $t_0 = k_0 T$ and $\varsigma_3 = \mu_{\varsigma_1} e^{\varsigma_2 T}$. Therefore, we conclude that the trivial solutions $x_d(t; t_0, x_0)$ to (12) with (14) are GES.

Remark 3. There are three points to be mentioned on Problem 1.

1. Conditions (17) and (18) may be solvable for K_d^i and E_d^i, if $p \geqslant n$ and $m \geqslant n$ and H_i is nonsingular. However, in control engineering these are usually overdetermined, hence one may suffer difficulty in having exact solutions.
2. It should be addressed that (17) and (18) are hardly solved in many cases of TS fuzzy model based control, since each variable K_d^i and E_d^i should satisfy r different equality constraints, respectively, for the global matching [19].
3. The stability property of the analog fuzzy control should be preserved in the IDR procedure. However securing the GES of (15) weights down solving K_d^i and E_d^i analytically.

In order to cope with the difficulties, an alternative approach is applied by relaxing Problem 1 and searching the digital fuzzy set-point regulator in a numerical manner. Our key idea is to find K_d^i and E_d^i in such a way that the norm distances between ϕ_i and $G_i + H_i K_d^j$, and ψ_{ij} and $H_i E_d^j$, respectively, are minimized by using a numerical optimization technique.

Problem 2. (γ-Suboptimal IDR Problem for a Fuzzy Set-Point Regulator) Given well-designed analog GES control gain matrices K_c^i and E_c^i for (10), find the digital control gain matrices K_d^i and E_d^i for (14) such that the followings are satisfied:

1. $\|\phi_i - G_i - H_i K_d^j\|$ and $\|\psi_{ij} - H_i E_d^j\|$ are minimized for all $(i, j) \in \mathcal{I}_R \times \mathcal{I}_R$, in the sense of the induced 2-norm measure.
2. The discretized closed-loop system (15) with $r(kT) = [0]_{p \times 1}$ is GES in the sense of Lyapunov.

Notice that Problem 2 becomes a convex optimization problem, hence can be numerically solved by formulating in terms of LMIs. The main results of this paper are summarized as follows:

Theorem 1. (γ-Suboptimal IDR for a Fuzzy Set-Point Regulator) *If there exist matrices $Q = Q^T \succ 0$, $X_{ij} = X_{ij}^T = X_{ji} = X_{ji}^T$, and F_d^i, E_d^i with compatible dimensions, and possibly small positive scalars $\gamma_1 \in \mathbb{R}_{>0}$, $\gamma_2 \in \mathbb{R}_{>0}$, such that the following minimization problem has solutions*

MP: $\underset{Q,X_{ij},F_d^i,E_d^i}{Minimize}$ trace$\{\gamma_1, \gamma_2\}$

subject to

$$\begin{bmatrix} -\gamma_1 Q & (\bullet)^{\mathrm{T}} \\ \phi_i Q - G_i Q - H_i F_d^j & -\gamma_1 I \end{bmatrix} \prec 0 \qquad (19)$$

$$\begin{bmatrix} -\gamma_2 I & (\bullet)^{\mathrm{T}} \\ \psi_{ij} - H_i E_d^j & -\gamma_2 I \end{bmatrix} \prec 0, \quad (i,j) \in \mathcal{I}_R \times \mathcal{I}_R \qquad (20)$$

$$\begin{bmatrix} -Q + X_{ij} & (\bullet)^{\mathrm{T}} \\ \left(\frac{G_i Q + H_i F_d^j + G_j Q + H_j F_d^i}{2} \right) & -Q \end{bmatrix} \prec 0 \qquad (21)$$

$$\left[X_{ij} \right]_{r \times r} \succ 0, \quad (i,j) \in \mathcal{I}_J \times \mathcal{I}_R \qquad (22)$$

then $x_d(kT)$ of the discrete-time representation (13) of (12) controlled by the intelligently redesigned digital set-point regulator (14) closely matches $x_c(kT)$ of the discrete-time representation (16) of the analog control system (11), and (15) is GES in the sense of Lyapunov, where $(\bullet)^{\mathrm{T}}$ denotes the transposed element in symmetric positions and $\mathcal{I}_J \times \mathcal{I}_R$ means all pairs $(i,j) \in \mathbb{Z}_{>0} \times \mathbb{Z}_{>0}$ such that $1 \leqslant i \leqslant j \leqslant r$.

Proof. First, consider the first constraint in the first objective of Problem 2. Introducing a free matrix variable W having a full column rank of n, then we have

$$\left\| \phi_i - G_i - H_i K_d^j \right\|_2 \leqslant \widehat{\gamma}_1$$

$$= \widehat{\gamma}_1 \frac{1}{\|W\|_2} \|W\|_2$$

$$= \gamma_1 \|W\|_2 \qquad (23)$$

for all $(i,j) \in \mathcal{I}_R \times \mathcal{I}_R$, where $\gamma_1 = \widehat{\gamma}_1 / \|W\|_2 \in \mathbb{R}_{>0}$. Without loss of generality, one can pick P such that $P \succ W^{\mathrm{T}} W$, which is reasonable since $P = P^{\mathrm{T}} \succ 0$ is definitely bounded from (19).

From the definition of the induced 2-norm, (23) holds if the following inequality are satisfied:

$$(\phi_i - G_i - H_i K_d^j)^{\mathrm{T}} (\phi_i - G_i - H_i K_d^j) \prec \gamma_1^2 P . \qquad (24)$$

Using Schur complement, (24) can be represented by LMIs of the form:

$$\begin{bmatrix} -\gamma_1 P & (\bullet)^{\mathrm{T}} \\ \phi_i - G_i - H_i K_d^j & -\gamma_1 I \end{bmatrix} \prec 0 . \qquad (25)$$

Further applying the congruence transformation to (25) with diag$\{P^{-1}, I\}$, and denoting $F_d^i = K_d^i P^{-1}$ yields (19). Next, $\|\psi_{ij} - H_i E_d^j\|_2$ is minimized whenever we minimize γ_2 over E_d^i subject to $(\psi_{ij} - H_i E_d^j)^{\mathrm{T}} (\psi_{ij} - H_i E_d^j) \prec \gamma_2^2 I$

or equivalently (20). LMI (21) and (22) directly follow from the standard Lyapunov GES theorem [21] details of which are shown in the Appendix. This completes the proof of the theorem.

Remark 4. The matrix constraints regarding ψ_i such that $\Gamma_i^h \cap \Gamma_j^h = \varnothing$, $h \in \mathcal{I}_P$ on U_{z_h}, and for all $t \in \mathbb{R}_{\geqslant 0}$ do not have to be solved in IDR procedure.

4 Examples

Two examples are included to visualize the theoretical analysis and design. More precisely, using the suggested technique digital control problems of two chaotic systems, the Duffing-like chaotic oscillator and the chaotic Chua circuit, are presented in this chapter.

4.1 IDR and Set-Point Regulation of the Duffing-like Chaotic Oscillator

Consider the following Duffing-like chaotic oscillator [32]:

$$\ddot{y}(t) - ay(t) + by(t)|y(t)| = -\epsilon(\zeta\dot{y}(t) - c\sin(\omega t)), \qquad (26)$$

where $a = 1.1, b = 1$, and $c = 21$ are some positive constants, and $\zeta = 3$ and $\epsilon = 0.1$ are small positive constants for chaos to emerge. The trajectory of this system is shown in Fig. 3, which is chaotic and irregular. Let the system state be $x_c(t) = [y(t), \dot{y}(t)]^\mathrm{T}$, and rewrite (26) as

$$\frac{\mathrm{d}}{\mathrm{d}t} x_c(t) = \begin{bmatrix} x_{c_2}(t) \\ ax_{c_1}(t) - bx_{c_1}(t)|x_{c_1}(t)| - \epsilon\zeta x_{c_2}(t) \end{bmatrix} + \begin{bmatrix} 0 \\ \epsilon c \sin(\omega t) \end{bmatrix}.$$

First, in order to construct the TS fuzzy system, the nonlinear term $x_{c_1}(t)|x_{c_1}(t)|$ should be expressed as a convex sum of the state as follows:

$$\begin{cases} x_{c_1}(t)|x_{c_1}(t)| = \Gamma_1^1(x_{c_1}(t))0 + \Gamma_1^2(x_{c_1}(t))Mx_{c_1}(t) \\ \Gamma_1^1(x_{c_1}(t)) + \Gamma_1^2(x_{c_1}(t)) = 1, \end{cases} \qquad (27)$$

where $x_{c_1}(t) \in U_y \subset (-M, M)$. M is reasonably chosen as 2.5, from Fig. 3. Solving (27) yields

$$\begin{cases} \Gamma_1^1(x_{c_1}(t)) = 1 - \frac{|x_{c_1}(t)|}{M} \\ \Gamma_1^2(x_{c_1}(t)) = \frac{|x_{c_1}(t)|}{M}. \end{cases}$$

Next, since we consider only the autonomous dynamical system for IDR, it is highly desirable that the sinusoidal function is included into the state vector. Hence, the state of the system can be redefined as

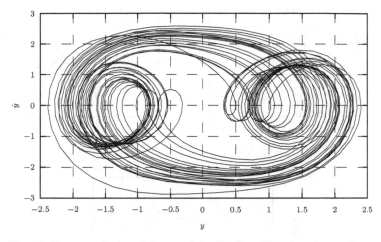

Fig. 3. Uncontrolled trajectory of the Duffing-like chaotic oscillator

$$x_c(t) = \left[y(t), \dot{y}(t), \sin(\omega t), \cos(\omega t) \right]^{\mathrm{T}} .$$

Now, the analytic TS fuzzy system of (26) is given by

$$R^1: \quad \text{IF } x_{c_1}(t) \text{ is about } \Gamma_1^1, \quad \text{THEN } \dot{x}_c(t) = A_1 x_c(t) ,$$
$$R^2: \quad \text{IF } x_{c_1}(t) \text{ is about } \Gamma_1^2, \quad \text{THEN } \dot{x}_c(t) = A_2 x_c(t) ,$$

where

$$A_1 = \begin{bmatrix} 0 & 1 & 0 & 0 \\ a & -\epsilon\zeta & \epsilon c & 0 \\ 0 & 0 & 0 & \omega \\ 0 & 0 & -\omega & 0 \end{bmatrix}, \qquad A_2 = \begin{bmatrix} 0 & 1 & 0 & 0 \\ a - bM & -\epsilon\zeta & \epsilon c & 0 \\ 0 & 0 & 0 & \omega \\ 0 & 0 & -\omega & 0 \end{bmatrix}.$$

Since the mathematical equations of the original Duffing-like chaotic oscillator and its TS fuzzy system are the same, one can easily expect that their trajectories are identical. For design of a suitable fuzzy-model-based control, the input and output matrices are assumed to be

$$B_1 = B_2 = \begin{bmatrix} 0 \\ 1 \\ 1 \\ 0 \end{bmatrix}, \qquad C = \begin{bmatrix} 1 & 0 & 0 & 0 \end{bmatrix},$$

which preserve the controllability of the system.

From Theorem 1 in [28] and [2], the well-constructed gain matrices for the analog fuzzy control are obtained as follows:

$$K_c^1 = \begin{bmatrix} -9.1263 & -2.6634 & -1.3199 & 0.0400 \end{bmatrix}, \quad E_c^1 = 8.0019$$
$$K_c^2 = \begin{bmatrix} -8.3869 & -3.5096 & -0.7655 & -0.3501 \end{bmatrix}, \quad E_c^1 = 9.5146 .$$

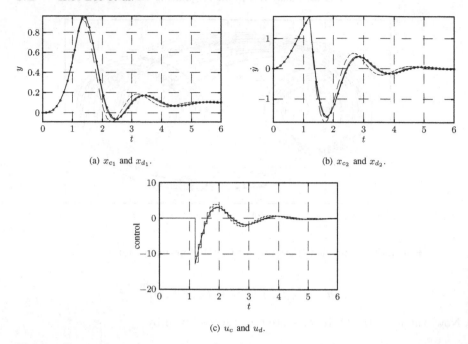

(a) x_{c_1} and x_{d_1}.

(b) x_{c_2} and x_{d_2}.

(c) u_c and u_d.

Fig. 4. Comparison of time responses of the controlled Duffing-like chaotic oscillator for $T = 0.1$ (s) (control input is activated at time $t = 1.2$ (s)): analog (solid line), proposed (solid line with circle), and simple digital implementation (dotted line)

Applying Theorem 1 yields the intelligently redesigned digital fuzzy control gain matrices, for the sampling period $T = 0.1$ (s), as

$$K_d^1 = \begin{bmatrix} -7.3545 & -2.4996 & -1.3297 & -0.0844 \end{bmatrix}, \qquad E_d^1 = \begin{bmatrix} 6.4246 \end{bmatrix}$$
$$K_d^2 = \begin{bmatrix} -6.6661 & -3.1038 & -0.9149 & -0.3562 \end{bmatrix}, \qquad E_d^1 = \begin{bmatrix} 7.6389 \end{bmatrix}.$$

To explicitly show the usefulness of the proposed method, we also simulate with $K_d^i = K_c^i$, $i \in \mathcal{I}_R$, which signifies naive digital implementation of analog control by simply inserting sample-and-hold devices into the original analog control loop.

The initial value is $x_c(0) = x_d(0) = x_0 = \begin{bmatrix} 1, 0, 0, 1 \end{bmatrix}^T$ and simulation time is 6 (s). For the quantitative comparison of the performance of the proposed method, the performance measure is defined as [16]

$$\mathcal{P} = \sum_{i=1}^{4} \left(\int_0^6 |x_{c_i}(t) - x_{d_i}(t)| \mathrm{d}t \right).$$

Time response of the simulation is shown in Fig. 4. Control input is activated at $t = 1.2$ (s). Before the control input is activated, $y_d(t)$ does not

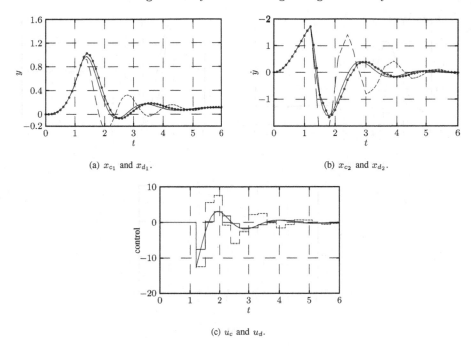

(a) x_{c_1} and x_{d_1}.

(b) x_{c_2} and x_{d_2}.

(c) u_c and u_d.

Fig. 5. Comparison of time responses of the controlled Duffing-like chaotic oscillator for $T = 0.3$ (s) (control input is activated at time $t = 1.2$ (s)): analog (solid line), proposed (solid line with circle), and simple digital implementation (dashed line)

converge to $r(t) = \begin{bmatrix} 1 \end{bmatrix}$. After the control input is activated, all output trajectories are guided to $r(t)$. Nevertheless, we see that our IDR method produces a closer state-matching performance than the other one.

In Fig. 5 the excellence of the proposed method is shown. For $T = 0.3$ (s), we intelligently redesign the digital control gains as follows:

$$K_d^1 = \begin{bmatrix} -2.0022 & -1.4249 & -1.1689 & -0.5963 \end{bmatrix}, \qquad E_d^1 = \begin{bmatrix} 1.4470 \end{bmatrix}$$
$$K_d^2 = \begin{bmatrix} -1.3277 & -1.2906 & -1.1066 & -0.5905 \end{bmatrix}, \qquad E_d^2 = \begin{bmatrix} 1.7318 \end{bmatrix}.$$

It should be strongly stressed that the trajectory controlled by the proposed method closely matches the original controlled trajectory, whereas the naive simple digital implementation of analog fuzzy control yields a poor state-matching. This is because the proposed method provides the global state-matching of the overall TS fuzzy system.

Another relatively longer sampling period $T = 0.6$ (s) is chosen so as to emphasize the superiority of the proposed method to the other in the angle of the stabilizability and control performance. Based on Theorem 1, the gain matrices for $T = 0.6$ (s) are obtained as follows:

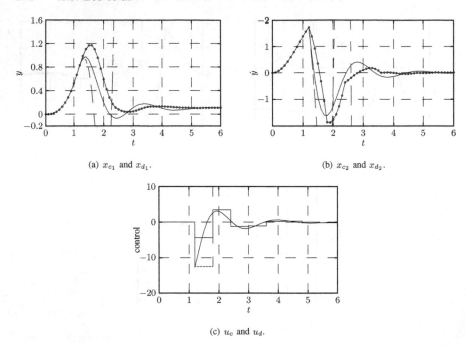

(a) x_{c_1} and x_{d_1}.

(b) x_{c_2} and x_{d_2}.

(c) u_c and u_d.

Fig. 6. Comparison of time responses of the controlled Duffing-like chaotic oscillator for $T = 0.6$ (s) (control input is activated at time $t = 1.2$ (s)): analog (solid line), proposed (solid line with circle), and simple digital implementation (dashed line)

$$K_d^1 = \begin{bmatrix} -2.0022 & -1.4249 & -1.1689 & -0.5963 \end{bmatrix}, \qquad E_d^1 = \begin{bmatrix} 1.4470 \end{bmatrix}$$
$$K_d^2 = \begin{bmatrix} -1.3277 & -1.2906 & -1.1066 & -0.5905 \end{bmatrix}, \qquad E_d^2 = \begin{bmatrix} 1.7318 \end{bmatrix}.$$

Figure 6 shows the time responses and the trajectories of two digitally controlled systems. As shown in the figures, the intelligently redesigned digital control by the proposed method not only drives $y(t)$ to $r(t) = \begin{bmatrix} 1 \end{bmatrix}$, but also relatively well matches the trajectories of the original system. However, the other digital control gives the deteriorated state-matching performance, and even instability. The performance comparison of these is shown in Table 1. It is noted that the proposed method guarantees the stability of the controlled system in much wider range of sampling period than does the simple digital implementation, which may fail to stabilize the system especially for relatively longer sampling period; this is another major advantage of the proposed method. This is because the proposed method incorporates the stability criterion in the IDR condition, whereas the other approach does not.

Table 1. Comparison of the performance $\mathcal{P}$ of the proposed method with that of other method according to the various values of sampling period T

	Sampling Period T (s)		
Method	0.1	0.3	0.6
Simple digital implementation	0.160447	0.557474	Unstable
Proposed	0.039535	0.183141	0.456691

4.2 Chaotic Chua Circuit

The chaotic Chua circuit, as shown in Fig. 7, is a simple electronic system that consists of one inductor (L), two capacitors (C_1,C_2), one linear resistor (R), and one piecewise linear resistor (g). The circuit diagram is illustrated in Fig. 8. Chua's circuit has been shown to possess very rich nonlinear dynamics such as bifurcations and chaos [33].

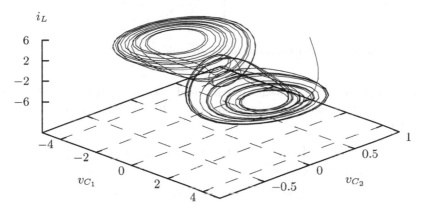

Fig. 7. Uncontrolled trajectory of the chaotic Chua's circuit oscillator

The dynamic equations of Chua's circuits are described by

$$\begin{cases} \dot{v}_{C_1} = \dfrac{1}{C_1}\left(\dfrac{1}{R}(v_{C_2} - v_{C_1}) - g(v_{C_1})\right) \\[2mm] \dot{v}_{C_1} = \dfrac{1}{C_2}\left(\dfrac{1}{R}(v_{C_1} - v_{C_2}) + i_L\right) \\[2mm] \dot{i}_L = -\dfrac{1}{L}(v_{C_1} + R_0 i_L) \,, \end{cases} \tag{28}$$

where R_0 is a constant and g denotes the nonlinear resistor, which is a function of the voltage across the terminals of C_1 defined as follows:

$$g(v_{C_1}) = m_b v_{C_1} + \frac{1}{2}(m_a - m_b)(|v_{C_1} + E| - |v_{C_1} - E|) \,,$$

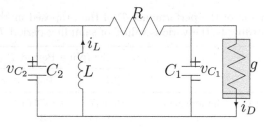

Fig. 8. Chua's circuit

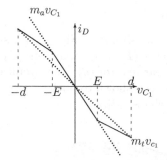

Fig. 9. Resistor characteristic with a piecewise linear function

where E is a constant voltage and m_a and m_b are negative, as shown in Fig. 9, which can also be represented by the three-segment piecewise linear function

$$g(v_{C_1}) = \begin{cases} m_b v_{C_1} + (m_a - m_b)E, & v_{C_1} \geqslant E \\ m_a v_{C_1}, & -E < v_{C_1} < E \\ m_b v_{C_1} - (m_a - m_b)E, & v_{C_1} \leqslant E \end{cases} \qquad (29)$$

We want to obtain a TS fuzzy system in the open-loop form (3) for Chua's circuit with characteristic (29). Assuming $v_{C_1} \in U_{v_{C_1}} \subset \mathbb{R}_{[-d,d]}$, $d > E$, then (29) can be represented by a nonlinear weighted sum of the form: $\chi(v_{C_1}) = m(\theta)v_{C_1}$, where $m(\theta) \in \text{co}\{m_a, m_t\}$ so that

$$\begin{cases} g(v_{C_1}) = m(\theta) \\ \theta_1 + \theta_2 = 1 \end{cases}$$

is fulfilled, where $m_t = m_a + (m_a - m_b)E/(d)$. Denote $x_c(t) = \left[v_{C_1}, v_{C_2}, i_L\right]^{\mathrm{T}}$. In order to build the intended fuzzy system, the parameter d must be chosen properly. Taking the above θ_1 and θ_2 as membership functions of fuzzy sets Γ_1^1 and Γ_1^2, $\chi(v_{C_1})$ can be viewed as a defuzzified output of a TS fuzzy system for $g(v_{C_1})$ with $\sup_{x \in U_{v_{C_1}}} \|g(v_{C_1}) - \chi(v_{C_1})\| \leqslant \delta = 0$. Then, the TS fuzzy system for Chua's circuit becomes

$$R^1: \quad \text{IF } v_{C_1} \text{ is about } \Gamma_1^1 \text{ THEN } \dot{x}_c(t) = A_1 x_c(t)$$

$$R^2: \quad \text{IF } v_{C_1} \text{ is about } \Gamma_1^2 \text{ THEN } \dot{x}_c(t) = A_2 x_c(t) ,$$

where

$$A_1 = \begin{bmatrix} -\frac{1}{C_1 R} - \frac{m_a}{C_1} & \frac{1}{C_1 R} & 0 \\ \frac{1}{C_2 R} & -\frac{1}{C_2 R} & \frac{1}{C_2} \\ 0 & -\frac{1}{L} & -\frac{R_0}{L} \end{bmatrix}, \quad A_2 = \begin{bmatrix} -\frac{1}{C_1 R} - \frac{m_t}{C_1} & \frac{1}{C_1 R} & 0 \\ \frac{1}{C_2 R} & -\frac{1}{C_2 R} & \frac{1}{C_2} \\ 0 & -\frac{1}{L} & -\frac{R_0}{L} \end{bmatrix} .$$

For set-point regulation purpose, we set the input and output matrices as

$$B_1 = B_2 = I_{3\times3}, \quad C = \begin{bmatrix} 0 & 0 & 1 \end{bmatrix} .$$

By borrowing the chaotic parameters as $R = \frac{10}{7}, R_0 = 0, C_1 = 1, C_2 = \frac{19}{2}, L = \frac{19}{14}, m_a = -\frac{8}{7}, m_b = -\frac{5}{7}, E = 1$ [5], and applying [28, Th. 1] and [2], and solving the corresponding LMIs, we obtain the following analog control gain matrices:

$$K_c^1 = \begin{bmatrix} -21.6190 & -5.1116 & -0.0140 \\ -4.8884 & -19.3333 & 6.5199 \\ 0.0140 & 6.7658 & -20.3333 \end{bmatrix}, \quad E_c^1 = \begin{bmatrix} 1.4008 \\ -7.2520 \\ 20.3333 \end{bmatrix}$$

$$K_c^2 = \begin{bmatrix} -18.5333 & -5.0111 & 0.0215 \\ -4.9889 & -19.3333 & 6.5723 \\ -0.0215 & 6.7134 & -20.3333 \end{bmatrix}, \quad E_c^2 = \begin{bmatrix} 1.4098 \\ -7.2958 \\ 20.3333 \end{bmatrix} .$$

Applying Theorem 1, the intelligently redesigned digital fuzzy control gains are computed for $T = 0.02$ (s):

$$K_d^1 = \begin{bmatrix} -17.5693 & -4.8727 & -0.2709 \\ -3.3367 & -16.6930 & 4.2035 \\ -0.1486 & 6.7931 & -16.7295 \end{bmatrix}, \quad E_d^1 = \begin{bmatrix} 1.4036 \\ -5.0302 \\ 16.3731 \end{bmatrix}$$

$$K_d^2 = \begin{bmatrix} -15.3289 & -4.7566 & -0.2407 \\ -3.5552 & -16.3462 & 4.2349 \\ -0.1892 & 6.8042 & -16.3663 \end{bmatrix}, \quad E_d^2 = \begin{bmatrix} 1.4129 \\ -5.0668 \\ 16.3705 \end{bmatrix} .$$

We set $r(t) = \begin{bmatrix} 0.1 \end{bmatrix}$ and measure the state-matching performance from $t = 0$ to 0.4 (s). A naive digital implementation of analog fuzzy control is also simulated. The initial states are set to be $x_c(0) = x_d(0) = x_0 = \begin{bmatrix} 0, 1, 0 \end{bmatrix}^T$. The simulated trajectories and their time responses by both methods are shown in Fig. 10. As one can immediately witness, the state trajectory by the proposed method is almost identical to that by the original analog control and i_{L_d} is guided to $r(t) = \begin{bmatrix} 0.1 \end{bmatrix}$. However, the compared method indeed heavily deteriorates the state-matching performance. In the other set of simulation run, we take $T = 0.04$ (s). The simulation results are depicted in Fig. 11. It is observed that the state-matching performances by the proposed method somewhat degraded yet the state trajectories have a strong resemblance to the original one, whereas the others do not.

Table 2. Comparison of the performance $\mathcal{P}$ of the proposed method with that of other method according to the various values of sampling period T

	Sampling Period T (s)	
Method	0.02	0.04
Simple digital implementation	0.014287	0.031420
Proposed	0.003869	0.005697

5 Closing Remarks

In this chapter, a novel and reliable IDR methodology has been presented for the digital set-point regulator for chaotic systems represented by TS fuzzy systems. The developed technique formulated the concerned IDR problem as constrained minimization problem and the related conditions are specified by LMIs, so as to realize the global state-matching on the whole U_x of chaotic systems as well as to preserve the GES by the intelligently redesigned digital set-point regulator through the proposed IDR algorithm. The simulation results on chaotic systems convincingly demonstrated the advantage of the developed method compared to the naive approach. It implies the reliable digital implementation of chaos control.

Appendix

We now show that (21) and (22) ensure the GES of (15) in the sense of Lyapunov. To this end, it suffices to prove that there exists a function $V(x_d(kT)) = x_d(kT)^{\mathrm{T}} P x_d(kT)$ such that it is indeed positive definite and radially unbounded, and its incremental difference $\Delta V(x_d(kT))$ is negative definite.

Clearly, $V(0) = 0$ and $V(x_d(kT)) > 0$ and radially unbounded in any neighborhood of $x_d(kT) = x_{d_{eq}} = [0]_{n \times 1}$. By (21), the rate of increases of $V(x_d(kT))$ is majorized by

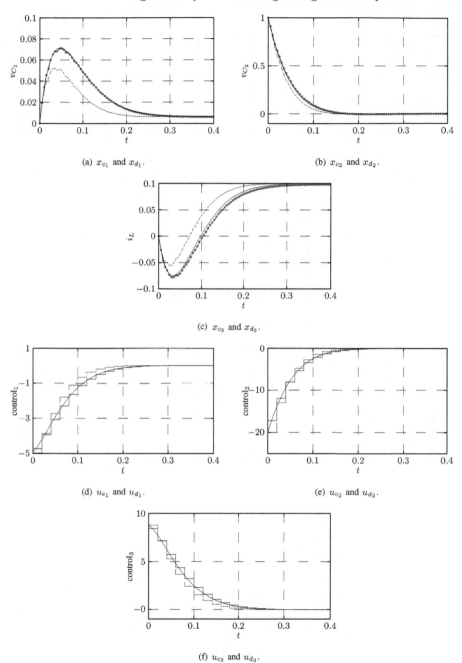

(a) x_{c_1} and x_{d_1}.

(b) x_{c_2} and x_{d_2}.

(c) x_{c_3} and x_{d_3}.

(d) u_{c_1} and u_{d_1}.

(e) u_{c_2} and u_{d_2}.

(f) u_{c_3} and u_{d_3}.

Fig. 10. Comparison of time responses of the controlled Chua's circuit for $T = 0.02$ (s): analog (solid line), proposed (solid line with circle), and simple digital implementation (dashed line)

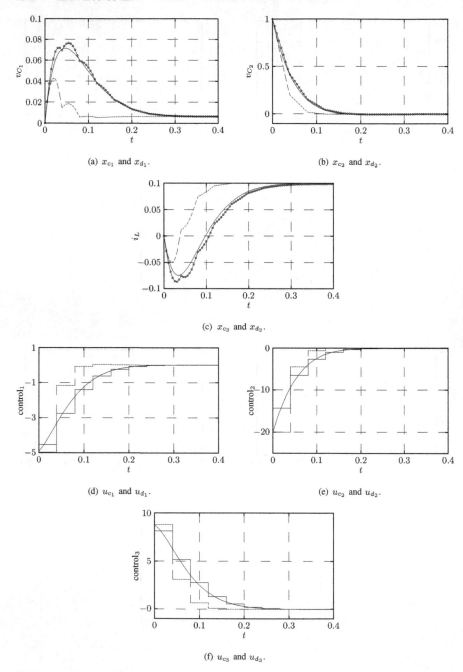

(a) x_{c_1} and x_{d_1}.

(b) x_{c_2} and x_{d_2}.

(c) x_{c_3} and x_{d_3}.

(d) u_{c_1} and u_{d_1}.

(e) u_{c_2} and u_{d_2}.

(f) u_{c_3} and u_{d_3}.

Fig. 11. Comparison of time responses of the controlled Chua's circuit for $T = 0.04$ (s): analog (solid line), proposed (solid line with circle), and simple digital implementation (dashed line)

$$\Delta V(x_d(kT)) = V(x_d(kT+T)) - V(x_d(kT))$$

$$= x_d(kT+T)^{\mathrm{T}} P x_d(kT+T) - x_d(kT)^{\mathrm{T}} P x_d(kT)$$

$$= \sum_{i=1}^{r}\sum_{j=1}^{r}\sum_{h=1}^{r}\sum_{g=1}^{r} \theta_i(z(kT))\theta_j(z(kT))\theta_h(z(kT))\theta_g(z(kT))$$

$$\times x_d(kT)^{\mathrm{T}}((G_i + H_i K_j)^{\mathrm{T}} P(G_h + H_h K_g) - P)x_d(kT)$$

$$= \frac{1}{4}\sum_{i=1}^{r}\sum_{j=1}^{r}\sum_{h=1}^{r}\sum_{g=1}^{r} \theta_i(z(kT))\theta_j(z(kT))\theta_h(z(kT))\theta_g(z(kT))$$

$$\times x_d(kT)^{\mathrm{T}}((G_i + H_i K_j + G_j + H_j K_i)^{\mathrm{T}} P$$

$$\times (G_h + H_h K_g + G_g + H_g K_h) - 4P)x_d(kT)$$

$$\leqslant \frac{1}{4}\sum_{i=1}^{r}\sum_{j=1}^{r} \theta_i(x_d(kT))\theta_j(x_d(kT))x_d(kT)^{\mathrm{T}}((G_i + H_i K_d^j + G_j$$

$$+ H_j K_d^i)^{\mathrm{T}} P(G_i + H_i K_d^j + G_j + H_j K_d^i) - 4P)x_d(kT)$$

$$= 2\sum_{i\leqslant j}^{r} \theta_i(x_d(kT))\theta_j(x_d(kT))x_d(kT)^{\mathrm{T}}$$

$$\times \left(\left(\frac{G_i + H_i K_d^j + G_j + H_j K_d^i}{2}\right)^{\mathrm{T}}\right.$$

$$\times P\left.\left(\frac{G_i + H_i K_d^j + G_j + H_j K_d^i}{2}\right) - P\right)x_d(kT)$$

$$\leqslant -2\sum_{i\leqslant j}^{r} \theta_i(x_d(kT))\theta_j(x_d(kT))x_d(kT)^{\mathrm{T}} P X_{ij} P x_d(kT)$$

$$= - \begin{bmatrix} \theta_1(z(kT))x_d(kT) \\ \theta_2(z(kT))x_d(kT) \\ \vdots \\ \theta_r(z(kT))x_d(kT) \end{bmatrix}^{\mathrm{T}} \begin{bmatrix} P & 0 & \cdots & 0 \\ 0 & P & \ddots & \vdots \\ \vdots & \ddots & \ddots & 0 \\ 0 & \cdots & 0 & P \end{bmatrix}$$

$$\times \begin{bmatrix} X_{11} & X_{12} & \cdots & X_{1r} \\ X_{12} & X_{22} & \cdots & X_{2r} \\ \vdots & \vdots & \ddots & \vdots \\ X_{1r} & X_{2r} & \cdots & X_{rr} \end{bmatrix} \begin{bmatrix} P & 0 & \cdots & 0 \\ 0 & P & \ddots & \vdots \\ \vdots & \ddots & \ddots & 0 \\ 0 & \cdots & 0 & P \end{bmatrix}$$

$$\times \begin{bmatrix} \theta_1(z(kT))x_d(kT) \\ \theta_2(z(kT))x_d(kT) \\ \vdots \\ \theta_r(z(kT))x_d(kT) \end{bmatrix},$$

where at the second majorization the congruence transformation with $\text{diag}\{P, I\}$ and Schur complement are sequentially applied to (21), and we let $P = Q^{-1}$ and $F_d^i = K_d^i P^{-1}$. Hence if (22) holds, $\Delta V(x_d(kT))$ is negative definite. Hence we conclude that (15) is GES in the sense of Lyapunov, whenever (21) is satisfied.

References

1. W. Chang, J.B. Park, Y.H. Joo: IEEE Trans. Fuzzy Syst. **11**(1), 35–44 (2003)
2. Y. H. Joo, L.S. Shieh, G. Chen: IEEE Trans. Fuzzy Syst. **7**(4), 394–408 (1999)
3. S.M. Guo, L.S. Shieh, G. Chen, C.F. Lin: IEEE Trans. Circuits Syst. I **47**(11), 1557–1570 (2000)
4. J. Xu, G. Chen, L.S. Shieh: IEEE Trans. Aero. Electr. **32**(8), 1488–1499 (1996)
5. S.M. Guo, L.S. Shieh, C.F. Lin, J. Chandra: Int. J. Birfurcation Chaos, **11**(4), 1079–1113 (2001)
6. W. Chang, J.B. Park, Y.H. Joo, G. Chen: IEEE Trans. Circuits Syst. I **49**(4), 509–517 (2002)
7. W. Chang, J.B. Park, H.J. Lee, Y.H. Joo: IEEE Proc., Control Theory Appl. **149**(4), 297–302 (2002)
8. H.J. Lee, J.B. Park, Y.H. Joo: IEEE Trans. Circuits Syst. I **50**(12), 1595–1601 (2003)
9. J.W. Sunkel, L.S. Shieh, J.L. Zhang: J. Guid. Control Dyn. **14**(4), 712–723 (1991)
10. L.S. Shieh, W.M. Wang, M.K. Appu Panicker: ISA Trans. **37**, 201–213 (1998)
11. L.S. Shieh, W.M. Wang, J. Bain, J.W. Sunkel: J. Guid. Control Dyn. **23**, 629–639 (2000)
12. C.A. Rabbath, N. Hori: IEEE Proc. Control Theory Appl. **150**(4), 335–346 (2003)
13. L.S. Shieh, W.M. Wang, J.B. Zheng: Int. J. Control **66**(1), 43–64 (1997)
14. L.S. Shieh, W.M. Wang, J.S.H. Tsai: Int. J. Control **70**(5), 665–683 (1998)
15. L.S. Shieh, Y.J. Wang, J.W. Sunkel: IEEE Proc. Control Theory Appl. **140**(2), 99–110 (1993)
16. B.C. Kuo: *Digital Control Systems* (Holt, Rinehart and Winston, New York, 1980)
17. G. Feng: IEEE Trans. Circuits Syst. I **48**(6), 760–779 (2001)
18. Z. Li, J.B. Park, Y.H. Joo: IEEE Trans. Circuits Syst. I **48**(10), 1237–1243 (2001)
19. H.J. Lee, H. Kim, Y.H. Joo, W. Chang, J.B. Park: IEEE Trans. Fuzzy Syst. **12**(2), 274–284 (2004)
20. K. Tanaka, T. Ikeda, H.O. Wang: IEEE Trans. Fuzzy Syst. **6**(2), 250–265 (1998)
21. E. Kim, H. Lee: IEEE Trans. Fuzzy Syst. **8**(5), 523–534 (2000)
22. K. Kiriakidis: IEEE Trans. Fuzzy Syst. **9**(2), 269–277 (2001)
23. C.S. Tseng, B.S. Chen, H.J. Uang: IEEE Trans. Fuzzy Syst. **9**(3), 381–392 (2001)
24. K. Tanaka, H.O. Wang: *Fuzzy Control Systems Design and Analysis: A Linear Matrix Inequality Approach* (Wiley, New York, 2001)
25. Z. Li, J.B. Park, G. Chen, Y.H. Joo, Y.H. Choi: Int. J. Bifurcation Chaos **12**(10), 2283–2291

26. Z. Li, W. Halang, G. Chen, L.F. Tian: Dyn. Contin. Discret. Impuls. Syst. B **10**, 813–832 (2003)
27. Z. Li, J.B. Park, Y.H. Joo, Y.H. Choi, G. Chen: IEEE Trans. Circuits Syst. I **49**(2), 249–253
28. H.J. Lee, J.B. Park, G. Chen: IEEE Trans. Fuzzy Syst. **9**(2), 369–379 (2001)
29. G. Chen, T. Ueta: Int. J. Bifurcation. Chaos **9**(7), 1465–1466 (1999)
30. O.E. Rössler: Phys. Lett. A **35**, 397–398 (1976)
31. P. Apkarian: Automatica **33**(4), 655–661 (1997)
32. K.S. Tang, K.F. Man, G.Q. Zhong, G. Chen: IEEE Trans. Circuits Syst. I **48**(5), 636–641 (2001)
33. L.O. Chua, M. Komuro, T. Matsumoto: IEEE Trans. Circuit Syst. **CS-33**, 1072–1118 (1986)

Anticontrol of Chaos for Takagi–Sugeno Fuzzy Systems

Zhong Li, Guanrong Chen, and Wolfgang A. Halang

Abstract. The current study on anticontrol of chaos for both discrete-time and continuous-time Takagi–Sugeno (TS) fuzzy systems is reviewed. To chaotifying discrete-time TS fuzzy systems, the parallel distributed compensation (PDC) method is employed to determine the structure of a fuzzy controller so as to make all the Lyapunov exponents of the controlled TS fuzzy system strictly positive. But for continuous-time ones, the chaotification approach is based on the fuzzy feedback linearization and a suitable approximate relationship between a time-delay differential equation and a discrete map. The time-delay feedback controller, chosen among several candidates, is a simple sinusoidal function of the delayed states of the system, which can have an arbitrarily small amplitude. These anticontrol approaches are all proved to be mathematically rigorous in the sense of Li and Yorke. Some examples are given to illustrate the effectiveness of the proposed anticontrol methodologies.

1 Introduction

Nowadays, it is well known that most conventional control methods and many special techniques can be used for chaos control [1], for which, no matter the purpose is to reduce harmful or undesirable chaos or to introduce useful or beneficial chaos, numerous control methodologies have been proposed, developed, tested, and applied. Similar to conventional systems control, the concept of "controlling chaos" is first to mean ordering or suppressing chaos in the sense of stabilizing chaotic system responses. In this pursuit, numerical and experimental simulations have convincingly demonstrated that chaotic systems respond well to these control strategies. These methods of ordering chaos include the now-familiar OGY method [2, 3], feedback controls [4, 5], and fuzzy control [6–9], to list just a few.

However, controlling chaos also encompasses many nontraditional tasks, particularly those of enhancing or generating chaos when it is beneficial. Thus, the process of chaos control is now understood as a transition between chaos and order, depending on the application of interest. The task of purposely creating chaos, sometimes called chaotification or anticontrol of chaos, has attracted increasing attention in recent years due to its great potential in

Z. Li et al.: *Anticontrol of Chaos for Takagi–Sugeno Fuzzy Systems*, StudFuzz **187**, 185–227 (2006)
www.springerlink.com

nontraditional applications such as those found within the context of physical, chemical, mechanical, electrical, optical, and particularly biological and medical systems [10–12]. Recently, there have been some successful reports on anticontrolling chaos [11–13]. Although these reports are essentially experimental or semianalytical, in the sense that no explicit and quantitative computational formula are provided with rigorous mathematical justification, especially for the continuous-time case, they are nevertheless interesting and promising.

One simple yet mathematically rigorous control method from the engineering feedback control approach has been developed [14–16], where a linear state feedback controller with an uniformly bounded control-gain sequence can be designed to make all Lyapunov exponents of the controlled system strictly positive and arbitrarily assigned. Moreover, such a controller can be designed for an arbitrarily given n-dimensional dynamical system that could be originally nonchaotic or even asymptotically stable. The goal of chaotification is finally achieved with a simple modulus operation or a sawtooth (or even a sine) function. The design criterion is to use the definition of chaos given by Devaney [18] or Li–Yorke [19], where for the n-dimensional case the Marotto theorem [20] was used for a proof. For the continuous-time case, a general approach to make an arbitrarily given autonomous system chaotic has also been proposed lately [21–24]. Here, the main tool to use is time-delay feedback perturbation on a system parameter or as an exogenous input [22].

The possible interactions between fuzzy logic and chaos theory have been explored since the late 1990s. The explorations have been carried mainly on linguistic descriptions [25, 26], fuzzy modeling of chaotic systems using the Takagi–Sugeno (TS) model [6–8, 27–29], and fuzzy control of chaos via an LMI-based fuzzy control system design [6, 7]. In these investigations, fuzzy modeling of chaotic systems in a typical approach was carried out in a linguistic manner based on the Mamdani fuzzy model. The stretching and folding features of the flow are responsible for the sensitivity to initial conditions, characterizing the chaotic behaviors. Besides, the TS fuzzy model was used to precisely represent chaotic systems, based on which a fuzzy controller is designed, and then the LMI-based design problems were defined and employed to find feedback gains of the fuzzy controllers that can satisfy some specifications such as stability, decay rate, and constraints on the control input and output of the overall fuzzy control systems.

In this chapter, the study of anticontrol of chaos for both continuous-time and discrete-time TS fuzzy systems is reviewed. Briefly, these anticontrol approaches are extensions of formerly developed methods for general systems to TS fuzzy systems, and are proved mathematically rigorous in the sense of Li and Yorke.

2 Chaotifying Discrete-Time TS Fuzzy Systems

2.1 The TS Fuzzy System

The TS fuzzy model [7, 8, 29] is described by a set of fuzzy implications, which characterize local relations of the underlying system in the state space. The main feature of a TS model is to express the local dynamics of each fuzzy implication (rule) by a linear state-space system model. The overall fuzzy system is then built up by fuzzy "blending" of these local linear system models.

A general discrete-time TS fuzzy system is described as follows:

Discrete-time TS fuzzy model:

Plant Rule i: IF $x_1(k)$ is Γ_1^i and $\ldots$ and $x_n(k)$ is Γ_n^i ,

THEN $x(k+1) = A_i x(k) + B_i u(k)$, $\qquad\qquad$ (1)

where

$$x^T(k) = [x_1(k), x_2(k), \ldots, x_n(k)] ,$$

$$u^T(k) = [u_1(k), u_2(k), \ldots, u_m(k)] ,$$

$i = 1, 2, \ldots, q$, in which q is the number of IF–THEN rules, Γ_j^i are fuzzy sets, and the equation $x(k+1) = A_i x(k) + B_i u(k)$ is the output of the ith IF–THEN rule.

Assume that A_i and B_i, $i = 1, 2, \ldots, q$, are uniformly bounded; that is, there are constants N and Q such that

$$\sup_{1 \leq i \leq q} \|A_i\| \leq N < \infty \quad \text{and} \quad \sup_{1 \leq i \leq q} \|B_i\| \leq Q < \infty$$

where $\| \cdot \|$ denotes the spectral norm of a finite-dimensional matrix, that is, the largest singular value of the matrix.

Now, given one pair of $(x(k), u(k))$, the final output of the fuzzy system at step k is inferred as follows:

$$x(k+1) = \frac{1}{\sum_{i=1}^{q} \omega_i(k)} \sum_{i=1}^{q} \omega_i(k)\{A_i x(k) + B_i u(k)\} , \qquad (2)$$

where

$$\omega_i(k) = \prod_{j=1}^{n} \Gamma_j^i(x_j(k)) .$$

in which $\Gamma_j^i(x_j(k))$ is the degree of membership of $x_j(k)$ in Γ_j^i, with

$$\begin{cases} \sum_{i=1}^{q} \omega_i(k) > 0, \\ \omega_i(k) \geq 0, \end{cases} \quad i = 1, 2, \ldots, q .$$

By introducing $h_i(k) = \frac{\omega_i(k)}{\sum_{i=1}^{q} \omega_i(k)}$ instead of $\omega_i(k)$, (2) is rewritten as

$$x(k+1) = \sum_{i=1}^{q} h_i(k)\{A_i x(k) + B_i u(k)\} \,. \tag{3}$$

Note that

$$\begin{cases} \sum_{i=1}^{q} h_i(k) = 1, \\ h_i(k) \geq 0, \end{cases} \quad i = 1, 2, \ldots, q \,, \tag{4}$$

in which $\{h_i(k)\}_{i=1}^{q}$ can be regarded as the normalized weight of the IF–THEN rules.

2.2 Chaos in the Sense of Li–Yorke: The Marotto Theorem

Consider a general n-dimensional discrete-time autonomous system of the form

$$x(k+1) = g(x(k)) \,, \tag{5}$$

where g is a C^1 nonlinear map. Let g^t denote the t times of compositions of g with itself. A point x^* is said to be a p-periodic point of g if $g^p(x^*) = x^*$ but $g^t(x^*) \neq x^*$ for $1 \leq t < p$. If $p = 1$, that is, $g(x^*) = x^*$, then x^* is called a *fixed point*. Let $g'(x)$ and $\det(g'(x))$ be the Jacobian of g at point x and its determinant, respectively, and let $B(x;r)$ denote a closed ball in $\mathbb{R}^n$ of radius r centered at the point x.

Definition 1. [20] *A fixed point x^* of (5) is said to be a snap-back repeller if*

1. *There exists a real number $r > 0$ such that g is continuously differentiable with all eigenvalues of $g'(x)$ exceeding the unity in absolute value for all $x \in B(x^*;r)$;*
2. *There exists a point $x^0 \in B(x^*;r)$, with $x^0 \neq x^*$, such that for some positive integer $m \geq 2$, $g^m(x^0) = x^*$ and $\det((g^m)'(x^0)) \neq 0$.*

Based on the above definition of the snap-back repeller, Marotto derived the following criterion [20].

Theorem 1. (Marotto Theorem) *If g has a snap-back repeller then system (5) is chaotic in the following generalized sense of Li–Yorke:*

(i)   *There is a positive integer N such that for each integer $p \geq N$, g has a point of period p;*
(ii)   *There is a scrambled set of g, i.e., an uncountable set S containing no periodic points of g, such that*

(a) $g(S) \subset S$,

(b) for every $x, y \in S$ with $x \neq y$,

$$\lim_{k \to \infty} \sup \|g^k(x) - g^k(y)\| > 0 \,,$$

(c) for every $x \in S$ and any periodic point y_{per} of g,

$$\lim_{k \to \infty} \sup \|g^k(x) - g^k(y_{\mathrm{per}})\| > 0 \,;$$

(iii) There is an uncountable subset S_0 of S such that for every $x, y \in S_0$:

$$\lim_{k \to \infty} \inf \|g^k(x) - g^k(y)\| = 0.$$

Straightforwardly, the definition of a chaotic TS fuzzy model can be given as follows.

Definition 2 (Chaotic TS Fuzzy Model [30]). *The TS fuzzy model (3) is said to be chaotic in the sense of Li and Yorke if it has a snap-back repeller.*

2.3 Anticontrol of Chaos via Parallel-Distributed Compensation (PDC)

2.3.1 Parallel-Distributed Compensation (PDC)

The parallel-distributed compensation (PDC) technique is employed to determine the structure of a fuzzy controller for a given TS fuzzy model. Each control rule in the PDC is constructed from the corresponding rule of the TS fuzzy model. The designed fuzzy controller shares the same fuzzy sets with the fuzzy model in the premise parts. The PDC provides the following fuzzy control rule structure from the fuzzy model (1):

> Control Rule i: IF $x_1(k)$ is Γ_1^i and ... and $x_n(k)$ is Γ_n^i ,
>
> THEN $u_i(k) = -F_i x(k)$, $i = 1, 2, \ldots, q$. $\qquad$ (6)

where $\{F_i\}_{i=1}^q$ are constant-gain matrices to be designed.

The fuzzy control rules have linear state feedback laws in the consequent parts. The overall fuzzy controller is represented by

$$u(k) = -\frac{1}{\sum_{i=1}^{q} \omega_i(k)} \sum_{i=1}^{q} \omega_i(k) F_i x(k) = -\sum_{i=1}^{q} h_i(k) F_i x(k) \,. \qquad (7)$$

To be practical, the control-gain matrices $\{F_i\}_{i=1}^q$ are required to be uniformly bounded:

$$\sup_{1 \le i \le q} \|F_i\| \le M < \infty \,, \qquad (8)$$

for some positive constant M.

Substituting (7) into (3) yields

$$x(k+1) = \sum_{i=1}^{q}\sum_{j=1}^{q} h_i(k)h_j(k)\{A_i - B_iF_j\}x(k) . \tag{9}$$

System (9) can also be written as

$$x(k+1) = \sum_{i=1}^{q} h_i(k)h_i(k)\{A_i - B_iF_i\}x(k)$$

$$+ 2\sum_{i<j}^{q} h_i(k)h_j(k)\frac{\{A_i - B_iF_j\} + \{A_j - B_jF_i\}}{2}x(k)$$

$$= \sum_{i=1}^{q} h_i(k)h_i(k)G_{ii}x(k)$$

$$+ 2\sum_{i>j}^{q} h_i(k)h_j(k)\left\{\frac{G_{ij} + G_{ji}}{2}\right\}x(k) , \tag{10}$$

where $G_{ij} = A_i - B_iF_j$ for all $i, j = 1, 2, \ldots, q$.

2.3.2 Anticontroller Design

For simplicity, assume that $B_i = B$, $i = 1, 2, \ldots, q$, so that (10) can be rewritten as

$$x(k+1) = \sum_{i=1}^{q} h_i(k)\{A_i - BF_i\}x(k) = \sum_{i=1}^{q} h_i(k)G_{ii}x(k) . \tag{11}$$

Then, one has the following result [8].

Lemma 1. *The TS fuzzy system (3) is exactly linearizable via the PDC fuzzy controller (6) if there exist feedback gains such that*

$$\{(A_1 - BF_1) - (A_i - BF_i)\}^{\mathrm{T}} \times \{(A_1 - BF_1) - (A_i - BF_i)\} = 0 , \tag{12}$$

for $i = 2, 3, \ldots, q$. The overall control system can be linearized as

$$x(k+1) = Gx(k) , \tag{13}$$

where $G = G_{ii}$, $i = 2, 3, \ldots, q$.

The jth Lyapunov exponent of the orbit of the controlled system (13), starting from the given x_0, is defined by

$$\lambda_j = \lim_{k \to \infty} \frac{1}{k} \ln |\mu_j(G^k)|, \quad j = 1, 2, \ldots, n, \tag{14}$$

where $\mu_j(G^k)$ is the jth singular value of matrix G^k.

In the controlled system (13), one is able to design the constant-gain matrices $\{F_i\}_{i=1}^q$, given in (6), such that the Lyapunov exponents of the controlled system orbit $\{x_k\}_{k=0}^\infty$ can be arbitrarily assigned:

$$\lambda_j(x_0) = \sigma_j, \quad j = 1, 2, \ldots, n, \tag{15}$$

where $\{\sigma_j\}_{j=1}^n$ are arbitrarily chosen by the user, which may be positive, zero, or negative (but all finite).

A convenient choice for the matrix G in (13) is, simply,

$$G = \text{diag}\{e^{\sigma_1}, e^{\sigma_2}, \ldots, e^{\sigma_n}\}.$$

It is clear that the eigenvalues of G are all larger than 1 if $\sigma_j > 0$, $j = 1, 2, \ldots, n$. The desired matrices F_i, $i = 1, 2, \ldots, q$, can then be obtained.

2.3.3 Verification of the Anticontrol Design with Mod-Operation

Theorem 2. [30] *The resulting overall controlled system (13), along with the mod-operation,*

$$x(k + 1) = Gx(k) \ (\text{mod-1}), \tag{16}$$

where $G = \text{diag}\{e^{\sigma_1}, e^{\sigma_2}, \ldots, e^{\sigma_n}\}$ and $\sigma_i > 0$, $i = 1, 2, \ldots, n$, is chaotic in the sense of Li and Yorke.

Proof. The controlled system (13) is

$$x(k + 1) = Gx(k) \ (\text{mod-1}) \equiv g(x(k)). \tag{17}$$

It is now to prove that the fixed point $x^* = 0$ of (13) is a snap-back repeller. To do so, define two n-dimensional vectors, $b = [1, 1, \ldots, 1]^T$ and

$$x^0 = G^{-2}b = \begin{bmatrix} e^{-2\sigma_1} & 0 & \cdots & 0 \\ 0 & 0 & \cdots & 0 \\ \vdots & \vdots & \ddots & \vdots \\ 0 & 0 & \cdots & e^{-2\sigma_n} \end{bmatrix} b \neq 0. \tag{18}$$

Since $\sigma_i > 0$, $i = 1, 2, \ldots, n$, $\|x^0\|_\infty < 1$. Clearly, after two-step iterations on (16) with $x^1 = g(x^0) = G^{-1}b$, one has

$$g^2(x^0) = g(x^1) = 0 = x^* .\tag{19}$$

Let r be a given constant satisfying $\|x^0\|_\infty \leq r \leq 1$. For any $x \in B_r \equiv \{x \in \mathbb{R}^n|\ \|x\|_\infty \leq r\}$, it is also clear that all the eigenvalues of G exceed unity. Therefore, the fixed point $x^* = 0$ of (13) is a snap-back repeller. By the Marotto theorem, the controlled system (11)–(13) is chaotic in the sense of Li and Yorke. Thus, it completes the proof.

2.4 A Simulation Example

Consider a nonchaotic discrete-time TS fuzzy model given as follows:

$$\text{Rule 1: IF } x(k) \text{ is } \Gamma_1, \text{ THEN } \begin{bmatrix} x(k+1) \\ y(k+1) \end{bmatrix} = A_1 \begin{bmatrix} x(k) \\ y(k) \end{bmatrix} + Bu(k),$$

$$\text{Rule 2: IF } x(k) \text{ is } \Gamma_2, \text{ THEN } \begin{bmatrix} x(k+1) \\ y(k+1) \end{bmatrix} = A_2 \begin{bmatrix} x(k) \\ y(k) \end{bmatrix} + Bu(k),$$

where

$$A_1 = \begin{bmatrix} d & 0.3 \\ 1 & 0 \end{bmatrix}, \qquad A_2 = \begin{bmatrix} -d & 0.3 \\ 1 & 0 \end{bmatrix},$$

$x(k) \in [-d, d]$ and $d > 0$, with membership functions

$$\Gamma_1 = \frac{1}{2}\left(1 - \frac{x(k)}{d}\right) \quad \text{and} \quad \Gamma_2 = \frac{1}{2}\left(1 + \frac{x(k)}{d}\right).$$

Without control, i.e., $\mathbf{u} \equiv 0$, the system is stable as shown in Fig. 1.

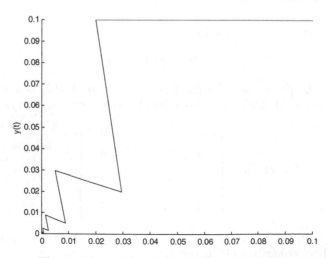

Fig. 1. The system orbit without control input

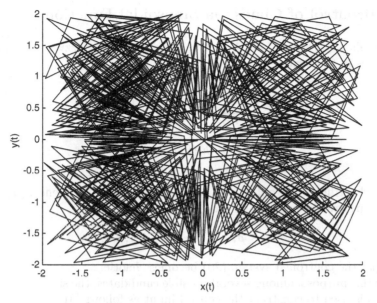

Fig. 2. The chaotic orbit of the anticontrolled system

Here, we choose two desired Lyapunov exponents, $\sigma_1 = \ln(1.9) = 0.6418539$ and $\sigma_2 = \ln(2.0) = 0.6931472$.

For simplicity, we assume that $B = \begin{bmatrix} 1 & 0 \\ 0 & 1 \end{bmatrix}$ and $d = 2$. We completed the design of the feedback controller by following the procedure described above. We obtained the controlled system output as shown in Figs. 2 and 3. The output trajectory is displayed in the phase plane after some mod-2 operations (they are obviously equivalent to mod-1 operations for anticontrol), which has the above-indicated Lyapunov exponents $\lambda_1 = \sigma_1$ and $\lambda_2 = \sigma_2$.

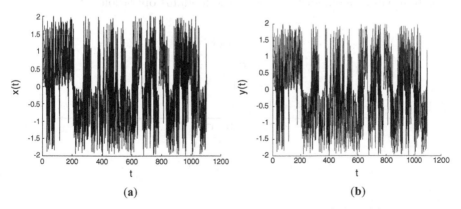

(a)   (b)

Fig. 3. The phase portraits of (a) $t - x(k)$; (b) $t - y(k)$

3 Anticontrol of Chaos via Sinusoidal Function

For simiplicity, assume that $B_i = I$ in the discrete TS fuzzy system (1), so that (3) is in the form of

$$x(k+1) = \sum_{i=1}^{q} h_i(k)\{A_i x(k) + u(k)\}$$

$$= \left[\sum_{i=1}^{q} h_i(k)A_i\right] x(k) + u(k) . \tag{20}$$

The chaotification problem is to design a control input sequence, $\{u(k)\}_{k=0}^{\infty}$, with an arbitrarily small magnitude, $\sigma > 0$, namely,

$$\|u(k)\|_{\infty} \le \sigma, \qquad \text{for all } k = 1, 2, \ldots, q , \tag{21}$$

such that the controlled system (20) becomes chaotic.

For this purpose, among several possible candidates, the simple sinusoidal function is used to construct the control input as follows [31]:

$$u(k) = \Phi(\beta x(k))$$

$$\equiv [\varphi(\beta x_1(k)), \varphi(\beta x_2(k)), \ldots, \varphi(\beta x_n(k))]^{\mathrm{T}} , \tag{22}$$

where $x(k) = [x_1(k), x_2(k), \ldots, x_n(k)]^{\mathrm{T}}$, β is a constant, and $\varphi : \mathbb{R} \to \mathbb{R}$ is a continuous sinusoidal function defined by (see Fig. 4)

$$\varphi(x) = \sigma \, \sin\left(\frac{\pi}{\sigma}x\right) . \tag{23}$$

Obviously, $|\varphi(x)| \le \sigma$ for all $x \in \mathbb{R}$, so that $\|u(k)\| \le \sigma$, where σ can be arbitrarily small, as required by condition (21). Here, the sinusoidal function actually serves as a smooth version of the modulus operation.

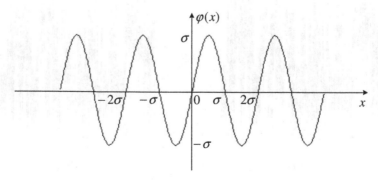

Fig. 4. The sinusoidal function used for chaotification

Lemma 2 (Boundedness). *The state vector of the controlled system* (20), *under the control of the controller* (22) *and* (23), *are uniformly bounded by a constant,* $\sigma(1-\alpha)^{-1}$.

Proof. The solution of the controlled system (20) can be written as

$$x(k) = \left[\sum_{i=1}^{q} h_i(k)A_i\right]^k x(0) + \sum_{j=1}^{k-1}\left[\sum_{i=1}^{q} h_i(k)A_i\right]^{k-1-j} u(j) .$$

Since $\max_{1\le i\le q}\{\|A_i\|\} = \alpha < 1$ and $\|u(k)\| \le \infty$, one has a decreasing sequence $\{x(k)\}_{k=0}^{\infty}$ with

$$\lim_{k\to\infty}\|x(k)\| \le \lim_{k\to\infty}\left\{\left[\left\|\sum_{i=1}^{q} h_i(k)A_i\right\|\right]^k \|x(0)\|\right.$$

$$+ \sum_{j=1}^{k-1}\left.\left[\left\|\sum_{i=1}^{q} h_i(k)A_i\right\|\right]^{k-1-j} \|u(j)\|\right\}$$

$$\le \lim_{k\to\infty}\alpha^k\|x(0)\| + \lim_{k\to\infty}\sum_{j=1}^{k-1}\alpha^{k-1-j}\|u(j)\|$$

$$\le \sigma \lim_{k\to\infty}\sum_{j=1}^{k-1}\alpha^{k-1-j}$$

$$= \frac{1}{1-\alpha}\sigma .$$

This means that the sinusoidal function folds an expanding trajectory back toward the origin when the trajectory becomes too large in magnitude, thus bounding the controlled system trajectory globally. On the other hand, it will be shown in the next section that if β is chosen to be large enough, the controller designed above can lead all eigenvalues of the controlled system Jacobian, at every time step, to exceed the unity in absolute value. Consequently, it can be proven that all the Lyapunov exponents of the controlled system are strictly positive, so that the system trajectory is locally expanding in all directions. The combination of these two effects, stretching and folding, will then yield complex chaotic dynamics within the bounded region of the controlled system trajectories.

In this case, the fuzzy control rule structure (6) is in the following form:

Control Rule i: IF $x_1(k)$ is Γ_1^i and ... and $x_n(k)$ is Γ_n^i ,

THEN $u(k) = \Phi(\beta x(k))$, $i = 1, 2, \ldots, q$.   (24)

The fuzzy control rules have linear state-feedback laws in the consequent parts. The overall fuzzy controller is represented by

$$u(k) = \frac{1}{\sum_{i=1}^{q} \omega_i(k)} \sum_{i=1}^{q} \omega_i(k)\Phi(\beta x(k)) = \Phi(\beta x(k)) . \tag{25}$$

In terms of the Marotto theorem, the theoretical result of the above controller design is summarized as follows [31].

Theorem 3. *Suppose that $h_i(k)$, $i = 1, 2, \ldots, q$, are continuously differentiable in the neighborhood of the fixed point, $x^* = 0$, of the controlled system (20). Then, there exists a positive constant $\bar{\beta}$ such that if $\beta > \bar{\beta}$ then the controlled TS fuzzy system (20) and (22) is chaotic in the sense of Li and Yorke.*

Proof. The controlled system (20) and (22) is

$$x(k+1) = \sum_{i=1}^{q} h_i(k)\{A_i x(k) + u(k)\}$$

$$= \left[\sum_{i=1}^{q} h_i(k)A_i\right] x(k) + \Phi(\beta x(k)) \equiv g(x(k)) . \tag{26}$$

Obviously, $x^* = 0$ is a fixed point of (26), which is now proven to be a snap-back repeller. Differentiating (26) at this fixed point yields

$$\|g'(0)\| = \left\| \left[\sum_{i=1}^{q} h_i(k)A_i\right]\bigg|_0 + \pi\beta I \right\| . \tag{27}$$

If $\beta > \frac{(1+\alpha)}{\pi}$, then $\|g'(0)\| > 1$. By the continuity of $g'(x)$ in the neighborhood of the fixed point, there exists a small positive constant r such that when $x \in B(x^*; r)$, $\|g'(x)\| > 1$.

Therefore, the Gerschgorin theorem [32] implies that all eigenvalues of $g'(x)$ exceed the unity in absolute value for all $x \in B(x_0; r)$.

Next, it is shown that there exists a point $x_0 \in B(x^*; r)$ such that $g^2(x_0) = 0 = x^*$ and $(g^2(x_0))' \neq 0$.

Indeed, it is easy to see that if $\beta > \frac{3\alpha}{2}$, then there exist $x_1 = [\frac{\sigma}{2\beta}, \ldots, \frac{\sigma}{2\beta}]^{\mathrm{T}}$ and $x_2 = [\frac{3\sigma}{2\beta}, \ldots, \frac{3\sigma}{2\beta}]^{\mathrm{T}}$, such that

$$g(x_1) > 0 \quad \text{and} \quad g(x_2) < 0 .$$

Therefore, by the Mean Value Theorem in Calculus, there exists a point $x_1 < x_1^* < x_2$ such that $g(x_1^*) = 0$.

Let $\tilde{x} = [r, r, \ldots, r]^{\mathrm{T}}$. It is clear that there exists a constant $\beta_1 > 0$ such that if $\beta > \beta_1$ then

$$g(0) = 0 < x_1^* \quad \text{and} \quad g(\tilde{x}) > x_1^* .$$

Using the Mean Value Theorem again, one concludes that there exists a point $x_0 \in B(x^*; r)$ such that $g(x_0) = x_1^*$. Therefore,

$$g^2(x_0) = g(x_1^*) = 0 .$$

On the other hand, there exists a constant $\beta_2 > 0$ such that $g'(x_1^*) < 0$, for $\cos(\frac{\pi}{\sigma \beta x_1^*}) < 0$. Therefore,

$$(g^2)'(x_0) = g'(x_1^*) g'(x_0) \neq 0 .$$

To conclude, if $\beta > \bar{\beta} \equiv \max\{(1+\alpha)/\pi, (3\sigma)/2, \beta_1, \beta_2\}$, then $x_0 = 0$ is a snap-back repeller of the map g defined in (26), so the controlled system (20) and (22) is chaotic in the sense of Li and Yorke.

3.1 A Simulation Example

To visualize the theoretical analysis and design, the same example as given above is used for illustration.

The controlled TS fuzzy system is described as follows:

$$x(k+1) = \sum_{i=1}^{q} h_i(k)\{A_i x(k) + u(k)\} = \sum_{i=1}^{q} h_i(k) A_i x(k) + u(k)$$

$$= \sum_{i=1}^{q} h_i(k) A_i x(k) + \sigma \, \sin\left(\frac{\pi}{\sigma}\beta x(k)\right) .$$

In the simulation, the magnitude of the control input is arbitrarily chosen to be $\sigma = 0.1$. Thus, $\|u(k)\| < \infty$, and β can also be regarded as a control parameter. Without control, the TS fuzzy model is stable, as shown in Fig. 1. When β takes values of 0.25, 0.4, 0.45, 0.5, and 1.3, the phase portraits, time evolutions, and bifurcation diagrams are shown in Figs. 5–11, respectively. These numerical results verify the theoretical analysis and the design of the chaos generator.

4 Anticontrol of Chaos for Continuous-Time TS Fuzzy Systems via Discretization

4.1 Continuous-Time TS Fuzzy Systems

Similar to model (1), a continuous-time TS fuzzy system is described as Continuous-time TS fuzzy model:

Plant Rule i: IF $x_1(t)$ is Γ_1^i and ... and $x_n(t)$ is Γ_n^i,

THEN $\dot{x}(t) = A_i x(t) + B_i u(t) ,$ (28)

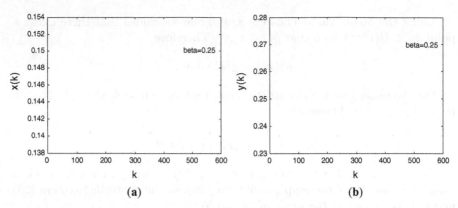

Fig. 5. Periodic orbits at $\beta = 0.25$

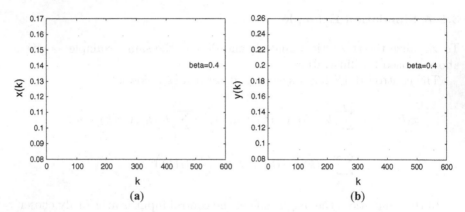

Fig. 6. Period-doubling bifurcation at $\beta = 0.4$

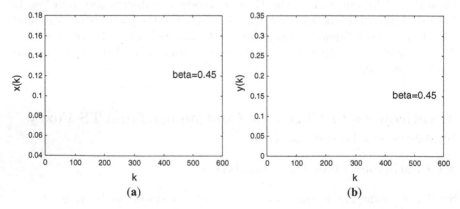

Fig. 7. Special period-doubling bifurcation at $\beta = 0.45$

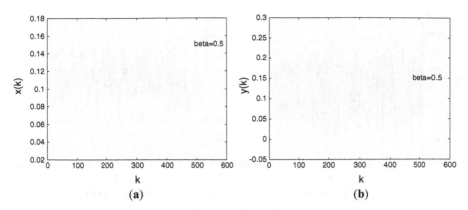

Fig. 8. Period-8 bifurcation at $\beta = 0.5$

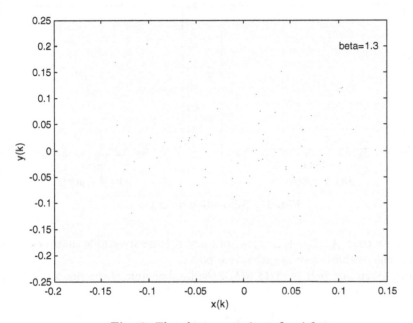

Fig. 9. The phase portrait at $\beta = 1.3$

where

$$x(t) = [x_1(t), x_2(t), \dots, x_n(t)]^{\mathrm{T}} ,$$

$$u(t) = [u_1(t), u_2(t), \dots, u_m(t)]^{\mathrm{T}} ,$$

$i = 1, 2, \dots, q$, in which q is the number of IF–THEN rules, Γ_j^i are fuzzy sets, and the equation $\dot{x}(t) = A_i x(t) + B_i u(t)$ is the output of the ith IF–THEN rule.

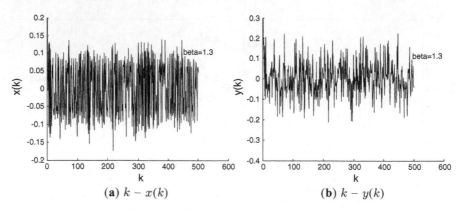

(a) $k - x(k)$ (b) $k - y(k)$

Fig. 10. The time evolution at $\beta = 1.3$

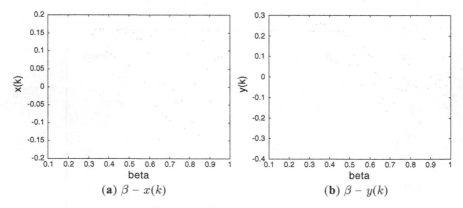

(a) $\beta - x(k)$ (b) $\beta - y(k)$

Fig. 11. Bifurcation diagrams

Assume that A_i, $i = 1, 2, \ldots, q$, are $n \times n$ Hurwitz stable matrices; that is, their eigenvalues have negative real parts.

Now, given one pair of $(x(t), u(t))$, the final output of the fuzzy system is inferred as follows:

$$\dot{x}(t) = \frac{1}{\sum_{i=1}^{q} \omega_i} \sum_{i=1}^{q} \omega_i \{A_i x(t) + B_i u(t)\} , \qquad (29)$$

where

$$\omega_i = \prod_{j=1}^{n} \Gamma_j^i(x_j(t)) ,$$

in which $\Gamma_j^i(x_j(t))$ is the degree of membership of $x_j(t)$ in Γ_j^i, with

$$\begin{cases} \sum_{i=1}^{q} \omega_i > 0, \\ \omega_i \geq 0, \end{cases} \qquad i = 1, 2, \ldots, q .$$

By introducing $h_i = \omega_i / \sum_{i=1}^{q} \omega_i$ instead of ω_i, (29) is rewritten as

$$\dot{x}(t) = \sum_{i=1}^{q} h_i \{A_i x(t) + B_i u(t)\}$$

$$= \left[\sum_{i=1}^{q} h_i A_i \right] x(t) + \left[\sum_{i=1}^{q} h_i B_i \right] u(t) . \tag{30}$$

Note that

$$\begin{cases} \sum_{i=1}^{q} h_i = 1, \\ h_i \geq 0, \end{cases} \quad i = 1, 2, \ldots, q , \tag{31}$$

in which $\{h_i\}_{i=1}^{q}$ can be regarded as the normalized weights of the IF–THEN rules.

In order to make a nonchaotic or even stable continuous-time TS fuzzy system chaotic, based on the above-proposed anticontrol approaches a natural and straightforward approach is to convert it to a discrete-time version, that is, to discretize the continuous-time TS fuzzy system. Conversion of the continuous-time controller to an equivalent digital controller is known as digital redesign. Digital redesign techniques were first considered in [33], and then developed by many others [27, 34–36]. Here, this digital redesign technique is adopted for discretization of a continuous-time TS fuzzy system.

4.2 Discretization of the Continuous-Time TS Fuzzy Model

There are a few methods for discretizing a linear time-invariant (LTI) continuous-time system. Unfortunately, these discretization methods cannot be directly applied to the discretization of the continuous-time TS fuzzy models, since the defuzzified system is not LTI but linear time-varying. It is very difficult to obtain the state transition matrix for their discretization. The following theorem gives a rigorous mathematical foundation for the discretization of a continuous-time TS fuzzy model [36, 37].

Theorem 4. (Discretization Theorem) *The continuous-time TS fuzzy model (28) can be converted to the following discrete counterpart:*
Discretized TS fuzzy model:

Plant Rule i: IF $x_1(k)$ is Γ_1^i and ... and $x_n(k)$ is Γ_n^i ,
THEN $x(k+1) = G_i x(k) + H_i u(k)$, $\tag{32}$

where

$$G_i = \exp(A_i T_{\mathrm{s}}) = I + A_i T_{\mathrm{s}} + A_i^2 \frac{T_{\mathrm{s}}^2}{2!} + \cdots,$$

$$H_i = \int_0^{T_{\mathrm{s}}} \exp(A_i \tau) B_i d\tau \equiv (G_i - I) A_i^{-1} B_i \,, \tag{33}$$

and T_{s} is the sampling time.

Here, it should be noted that (33) is only a short-hand notation, in which A_i does not have to be invertible.

Proof. The exact solution of (30) at $t = kT_{\mathrm{s}} + T_{\mathrm{s}}$, where $T_{\mathrm{s}} > 0$ is the sampling period, is given by

$$x(kT_{\mathrm{s}} + T_{\mathrm{s}}) = \Phi\left(\sum_{i=1}^q h_i(kT_{\mathrm{s}} + T_{\mathrm{s}}), \sum_{i=1}^q h_i(kT_{\mathrm{s}})\right) x(kT_{\mathrm{s}})$$

$$+ \int_{kT_{\mathrm{s}}}^{kT_{\mathrm{s}} + T_{\mathrm{s}}} \Phi\left(\sum_{i=1}^q h_i(kT_{\mathrm{s}} + T_{\mathrm{s}}), \sum_{i=1}^q h_i(\tau)\right)$$

$$\times \left[\sum_{i=1}^q (h_i B_i)\right] u(\tau) \, d\tau \,, \tag{34}$$

where

$$\Phi\left(\sum_{i=1}^q h_i(kT_{\mathrm{s}} + T_{\mathrm{s}}), \sum_{i=1}^q h_i(kT_{\mathrm{s}})\right) = \Psi\left(\sum_{i=1}^q h_i(kT_{\mathrm{s}} + T_{\mathrm{s}})\right) \Psi\left(\sum_{i=1}^q h_i(kT_{\mathrm{s}})\right)$$

is the state transition matrix of (30), and Ψ is the fundamental matrix of the uncontrolled TS fuzzy system (with $u(t) = 0$), and is nonsingular for all t.

For a sufficiently small T_{s}, the input $u(t)$ can be regarded approximately as piecewise constant over the integration interval; namely, $u(t) = u(kT_{\mathrm{s}})$ for $kT_{\mathrm{s}} \le t < kT_{\mathrm{s}} + T_{\mathrm{s}}$. Then, (34) can be rewritten as

$$x(kT_{\mathrm{s}} + T_{\mathrm{s}}) = \bar{G}x(kT_{\mathrm{s}}) + \bar{H}u(kT_{\mathrm{s}}) \,, \tag{35}$$

where

$$\bar{G} = \Phi\left(\sum_{i=1}^q h_i(kT_{\mathrm{s}} + T_{\mathrm{s}}), \sum_{i=1}^q h_i(kT_{\mathrm{s}})\right),$$

$$\bar{H} = \int_{kT_{\mathrm{s}}}^{kT_{\mathrm{s}} + T_{\mathrm{s}}} \Phi\left(\sum_{i=1}^q h_i(kT_{\mathrm{s}} + T_{\mathrm{s}}), \sum_{i=1}^q h_i(\tau)\right) \times \left[\sum_{i=1}^q (h_i B_i)\right] d\tau \,.$$

The exact evaluation of the state transition matrix $\Phi(\cdot, \cdot)$ is very difficult, if not impossible, since the continuous-time TS fuzzy model (30) is time-varying. To solve this problem, we select a set of discrete-time points, kT_{s}, such that $\sum_{i=1}^{q} h_i(t)A_i$ and $\sum_{i=1}^{q} h_i(t)B_i$ can be approximated by constant matrices $\sum_{i=1}^{q} h_i(kT_s)A_i$ and $\sum_{i=1}^{q} h_i(kT_s)B_i$, respectively, over each interval $[kT_s, kT_s + T_s]$. Then, a set of difference equations can be used to describe the discrete-time TS fuzzy model at each kT_s [38].

In the time interval $kT_s \le t < kT_s + T_s$, $\bar{G}$ and $\bar{H}$ have the following representations:

$$\bar{G}(kT_s) = \exp\left(\left(\sum_{i=1}^{q} h_i(kT_s)A_i\right)T_s\right),$$

$$\begin{aligned}
\bar{H}(kT_s) &= \int_{kT_s}^{kT_s+T_s} \exp\left(\left(\sum_{i=1}^{q} h_i(kT_s)A_i\right)\right. \\
&\quad \times \left.(kT_s + T_s - \tau)\right)\left(\sum_{i=1}^{q} h_i(kT_s)B_i\right) d\tau \\
&= \left(\exp\left(\left(\sum_{i=1}^{q} h_i(kT_s)A_i\right)T_s\right) - I\right)\left(\sum_{i=1}^{q} h_i(kT_s)A_i\right)^{-1} \\
&\quad \times \left(\sum_{i=1}^{q} h_i(kT_s)B_i\right).
\end{aligned} \tag{36}$$

For a sufficiently small sampling period $T_s > 0$, and by using a power series expansion, one has

$$\begin{aligned}
\bar{G}(kT_s) &= \exp\left(\left(\sum_{i=1}^{q} h_i(kT_s)A_i\right)T_s\right) \\
&= I + \sum_{i=1}^{q} h_i(kT_s)A_iT_s + O(T_s^2) \\
&\approx \sum_{i=1}^{q} h_i(kT_s)(I + A_iT_s) \\
&\approx \sum_{i=1}^{q} h_i(kT_s) \exp(A_iT_s) \\
&= \sum_{i=1}^{q} h_i(kT_s)G_i,
\end{aligned}$$

and by using the short-hand notation $(\exp(x) - I)x^{-1} = I + (1/2!)x + (1/3!)x^2 + \cdots$, one also has

$$\bar{H}(kT_s) = \int_{kT_s}^{kT_s+T_s} \exp\left(\left(\sum_{i=1}^{q} h_i(kT_s)A_i\right) \times (kT_s + T_s - \tau)\right)$$

$$\times \left(\sum_{i=1}^{q} h_i(kT_s)B_i\right) d\tau$$

$$= \left(\exp\left(\left(\sum_{i=1}^{q} h_i(kT_s)A_i\right) T_s\right)_I\right)$$

$$\times \left(\sum_{i=1}^{q} h_i(kT_s)A_i\right)^{-1} \left(\sum_{i=1}^{q} h_i(kT_s)B_i\right)$$

$$= \left[\left(\sum_{i=1}^{q} h_i(kT_s)A_i\right) T_s + O(T_s^2)\right]$$

$$\times \left(\sum_{i=1}^{q} h_i(kT_s)A_i\right)^{-1} \left(\sum_{i=1}^{q} h_i(kT_s)B_i\right)$$

$$\approx \sum_{i=1}^{q} h_i(kT_s)B_i T_s$$

$$= \sum_{i=1}^{q} h_i(kT_s)H_i \ ,$$

where

$$G_i = \exp(A_i T_s) = I + A_i T_s + A_i^2 \frac{T_s^2}{2!} + \cdots \approx I + A_i T_s$$

and

$$H_i = B_i T_s \approx \int_0^{T_s} \exp(A_i \tau) \, d\tau = (G_i - I)A_i^{-1} B_i \ .$$

Denoting kT_s by k gives (33).

If the subsystems are stable in each local subspace, and $h_i(x(t))$, $i = 1, 2, \ldots, q$, are continuously differentiable in a neighborhood of the origin, then the local stability of the overall system is guaranteed by the following theorem.

Theorem 5 (Local Stability Theorem). *In (28), if A_i, $i = 1, 2, \ldots, q$, are all $n \times n$ Hurwitz stable matrices, then the uncontrolled system (30) (with $u(t) = 0$) and uncontrolled discretized system (33) (with $u(k) = 0$) are both stable in the neighborhood of the origin.*

Proof. We first note that A_i, $i = 1, 2, \ldots, q$, are all $n \times n$ Hurwitz stable matrices; that is, all of their eigenvalues have negative real parts.

1. For the uncontrolled system (30), the origin $x_0 = 0$ is obviously the fixed point of the uncontrolled system (30), and its Jacobian at the origin is $\sum_{i=1}^{q} h_i(x_0)A_i$. The characteristic equation is

$$\left| \lambda I - \sum_{i=1}^{q} h_i(x_0)A_i \right| = \left| \sum_{i=1}^{q} h_i(x_0)(\lambda I - A_i) \right|$$

$$= \sum_{i=1}^{q} h_i(x_0) \prod_{j=1}^{n} (\lambda - \lambda_{ij})$$

$$= 0 ,$$

where λ_{ij}, $j = 1, 2, \ldots, n$, are the eigenvalues of A_i, $i = 1, 2, \ldots, q$, which all have negative real parts. If $\lambda \geq 0$, then $|\lambda I - \sum_{i=1}^{q} h_i A_i| > 0$. So $|\lambda I - \sum_{i=1}^{q} h_i A_i| = 0$ holds only when $\lambda < 0$. This means that $\sum_{i=1}^{q} h_i(x_0)A_i$ is a Hurwitz stable matrix, and hence the uncontrolled system (30) is stable in a neighborhood of the origin.

2. For the uncontrolled discretized system (33): Since A_i, $i = 1, 2, \ldots, q$, are Hurwitz stable matrices, $G_i = \exp(A_i T_s)$ are Schur stable matrices, that is, $\rho(G_i) < 1$, therefore, there exists a certain norm, $\| \cdot \|$, such that $\|G_i\| < 1$. From (31) and the convexity of the matrix norm, it follows that

$$\left\| \sum_{i=1}^{q} h_i G_i \right\| \leq \sum_{i=1}^{q} h_i \|G_i\|$$

$$\leq \left[\sum_{i=1}^{q} h_i \right] \max\{\|G_i\|\} ,$$

$$= \max\{\|G_i\|\}$$

$$\triangleq \alpha < 1 .$$

Thus,

$$\|x(k+1)\| \leq \left\| \sum_{i=1}^{q} h_i G_i \right\| \|x(k)\| \leq \alpha \|x(k)\| .$$

By the contraction mapping theorem, it is concluded that the uncontrolled discretized system (33) is stable in the neighborhood of the origin.

After converting the continuous-time TS fuzzy system to a discrete-time counterpart, we can design a controller for the discretized TS fuzzy system to make it chaotic. The anticontroller design and the verification of chaos in the controlled TS fuzzy system are similar to that discussed in Sect. 4.1, so they are omitted.

4.3 A Simulation Example

Consider a continuous-time TS fuzzy system, which is the fuzzy model of the Lorenz equation, given by

$$\text{Rule 1: IF } x_1(t) \text{ is } \Gamma_1, \text{ THEN } \frac{d}{dt}\begin{bmatrix} x_1(t) \\ x_2(t) \\ x_3(t) \end{bmatrix} = A_1 \begin{bmatrix} x_1(t) \\ x_2(t) \\ x_3(t) \end{bmatrix}$$

$$\text{Rule 2: IF } x_1(t) \text{ is } \Gamma_2, \text{ THEN } \frac{d}{dt}\begin{bmatrix} x_1(t) \\ x_2(t) \\ x_3(t) \end{bmatrix} = A_2 \begin{bmatrix} x_1(t) \\ x_2(t) \\ x_3(t) \end{bmatrix}$$

where

$$A_1 = \begin{bmatrix} -d & d & 0 \\ r & -1 & -x_{1\min} \\ 0 & x_{1\min} & -b \end{bmatrix} \quad \text{and} \quad A_2 = \begin{bmatrix} -d & d & 0 \\ r & -1 & -x_{1\max} \\ 0 & x_{1\max} & -b \end{bmatrix}$$

and the membership functions are

$$\Gamma_1 = \frac{-x_1 + x_{1\max}}{x_{1\max} - x_{1\min}} \quad \text{and} \quad \Gamma_2 = \frac{x_1 - x_{1\min}}{x_{1\max} - x_{1\min}},$$

where Γ_i, $i = 1, 2$, are positive semidefinite for all $x_1 \in [x_{1\min}, x_{1\max}]$, and d, r, and b are parameters.

With the parameters $(d, r, b) = (10, 28, 8/3)$ and initial values $(0.1, 0.1, 0.1)$, the trajectory of the TS fuzzy system of the Lorenz system is shown in Fig. 12a.

In terms of Theorem 4, the discretized TS fuzzy model of the original continuous-time TS fuzzy Lorenz system is obtained as follows:

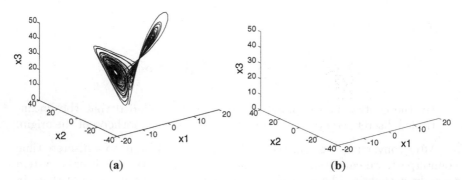

(a)   (b)

Fig. 12. Trajectories of the original (a) and discretized (b) TS fuzzy Lorenz system

$$\text{Rule 1: IF } x_1(k) \text{ is } \Gamma_1, \text{ THEN } \begin{bmatrix} x_1(k+1) \\ x_2(k+1) \\ x_3(k+1) \end{bmatrix} = G_1 \begin{bmatrix} x_1(k) \\ x_2(k) \\ x_3(k) \end{bmatrix}$$

$$\text{Rule 2: IF } x_1(k) \text{ is } \Gamma_2, \text{ THEN } \begin{bmatrix} x_1(k+1) \\ x_2(k+1) \\ x_3(k+1) \end{bmatrix} = G_2 \begin{bmatrix} x_1(k) \\ x_2(k) \\ x_3(k) \end{bmatrix}$$

where

$$G_1 = \begin{bmatrix} 1 - dT_s & dT_s & 0 \\ rT_s & 1 - T_s & -x_{1\min}T_s \\ 0 & x_{1\min}T_s & 1 - bT_s \end{bmatrix} \quad \text{and}$$

$$G_2 = \begin{bmatrix} 1 - dT_s & dT_s & 0 \\ rT_s & 1 - T_s & -x_{1\max}T_s \\ 0 & x_{1\max}T_s & 1 - bT_s \end{bmatrix}.$$

Figure 12b shows the trajectory of the discrete-time version of the continuous-time TS fuzzy Lorenz model, with $T_s = 0.004$. It can be seen that the overall shape of the trajectory is very similar to that in Fig. 12a.

When the parameters are chosen as $d = 200$, $r = 40$, and $b = 200$, the eigenvalues of A_i are -132.1267, -68.8733, and -200.0000, and so they are Hurwitz stable matrices. Hence, by Theorem 5, the continuous-time and the corresponding discretized TS fuzzy Lorenz system are both stable. The anticontroller is designed as described in (22)–(23). The controlled TS fuzzy Lorenz system is described as follows:

$$x(k+1) = \sum_{i=1}^{q} h_i \{G_i x(k) + u(k)\}$$

$$= \sum_{i=1}^{q} h_i G_i x(k) + u(k)$$

$$= \sum_{i=1}^{q} h_i G_i x(k) + \sigma \sin\left(\frac{\pi}{\sigma}\beta x(k)\right).$$

In the simulation, the magnitude of the control input is arbitrarily chosen to be $\sigma = 0.1$. Thus, $\|u(k)\| < \sigma$, and β can also be regarded as a control parameter. For β values of 0.6 and 2.1, the time evolutions, phase portraits, and bifurcation diagrams are shown in Figs. 13–16, respectively. These numerical results verify the theoretical analysis and the design of the chaotifying controller developed above.

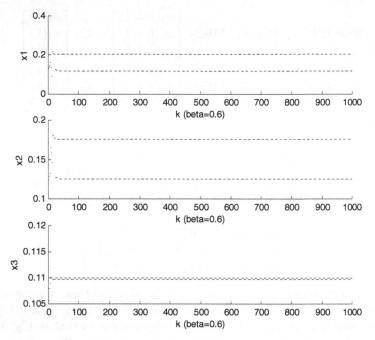

Fig. 13. Period-doubling bifurcations at $\beta = 0.6$

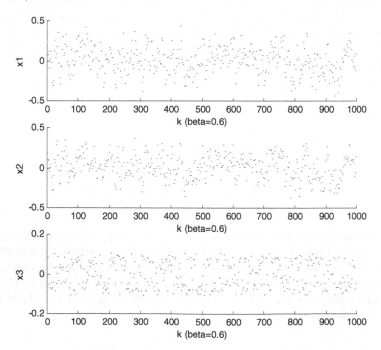

Fig. 14. Chaotic time evolutions

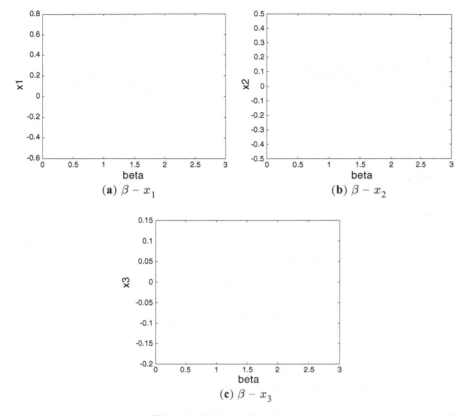

(a) $\beta - x_1$ (b) $\beta - x_2$

(c) $\beta - x_3$

Fig. 15. Bifurcation diagrams

5 Anticontrol of Chaos for Continuous-Time TS Fuzzy Systems via Time-Delay Feedback

A new technique was proposed for chaotification, which is general and yet by nature very different from the aforementioned approaches [39]. It can make an arbitrarily given continuous-time TS fuzzy system chaotic via fuzzy feedback linearization, differential geometric control theory, and time-delay feedback perturbations [40–42]. In this approach, asymptotic analysis is used to establish an approximate relationship between a time-delay differential equation and a discrete chaotic map, so that a time-delay feedback control term can be constructed to make the controlled TS fuzzy system chaotic, where the generated chaos is in the sense of Li and Yorke [19]. Systems with time-delay are inherently infinite-dimensional; therefore, it is possible to produce complicated dynamics such as bifurcation and chaos, even in a very simple first-order system [43].

Specifically, consider a general *single-input* TS fuzzy system in the following form:

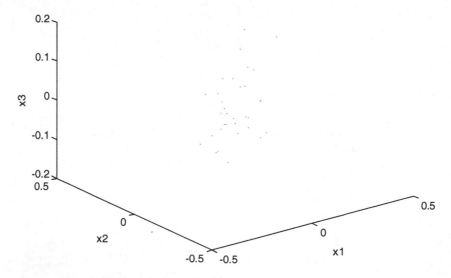

Fig. 16. Chaotic phase portrait at $\beta = 2.1$

Plant Rule i: IF $x_1(t)$ is Γ_1^i ... and $x_n(t)$ is Γ_n^i,

$$\text{THEN } \dot{x}(t) = A_i x(t) + B_i u(t) , \tag{37}$$

where $x(t) \in \mathbb{R}^n$, $u(t) \in \mathbb{R}$, $A_i \in \mathbb{R}^{n \times n}$, $B_i \in \mathbb{R}^n$, $i = 1, 2, \ldots, q$, in which r is the number of IF–THEN rules, Γ_j^i are fuzzy sets, and the equation $\dot{x}(t) = A_i x(t) + B_i u(t)$ is the output from the ith IF–THEN rule.

Similarly, by using $\mu_i = \omega_i / \sum_{j=1}^r \omega_j$ instead of ω_i, for a given pair of $(x(t), u(t))$, the final output of the fuzzy system is inferred by

$$
\begin{aligned}
\dot{x}(t) &= \frac{1}{\sum_{i=1}^r \omega_i} \sum_{i=1}^r \omega_i \{ \mathbf{A}_i \mathbf{x}(t) + \mathbf{B}_i u(t) \} \\
&= \sum_{i=1}^q \mu_i \{ A_i x(t) + B_i u(t) \} \\
&= \left[\sum_{i=1}^q \mu_i A_i \right] x(t) + \left[\sum_{i=1}^q \mu_i B_i \right] u(t) \\
&= \bar{A} x(t) + \bar{B} u(t) ,
\end{aligned}
\tag{38}
$$

where $\bar{A} = \sum_{i=1}^q \mu_i A_i$ and $\bar{B} = \sum_{i=1}^q \mu_i B_i$.

Definition 3 (Local and Global Controllability). *If (A_i, B_i), $i = 1, 2, \ldots, q$, are controllable pairs, then the fuzzy model (38) is called locally controllable. If $(\bar{A}, \bar{B})$ is a controllable pair, then the fuzzy model (38) is called globally controllable.*

For a special kind of locally controllable TS fuzzy systems and a more general type of TS fuzzy systems, fuzzy feedback linearization technique, nonlinear control theory, and the PDC with time-delay can be employed to make them chaotic.

5.1 PDC Controller for Locally Controllable TS Fuzzy Systems

The PDC controller discussed in Sect. 2.3.1 is employed here to determine the structure of a fuzzy controller from a given TS fuzzy model [6, 44–47]. The PDC provides the following fuzzy-control rule structure from the fuzzy model (38):

$$\text{Control Rule } i: \quad \text{IF } x_1(t) \text{ is } \Gamma_1^i \ldots \text{ and } x_n(t) \text{ is } \Gamma_n^i,$$
$$\text{THEN } u(t) = -K_i x(t) + \nu, \qquad i = 1, 2, \ldots, q, \tag{39}$$

where K_i are feedback control gains and ν is a time-delay feedback term of the form

$$\nu(t - \tau) = \omega(h(x(t - \tau)))\,,$$

in which h is a scalar function needed to be further determined.

The fuzzy control rules have linear state feedback laws and a scalar function in the consequent parts. The overall fuzzy controller is represented by

$$
\begin{aligned}
u(t) &= -\frac{1}{\sum_{i=1}^{q} \omega_i} \sum_{i=1}^{q} \omega_i \left(K_i x(t) + \nu \right) \\
&= -\sum_{i=1}^{q} \mu_i K_i x(t) + \nu\,.
\end{aligned}
\tag{40}
$$

Chaotifying the TS fuzzy system (38) is generally a very difficult task. However, in some special cases, this can be done. Consider the case of a common B in (38), i.e., $B_1 = B_2 = \cdots = B_q = B$.

Definition 4. *The TS fuzzy system (38) is called exactly linearizable via the feedback controller (40) if there exist feedback gains K_i such that*

$$A_i - BK_i = G\,, \tag{41}$$

for $i = 2, 3, \ldots, q$.

Notice that having a common G might not always be possible even if (A_i, B_i), $i = 1, 2, \ldots, q$, are controllable.

If the locally controllable TS fuzzy system (38) is exactly linearizable, one can choose K_i such that $A_i - BK_i = G$ for $i = 1, 2, \ldots, q$, where G is a Hurwitz stable matrix with desired eigenvalues of μ_1, μ_2, $\ldots$, μ_n, and (G, B) is also controllable. Thus, the global fuzzy model (38) is reduced to the following form:

$$\dot{x}(t) = Gx(t) + B\nu .$$ (42)

The feedback control law (40) can then be designed by using conventional linear system theory [40] so that the eigenvalues of $A_i - BK_i$ are the specified ones. The feedback gain K_i can be obtained by using the Ackerman's formula [40], in the case of a single-input system, as follows:

$$K_i = [0\,0 \cdots 0\,1]\,\big[B\,|\,A_iB\,|\,\cdots\,|\,A_i^{n-1}B\big]^{-1}\phi(A_i) ,$$ (43)

where $\phi(A_i) = A_i^n + a_1 A_i^{n-1} + \cdots + a_{n-1} A_i + a_n I$, $i = 1, 2, \ldots, q$, are Hurwitz stable polynomials, and $a_1, a_2, \ldots, a_n$ are coefficients of the characteristic polynomial

$$|\lambda I - G| = (\lambda - \mu_1)(\lambda - \mu_2) \cdots (\lambda - \mu_n)$$
$$= \lambda^n + a_1 \lambda^{n-1} + \cdots + a_{n-1}\lambda + a_n .$$ (44)

Define a transformation matrix T by

$$T = MW ,$$ (45)

where M is the controllability matrix of the form

$$M = \big[B\,|\,GB\,|\,\ldots\,|\,G^{n-1}B\big]$$

and

$$W = \begin{bmatrix} a_{n-1} & a_{n-2} & \cdots & a_1 & 1 \\ a_{n-2} & a_{n-3} & \cdots & 1 & 0 \\ \vdots & \vdots & \ddots & \vdots & \vdots \\ a_1 & 1 & \cdots & 0 & 0 \\ 1 & 0 & \cdots & 0 & 0 \end{bmatrix} .$$

Also define a new state vector $\hat{x}$ by

$$x = T\hat{x} .$$

The rank of the controllability matrix M is n because (G, B) is controllable. Then, the inverse of matrix T exists and (43) can be modified to

$$\dot{\hat{x}} = T^{-1}GT\hat{x} + T^{-1}B\nu ,$$ (46)

where

$$T^{-1}GT = \begin{bmatrix} 0 & 1 & 0 & \cdots & 0 \\ 0 & 0 & 1 & \cdots & 0 \\ \vdots & \vdots & \vdots & \ddots & \vdots \\ 0 & 0 & 0 & \cdots & 1 \\ -\alpha_n & -\alpha_{n-1} & -\alpha_{n-2} & \cdots & -\alpha_1 \end{bmatrix}$$

and

$$T^{-1}B = \begin{bmatrix} 0 \\ 0 \\ \vdots \\ 0 \\ 1 \end{bmatrix},$$

with

$$\nu = \omega(\hat{x}_1(t - \tau)) = \omega((T^{-1}x)_1(t - \tau)), \tag{47}$$

in which $\hat{x}_1$ means the first component of vector $\hat{x}$.

This will be further determined in the following.

5.2 Controller Design for General TS Fuzzy Systems

Two different approaches are employed to study the problem of chaotification for a more general TS fuzzy system.

5.2.1 Approximate Linearization Approach

Obviously, the origin is an equilibrium of the uncontrolled system (38) with $u = 0$ therein, and by Theorem 5 it is asymptotically stable within its sufficiently small neighborhood if the subsystems (37) is stable.

In a small neighborhood of the origin, system (38) can be represented by its linearization, evaluated at the origin, as follows:

$$\dot{x}(t) = \bar{A}_0 x(t) + \bar{B}_0 u(t) . \tag{48}$$

Suppose $(\bar{A}_0, \bar{B}_0)$ is controllable, a feedback control law similar to the above-proposed formula (45) can be used to convert it to the controllable canonical form (46).

5.2.2 Globally Exact Linearization Approach

To achieve the intended chaotification, differential geometric control theory can be applied [41, 48, 49]. Assume that $y = h(x)$ is a scalar output of system (38), where h is a smooth function satisfying $h(x) = 0$. Combined with (38), it can be described as a SISO affine nonlinear system as follows:

$$\begin{cases} \dot{x}(t) = \bar{A}x(t) + \bar{B}u(t) := f(x) + g(x) \\ \quad y = h(x) \, . \end{cases} \tag{49}$$

Let $L_f^i h(x)$ denote the ith *Lie derivative* of the smooth function $h(x)$ with respect to a vector field $f(x)$, and $ad_f^i g(x)$ the ith *Lie bracket* of the two smooth vector fields $f(x)$ and $g(x)$.

Definition 5. *The SISO affine system* (49) *is said to have a relative degree r at x^* if there exists a neighborhood U of x^* such that*

$$L_g L_f^k h(x^*) = 0, \forall \, x \in U, \quad 0 \le k < r - 1 \, ;$$

$$L_g L_f^{r-1} h(x) \neq 0 \, .$$

Definition 6. (Involutive Distributions) *A distribution $\Delta = \text{span}$ $\{f_1, \dots, f_m\}$ is called involutive if for any two vector fields $\tau_1, \tau_2 \in \Delta$, their Lie bracket $ad_{\tau_1} \tau_2 \in \Delta$.*

The following result is well known [41].

Lemma 3. *The SISO system* (49) *has relative degree n at x^* if and only if there exists a neighborhood U of x^* such that*

(i) $\text{rank} \, [g(x), ad_f g(x), \dots, ad_f^{n-1} g(x)] = n$ *for all $x \in U$,*
(ii) $\text{span} \, \{g(x), ad_f g(x), \dots, ad_f^{n-2} g(x)\}$ *is involutive in U.*

Definition 7. *The SISO system* (49) *is called globally feedback linearizable if it has relative degree n at x^*.*

In this case, the output $y = h(x)$ is a solution to the following partial differential equation:

$$\frac{\partial h(x)}{\partial x} \left[g(x), \, ad_f g(x), \, \dots, \, ad_f^{n-2} g(x) \right] = 0 \, .$$

If the output $y = h(x)$ results in a relative degree of n, then one may use

$$y = \Phi(x) = \begin{pmatrix} h(x) \\ L_f h(x) \\ L_f^2 h(x) \\ \vdots \\ L_f^{n-1} h(x) \end{pmatrix} \tag{50}$$

and the control law

$$u(x) = \frac{1}{L_g L_f^{n-1} h(x)} (-L_f^n h(x) + v) \tag{51}$$

to yield the linear system

$$\dot{\mathbf{y}} = \begin{bmatrix} 0 & 1 & 0 & \cdots & 0 \\ 0 & 0 & 1 & \cdots & 0 \\ \vdots & \vdots & \vdots & \ddots & \vdots \\ 0 & 0 & 0 & \cdots & 0 \\ 0 & 0 & 0 & \cdots & 0 \end{bmatrix} \mathbf{y} + \begin{bmatrix} 0 \\ 0 \\ \vdots \\ 0 \\ 1 \end{bmatrix} u, \tag{52}$$

where $\mathbf{y} = \left[y, \dot{y}, \ldots, y^{(n-1)}\right]^T$, and

$$\begin{aligned} u &= \frac{1}{L_g L_f^{n-1}} \left(-L_f^n h(x) + \omega(y(t-\tau))\right) \\ &= \frac{1}{L_g L_f^{n-1}} \left(-L_f^n h(x) + \omega(h(x(t-\tau)))\right), \end{aligned} \tag{53}$$

in which the time-delay term also remains to be further determined in the following.

5.2.3 Verification of Chaos

The controlled TS fuzzy system (48) with time-delay term (52) can be recast in the following n-dimensional state-space form:

$$\dot{x}(t) = Gx(t) + Bv, \tag{54}$$

where G and B are in the controllable canonical form, namely,

$$G = \begin{bmatrix} 0 & 1 & \cdots & 0 \\ \vdots & \vdots & \ddots & \vdots \\ 0 & 0 & \cdots & 1 \\ -\alpha_n & -\alpha_{n-1} & \cdots & -\alpha_1 \end{bmatrix} \quad \text{and} \quad B = \begin{bmatrix} 0 \\ 0 \\ \vdots \\ 0 \\ 1 \end{bmatrix}.$$

Since G is a Hurwitz stable matrix and $v(t)$ is uniformly bounded, the solution of (54) is bounded for any bounded initial condition and can be computed iteratively on each τ-time interval $(m\tau, (m+1)\tau]$ for $m = 0, 1, \ldots$.

Denote $x(t) \equiv x(m\tau + \hat{t}) \equiv x(m, \hat{t})$ for $t \equiv m\tau + \hat{t}$, $\hat{t} \in (0, \tau]$. It follows that

$$x(t) = x(m, \hat{t}) = e^{G\hat{t}} x(m-1, \hat{t}) + \int_0^{\hat{t}} e^{G(\hat{t}-t')} B\omega(x_1(m-1, t'))dt'. \tag{55}$$

Lemma 4. *Let $\delta(t - t_0)$ be the scalar-valued Dirac distribution centered at $t_0 \geq 0$, and let $d\xi(t, t_0) = e^{G(t_0-t)}\,dt$ be a matrix-valued measure defined on $[0, \tau]$. If it is imposed that $d\xi(t, t_0) = C(t, t_0)\delta(t - t_0)\,dt$, then $C(t, t_0) \approx -G^{-1}e^{G(t_0-t)}$ for a sufficiently large $\tau > 0$. Moreover, $C(t, t_0) \to -G^{-1}$ as $t \to t_0$.*

Proof. See [22, 50].

Lemma 5. *For a sufficiently large τ and a large $\hat{t} \in (t_0, \tau]$,*

$$x_1(m, \hat{t}) \approx \omega(x_1(m - 1, \hat{t})) \quad \text{and} \quad x_i(m, \hat{t}) \approx 0 , \tag{56}$$

for $m = 0, 1, \ldots$ and $i = 2, \ldots, n$.

Proof. Note that for any given bounded initial condition, $x(t)$ is uniformly bounded and $e^{G\hat{t}}x(m-1, \hat{t})$ tends to 0 rapidly as $\hat{t} \to \infty$. Therefore, it follows from (49) that

$$
\begin{aligned}
x(t) &\approx \int_0^{\hat{t}} e^{G(\hat{t}-t')} B\omega(x_1(m - 1, t'))\,dt' \\
&\approx \int_0^{\hat{t}} C(t', \hat{t})\delta(t' - \hat{t}) B\omega(x_1(m - 1, t'))\,dt' \\
&\approx C(t', \hat{t}) B\omega(x_1(m - 1, \hat{t})) \\
&\approx -G^{-1}\omega(x_1(m - 1, t')) .
\end{aligned}
$$

Since

$$
G^{-1} = \begin{bmatrix}
-\dfrac{\alpha_1}{\alpha_0} & -\dfrac{\alpha_2}{\alpha_0} & \cdots & -\dfrac{\alpha_{n-1}}{\alpha_0} & \dfrac{1}{\alpha_0} \\
1 & 0 & \cdots & 0 & 0 \\
0 & 1 & \ddots & 0 & 0 \\
0 & 0 & \cdots & 0 & 0 \\
0 & 0 & \cdots & 1 & 0
\end{bmatrix} ,
$$

one has

$$x(m, \hat{t}) \approx \left(\alpha_0^{-1}\omega(x_1(m - 1, \hat{t})), 0, \ldots, 0\right)^{\mathrm{T}} .$$

The proof is thus completed.

Lemma 5 establishes an asymptotically approximate relationship between the time-delay equation (54) and the difference equation (56). Although there is an essential difference between the dynamics of a time-delay equation and that of its associated difference equation [51], it is reasonable to expect that the first state component of the time-delay equation (54) is chaotic if $\omega(\cdot)$ is a bounded chaotic map and the delay time is sufficiently large. This implies that one can use time-delay feedback to drive system (54) to be chaotic.

Theorem 6. *If $\omega(\cdot)$ is a bounded chaotic map and if the delay time is sufficiently large, then the TS fuzzy system (38) controlled by the time-delayed feedback control law is chaotic.*

Proof It follows directly from Lemma 5.

There are many well-known chaotic maps, such as the Logistic map, Henon map, Baker's map, which can be used to construct the time-delay feedback $\omega(\cdot)$, thus making the controlled TS fuzzy system (38) chaotic. One simple choice is

$$\nu(t) = \omega(x_1(t - \tau)) = \sigma \, \sin\left(\frac{\pi}{\sigma}\beta x_1(t - \tau)\right) , \qquad (57)$$

which is shown in Fig. 4.

Obviously, $\mid \nu(t) \mid \leq \sigma$ for all $x_1 \in \mathbb{R}$, which can be arbitrarily small.

As mentioned above, if the map (57) is chaotic, then one can expect that the time-delay PDC will make the TS fuzzy system (38) chaotic, provided that the delay time τ is sufficiently large. Mathematically, we can show that the map (57) is indeed chaotic in the sense of Li and Yorke by arguments similar to that given in [18, 19]. The bifurcation diagram of map (57) is shown in Fig. 17, with $\sigma = 0.1$, which reveals the chaotic nature of the map (57).

5.3 Simulation Examples

To visualize the theoretical analysis and design, two examples are included here for illustration.

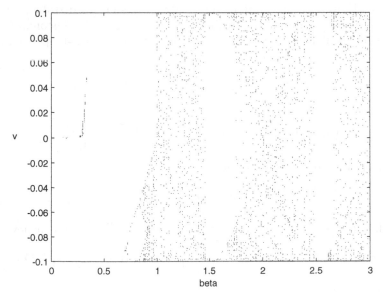

Fig. 17. Bifurcation diagram of map (57) with $\sigma = 0.1$

Example 1. Consider the fuzzy system

$$\text{Rule 1: IF } x_2(t) \text{ is } \Gamma_1, \text{ THEN } \dot{x} = A_1 x + Bu ,$$
$$\text{Rule 2: IF } x_2(t) \text{ is } \Gamma_2, \text{ THEN } \dot{x} = A_2 x + Bu ,$$

where

$$A_1 = \begin{bmatrix} 1 & -0.5 \\ 1 & 0 \end{bmatrix}, \ A_2 = \begin{bmatrix} -1 & -0.5 \\ 1 & 0 \end{bmatrix}, \text{ and } B = \begin{bmatrix} 1 \\ 0 \end{bmatrix}.$$

Figure 18 shows the membership functions of Γ_1 and Γ_2.

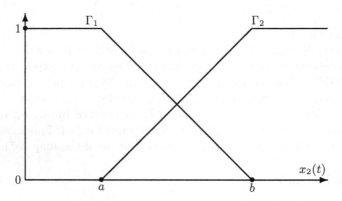

Fig. 18. Membership functions of Example 1

Since A_1 and A_2 are stable, the linear subsystems are stable, and the overall system is stable in a neighborhood of the origin by Theorem 5. Employing the feedback controller (40) and choosing the closed-loop eigenvalues as $[-0.35, -0.5]$, we obtain

$$K_1 = [1.85, -0.325], \quad K_2 = [-0.15, -0.325] ,$$

and

$$A_1 - BK_1 = A_2 - BK_2 = G = \begin{bmatrix} -1.85 & -0.175 \\ 1 & 0 \end{bmatrix}.$$

The closed-loop system becomes

$$\dot{x} = Gx + B\nu .$$

Define a transformation matrix, $T = MW = \begin{bmatrix} 0 & 1 \\ 1 & 0 \end{bmatrix}$, and a new state vector $\hat{x}$ by $x = T\hat{x}$, where $M = \begin{bmatrix} 1 & -0.85 \\ 0 & 1 \end{bmatrix}$ and $W = \begin{bmatrix} 0.85 & 1 \\ 1 & 0 \end{bmatrix}$. One has

$$\dot{\hat{x}} = \begin{bmatrix} 0 & 1 \\ -0.175 & -0.85 \end{bmatrix} \hat{x} + \begin{bmatrix} 0 \\ 1 \end{bmatrix} \nu ,$$

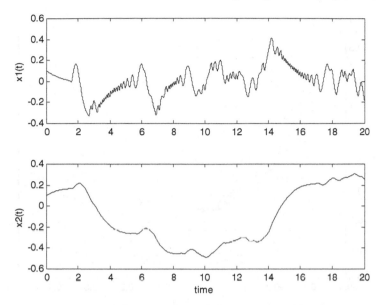

Fig. 19. Time response of the controlled TS fuzzy system (the first component is chaotic)

where

$$\nu = \sigma \sin\left(\frac{\pi}{\sigma}\beta\hat{x}_1(t-\tau)\right) = \sigma \sin\left(\frac{\pi}{\sigma}\beta(T^{-1}x)_1(t-\tau)\right) = \sigma \sin\left(\frac{\pi}{\sigma}\beta x_2(t-\tau)\right).$$

When $\sigma = 1$, $\beta = 141$, and $\tau = 1$, and the control starts at $t = 3$, the time response and chaotic attractor of the controlled TS fuzzy system are obtained as shown in Figs. 19 and 20, respectively.

Example 2. Consider a continuous-time TS fuzzy system, the fuzzy model of Chen's system [36], given by

$$\text{Rule 1: IF } x_1(t) \text{ is } \Gamma_1, \text{ THEN } \frac{d}{dt}\begin{bmatrix} x_1(t) \\ x_2(t) \\ x_3(t) \end{bmatrix} = A_1 \begin{bmatrix} x_1(t) \\ x_2(t) \\ x_3(t) \end{bmatrix} + Bu$$

$$\text{Rule 2: IF } x_1(t) \text{ is } \Gamma_2, \text{ THEN } \frac{d}{dt}\begin{bmatrix} x_1(t) \\ x_2(t) \\ x_3(t) \end{bmatrix} = A_2 \begin{bmatrix} x_1(t) \\ x_2(t) \\ x_3(t) \end{bmatrix} + Bu,$$

where

$$A_1 = \begin{bmatrix} -a & a & 0 \\ c-a & c & -x_{1\,\text{min}} \\ 0 & x_{1\,\text{min}} & -b \end{bmatrix}, \quad A_2 = \begin{bmatrix} -a & a & 0 \\ c-a & c & -x_{1\,\text{max}} \\ 0 & x_{1\,\text{max}} & -b \end{bmatrix},$$

with membership functions

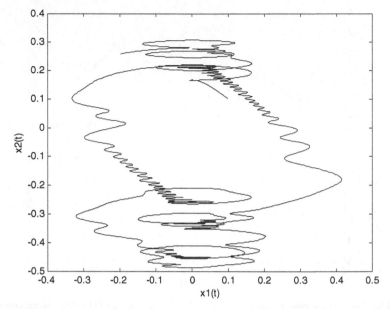

Fig. 20. Chaotic attractor of the controlled TS fuzzy system (phase portrait)

$$\Gamma_1 = \frac{-x_1 + x_{1\max}}{x_{1\max} - x_{1\min}} \quad \text{and} \quad \Gamma_2 = \frac{x_1 - x_{1\min}}{x_{1\max} - x_{1\min}},$$

where Γ_i, $i = 1, 2$, are positive semidefinite for all $x_1 \in [x_{1\min}, x_{1\max}] = [-30, 30]$, and a, b, and c are parameters.

We fixed $a = 35$, $b = 3$, and let c vary. For $c < 12.8$, the origin of the uncontrolled subsystems is a globally exponentially stable equilibrium point. For $17.5061 \approx c_{H_1} < c < 20$, the uncontrolled Chen's system has two locally exponentially stable equilibrium points given by the following formula: $P_\pm = \left(\pm\sqrt{b(a - 2c)}, \pm\sqrt{b(a - 2c)}, a - 2c \right)$, and the system is chaotic only if $c > c_{H_2} \approx 37.9868$ or $12.8 < c < c_{H_1} \approx 17.5061$.

5.3.1 Approximate Linearization Approach

In the simulation, we took $c = 10$, so the origin of the uncontrolled overall system is locally stable. Its linearization is as follows:

$$\dot{x} = \begin{bmatrix} -35 & 35 & 0 \\ -25 & 10 & 0 \\ 0 & 0 & -3 \end{bmatrix} x + \begin{bmatrix} 1 \\ 1 \\ 1 \end{bmatrix} u \,.$$

Feedback gain $K = [-20.2, -1.8, 0]$ can be chosen to make the system have the desired eigenvalues -1, -2, and -3. The system becomes

$$\dot{x} = \begin{bmatrix} -14.8 & 36.8 & 0 \\ -4.8 & 11.8 & 0 \\ 20.2 & 1.8 & -3 \end{bmatrix} x + \begin{bmatrix} 1 \\ 1 \\ 1 \end{bmatrix} \nu .$$

Define a transformation matrix T by

$$T = \begin{bmatrix} -54.69 & 22.6 & 1 \\ -18.69 & 7.6 & 1 \\ 411.51 & 19.6 & 1 \end{bmatrix} ,$$

and define a new state vector $\hat{x}$ by

$$x = T\hat{x} .$$

The controller is designed as

$$u = Kx + \sigma \sin\left(\frac{\pi}{\sigma}\beta(T^1 x)_1(t - \tau)\right)$$

$$= Kx + \sigma \sin\left(\frac{\pi}{\sigma}\beta(-0.00170x_1(t - \tau) - 0.0004x_2(t - \tau)\right.$$

$$\left. + 0.0022x_3(t - \tau))\right) .$$

In the simulation, we fixed $\tau = 1$. Figure 21 shows the chaotic attractor of the controlled Chen's TS fuzzy system, with $\sigma = 1$ and $\beta = 31$.

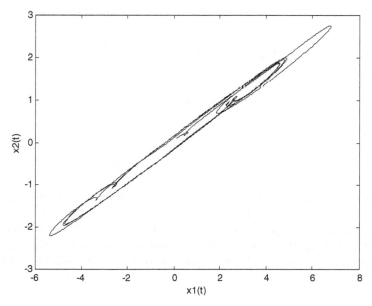

Fig. 21. Chaotic attractor of the controlled Chen's TS fuzzy system

5.3.2 Globally Exact Linearization Approach

In the simulation, we took $c = 19$. In this setting, the uncontrolled TS fuzzy model of Chen's system has two locally exponentially stable equilibrium points. Here, we use c as a control parameter, and denote $c = c_0 + \delta c$. The controlled Chen's TS fuzzy system becomes

$$\dot{x} = \begin{bmatrix} a(x_2 - x_1) \\ (c - a)x_1 - x_1 x_3 - cx_2 \\ x_1 x_2 - bx_3 \end{bmatrix} + \begin{bmatrix} 0 \\ x_1 + x_2 \\ 0 \end{bmatrix} \delta c$$

with

$$ad_f g = \begin{bmatrix} -a(x_1 + x_2) \\ ax_2 - 2ax_1 - x_1 x_3 \\ -x_1(x_1 + x_3) \end{bmatrix} .$$

We still fixed $a = 35$, $b = 3$, and $c = 19$, but let $0 < \delta c < 1.48$. It can be verified that conditions (i) and (ii) in Lemma 3 are satisfied for all $\mathbf{x} \neq 0$, and so the relative degree of the system at the nontrivial equilibrium points is 3. It follows from (50) that

$$\frac{\partial h}{\partial x} g(x) = \frac{\partial h}{\partial x_2}(x_1 + x_2) = 0 \,,$$

$$\frac{\partial h}{\partial x} ad_f g(x) = 0 \,,$$

so that

$$\frac{\partial h}{\partial x_2} = 0, \quad a\frac{\partial h}{\partial x_1} + x_1 \frac{\partial h}{\partial x_3} = 0 \,.$$

The solution is given by

$$y = h(x) = 0.5x_1^2 - ax_3 \,.$$

Therefore, we may take

$$\delta c = \sigma \sin\left(\frac{\pi}{\sigma}\beta\left(0.5x_1^2(t - \tau) - ax_3(t - \tau)\right)\right) \,.$$

In the simulation, we fixed $\tau = 1$. For $\sigma = 1$ and $\beta = 31$, the controlled system has two separated chaotic attractors; each is near to one of the two originally stable fixed points, as shown in Figs. 22 and 23, respectively.

For $\sigma = 5$, Fig. 24 shows that the two separated attractors merge into one chaotic attractor with β unchanged, and its corresponding 3-D phase portrait is shown in Fig. 25.

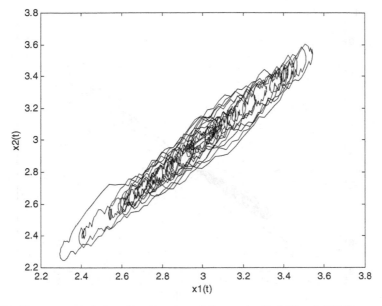

Fig. 22. One separated chaotic attractor of the controlled Chen's TS fuzzy system

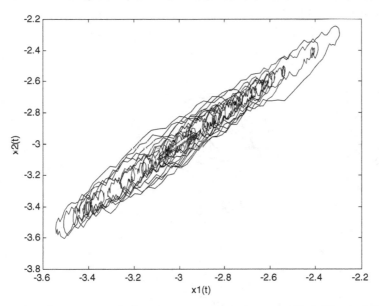

Fig. 23. Another separated chaotic attractor of the controlled Chen's TS fuzzy system

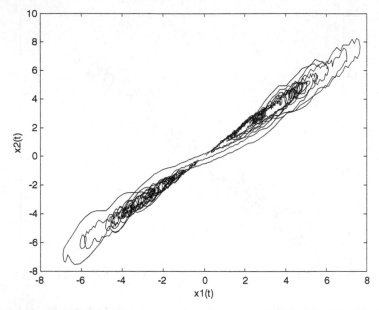

Fig. 24. Projection of the chaotic attractor of the controlled Chen's TS fuzzy system on $x_1 - x_2$ plane

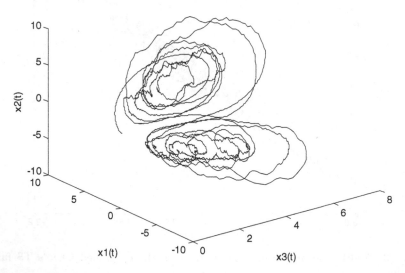

Fig. 25. The 3-D chaotic attractor of the controlled Chen's TS fuzzy system

5.4 A Remark on the Time-Delay Anticontrol Approach

Although the fundamental idea of the anticontrol algorithm used in this section to chaotify continuous-time TS fuzzy systems is correct and insightful, the time-delay differential equation is only approximated by a related discrete map. Inspired by this fundamental idea, [23] improved the technical contents by deriving a similar yet rigorous design method for chaotification, which may be modified to apply to the TS fuzzy systems setting.

6 Concluding Remarks

The study on anticontrol of chaos for both discrete-time and continuous-time Takagi–Sugeno (TS) fuzzy systems has been reviewed in this chapter. The anticontrol approaches are proved to be mathematically rigorous and the chaos generated by this method has been proved to be in the sense of Li and Yorke. An open problem is how to generate a chaotic attractor, which has preferred features such as multiscroll attractors. The Mamdani fuzzy modeling and control approaches may be able to provide a possible solution, which needs to be further studied in the future.

References

1. G. Chen, X. Dong: Int. J. Bifur. Chaos **3**, 1363–1409 (1993)
2. E. Ott, C. Grebogi, J.A. Yorke: Phys. Rev. Lett. **64**, 1196–1199 (1990)
3. T. Shinbrot, C. Grebogi, E. Ott, J.A. Yorke: Nature **363**, 411–417 (1993)
4. G. Chen, X. Dong: *From Chaos to Order: Methodologies, Perspectives and Applications* (World Scientific, Singapore, 1998)
5. G. Chen, X. Dong: IEEE Trans. Circuits Syst. I **40**(9), 591–601 (1993)
6. H. Wang, K. Tanaka, M.F. Griffin: IEEE Trans. Fuzzy Syst. **4**(1), 14–23 (1996)
7. H. Wang, K. Tanaka, T. Ikeda: IEEE Int. Symp. Circuits Syst. **3**, 209–212 (1996)
8. K. Tanaka, T. Ikeda, H. Wang: IEEE Trans. Circuits Syst. **45**(10), 1021–1040 (1998)
9. O. Calvo, J.H.E. Cartwright: Int. J. Bifurc. Chaos **8**(8), 1743–1747 (1998)
10. G. Chen: IEEE Circuits Syst. Soc. Newslett. **9**(1), 1–5 (1998)
11. S.J. Schiff, K. Jerger, D.H. Duong, T. Chang, M.L. Spano, W.L. Ditto: Nature **370**, 615–620 (1994)
12. W. Yang, M. Ding, A.J. Mandell, E. Ott: Phys. Rev. E **51**, 102–110 (1995)
13. I. Triandaf, I.B. Schwartz: Phys. Rev. E **62**, 3529–2534 (2000)
14. G. Chen, D. Lai: Int. J. Bifurc. Chaos **6**, 1341–1349 (1996)
15. G. Chen, D. Lai: "Anticontrol of chaos via feedback" In: *Proc. IEEE Conf. Decision and Control*, San Diego, CA, Dec. **10–12**, 1997 pp 367–372
16. G. Chen, D. Lai: Int. J. Bifurc. Chaos **8**, 1585–1590 (1998)
17. Y. Shi, G. Chen, "Chaotification of discrete dynamical systems governed by continuous maps, Int. J. Bifurc. Chaos, 2005, to appear.

18. R.L. Devaney, *An Introduction to Chaotic Dynamical Systems* (Addison-Wesley, New York, 1987)
19. T.Y. Li, J.A. Yorke: Am. Math. Monthly **82**, 481–485 (1975)
20. F.R. Marotto: J. Math. Anal. Appl. **63**, 199–223 (1978)
21. X.F. Wang, G. Chen: IEEE Trans. Circuits Syst. I **47**(3), 410–415 (2000)
22. X.F. Wang, G. Chen, X. Yu: Chaos **10**(4), 1–9 (2000)
23. T.S. Zhou, G. Chen, Q.G. Yang: Chaos **14**, 662–668 (2004)
24. H.G. Zhang, Z.L. Wang, D.R. Liu: Int. J. Bifurc. Chaos **14**(10), 3505–3517 (2004)
25. M. Porto, P. Amato: FUZZ-IEEE **1**, 435–440 (2000)
26. S. Baglio, L. Fortuna, G. Manganaro: Electron. Lett. **32**(4), 292–293 (1996)
27. Y.H. Joo, L.S. Shieh, G. Chen: IEEE Trans. Fuzzy Syst. **7**(4), 394–408 (1999)
28. H.J. Lee, J.B. Park, G. Chen: IEEE Trans. Fuzzy Syst. **9**(2), 369–379 (2001)
29. T. Takagi, M. Sugeno: IEEE Trans. Syst. Man, Cybern. **15**(1), 116–132 (1985)
30. Z. Li, J.B. Park, Y.H. Joo, Y.H. Choi, G. Chen: IEEE Trans. Circuits Syst. **49**(2), 249–253 (2002)
31. Z. Li, J.B. Park, G. Chen, Y.H. Joo, Y.H. Choi: Int. J. Bifurc. Chaos **12**(10), 2283–2291 (2002)
32. G.H. Golub, C.F. Van Loan: *Matrix Computations* (Johns Hopkins University Baltimore, MD, Press, 1983)
33. B.C. Kuo: Digital Control Systems (Holt, Rinehart and Winston, New York, 1980)
34. L.S. Shieh, J.L. Zhang, J.W. Sunkel: Control Theory Adv. Technol. **8**, 37–57 (1992)
35. L.S. Shieh, J.L. Zhang, W. Wang: J. Guid. Control Dyn. **20**(4), 721–728 (1997)
36. W. Chang, J.B. Park, Y.H. Joo, G. Chen: IEEE Trans. Circuits Syst. I **49**(4), 509–517 (2002)
37. Z. Li, J.B. Park, Y.H. Joo: IEEE Trans. Circuits. Syst. I **48**(10), 1237–1243 (2001)
38. W.L. Brogan: *Modern Control Theory* (Prentice Hall, Englewood Cliffs, NJ, 1991)
39. Z. Li, W. Halang, G. Chen, L.F. Tian: Dyn. Contin. Discrete Impuls. Syst., **10**, 813–832 (2003)
40. K. Ogata: *Modern Control Engineering* (Prentice Hall, Englewood Cliffs, NJ, 1990)
41. S. Sastry: *Nonlinear Systems: Analysis, Stability, and Control* (Springer, Berlin Heidelberg New York, 1999)
42. A. Isidori: Nonlinear Control Systems (Springer, Berlin Heidelberg New York, 1995)
43. H. Lu, Z. He: IEEE Trans. Circuits Syst. Video Tech. **43**, 700–702 (1996)
44. J. Huffman: Dealing with chaos: Neural nets and fuzzy logic – Tools for building products in the information age, In: *WESCON/94. Idea/Microelectronics. Conference Record*, 1994, pp 332–336
45. K. Hirota: IEEE Int. Conf. Syst. Man Cybern. **3**, 2446–2459 (1995)
46. T. Akagi, M. Sugeno: IEEE Trans. Syst. Man, Cybern. **15**(1), 116–132 (1985)
47. H.O. Wang, K. Tanaka, M.F. Griffin: Parallel distributed compensation of nonlinear systems by Takagi–Sugeno fuzzy model. In: *Proc. Fuzz IEEE/IFES '95*, 1995, pp 531–538

48. C.I. Byrnes, A. Isidori: Exact linearization and zero dynamics. In: *Proc. of the 29th Conference on Decision and Control*, Honolulu, Hawaii, 1990, pp 2080–2084
49. S.S. Sastry, A. Isidori: IEEE Trans. Auto. Contr. **34**(11), 1123–1131 (1989)
50. P. Celka: Physica D **104**, 127–147 (1997)
51. J.D. Farmer: Physica D **4**, 366–393 (1982)

Chaotification of the Fuzzy Hyperbolic Model

Huaguang Zhang, Zhiliang Wang, and Derong Liu

Abstract. In this chapter, the problem of chaotifying the continuous-time fuzzy hyperbolic model (FHM) is studied. We first use impulsive and nonlinear feedback control methods to chaotify the FHM and we show that the chaos produced by the present methods satisfy the three criteria of Devaney. We then design a controller based on inverse optimal control and adaptive parameter tuning methods to chaotify the FHM by tracking the dynamics of a chaotic system. Computer simulation results show that for any initial value the FHM can track a chaotic system asymptotically.

1 Introduction

As an intersection of chaos theory and control theory, chaos control has attracted more and more attention since the seminal work by Ott et al. [1]. In the past decade, many researchers have studied control methods with the purpose of either reducing "bad" chaos or introducing "good" chaos [2]. Due to its great potential in nontraditional applications such as those found in the fields of physical, chemical, mechanical, electrical, optical, and particularly, biological and medical systems [3–5], making a nonchaotic system chaotic or maintaining existing chaos, known as "chaotification" or "anticontrol," has attracted increasing attention in recent years. The process of chaos control is now understood as a transition from chaos to order and sometimes from order to chaos, depending on the purposes of different applications. Studies have shown that discrete maps can be chaotified in the sense of Devaney or Li–York by a state feedback controller with a control sequence of uniformly bounded gain designed to make all Lyapunov exponents of the controlled system strictly positive or arbitrarily assigned [6–11]. Even though there are also some research works showing that certain continuous stable systems can be chaotified [12–18], the problems of how to chaotify unlinearizable nonlinear systems as well as how to make nonchaotic systems produce expected chaotic states, are still open problems.

In this chapter, we make an effort to solve the above problems. Instead of directly chaotifying a nonlinear system, we will first design a controller to chaotify a fuzzy hyperbolic model (FHM), which is used to describe the original nonlinear system. After that, we impose the designed controller on the original nonlinear system. Due to the universal approximation property of

H. Zhang et al.: *Chaotification of the Fuzzy Hyperbolic Model*, StudFuzz **187**, 229–257 (2006)
www.springerlink.com

the FHM, it is reasonable to believe that chaos will emerge when controlling the original nonlinear system. There are two approaches used in this chapter for the chaotification of the FHM. First the impulsive control method will be investigated, which is shown to produce chaos in the sense of Devaney when controlling the FHM. Next, inverse optimal control theory will be used to design a chaotifying controller for the same model. Finally, we apply the controllers designed for the chaotification of the FHM to the chaotification of a piecewise-linear continuous system to show the effectiveness of the present chaotifying controllers.

2 Chaotification of the Fuzzy Hyperbolic Model by Impulsive Control Method

Commonly speaking, a system's trajectory can have three kinds of states: convergent, periodic, and divergent. To make a continuous nonchaotic system chaotified, one intuitive idea is to make the system's trajectory contain irregular jumps or oscillations. Furthermore, if the system's trajectory with such jumps or oscillations can be kept in a bounded region, it is reasonable to infer that the system maybe in a chaotic state. To verify the inference, we must prove that the system indeed does move in a chaotic manner. In this section, we take Devaney's definition of chaos as our theoretical foundation and use the FHM to test the above idea.

2.1 Preliminaries

First, we review Devaney's definition of chaos.

Definition 1. (cf. [19]) *A map* $\phi : S \to S$, *where* S *is a set, is chaotic if*

(i) ϕ *has sensitive dependence on initial conditions, in that for any* $x \in S$ *and any neighborhood* N *of* x *in* S, *there exists a* $\delta > 0$ *such that* $\|\phi^m(x) - \phi^m(y)\| > \delta$ *for some* $y \in N$ *and* $m > 0$, *where* ϕ^m *is the mth-order iteration of* ϕ, *i.e.,* $\phi^m \overset{\Delta}{=} \phi \circ \phi \circ \cdots \circ \phi$ (*m times*).

(ii) ϕ *is topological transitive, in that for any pair of subsets* $U, V \subset S$, *there exists an integer* $m > 0$ *such that* $\phi^m(U) \cap V \neq \varnothing$.

(iii) *the periodic points of* ϕ *are dense in* S.

Remark 1. Definition 1 is for discrete-time systems. For continuous-time systems, we need to construct Poincaré section to get a Poincaré map. Details will be provided later.

In this chapter, we use FHM to refer to the fuzzy hyperbolic model as well as the generalized fuzzy hyperbolic model introduced in previous chapters. We will concern ourselves primarily with the kind of FHM defined as follows.

Definition 2. *Given a plant with n input variables $x = (x_1, \ldots, x_n)^{\mathrm{T}}$ and the output $\dot{x}$, we call the fuzzy rule base "hyperbolic type fuzzy rule base" if it satisfies the following conditions:*

(i)   Every fuzzy rule has the following form:

R^l: IF x_1 is F_{x_1}, x_2 is F_{x_2}, ..., and x_n is F_{x_n},
 THEN $\dot{x}_j = \pm c_{x_1} \pm c_{x_2} \pm \cdots \pm c_{x_n}$, $j = 1, \ldots, n$, $l = 1, \ldots, 2^n$,

where F_{x_i} $(i = 1, \ldots, n)$ are fuzzy sets of x_i, which include P_{x_i} (positive) and N_{x_i} (negative), c_{x_i} $(i = 1, \ldots, n)$ are positive constants corresponding to F_{x_i}, and $\pm$ stands for either the plus or the minus sign. The actual signs in the THEN part are determined in the following manner: If in the IF part the term characterizing F_{x_i} is P_{x_i}, then in the THEN part c_{x_i} appears with a plus sign; otherwise, c_{x_i} appears with a minus sign.

(ii) The constant terms c_{x_i} in the THEN part correspond to F_{x_i} in the IF part; i.e., if there is an F_{x_i} term in the IF part, c_{x_i} must appear in the THEN part. Otherwise, c_{x_i} does not appear in the THEN part.

(iii) There are 2^n fuzzy rules in the rule base; i.e., there are a total of 2^n possible P_{x_i} and N_{x_i} combinations of input variables in the IF part, or a total of 2^n sign combinations of constants in the THEN part.

The following theorem explains how a fuzzy hyperbolic model is constructed.

Theorem 1. *(cf. [20]) Given a hyperbolic type rule base, if we define the membership function of P_{x_i} and N_{x_i} as*

$$\mu_{P_{x_i}}(x_i) = e^{-\frac{1}{2}(x_i - k_{x_i})^2}, \qquad \mu_{N_{x_i}}(x_i) = e^{-\frac{1}{2}(x_i + k_{x_i})^2},$$

where $k_{x_i} > 0$, then we can always derive the following model:

$$\dot{x} = A \tanh(Kx), \tag{1}$$

where $K = \mathrm{diag}[k_{x_1}, \ldots, k_{x_n}]$ and A is a constant matrix.

Definition 3. *(cf. [21]) A matrix A is said to be a Hurwitz diagonally stable matrix if there exits a diagonal matrix $Q > 0$ such that $A^{\mathrm{T}}Q + QA < 0$.*

We use the notation $Q > 0$ to indicate that Q is a positive definite matrix and we use $Q < 0$ to indicate that Q is a negative definite.

Lemma 1. *(cf. [21]) For system (1), if A is a Hurwitz diagonally stable matrix, then the equilibrium $x = 0$ of (1) is globally and asymptotically stable.*

2.2 Stabilization of the Fuzzy Hyperbolic Model

Consider the following control system:

$$\dot{x} = Af(x) + Bu \,, \tag{2}$$

where $x \in \mathbb{R}^n$, $f(x) = \tanh(Kx)$ is a n-dimensional vector function, $u = u(x(t)) \in \mathbb{R}^m$ is a vector function of state x, and $A \in \mathbb{R}^{n \times n}$ and $B \in \mathbb{R}^{n \times m}$ are both constant matrices.

Assumption 1 (A, B) is completely controllable.

Under this assumption, we have the following lemma.

Lemma 2. *For system* (2), *if the controller u is chosen as*

$$u(x(t)) = L \, \tanh(Kx(t)) \,, \tag{3}$$

where $L \in \mathbb{R}^{m \times n}$ is a constant matrix such that $(A + BL)$ is a Hurwitz diagonally stable matrix, then the origin of the closed-loop system (2) *is globally asymptotically stable.*

Proof. The result can directly be obtained from Lemma 1 by replacing A in Lemma 1 with $A + BL$.

Remark 2. System (1) and the closed-loop system (2) can be regarded as recurrent neural networks. They are also special cases of Lur'e systems [22].

Remark 3. Lemma 2 implies that solutions of the control system (2) are defined in a bounded region $D \subset \mathbb{R}^n$.

Remark 4. If we let $\tilde{A} = A + BL$, system (2) becomes

$$\dot{x} = \tilde{A}f(x) \,. \tag{4}$$

Because of the form of $f(x)$, there exits a constant γ such that for all different $x, y \in D$, the following inequality holds:

$$\|f(y) - f(x)\| \leq \gamma \|y - x\| \,. \tag{5}$$

That is to say, $f(x)$ is a Lipschitz function over D. From the theory of ordinary differential equations, we know that system (2) has a unique continuous solution $\phi(t, t_0, x_0)$ through a given initial point, (t_0, x_0), which is also continuously dependent on x_0 [23, 24].

2.3 Chaotification of the Closed-Loop Fuzzy Hyperbolic Model

Now, let us consider the following impulsive control system:

$$\begin{cases} \dot{x} = \tilde{A}f(x), & t \neq \tau_k \\ \Delta x = I_k(x(t)), & t = \tau_k, \quad k = 1, 2, 3, \ldots \\ x(t_0+) = x_0, & t_0 \geq 0 \end{cases} \qquad (6)$$

where $\tilde{A}$ is defined in (4), $I_k(x(t))$, $k = 1, 2, 3, \ldots$, is impulsive control defined in $D \subset \mathbb{R}^n$, $t_0 = \tau_0 < \tau_1 < \cdots < \tau_k$, $\tau_k - \tau_{k-1} = T$, $\lim \tau_k {}_{k \to +\infty} = +\infty$, $\Delta x = x(\tau_k+) - x(\tau_k-)$, and $x(\tau_k+)$ and $x(\tau_k-)$ are the right and left limit of $x(t)$ at $t = \tau_k$.

For system (6), we have the following lemma.

Lemma 3. *If $I_k(x(t))$ is chosen as*

$$I_k(x(t)) = y_k - x(t-), \quad t = \tau_k, \quad k = 1, 2, 3, \ldots \qquad (7)$$

where $y_k = g(y_{k-1})$ and $g \colon Y \to Y$, $Y \subset D$. Then, when $g(\cdot)$ is chaotic and satisfies the Devaney's three criteria, the control system (6) is also chaotic and satisfies the Devaney's criteria.

Proof. First, it should be emphasized that for system (6) to display chaotic dynamics, its phase space must be a finite region, i.e., there exists an $M > 0$ such that the trajectory of system (6), $\phi(t, t_0, x_0)$, satisfies $\|\phi(t, t_0, x_0)\| \leq M$ for all t in the domain. In fact, for $t \in (\tau_0, \tau_1)$, system (6) is under the action of $\tilde{A}f(x)$. By Lemma 1, its trajectory, $\phi(t, t_0+, x_0)$, is asymptotically stable. When $t = \tau_1$, $\tilde{A}f(x)$ turns off and $I_1(x(\tau_1))$ acts on system (6). The trajectory of system (6) jumps from $\phi(\tau_1-, t_0+, x_0)$ to x_0^1. We denote the trajectory at each instant τ_k as x_0^k, $k = 0, 1, 2, \ldots$, where $x_0^0 = x_0$. Since $I_1(x(\tau_1))$ is a finite impulse, x_0^1 is also finite. For $t \in (\tau_1, \tau_2)$, $I_1(x(\tau_1))$ turns off and $Af(x)$ turns on with initial point (τ_1+, x_0^1). The trajectory of system (6) in this time span is also asymptotically stable, denoted by $\phi(t, \tau_1+, x_0^1)$. The analysis in other time spans, (τ_k, τ_{k+1}), $k = 2, 3, 4, \ldots$, are the same as that in (τ_1, τ_2), and situations at other time instants, τ_k, $k = 2, 3, 4, \ldots$, are analogous to that at τ_1. It is easy to see that although the trajectory may be discontinuous at τ_k, $k = 1, 2, 3, \ldots$, the whole trajectory is indeed in a bounded region.

In the following we will prove that system (6) is chaotic and satisfies Devaney's three criteria.

To prove that a continuous system is chaotic, one method is to show that its Poincaré map is chaotic [24]. Suppose that $\phi(t, t_0+, x_0)$ is a solution of system (4) with initial value $x(t_0+) = x_0$. According to Remark 4, $\phi^{-1}(t, t_0+, x_0)$ exists. Define a set of Poincaré sections S_k as

$$S_k = \{(t, x) \mid x \in D, t = \tau_k\}, \quad k = 1, 2, 3, \ldots.$$

The Poincaré map is defined as

$$P: V \to V, \qquad P = \psi \circ g \circ \psi^{-1},$$

where $V = \psi(Y)$ and

$$\psi(x) = \phi(\tau_k + T, \tau_k, x) = \phi(\tau_0 + T, \tau_0, x).$$

We will first prove that P is extremely sensitive to initial values, i.e., for any proper constant $\varepsilon > 0$ and any two points $x_0, \bar{x}_0 \in V$, there exists an $N > 0$, such that

$$\left\| P^N(x_0) - P^N(\bar{x}_0) \right\| > \varepsilon. \tag{8}$$

Since

$$P^N = \underbrace{(\psi \circ g \circ \psi^{-1}) \circ \cdots \circ (\psi \circ g \circ \psi^{-1})}_{N\text{th}-\text{order iteration of } P} = \psi \circ g^N \circ \psi^{-1}, \tag{9}$$

to prove that (8) holds is equivalent to prove that

$$\left\| \psi(g^N(y_0)) - \psi(g^N(\bar{y}_0)) \right\| > \varepsilon \tag{10}$$

holds, where $y_0 = \psi^{-1}(x_0)$ and $\bar{y}_0 = \psi^{-1}(\bar{x}_0)$.

If (10) does not hold, i.e., for all $n \geq 0$, there exists an $\varepsilon_0 > 0$ such that

$$\left\| \psi(g^n(y_0)) - \psi(g^n(\bar{y}_0)) \right\| \leq \varepsilon_0,$$

from the fact that $\psi(x)$ is continuous we know that for the above ε_0, there exists a δ_0 such that

$$\left\| g^n(y_0) - g^n(\bar{y}_0) \right\| < \delta_0 \quad \text{for any } n \geq 0.$$

But this contradicts the fact that g is a chaotic map satisfying Devaney's definition. Therefore, P satisfies the first criterion of Devaney.

Next, we show that P is topologically transitive; that is, for any pair of subsets $E, F \subset V$, there exists an integer $N > 0$ such that

$$P^N(E) \cap F \neq \varnothing. \tag{11}$$

It is known that for any two subsets $\psi^{-1}(E), \psi^{-1}(F) \subset Y$, there exists a number $N > 0$ such that

$$g^N[\psi^{-1}(E)] \cap \psi^{-1}(F) \neq \varnothing. \tag{12}$$

Acting on both sides of (12) by ψ, we get

$$\psi\{g^N[\psi^{-1}(E)]\} \cap F \neq \varnothing.$$

Noticing (9), we therefore proved (11).

Finally, we will prove that the periodic points of P are dense in V. Denote the set of periodic points of a map f as $\overline{\text{Per}(f)}$. Since $\overline{\text{Per}(g)} = Y$ and ψ is one to one, we have $\overline{\text{Per}(P)} = V$.

Thus, we have proved the lemma.

Summarizing the results above, we have the following theorem.

Theorem 2. *For the following system*

$$\begin{cases} \dot{x} = Af(x) + Bu, & t \neq \tau_k \\ \Delta x = I_k(x(t)), & t = \tau_k, \quad k = 1, 2, 3, \ldots \\ x(t_0+) = x_0, & t_0 \geq 0 \end{cases} \tag{13}$$

if u is designed according to (3) and $I_k(x(t))$ is chosen as in (7), the system (13) will display chaotic dynamics and satisfies the three criteria of Devaney.

Remark 5. The conclusion of Theorem 2 can also be kept if the impulsive control is chosen as $I_k(x(t)) = h(y_k) - x(t-)$, where $h : Y \rightarrow Y$ is a homeomorphism. A special case is to choose h as a constant diagonal matrix $\Lambda = \text{diag}[\lambda_1, \ldots, \lambda_n]$ with $\lambda_i > 0, i = 1, \ldots, n$, which means that we can chaotify the original system with arbitrarily small impulsive energy.

2.4 Simulation Results

Suppose that we have the following fuzzy rule base:

IF x_1 is P_{x_1} and x_2 is P_{x_2}, THEN $\dot{x}_3 = C_{x1} + C_{x2}$;
IF x_1 is N_{x_1} and x_2 is P_{x_2}, THEN $\dot{x}_3 = -C_{x1} + C_{x2}$;
IF x_1 is P_{x_1} and x_2 is N_{x_2}, THEN $\dot{x}_3 = C_{x1} - C_{x2}$;
IF x_1 is N_{x_1} and x_2 is N_{x_2}, THEN $\dot{x}_3 = -C_{x1} - C_{x2}$;

IF x_1 is P_{x_1} and x_3 is P_{x_3}, THEN $\dot{x}_2 = C_{x1} + C_{x3}$;
IF x_1 is N_{x_1} and x_3 is P_{x_3}, THEN $\dot{x}_2 = -C_{x1} + C_{x3}$;
IF x_1 is P_{x_1} and x_3 is N_{x_3}, THEN $\dot{x}_2 = C_{x1} - C_{x3}$;
IF x_1 is N_{x_1} and x_3 is N_{x_3}, THEN $\dot{x}_2 = -C_{x1} - C_{x3}$;

IF x_2 is P_{x_2} and x_3 is P_{x_3}, THEN $\dot{x}_1 = C_{x2} + C_{x3}$;
IF x_2 is N_{x_2} and x_3 is P_{x_3}, THEN $\dot{x}_1 = -C_{x2} + C_{x3}$;
IF x_2 is P_{x_2} and x_3 is N_{x_3}, THEN $\dot{x}_1 = C_{x2} - C_{x3}$;
IF x_2 is N_{x_2} and x_3 is N_{x_3}, THEN $\dot{x}_1 = -C_{x2} - C_{x3}$.

Here, we choose fuzzy membership functions P_{x_i} and N_{x_i} as follows:

$$\mu_{P_{x_i}}(x) = e^{-\frac{1}{2}(x_i - k_i)^2}, \qquad \mu_{N_{x_i}}(x) = e^{-\frac{1}{2}(x_i + k_i)^2}.$$

Then we have the following three-dimensional model:

$$\dot{x} = Af(x) = A\tanh(Kx) \tag{14}$$

where $x = [x_1 \ x_2 \ x_3]^{\mathrm{T}}$, $A = \begin{bmatrix} 0 & C_{x2} & C_{x3} \\ C_{x1} & 0 & C_{x3} \\ C_{x1} & C_{x2} & 0 \end{bmatrix} = \begin{bmatrix} 0 & 1 & 3 \\ 2 & 0 & 3 \\ 2 & 1 & 0 \end{bmatrix}$, and $K = $

$\text{diag}\,[k_1 \ k_2 \ k_3] = \text{diag}\,[2 \ 3 \ 1]$. According to Theorem 2, the control system is

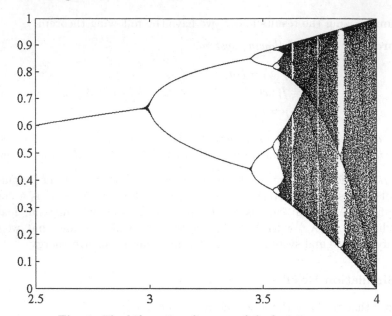

Fig. 1. The bifurcation diagram of the logistic map

$$\begin{cases} \dot{x} = (A + BL)f(x), & t \neq \tau_k \\ \Delta x = I_k(x(t)), & t = \tau_k, \quad k = 1, 2, 3, \ldots \\ x(t_0+) = x_0, & t \geq 0 . \end{cases} \tag{15}$$

We select $B = [1 \ 1 \ 1]^{\mathrm{T}}$ and $L = [-2 \ -1 \ -2]$. It is easy to verify that matrix $A + BL$ is Hurwitz diagonally stable by choosing $Q = \mathrm{diag}\,[q_1 \ q_2 \ q_3] = \mathrm{diag}\,[2 \ 2 \ 2]$. The initial value is chosen as $x_0 = [5 \ 2 \ -1]^{\mathrm{T}}$ and the impulsive control is

$$I_k(x) = \Lambda g(y_{k-1}) - x(t-),$$

where $\Lambda = \mathrm{diag}[\lambda_1 \ \lambda_2 \ \lambda_3]$ and $g(y_{k-1}) = [g_1(y_{k-1}^1) \ g_2(y_{k-1}^2) \ g_3(y_{k-1}^3)]^{\mathrm{T}}$ with $g_i(\cdot)$ the logistic map:

$$y_k^i = g_i(y_{k-1}^i) = a y_{k-1}^i (1 - y_{k-1}^i), \quad i = 1, 2, 3 .$$

When $a = 4.0$, the logistic map is chaotic; see Fig. 1 for an illustration. Without impulsive control, the following system

$$\dot{x} = (A + BL)f(x) \tag{16}$$

is globally and asymptotically stable; its trajectories are shown in Fig. 2.

With impulsive control and $T = \tau_k - \tau_{k-1} = 0.01$, from Figs. 3–6, we can see that the control system is chaotic for different Λ. These results verify the claim of Theorem 2 and Remark 5.

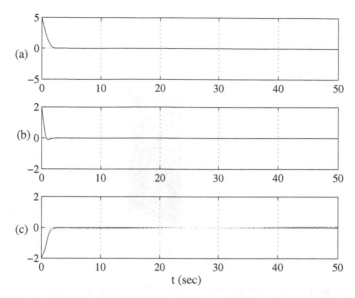

Fig. 2. State trajectories of (16): (a) the state x_1; (b) the state x_2; (c) the state x_3

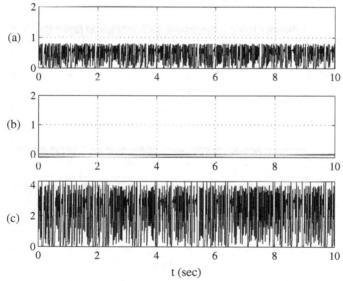

Fig. 3. State trajectories of system (15) when $\Lambda = \mathrm{diag}[0.8 \quad 0 \quad 4]$: (a) the state x_1; (b) the state x_2; (c) the state x_3

3 Chaotification of the Fuzzy Hyperbolic Model by Inverse Optimal Control Method

We have shown that with impulsive control a FHM can be chaotified. But in some applications, we want to control a nonchaotic system to produce

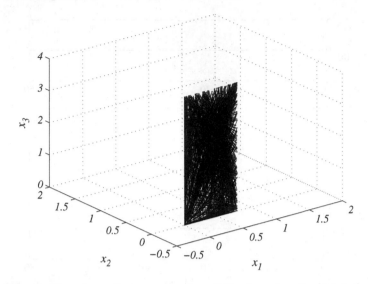

Fig. 4. The phase diagram of system (15) when $\Lambda = \text{diag}[0.8\ 0\ 4]$

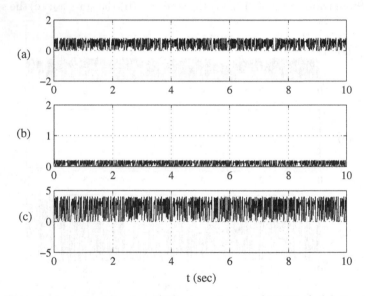

Fig. 5. State trajectories of system (15) when $\Lambda = \text{diag}[0.8\ 0.2\ 4]$: (a) the state x_1; (b) the state x_2; (c) the state x_3

specified chaos. It is reasonable to believe that to achieve this goal, the controller must have more complex structure than that proposed in the previous section, and the amplitude of control signals should also be increased. However, due to actuator saturation, the power and energy of control actions should be always limited to certain range in practice. The goal of this

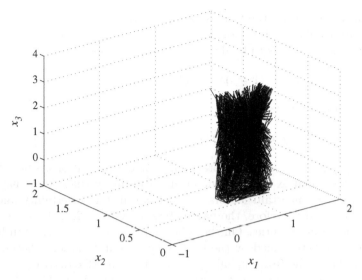

Fig. 6. The phase diagram of system (15) when $\Lambda = \mathrm{diag}[0.8\ 0.2\ 4]$

section is to develop a method that utilizes finite energy to make a nonchaotic system produce specified chaos. In this section, we develop a method under the framework of inverse optimal control theory to design such a controller for the fuzzy hyperbolic model.

3.1 Problem Formulation

Suppose that there exists a chaotic system having the following form:

$$\dot{x}_r = f_r(x_r), \quad x_r \in \mathbb{R}^n, \quad f_r(\cdot) \in \mathbb{R}^n . \tag{17}$$

Consider the following fuzzy hyperbolic model

$$\dot{x} = Af(x) = A \tanh(Kx) , \tag{18}$$

where x is the state and $A = [a_{ij}]_{n \times n}$ is a matrix. Because of the hyperbolic tangent form of $f(x)$, we know that $f^{\mathrm{T}}(x)x \geq 0$ for all x, $f(x) = 0$ only at $x = 0$, and $\lim\limits_{\|x\|_2 \to \infty} f^{\mathrm{T}}(x)x = +\infty$. Therefore, there exist positive constants γ_1 and γ_2 such that $\gamma_1 \|x\|_2^2 \leq f^{\mathrm{T}}(x)x \leq \gamma_2 \|x\|_2^2$ [24].

The idea here is to design a controller u such that the controlled fuzzy hyperbolic model

$$\dot{x} = Af(x) + u \tag{19}$$

can track the dynamics of system (17), i.e.,

$$\lim_{t \to \infty} \|e(t)\|_2 = \lim_{t \to \infty} \|x(t) - x_r(t)\|_2 = 0 .$$

Suppose that u has the form of $u(x,t) = v(x) + w(x,t)$, where $v(x) = \Lambda x$ is the state feedback, $\Lambda = -\text{diag}[\lambda_i]_{n \times n} \in \mathbb{R}^{n \times n}$ with λ_i $(i = 1, \ldots, n)$ real numbers, and $w(x,t)$ is to be designed. Then, (19) becomes

$$\dot{x} = \Lambda x + A f(x) + w(x,t) . \tag{20}$$

From (17) and (20), we obtain

$$\dot{e} = \Lambda x + A f(x) + w(x,t) - f_r(x_r) , \tag{21}$$

where $e = [e_1, e_2, \ldots, e_n]^T$. In practical control applications, since Λ is the state feedback matrix and A is determined by fuzzy rules, they may be affected by some uncertain factors such as parameter uncertainty and modelling errors, and therefore, the parameters of system (20), Λ and A, would include some uncertainties. On the other hand, k_i $(i = 1, \ldots, n)$ can be fixed since they are determined by the fuzzy membership functions chosen during modeling. Once the fuzzy membership functions are fixed, k_i $(i, = 1, \ldots, n)$ are determined. In this section, we assume Λ and A are tunable, and k_i $(i = 1, \ldots, n)$ are constants.

For system (20) to track the system (17), the following solvability assumption is needed [26].

Assumption 2. There exist functions $\rho(t)$ and $\alpha(t)$ such that

$$\dot{\rho}(t) = \Lambda_0 \rho(t) + A_0 f(\rho(t)) + \alpha(t) ,$$
$$\rho(t) = x_r(t) ,$$

where $A_0 = [a_{0ij}]_{n \times n}$ and $\Lambda_0 = -\text{diag}[\lambda_{0i}]_{n \times n}$ are known constant matrices and λ_{0i} are positive real numbers for $i = 1, \ldots, n$.

From (17) and Assumption 2, the following equation can then be derived:

$$\Lambda_0 x_r + A_0 f(x_r) + \alpha(t) = f_r(x_r) . \tag{22}$$

Substituting (22) into (21), we have

$$\dot{e} = \Lambda_0 e + A_0 [f(e + x_r) - f(x_r)] + [w - \alpha(t)] + \tilde{\Lambda} x + \tilde{A} f(x) , \tag{23}$$

where $\tilde{\Lambda} = \Lambda - \Lambda_0 = -\text{diag}[\lambda_i - \lambda_{0i}]_{n \times n}$ and $\tilde{A} = A - A_0 = [a_{ij} - a_{0ij}]_{n \times n} = [\tilde{a}_{ij}]_{n \times n}$. Let $\phi(e, x_r) = f(e + x_r) - f(x_r)$ and $\tilde{u} = w - \alpha(t)$. Then (23) can be rewritten as

$$\dot{e} = \Lambda_0 e + A_0 \phi(e, x_r) + \tilde{\Lambda} x + \tilde{A} f(x) + \tilde{u} . \tag{24}$$

Remark 6. It is clear that $\phi(e, x_r) = 0$, if $e = 0$. Moreover, $f(e + x_r) = A \tanh(K(e + x_r))$ is monotonically increasing (or decreasing) for each component e_i of e. Since $e_i > 0$ (or $e_i < 0$) implies that $e_i + x_{ri} > x_{ri}$ (or $e_i + x_{ri} < x_{ri}$) for all x_{ri}, $f_i(e_i + x_{ri}) > f_i(x_{ri})$ (or $f_i(e_i + x_{ri}) < f_i(x_{ri})$).

This means that $\phi^{\mathrm{T}}(e, x_r)e = (f(e + x_r) - f(x_r))^{\mathrm{T}}e > 0$. Therefore, there exist positive constants γ_1, γ_2, and L_ϕ such that

$$\gamma_1\|e\|_2^2 \leq \phi^{\mathrm{T}}(e, x_r)e \leq \gamma_2\|e\|_2^2 \tag{25}$$

and

$$\|\phi(e, x_r)\|_2 < L_\phi\|e\|_2 . \tag{26}$$

Therefore, $\phi(e, x_r)$ is Lipschitz with respect to e.

3.2 Controller Design

We first state the following lemma that is required in our controller design.

Lemma 4. *For matrices X, $Y \in \mathbb{R}^{n \times k}$ and $Q \in \mathbb{R}^{n \times n}$ with $Q = Q^{\mathrm{T}} > 0$, the following inequality holds:*

$$X^{\mathrm{T}}Y + Y^{\mathrm{T}}X \leq X^{\mathrm{T}}QX + Y^{\mathrm{T}}Q^{-1}Y .$$

Proof. Set $G_1 = Q^{-1/2}Y - Q^{1/2}X$, then the lemma can be obtained directly from $G_1^{\mathrm{T}}G_1 \geq 0$.

Theorem 3. *For system (18), if the controller is chosen as*

$$\begin{aligned} w = &- (A_0^{\mathrm{T}}A_0 + I)\phi(e, x_r) - A_0 x_r \\ &- A_0 f(x_r) + f_r(x_r) \end{aligned}$$

and the parameter adaptive update laws are chosen as

$$\begin{aligned} \dot{\lambda}_i &= -\phi_i(e, x_r)x_i , \\ \dot{a}_{ij} &= -\phi_i(e, x_r)f_j(x) , \end{aligned} \tag{27}$$

for $i = 1, \ldots, n$, $j = 1, \ldots, n$, where $\phi_i(e, x_r)$ and x_i are the ith component of $\phi(e, x_r)$ and x, respectively, and $f_j(x)$ is the jth component of $f(x)$, then the system (24) is globally asymptotically stable, i.e.,

$$\lim_{t \to \infty} \|e(t)\|_2 = 0 .$$

Proof. Using (22), we can get $\alpha(t) = f_r(x_r) - A_0 x_r - A_0 f(x_r)$. Thus,

$$\tilde{u} = w(x, t) - \alpha(t) = -(A_0^{\mathrm{T}}A_0 + I)\phi(e, x_r) . \tag{28}$$

Define

$$\begin{aligned} \varepsilon &\triangleq \left[e^{\mathrm{T}}(t), \ \theta^{\mathrm{T}}(t)\right]^{\mathrm{T}} \\ &= \left[e^{\mathrm{T}}(t), \ \tilde{\lambda}_1(t), \ \ldots, \tilde{\lambda}_n(t), \ \tilde{a}_{11}(t), \ \ldots, \tilde{a}_{1n}(t), \tilde{a}_{21}(t), \ldots, \tilde{a}_{nn}(t)\right]^{\mathrm{T}}. \end{aligned}$$

We choose a Lyapunov function as follows:

$$V(\varepsilon) = \sum_{i=1}^{n} \int_{0}^{e_i} \phi_i(\eta, x_r) \mathrm{d}\eta_i + \frac{1}{2} \sum_{i=1}^{n} \tilde{\lambda}_i^2 + \frac{1}{2} \sum_{i=1,j=1}^{n} \tilde{a}_{ij}^2 \,, \qquad (29)$$

where η_i is the ith element of η.

Because of (25), we know that $V(\varepsilon)$ is radially unbounded, i.e., $V(\varepsilon) > 0$ for all ε and $V(\varepsilon) \to \infty$ as $\|\varepsilon\|_2 \to \infty$. Along the solutions of (23), the time-derivative of $V(\varepsilon)$ is derived as follows:

$$\dot{V}(\varepsilon) = \phi^{\mathrm{T}}(e, x_r)(\Lambda_0 e + A_0 \phi(e, x_r) + \tilde{\Lambda} x + \tilde{A} f(x) + \tilde{u})$$
$$+ \sum_{i=1}^{n} \tilde{\lambda}_i \dot{\lambda}_i + \sum_{i=1,j=1}^{n} \tilde{a}_{ij} \dot{a}_{ij}$$
$$= \phi^{\mathrm{T}}(e, x_r) \Lambda_0 e + \phi^{\mathrm{T}}(e, x_r) A_0 \phi(e, x_r) + \phi^{\mathrm{T}}(e, x_r)(\tilde{\Lambda} x + \tilde{A} f(x))$$
$$+ \phi^{\mathrm{T}}(e, x_r) \tilde{u} + \sum_{i=1}^{n} \tilde{\lambda}_i \dot{\lambda}_i + \sum_{i=1,j=1}^{n} \tilde{a}_{ij} \dot{a}_{ij}$$
$$\triangleq L_{\tilde{f}} V + (L_g V) \tilde{u} \,, \qquad (30)$$

where

$$L_{\tilde{f}} V \triangleq \phi^{\mathrm{T}}(e, x_r) \Lambda_0 e + \phi^{\mathrm{T}}(e, x_r) A_0 \phi(e, x_r) + \phi^{\mathrm{T}}(e, x_r)(\tilde{\Lambda} x + \tilde{A} f(x))$$
$$+ \sum_{i=1}^{n} \tilde{\lambda}_i \dot{\lambda}_i + \sum_{i=1,j=1}^{n} \tilde{a}_{ij} \dot{a}_{ij} \,, \qquad (31)$$

and

$$L_g V \triangleq \phi^{\mathrm{T}}(e, x_r) \,. \qquad (32)$$

Applying Lemma 4 with $Q = I$, we obtain

$$\dot{V}(\varepsilon) = \phi^{\mathrm{T}}(e, x_r) \Lambda_0 e + \frac{1}{2} \phi^{\mathrm{T}}(e, x_r) \phi(e, x_r) + \frac{1}{2} \phi^{\mathrm{T}}(e, x_r) A_0^{\mathrm{T}} A_0 \phi(e, x_r)$$
$$+ \phi^{\mathrm{T}}(e, x_r) \tilde{u} + \phi^{\mathrm{T}}(e, x_r)(\tilde{\Lambda}(t) x + \tilde{A}(t) f(x))$$
$$+ \sum_{i=1}^{n} \tilde{\lambda}_i \dot{\lambda}_i + \sum_{i=1,j=1}^{n} \tilde{a}_{ij} \dot{a}_{ij}$$
$$= \phi^{\mathrm{T}}(e, x_r) \Lambda_0 e + \frac{1}{2} \phi^{\mathrm{T}}(e, x_r) \phi(e, x_r) + \frac{1}{2} \phi^{\mathrm{T}}(e, x_r) A_0^{\mathrm{T}} A_0 \phi(e, x_r)$$
$$+ \phi^{\mathrm{T}}(e, x_r) \tilde{u} + \sum_{i=1}^{n} \tilde{\lambda}_i (\dot{\lambda}_i + \phi_i(e, x_r) x_i)$$
$$+ \sum_{i=1,j=1}^{n} \tilde{a}_{ij} (\dot{a}_{ij} + \phi_i(e, x_r) f_j(x)) \,.$$

Substituting (27) into the above equality and using inequalities (25) and (26), we obtain

$$\dot{V}(\varepsilon) \leq -\left(\lambda^{*}\gamma_{1} - \frac{1}{2}L_{\phi}^{2}\right)\|e\|_{2}^{2} + \frac{1}{2}\phi^{\mathrm{T}}(e, x_{r})A_{0}^{\mathrm{T}}A_{0}\phi(e, x_{r}) + \phi^{\mathrm{T}}(e, x_{r})\tilde{u}\ ,$$

$$(33)$$

where $\lambda^{*} = \min\{\lambda_{0i}; i = 1,\ldots, n\}$.

If we let $R^{-1}(\varepsilon) = \frac{1}{\beta}(A_{0}^{\mathrm{T}}A_{0} + I)$, where $\beta \geq 2$ is a constant, we have

$$-\beta R^{-1}(\varepsilon)(L_{g}V)^{\mathrm{T}} = \tilde{u} = -(A_{0}^{\mathrm{T}}A_{0} + I)\phi(e, x_{r})\ . \qquad (34)$$

In general, $R^{-1}(\varepsilon)$ is a function of ε, but for our purpose it is chosen as a constant matrix. The motivation for this operation will be seen from the inverse optimization problem to be discussed later.

Substituting (28) into (33), we get

$$\dot{V}(\varepsilon) \leq -\left(\lambda^{*}\gamma_{1} - \frac{1}{2}L_{\phi}^{2}\right)\|e\|_{2}^{2} - \frac{1}{2}\left\|A_{0}^{\mathrm{T}}A_{0}\right\|L_{\phi}^{2}\|e\|_{2}^{2} - L_{\phi}^{2}\|e\|_{2}^{2}$$

$$= -\left(\lambda^{*}\gamma_{1} + \frac{1}{2}\left\|A_{0}^{\mathrm{T}}A_{0}\right\|L_{\phi}^{2} + \frac{1}{2}L_{\phi}^{2}\right)\|e\|_{2}^{2} \leq 0\ . \qquad (35)$$

By LaSalle's invariance principle [27], we know that the invariant set of (24) and (27), I_{S}, has the following form:

$$I_{S} = \{\varepsilon \mid \dot{V} = 0\} = \{\varepsilon|(\mathbf{0}, \theta^{\mathrm{T}})\}\ .$$

This completes the proof of the theorem.

To avoid the heavy computation burden that the Hamilton–Jacobi–Bellman (HJB) equation imposes upon the problem of optimal control of nonlinear systems, inverse optimal control theory has been developed recently. The difference between the traditional optimal control and inverse optimal control is that, the former seeks a controller that minimizes a given cost, while the latter is concerned with finding a controller that minimizes some "meaningful" cost.

According to literature [26, 28], for the inverse optimal control problem of system (24) to be solvable under the control of (27) and (28), we need to find a positive real-valued function $R(\varepsilon)$ and a positive definite function $l(\varepsilon)$ such that the following cost functional

$$J(\tilde{u}) = \lim_{t \to \infty}\left\{2\beta V(\varepsilon(\tau)) + \int_{0}^{\mathrm{T}}(l(\varepsilon(\tau))\right.$$

$$\left. + \tilde{u}(\tau)^{\mathrm{T}}R(\varepsilon(\tau))\tilde{u}(\tau))\,\mathrm{d}\tau\right\}, \quad \beta \geq 2 \qquad (36)$$

is minimized.

In the following, we will show that the controller we have designed can indeed solve the inverse optimal control problem.

Theorem 4. *If we choose*

$$l(\varepsilon) = -2\beta L_{\tilde{f}}V + 2\beta(L_g V)R^{-1}(\varepsilon)(L_g V)^{\mathrm{T}}$$
$$+ \beta(\beta - 2)((L_g V)R^{-1}(\varepsilon)(L_g V)^{\mathrm{T}})$$
$$= -2\beta L_{\tilde{f}}V + \beta(L_g V)(\beta R^{-1})(\varepsilon)(L_g V)^{\mathrm{T}} \tag{37}$$

and

$$R(\varepsilon) = \beta(A_0^{\mathrm{T}} A_0 + I)^{-1}, \quad \beta \geq 2, \tag{38}$$

the cost functional (36) *for system* (24) *under the parameter update laws* (27) *and the state feedback law* (28) *will be minimized.*

Proof. To prove this theorem, first we should prove that $R(\varepsilon)$ is positive and symmetric, and $l(\varepsilon)$ is radially unbounded, i.e., $R(\varepsilon) = R^{\mathrm{T}}(\varepsilon) > 0$ and $l(\varepsilon) > 0$ for all $\varepsilon \neq 0$ and $l(\varepsilon) \to +\infty$ as $\varepsilon \to \infty$. It is clear that $R(\varepsilon)$ chosen according to (38) satisfies this requirement. Using (27), (28), (31), (32), and (38), we obtain

$$l(\varepsilon) = 2\beta\phi^{\mathrm{T}}(e, x_r)\Lambda_0 e - 2\beta\phi^{\mathrm{T}}(e, x_r)A_0\phi(e, x_r)$$
$$+ \beta\phi^{\mathrm{T}}(e, x_r)(A_0^{\mathrm{T}} A_0 + I)\phi(e, x_r). \tag{39}$$

Applying Lemma 4 to the second term on the right-hand side of (39) with $X = \phi(e, x_r)$ and $Y = A_0\phi(e, x_r)$, we get

$$l(\varepsilon) \geq 2\beta\lambda^*\phi^{\mathrm{T}}(e, x_r)e - \beta\phi^{\mathrm{T}}(e, x_r)\phi(e, x_r)$$
$$- \beta\phi^{\mathrm{T}}(e, x_r)A_0^{\mathrm{T}} A_0\phi(e, x_r)$$
$$+ \beta\phi^{\mathrm{T}}(e, x_r)(A_0^{\mathrm{T}} A_0 + I)\phi(e, x_r)$$
$$\geq 2\beta\lambda^*\phi^{\mathrm{T}}(e, x_r)e.$$

This means that $l(\varepsilon)$ is radially unbounded. Substituting (34) into (30), we get

$$\dot{V} = L_{\tilde{f}}V + (L_g V)(-\beta R^{-1}(\varepsilon))(L_g V)^{\mathrm{T}}.$$

Multiplying it by -2β, we obtain

$$-2\beta\dot{V}(\varepsilon(t)) = -2\beta L_{\tilde{f}}V + 2\beta^2(L_g V)R^{-1}(\varepsilon)(L_g V)^{\mathrm{T}}.$$

Considering (34) and (37), we get

$$l(\varepsilon) + \tilde{u}^{\mathrm{T}} R(\varepsilon)\tilde{u} = -2\beta\dot{V}(\varepsilon(t)). \tag{40}$$

Substituting (40) into (36), we have

$$J(\tilde{u}) = \lim_{t \to \infty} \left\{ 2\beta V(\varepsilon(t)) + \int_0^t -2\beta\dot{V}(\varepsilon(\tau))\,\mathrm{d}\tau \right\}$$
$$= \lim_{t \to \infty} \left\{ 2\beta V(\varepsilon(t)) - 2\beta V(\varepsilon(t)) + 2\beta V(\varepsilon(0)) \right\}$$
$$= 2\beta V(\varepsilon(0)).$$

Thus, the minimum of the cost functional is $J(\tilde{u}) = 2\beta V(\varepsilon(0))$ for system (24) with the control law (28) and parameter update law (27). This completes the proof of the theorem.

We now summarize the results presented above in the next theorem.

Theorem 5. *If we choose feedback control law $u = v + w$ in which $v = \Lambda x$ is the linear state feedback with Λ a diagonal constant matrix and $w = -(A_0^{\mathrm{T}} A_0 + I)\phi(e, x_r) - \Lambda_0 x_r - A_0 f(x_r) + f_r(x_r)$ is the nonlinear feedback, and at the same time we choose parameter update laws for Λ and A according to (27), the controlled fuzzy hyperbolic model (19) will be chaotified through minimizing the cost functional (36).*

Remark 7. We note that the designed controller is not unique since we have freedom to select Λ_0 and A_0. In fact, it is necessary to select proper Λ_0 and A_0 so that the controller's energy satisfies requirements in practical applications. Once Λ_0 and A_0 are fixed, the controller is optimal in minimizing some "meaningful" cost. Theorem 5 also indicates that we can use as a general device to produce various chaotic dynamics.

3.3 Simulation Results

In this section, we take model (14) as our example and rewrite it as follows:

$$\dot{x} = Af(x).$$

Then the controlled system is

$$\dot{x} = Af(x) + u = \Lambda x + Af(x) + w , \tag{41}$$

where $\Lambda = \mathrm{diag}\,[\lambda_1,\ \lambda_2,\ \lambda_3]$. Here, because of the special form of A, only three parameter updating laws will be required.

Suppose that the chaotic system we want to track is the Lorenz system:

$$\dot{x}_r = f_r(x_r) , \tag{42}$$

where $x_r = [x_{1r},\ x_{2r},\ x_{3r}]^{\mathrm{T}}$ and

$$f_r(x_r) = [a(x_{2r} - x_{1r}),\ cx_{1r} - x_{1r}x_{3r} - x_{2r},\ x_{1r}x_{2r} - bx_{3r}]^{\mathrm{T}} .$$

When $a = 10$, $b = 8/3$, and $c = 28$, the Lorenz system has a chaotic attractor shown in Fig. 7.

In this example, we choose

$$\Lambda_0 = \mathrm{diag}\,[-2,\ -2,\ -2]^{\mathrm{T}} , \qquad \Lambda(0) = \mathrm{diag}\,[-1,\ -1,\ -2]^{\mathrm{T}} ,$$

$$A_0 = \begin{bmatrix} 0 & 3 & 4 \\ 3 & 0 & 4 \\ 3 & 3 & 0 \end{bmatrix} , \qquad A(0) = \begin{bmatrix} 0 & 2.8 & 3.7 \\ 2.8 & 0 & 3.7 \\ 2.8 & 2.8 & 0 \end{bmatrix} ,$$

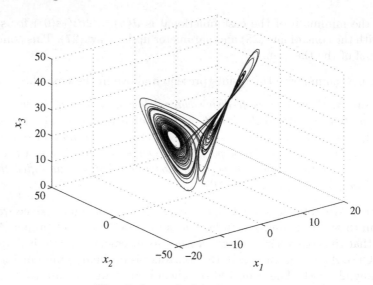

Fig. 7. Lorenz's chaotic attractor

$[k_1,\ k_2,\ k_3]^{\mathrm{T}} = [2,\ 3,\ 1]^{\mathrm{T}}$, $x_r(0) = [2,\ 1,\ 3]^{\mathrm{T}}$, and $x(0) = [0,\ 0,\ 0]^{\mathrm{T}}$. We choose these matrices in our simulation according to the following guidelines: (1) Λ_0 is a diagonal matrix with negative diagonal elements; (2) $\Lambda(0)$ is a perturbation of Λ_0; (3) from the process of fuzzy modelling, we know that each element of matrix A_0 is either positive or zero; (4) $A(0)$ is a perturbation of A_0.

The simulation results are shown in Figs. 8–12. From these figures, we can see that the controlled system (41) produces chaotic dynamics that have the same topological structure as system (42) and the two systems' states become indistinguishable after a short period of time. Figures 11 and 12 show that the parameters also approach some constants, which is in accordance with Remark 7.

4 Chaotification of the Original System

In this part, we want to verify our conjecture, i.e., the control method proposed in the preceding sections can make the original system chaotic. Before establishing our main theorem of this section, the following lemma is needed:

Lemma 5. (cf. [25]) *If $u(t)$, $v(t)$ and $c(t) \geq 0$ on $[0,t]$, c is differentiable and*

$$v(t) \leq c(t) + \int_0^t u(s)v(s)\,\mathrm{d}s\,,$$

then

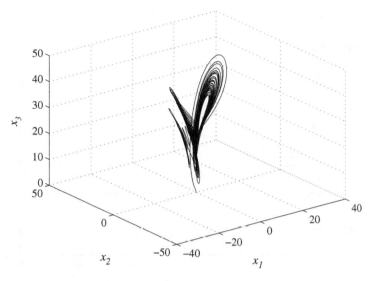

Fig. 8. The phase diagram of system (41) when the tracking object is system (42)

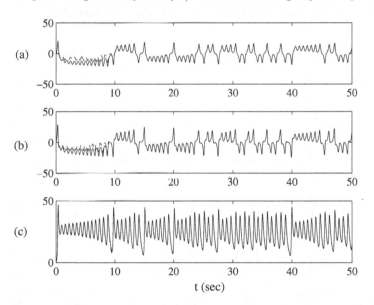

Fig. 9. State trajectories of system (41) when tracking object is system (42). The solid lines are the trajectories of system (41) and the dashed lines are the trajectories of system (42)

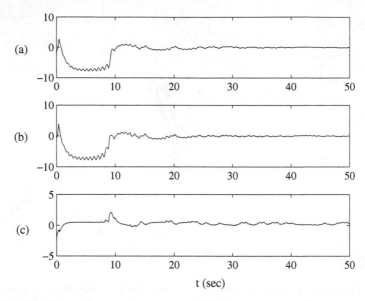

Fig. 10. (a) The trajectory of $x_1 - x_{r1}$; (b) the trajectory of $x_2 - x_{r2}$; (c) the trajectory of $x_3 - x_{r3}$

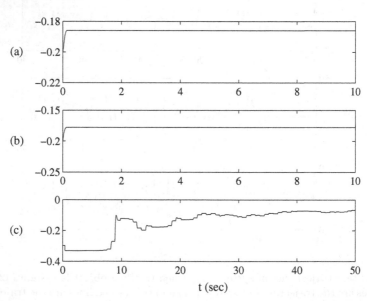

Fig. 11. (a) The plot of $C_{x_1} - C_1^0$; (b) the plot of $C_{x_2} - C_2^0$; (c) the plot of $C_{x_3} - C_3^0$

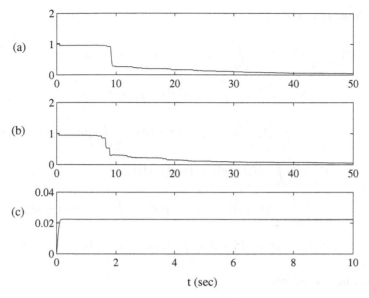

Fig. 12. (a) The plot of $\lambda_1 - \lambda_{10}$; (b) the plot of $\lambda_2 - \lambda_{20}$; (c) the plot of $\lambda_3 - \lambda_{30}$

$$v(t) \le c(0) \exp\left(\int_0^t u(s)\,ds\right) + \int_0^t c'(s)\left[\exp\left(\int_s^t u(\tau)\,d\tau\right)\right]ds\,.$$

Now we consider systems

$$\dot{x} = f_1(x) + u(x) \overset{\triangle}{=} F_1(x), \quad x(0) = x^0 \tag{43}$$

and

$$\dot{y} = f_2(y) + u(y) \overset{\triangle}{=} F_2(y), \quad y(0) = y^0\,, \tag{44}$$

where $x, y \in D \subset \mathbb{R}^n$ and $f_1, f_2 : \mathbb{R}^n \to \mathbb{R}^n$ are maps. If f_1 is integrable in the sense of Lebesgue, u and f_2 satisfy Lipschitz's condition, i.e., for any $x, y \in D$,

$$\|f_2(x) - f_2(y)\| \le L_1 \|x - y\|$$

and

$$\|u(x) - u(y)\| \le L_2 \|x - y\|\,,$$

then we have the following theorem.

Theorem 6. *If $\|f_2 - f_1\| < \varepsilon_1$ and $\|x^0 - y^0\| < \varepsilon_2$, then for any $t \in [0, T]$, there exists a constant $M(T, L_1, \varepsilon_1, \varepsilon_2) > 0$ such that*

$$\|x(t) - y(t)\| < M(L, T, \varepsilon, \varepsilon_2)\,, \tag{45}$$

where $L = L_1 + L_2$.

Proof. From (43) and (44),we have

$$\|x(t) - y(t)\| = \left\| x^0 - y^0 + \int_0^t [F_1(x(s)) - F_2(y(s))] \, ds \right\|$$

$$\leq \|x^0 - y^0\| + \left\| \int_0^t [f_1(x(s)) - f_2(x(s))] ds \right\|$$

$$+ \left\| \int_0^t [f_2(x(s)) - f_2(y(s))] \, ds \right\| + \left\| \int_0^t [u(x(s)) - u(y(s))] \, ds \right\|$$

$$< \varepsilon_2 + \int_0^t \|f_1(x(s)) - f_2(x(s))\| \, ds$$

$$+ \int_0^t \|f_2(x(s)) - f_2(y(s))\| \, ds + \int_0^t \|u(x(s)) - u(y(s))\| \, ds$$

$$< \varepsilon_2 + \varepsilon_1 t + (L_1 + L_2) \int_0^t \|x(s) - y(s)\| ds \ .$$

Using Lemma 5, we get

$$\|x(t) - y(t)\| < \left(\frac{\varepsilon_1}{L} + \varepsilon_2\right) e^{Lt} - \frac{\varepsilon_1}{L}$$

$$< \left(\frac{\varepsilon_1}{L} + \varepsilon_2\right) e^{Lt}$$

$$< \left(\frac{\varepsilon_1}{L} + \varepsilon_2\right) e^{Lt}$$

Thus, letting $M(T, L, \varepsilon_1, \varepsilon_2) = (\varepsilon_1/L + \varepsilon_2)e^{Lt}$, the theorem is proved.

Remark 8. From Theorem 6, we can conclude that the original system can produce almost the same state as that of the model as long as the modeling error is small enough.

Next, we will verify our claims by computer simulations. Suppose the original system has the following form:

$$\dot{x} = p(x) = \begin{cases} Ax, & \text{if } \|x\| < 1 \ ; \\ A\text{sign}(x), & \text{if } \|x\| \geq 1 \ . \end{cases} \tag{46}$$

Here, $x \in \mathbb{R}^3$, $\text{sign}(x) = [\text{sign}(x_1) \ \ \text{sign}(x_2) \ \ \text{sign}(x_3)]^{\mathrm{T}}$, and

$$A = \begin{bmatrix} 0 & 1 & 3 \\ 2 & 0 & 3 \\ 2 & 1 & 0 \end{bmatrix} \ .$$

It is easy to know that the system's trajectories are divergent. System (46) can be modeled using the FHM (14). The state trajectories of system (46) and system (14) are shown in Figs. 13 and 14. From these two figures we can conclude that model (14) can describe system (46) quite well.

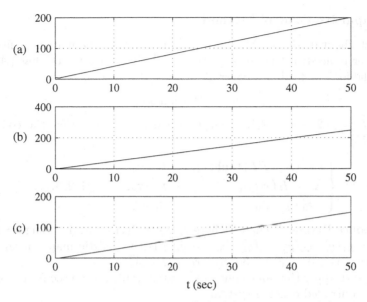

Fig. 13. The state trajectories of system (46): (a) the state x_1; (b) the state x_2; (c) the state x_3

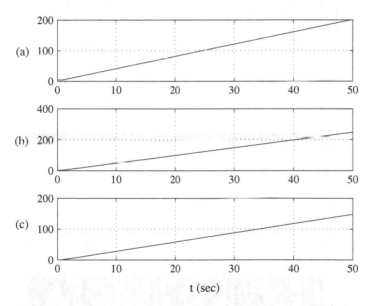

Fig. 14. The state trajectories of system (14): (a) the state x_1; (b) the state x_2; (c) the state x_3

4.1 Impulsive Control Method

We first chaotify the FHM. Since A is not a Hurwitz matrix, a feedback controller is needed. Choose $B = [1, \ 1, \ 1]^{\mathrm{T}}$. It is easy to see that (A, B) is controllable. The feedback controller can be designed as

$$u = BL \tanh(x) , \qquad (47)$$

where $L = [-2 \ -1 \ -2]^{\mathrm{T}}$. From Theorem 2, we know that the controlled FHM (15)

$$\begin{cases} \dot{x} = (A + BL)\tanh(x), & t \neq \tau_k \\ \Delta x = I_k(x(t)), & t = \tau_k, \quad k = 1, 2, 3, \ldots \\ x(t_0+) = x_0, & t_0 \geq 0 \end{cases} \qquad (48)$$

can produce Devaney's chaos, where $I_k(x) = \Lambda(y_k) - x, \Lambda = \mathrm{diag}[0.8\ 0.2\ 1]$, $g(y_k) = [g_1(y_k^1)\ g_2(y_k^2)\ g_3(y_k^3)]^{\mathrm{T}}$, and $g_i(\cdot)$ is the logistic map. Initial values are chosen as $g(0) = [0.1\ 0.3\ 0.8]^{\mathrm{T}}, x_0 = [2\ 2\ 1]^{\mathrm{T}}$.

We next apply the same control to system (46), we can see from Figs. 15–18 that controlled original system

$$\begin{cases} \dot{x} = p(x) + BL \tanh(x), & t \neq \tau_k \\ \Delta x = I_k(x(t)), & t = \tau_k, \quad k = 1, 2, 3, \ldots \\ x(t_0+) = x_0, & t_0 \geq 0 \end{cases} \qquad (49)$$

also produces similar state to that of the controlled FHM.

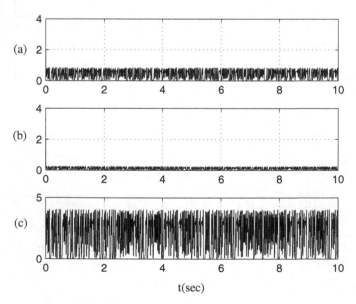

Fig. 15. The state trajectories of controlled FHM (48): (**a**) the state x_1; (**b**) the state x_2; (**c**) the state x_3

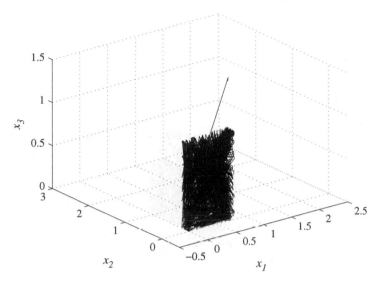

Fig. 16. The phase diagram of controlled FHM (48)

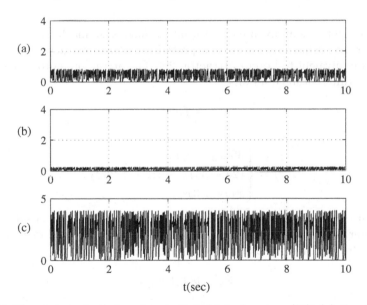

Fig. 17. The state trajectories of controlled original system (49): (**a**) the state x_1; (**b**) the state x_2; (**c**) the state x_3

4.2 Inverse Optimal Control Method

Although the system (46) can produce chaos under the state feedback and impulsive control, in some special cases we want the nonchaotic system to produce an attractor in some degree analogous to a predesigned chaotic

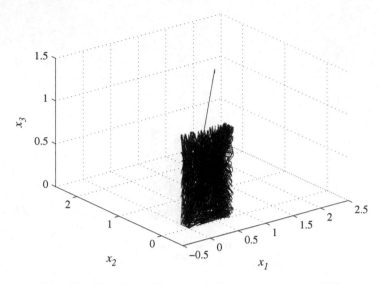

Fig. 18. The phase diagram of the original system (49)

attractor. Here, as an example, we want to make system (46) produce Chen attractor-like behavior. It has been proved that Chen attractor is not topologically equivalent to Lorenz attractor [29]. The attractor is produced by the following differential equations:

$$\dot{x} = f_s(x) \,, \tag{50}$$

where

$$f_s(x) = \begin{cases} 35(x_2 - x_1) \,, \\ (28 - 35)x_1 - x_1x_3 + 28x_2 \,, \\ x_1x_2 - 3x_3 \,. \end{cases}$$

Figure 19 shows the diagram of Chen attractor. Without considering the parameter disturbances, we apply the control scheme designed in Sect. 3.3 to system (46) and system (14), respectively, to obtain

$$\dot{x}_s = p(x_s) + u \,, \tag{51}$$

$$\dot{x}_m = A\tanh(x_m) + u \,. \tag{52}$$

From Fig. 20, we can see that state errors between Chen system (50) and model (52) are convergent. Figure 21 shows the phase diagram of controlled model (52), which shows that the FHM (52) is indeed chaotified and produces an attractor similar to the Chen attractor.

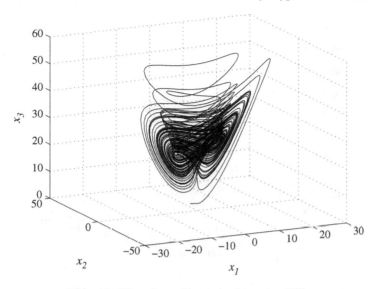

Fig. 19. The phase diagram of system (50)

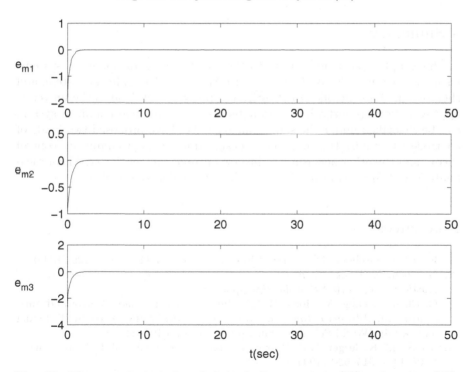

Fig. 20. The error trajectories of states between system (52) and system (50), where $e_{mi} = x_{mi} - x_i$, $i = 1, 2, 3$

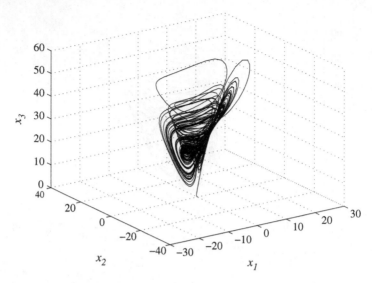

Fig. 21. The phase diagram of system (52)

5 Summary

In this chapter, we made efforts to the chaotification of nonlinear systems with unknown models. We first used a FHM to model a nonlinear system, and then designed a controller to chaotify the fuzzy model. Due to the universal approximation property, it is reasonable to expect the real nonlinear system can be chaotified under the same controller. We have proposed two kinds of controllers to realize this goal: one is designed according to impulsive control theory and the other according to inverse optimal control theory. Theoretical analysis and simulation results show that the methods are effective.

References

1. E. Ott, C. Grebogi, J.A. Yorke: Phys. Rev. Lett. **64**(11), 1196–1199 (1990)
2. G. Chen, X. Dong: *From Chaos to Order: Methodologies, Perspectives, and Applications* (World Scientific, Singapore, 1998)
3. G. Chen, J. Fang, Y. Hong, H. Qin: Introduction to chaos control and anti-control. In: *Advanced Topics in Nonlinear Control Systems*, ed by T. Leung and H. Qin (World Scientific, Singapore, 2001) pp 193–245
4. S.J. Schiff, K. Jerger, D.H. Duong, T. Chang, M.L. Spano, W.L. Ditto: Nature **370**(6491), 615–620 (1994)
5. W. Yang, M. Ding, A.J. Mandell, E. Ott: Phys. Rev. E **51**(1), 102–110 (1995)
6. G. Chen, D. Lai: IEEE Trans. Circuits Syst. I **44**(3), 250–253 (1997)
7. G. Chen, D. Lai: Int. J. Bifurcation Chaos **8**(7), 1585–1590 (1998)
8. X.F. Wang, G. Chen: Int. J. Bifurcation Chaos **9**(7), 1435–1441 (1999)

9. X.F. Wang, G. Chen: Int. J. Circuit Theory Appl. **28**(3), 305–312 (2000)
10. X.F. Wang, G. Chen: IEEE Trans. Circuits Syst. I **47**(3), 410–415 (2000)
11. X.F. Wang, G. Chen: Int. J. Bifurcation Chaos **10**(3), 549–570 (2000)
12. J.H. Lü, T.S. Zhou, G. Chen, X.S. Yang: Chaos **12**(2), 344–349 (2002)
13. E.N. Sanchez, J.P. Perez, G. Chen: Int. J. Bifurcation Chaos **113**, 857–863 (2001)
14. K.S. Tang, K.F. Man, G.Q. Zhong, G. Chen: IEEE Trans. Circuits Syst. I **48**(5), 636–641 (2001)
15. X.F. Wang, G. Chen, K.F. Man: IEEE Trans. Circuits Syst. I **48**(5), 641–645 (2001)
16. X.F. Wang, G. Chen, X. Yu: Chaos **10**(4), 771–779 (2000)
17. L. Yang, Z. Liu, G. Chen: Int. J. Bifurcation Chaos **12**(5), 1121–1128 (2002)
18. G.Q. Zhong, K.F. Man, G. Chen: Int. J. Bifurcation Chaos **11**(3), 865–869 (2001)
19. R.L. Devaney: *An Introduction to Chaotic Dynamical Systems* (Addison-Wesley, New York, 1987)
20. H. Zhang Y. Quan: IEEE Trans. Fuzzy Syst. **9**(2), 349–354 (2001)
21. E. Kaszkurewicz, A. Bhaya: *Matrix Diagonal Stability in Systems and Computation* (Birkhaüser, Boston, 2000)
22. H.K. Khalil: *Nonlinear Systems*, 3rd edn (Prentice Hall, Upper Saddle River, NJ, 2002)
23. M.W. Hirsch, S. Smale: *Differential Equations, Dynamical Systems, and Linear Algebra* (Academic Press, New York, 1974)
24. R.K. Miller, A.N. Michel: *Ordinary Differential Equations* (Academic Press, New York, 1982)
25. J. Guckenheimer, P. Holmes: *Nonlinear Oscillations, Dynamical Systems, and Bifurcation of Vector Fields* (Spriger, Berlin Heidelberg New York, 1983)
26. Z.-H. Li, M. Krstic: Automatica **33**(8), 1459–1473 (1997)
27. J.E. Slotine, W. Li: *Applied Nonlinear Control* (Prentice Hall, Englewood Cliffs, NJ, 1991)
28. M. Krstic, Z.-H. Li: IEEE Trans. on Automatic Control **43**(3), 336–351 (1998)
29. G. Chen, T. Ueta: Int. J. Bifurcation Chaos **9**(7), 1465–1466 (1999)

Fuzzy Chaos Synchronization via Sampled Driving Signals

Juan Gonzalo Barajas-Ramírez

Abstract. In this chapter the Tagaki–Sugeno fuzzy model representation of a chaotic system is used to find an alternative solution to the chaos synchronization problem. One of the advantages of the proposed approach is that it allows to express the synchronization problem as a fuzzy logic observer design in terms of linear matrix inequalities, which can be solved numerically using readily advailable software packages. Also, given the linear nature of this fuzzy representation, it is possible to use sophisticated methodologies to consider the more practical problem of digital implementation of a synchronization design. In particular, in this contribution the problem of a *master–slave* chaos synchronization design from sampled drive signals is considered and a solution is proposed as the state-matching digital redesign of the fuzzy logic observer designed to solve the continuous-time synchronization problem. The effectiveness of the proposed synchronization method is illustrated through numerical simulations of three well-known benchmark chaotic system, namely, Chua's circuit, Chen's equation, and the Duffing oscillator.

1 Introduction

Fuzzy logic was develop as a way to mimic the capacity of the human mind to operate with vague concepts and approximated reasoning. Therefore, one of the most significant advantages of using fuzzy logic in practical applications is the capacity to describe a systems in a way that tolerates imprecision and uncertainty. Using fuzzy logic a real-world system can be modeled in terms of linguistic variables, fuzzy sets, and membership functions, this adds the expert knowledge from human operators to the modeling process. The resulting "intelligent"model is an alternative representation of the original system, which usually is much simpler in structure than a model obtained using a conventional approach.

It has been shown that with a sufficiently large number of fuzzy sets and associated membership functions any bounded nonlinearity can be approximated to an arbitrary accuracy [1]. This is particularly significant in the case of chaotic systems given that they evolve on bounded area of state space, namely their strange attractor, which makes then very well suitable for this type of representation.

In this chapter, the fuzzy representation of a chaotic system is used as bases for an alternative solution to the chaos synchronization problem. Since

J.G. Barajas-Ramírez: *Fuzzy Chaos Synchronization via Sampled Driving Signals*,
StudFuzz **187**, 259–283 (2006)
www.springerlink.com

the appearance of the seminal paper by Pecora and Carol [2] in 1990, chaos synchronization has attracted increasing attention from the scientific community, as evidenced by the large numbers of research results published in a variety of journals, conferences [3–6], as well as many excellent books on the subject [7–10].

Significant advances on the study of synchronization of nonlinear dynamical systems has come from the reformulation of the synchronization problem in terms of well-established results from classical control theory, see for example the works of Nijmeijer et al. [11, 12], Mogül [13], and many others [14, 15]. In general terms, an effective and systematic solution to the *master–slave* synchronization problem can be found by expressing the synchronization objective as an observer design problem, where the output of the *master* system is interpreted as the driving signal and the objective is to design the *slave* system such that the observation error vanish at least asymptotically.

From the control theory point of view, some attractive results have been reported using alternative representations of chaotic systems. A particularly convenient representation to solve the synchronization problem is the Takagi–Sugeno (TS) fuzzy model, where the nonlinear dynamics are represented by local linear subsystems coupled nonlinearly, from which a fuzzy observer can be designed using linear techniques to achieve chaos synchronization [16–21].

Unlike most of the previously cited references, where synchronization design is carried out either in continuous or discrete-time, in this chapter the case of a hybrid configuration is considered. The motivation for this hybrid expression of the synchronization problem comes from the fact that most analyzers and controllers of complex analog systems are physically implemented by digital computers, which means that at least one part of the synchronized system is discrete-time in nature. In particular, the case of two continuous-time chaotic systems connected unidirectionally through a digital system is considered, that is, a *master–slave* configuration where the driving signal is a sampled version of the output from the *master* system.

A common practice when dealing with digital implementations is to follow the argument that with a sufficiently small sampling period the performance of the continuous-time design will be maintained in the digital device. Unfortunately, this is not always possible, since sampling affects the stability of the resulting system and there are physical limits to the sampling frequency for real systems. Furthermore, even if these restrictions are not present, the CPU time consumption in implementing a very small sampling period can be costly in many cases, so the idea becomes impractical in applications.

The problem of digital implementation is further complicated in the case of chaotic systems due to the extreme sensibility of chaotic dynamics to small changes. Nevertheless, some advances in this topic have been achieved, as can be found in the works of Shieh and his collaborators [22–24] where the discrete-time controller is derived from the continuous-time design via state-matching. This is called digital redesign, and it works well for many chaotic systems. Digital redesign consists in deriving a digital controller for a continuous-time plant by first designing a continuous-time controller to satisfy a set of control specifications, and then converting it to an equivalent digital controller such that the states of the continuous and discrete-time controlled systems are matched at least at each sampling instant. Digital redesign was originally proposed by Kuo [25] in 1980. During the last decades, Shieh et al. have thoroughly investigated this topic and developed several digital redesign methods that allow for (sub)-optimal control performance even for relatively large sampling periods, while at the same time requiring a significantly smaller control energy [22, 26, 27].

In what follows, the methodologies for digital redesign of controllers available in the literature are used, via the dual-system approach, to redesign a TS fuzzy observer designed to synchronize two chaotic systems through a sampled driving signal.

2 Fuzzy Modeling of Dynamical Systems

The fuzzy model representation of a dynamical system is an inference machine that relates inputs to outputs as a set of fuzzy If–Then rules, with the following general structure:

$$\text{System rule } i: \quad i = 1, 2, \ldots, r$$
$$\text{If} \quad z_1(t) \text{ is } M_{i,1} \text{ and } \cdots \text{ and } z_p(t) \text{ is } M_{i,p} \quad \text{Then} \quad \chi_r(t) \text{ is } \Psi_r \quad (1)$$

where r is the number of rules in the model, $z(t)$ are the permise variables, p is the number of premise variables, $\chi(t)$ are the conclusion variables, and "$z_1(t)$ is M" is a short form notation of "$z_1(t)$ belongs to the fuzzy set M, with a value $\mu_M(z_1(t))$ given by the associated membership function μ_M."

In the fuzzy If–Then rule set (1) the premise and conclusion variables represent inputs, states, and outputs of the dynamical system defined in terms of their corresponding fuzzy sets. The process of redefining these variables in terms of fuzzy sets and membership functions is called fuzzification.

The results of the logical operations on the fuzzy variables are significant only within the fuzzy model representation of the system; to express the conclusions of the inference machine in terms of the original input, output, and state variables of the system, it is necessary to find the overall result of the fuzzy model, which is obtained converting the fuzzy conclusion variables to their *crisp* or clear value—a process called defuzzification.

Depending on the method used for defuzzification of the inferred overall result, fuzzy models can be additive or nonadditive. In the Mamdani model [28], which corresponds to the nonadaptive approach, one assumes that no explicit model of the systems is available either because the system is partially unknown or too complex, and is necessary to model the system from informal knowledge available from human experts on the system's operation. Then, the overall result is inferred directly from the fuzzy rules activated at each instance in a *min–max* format.

Two types of fuzzy logic models correspond to the additive approach: standard-additive-model (SAM) [29] and Takagi–Sugeno (TS) models [17]. In the SAM model, the overall result is inferred from all the fuzzy rules activated by a given instance of the fuzzy variables, summing all the conclusion values using the center of gravity formula. The most significant characteristic of the TS fuzzy model is that in this representation, the dynamical system is expressed as a rule set, where each rule defines a region of action for a linear differential/difference algebraic representation of a local subsystem.

2.1 Takagi–Sugeno Fuzzy Model

The TS model was originally proposed for the fuzzy representation of systems with an explicit mathematical model [30] in 1985. Later Sugeno and Kang extended it to practical systems requiring identification [31]. It has been shown that with a sufficiently large number of fuzzy rules, the TS model serves as a universal approximator for an arbitrary continuous function defined on a compact set [1]. In particular, for chaotic systems that evolve within a bounded region of the state space, the TS fuzzy model can represent "exactly"the nonlinear dynamics by a small set of linear subsystems coupled by linguist variables [19].

A dynamical system can be expressed by a TS fuzzy model with the form

$$\text{System rule } i: \quad i = 1, 2, \ldots, r$$

$$\text{If} \quad z_1(t) \text{ is } M_{i,1} \text{ and } \cdots \text{ and } z_p(t) \text{ is } M_{i,p} \tag{2}$$

$$\text{Then} \quad \begin{aligned} \dot{x}(t) &= A_i x(t) + B_i u(t) \\ y(t) &= C_i x(t) \end{aligned}$$

where $M_{i,j}$ represents the fuzzy sets; $x(t) \in \mathbf{R}^n$, $u(t) \in \mathbf{R}^m$, and $y(t) \in \mathbf{R}^q$ are the state, input, and output variables of the system, respectively; A_i, B_i, and C_i are constant matrices of appropriate dimensions; and $z_1(t), z_2(t), \ldots, z_p(t)$ are the premise variables, which may be functions of the states, external disturbances and/or time (but here for simplicity they are assumed independent of the input).

The overall result of the fuzzy representation in (2) for a given instance is found using

$$\dot{x}(t) = \sum_{i=1}^{r} h_i(z(t)) \left[A_i \, x(t) + B_i \, u(t) \right] \tag{3}$$

$$y(t) = \sum_{i=1}^{r} h_i(z(t)) \, C_i \, x(t) \tag{4}$$

with $h_i(z(t)) = \frac{\omega_i(z(t))}{\sum_{i=1}^{r} \omega_i(z(t))}$, where $\omega_i(z(t)) = \prod_{j=1}^{p} M_{i,j}(z_j(t))$ and $z(t) = [z_1(t), \ldots, z_p(t)]$.

As usual, the term $M_{i,j}(z_j(t))$ is the grade of membership of the premise variable $z_j(t)$ belonging to the fuzzy set $M_{i,j}$, and its value is given by the associated membership function, $\mu_{M_{i,j}}$, defined for each fuzzy set.

For a fuzzy If–Then rule set to be an appropriate model of a dynamical system it must satisfy the 3C's principle: It must be complete, that is, all the possible conditions from the premise variables must be present; it must be consistent, logical contradictions must be avoided; and it must be concise, with no redundant rules on the set of conditions. In particular, the TS fuzzy model in (2) is well-defined if it satisfies the 3C's principle and the following conditions:

$$\sum_{i=1}^{r} \omega_i(z(t)) > 0 \quad \text{and} \quad \omega_i(z(t)) \geq 0 \quad \text{for } i = 1, 2, \ldots, r \tag{5}$$

$$\sum_{i=1}^{r} h_i(z(t)) = 1 \quad \text{and} \quad h_i(z(t)) \geq 0 \quad \text{for } i = 1, 2, \ldots, r \tag{6}$$

3 Fuzzy Logic Controller Design

The design of controllers for systems described in terms of fuzzy If–Then rule sets is important from the point of view of control engineering practice, because this representation allows inclusion of knowledge from human experts in modeling and control of complex systems.

A natural connection between fuzzy logic and control theory is accomplished using the TS fuzzy representation, which make it possible to use conventional linear techniques to design controllers and observers for dynamical systems from their fuzzy representation.

Two basic approaches can be taken when designing a fuzzy logic controller (FLC): replacement and combination. In the first approach, the idea is to replace a conventional controller by an inference machine designed in terms of a fuzzy logic description of the system, the most representative example of this approach is the Mamdani controller, which is the oldest and most widely used FLC design method for practical applications [28]. In the latter approach, fuzzy logic is used to extend or somehow improve a conventional controller, and examples of this complementary or combinatorial approach to FLC design are the parallel distributed compensator and the fuzzy-PID controllers [32].

3.1 Parallel Distributed Compensation

The simplest most intuitive manner to construct a controller for a system described by a fuzzy TS model is to designed a linear controller for each linear subsystem of the conclusion parts of the rule set. In 1986, Sugeno and Kant proposed, without stability analysis, the control of a fuzzy model by the

design of a static state feedback controller for each linear subsystem in the
model [33]. Then, in [34] the basic stability analysis was introduced and the
procedure began to be called parallel distributed compensation (PDC) in
[35]. In later publications the PDC controller design was extended to the
tracking problem for sampled-data systems [36] and its stability analysis was
expressed as an LMI problem in [17, 37].

A PDC controller is composed of the same rule set of the TS fuzzy model
of the system, and shares the same premise variables and fuzzy sets. For the
TS fuzzy model in (2) the PDC controller will have the form

Controller rule j: $j = 1, 2, \ldots, r$

If $z_1(t)$ is $M_{j,1}$ and $\cdots$ and $z_p(t)$ is $M_{j,p}$ Then $u(t) = -K_j x(t)$ (7)

where K_j are the controller gain matrices to be designed with conventional
linear control techniques like pole placement, optimal, or robust control.

The overall fuzzy controller is given by

$$u(t) = -\sum_{j=1}^{r} h_j(z(t)) K_j x(t) \tag{8}$$

Substituting (8) into (3), we find the overall closed-loop system as

$$\dot{x}(t) = \sum_{i=1}^{r} \sum_{j=1}^{r} h_i(z(t)) h_j(z(t)) \left[A_i - B_i \, K_j \right] x(t) \tag{9}$$

Defining $G_{i,j} := A_i - B_i \, K_j$, the closed-loop system can be rewriten as

$$\dot{x}(t) = \sum_{i=1}^{r} h_i(z(t)) h_i(z(t)) G_{i,i} x(t) \tag{10}$$

$$+ 2 \sum_{i=1}^{r} \sum_{i<j}^{r} h_i(z(t)) h_j(z(t)) \left[\frac{G_{i,j} + G_{j,i}}{2} \right] x(t)$$

where the second term on the right side is formed by all pairs (i, j) such that
$h_i(z(t)) \, h_j(z(t)) \neq 0$. That is, where the fuzzy sets $h_i(z(t))$ and $h_j(z(t))$ do
not overlap: $h_i(z(t)) \cap h_j(z(t)) \neq \emptyset$, for $i \neq j$.

The stability of system (9) can be determined using the direct Lyapunov
method, which for the zero input case result in the following.

Theorem 1 (Stability of TS Fuzzy Systems). *The zero equilibrium of
the TS fuzzy system*

$$\dot{x}(t) = \sum_{i=1}^{r} h_i(z(t)) A_i x(t) \tag{11}$$

*is globally asymptotically stable if there exists a common positive definite
matrix $P = P^T > 0$ such that*

$$A_i^{\mathrm{T}} P + P A_i < 0 \quad for \quad i = 1, 2, \ldots, r \tag{12}$$

is satisfied.

This result comes directly from the choice of a quadratic Lyapunov candidate function $V(x(t)) = x(t)^{\mathrm{T}} P x(t)$, and is usually referred to as the *quadratic stability theorem* for fuzzy systems.

An important remark is that the condition of a common matrix $P > 0$ is necessary and cannot be relaxed. It can be shown that simply requiring that all linear subsystems in (11) be stable is not sufficient to have stability of the overall fuzzy system. Furthermore, if no such common P matrix exists, the fuzzy system is unstable at least for a set of initial conditions [17, 37].

Applying the quadratic stability theorem to the closed-loop system (10) yields the following result.

Theorem 2 (Stability of Controlled TS Fuzzy Systems). *The zero equilibrium of the continuous-time controlled TS fuzzy system (10) is globally asymptotically stable if there exists a common positive definite matrix $P = P^{\mathrm{T}} > 0$ such that*

$$A_i^{\mathrm{T}} P + P A_i < 0 \quad for \quad i = 1, 2, \ldots, r \tag{13}$$

$$\left(\frac{G_{i,j} + G_{j,i}}{2}\right)^{\mathrm{T}} P + P \left(\frac{G_{i,j} + G_{j,i}}{2}\right) < 0$$

$$for \quad i < j \quad where \quad h_i \cap h_j = \emptyset \tag{14}$$

are both satisfied.

A particularly interesting simplification of the above result is obtained if all B_i matrices are identical. In this case, the conditions of Theorem 2 reduce to the following:

Corollary 1. *Assume $B_1 = B_2 = \cdots = B_r = B$ in (10). Then, the zero equilibrium point of the Controlled TS Fuzzy System is globally asymptotically stable if there exists a common positive definite matrix $P = P^{\mathrm{T}} > 0$ such that*

$$(A_i - B K_i)^{\mathrm{T}} P + P (A_i - B K_i) < 0 \tag{15}$$

is satisfied for $i = 1, 2, \ldots, r$.

The main problem to design a fuzzy PDC controller such that the conditions on (13) and (14) be satisfied is finding a common P matrix. Different methods can be used to solve this problem, the simplest approach is trail and error, but is difficult to find a solution this way, especially if the model has a large number of rules. A constructive method to design a stable fuzzy controller can be derived by expressing the inequalities of the stability theorems as a LMI problems. The LMI approach has the advantage that a solution can be found using numerical methods that are readily available. Also, it is possible to determine if a fuzzy controller exists, by simply determining the feasibility of the LMI problem.

The conditions of Theorem 1 and Theorem 2 cannot be solved as LMIs directly for P and K_i since they are not jointly convex. To express these conditions as LMIs, the inequalities on (13) and (14) can be multiplied by P^{-1} on both sides, and then defining the new variables $X = P^{-1}$ and $N_i = K_i\, P^{-1}$ the controller gains are found from the solutions of the LMI problem:

$$X\, A_i^{\mathrm{T}} + A_i\, X - N_i^{\mathrm{T}} B_i^{\mathrm{T}} - B_i\, N_i \quad < 0$$

$$X\, A_i^{\mathrm{T}} + A_i\, X + X\, A_j^{\mathrm{T}} + A_j\, X \tag{16}$$

$$-N_i^{\mathrm{T}} B_i^{\mathrm{T}} - B_i\, N_j - N_i^{\mathrm{T}} B_j^{\mathrm{T}} - B_j\, N_i \le 0$$

with $X = X^{\mathrm{T}} > 0$, for $i = 1, 2, \ldots, r$, where $h_i \cap h_j = \emptyset$.

3.2 Fuzzy H_∞ Robust Controller Design

Consider the following TS fuzzy system with an extra input $\nu(t) \in \mathbf{R}^d$ which represents the disturbances and noise affecting the system:

$$\text{Disturbed system rule } i: \quad i = 1, 2, \ldots, r$$

$$\text{If} \quad z_1(t) \text{ is } M_{i,1} \text{ and } \cdots \text{ and } z_p(t) \text{ is } M_{i,p} \tag{17}$$

$$\text{Then} \quad \begin{aligned} \dot{x}(t) &= A_i x(t) + B_i u(t) + E_i\, \nu(t) \\ y(t) &= C_i x(t) \end{aligned}$$

where $E_i \in \mathbf{R}^{n \times d}$ is a constant input matrix for the disturbance.

The objective of the fuzzy H_∞ robust controller design is to attenuate the effects of the disturbance input $\nu(t)$ on the overall system output $y(t)$ to a prescribed level in terms of the H_∞ norm of the transfer function from the disturbance input to the overall output $T_{y,\nu}(s)$.

In time-domain this condition can be described more precisely in the following manner:

Fuzzy H_∞ Controller Problem Given the fuzzy TS system (17), find a PDC controller (7), which in closed-loop makes the overall controlled fuzzy system internally stable and satisfy the performance index

$$\sup_{\|\nu(t)\|_2 \neq 0} \frac{\|y(t)\|_2}{\|\nu(t)\|_2} \le \gamma \tag{18}$$

for a prescribed constant bound $\gamma > 0$.

A solution to the above H_∞ disturbance attenuation problem can be derived from the quadratic Lyapunov stability conditions by requiring that

$$\dot{V}(x(t)) + y^{\mathrm{T}}(t)\, y(t) - \gamma^2 \nu(t)^{\mathrm{T}} \nu(t) \le 0 \qquad (19)$$

where the Lyapunov function is given by $V(x(t)) = x^{\mathrm{T}}(t)\, P\, x(t)$, with $P = P^{\mathrm{T}} > 0$.

This can be verified by integrating (19), which gives

$$V(x(t)) + \int_0^{\tau} \left[y^{\mathrm{T}}(t)\, y(t) - \gamma^2 \nu(t)^{\mathrm{T}} \nu(t) \right]\, \mathrm{d}t \le 0 \qquad (20)$$

From the stability condition (18) for the overall results of the disturbed TS fuzzy system (17) with the PDC controller (7) the design can be express in the following result:

Theorem 3 (Fuzzy H_∞ Robust Controller). *The controller gain matrices K_i of the PDC controller (7) that solve the Fuzzy H_∞ Controller Problem are obtained from the solutions of the LMI problem*

Minimize $\gamma^2 > 0$,

subject to $X = X^{\mathrm{T}} > 0$

$$\begin{bmatrix} \frac{1}{2}\begin{pmatrix} X\,A_i^{\mathrm{T}} + A_i\,X \\ +X\,A_j^{\mathrm{T}} + A_j\,X \\ -N_j^{\mathrm{T}} B_i^{\mathrm{T}} - B_i\,N_j \\ -N_i^{\mathrm{T}} B_j^{\mathrm{T}} - B_j\,N_i \end{pmatrix} & -\frac{1}{2}\,(E_i + E_j) & \frac{1}{2} X\,(C_i + C_j)^{\mathrm{T}} \\ -\frac{1}{2}\,(E_i + E_j)^{\mathrm{T}} & \gamma^2 I & 0 \\ \frac{1}{2}\,(C_i + C_j)\,X & 0 & I \end{bmatrix} \le 0 \quad (21)$$

for $i = 1, 2, \ldots, r$, where $h_i \cap h_j = \emptyset$

then, the controller gains are finally given by $K_i = X^{-1} N_i$.

4 Fuzzy Logic Observer Design

An observer for the TS fuzzy system (2) can be designed following the same procedure as for the PDC controller design. That is, designing, in terms of the same premise variables and the same fuzzy sets as the TS representation of the system, a linear observer for each fuzzy rule. This yields the following results:

Observer rule j: $j = 1, 2, \ldots, r$

If $\quad \hat{z}_1(t)$ is $M_{j,1}$ and $\cdots$ and $\hat{z}_p(t)$ is $M_{j,p}$ $\qquad (22)$

Then $\quad \begin{aligned} \dot{\hat{x}}(t) &= A_j \hat{x}(t) + B_j u(t) + L_j \left[y(t) - \hat{y}(t) \right] \\ \hat{y}(t) &= C_j \hat{x}(t) \end{aligned}$

where $L_j \in \mathbf{R}^{n \times q}$ are the observer gain matrices, which must be designed such that the observation error, $e(t) = x(t) - \hat{x}(t)$, be asymptotically stable.

The overall state dynamics and output of the fuzzy TS observer are inferred as before by

$$\dot{\hat{x}}(t) = \sum_{j=1}^{r} h_j(\hat{z}(t)) \left[A_j \hat{x}(t) + B_j \, u(t) + L_j \left(y(t) - \hat{y}(t) \right) \right] \tag{23}$$

$$\hat{y}(t) = \sum_{j=1}^{r} h_j(\hat{z}(t)) C_j \hat{x}(t) \tag{24}$$

The error dynamics are obtained from (23) and (3) as

$$\begin{aligned} \dot{e}(t) &= \dot{x}(t) - \dot{\hat{x}}(t) \\ &= \sum_{i=1}^{r} \sum_{j=1}^{r} h_i(z(t)) \, h_j(\hat{z}(t)) \{ [A_i x(t) + B_i \, u(t)] \\ &\quad - [A_j \hat{x}(t) + B_j \, u(t) + L_j \left(y(t) - \hat{y}(t) \right)] \} \end{aligned} \tag{25}$$

For simplicity, assume that input and output matrices are equal, that is, $B_i = B_j = B$ and $C_i = C_j = C$. Then, the error dynamics become

$$\dot{e}(t) = \sum_{i=1}^{r} \sum_{j=1}^{r} h_i(z(t)) \, h_j(\hat{z}(t)) \{ [A_i - L_j \, C] \, x(t) - [A_j - L_j C] \, \hat{x}(t) \} \tag{26}$$

Similarly, assuming that $\hat{z}(t)$ is independent of the observer states, one has $z(t) = \hat{z}(t)$ and the error dynamics are reduced to

$$\dot{e}(t) = \sum_{j=1}^{r} h_j(z(t)) \left[A_j - L_j \, C \right] e(t) \tag{27}$$

Under these conditions, the observer gains L_j can be designed as a PDC fuzzy controller for the dual system of (27), with $A_i = A_j^{\mathrm{T}}$ and $B = C^{\mathrm{T}}$, using the results presented above, the fuzzy observer design is found from the solutions to the following LMI problem:

Theorem 4 (TS Fuzzy Observer Design: Common B and C). *The error dynamics of the TS fuzzy observer (22) for the TS fuzzy system (2), where $B_i = B_j = B$, $C_i = C_j = C$, and $z(t) = \hat{z}(t)$, are given by (27) and will be globally asymptotically stable if there exist a common possitive matrix $X = X^{\mathrm{T}} > 0$ and the matrices N_j that satisfy the LMI*

$$X \, A_j + A_j^{\mathrm{T}} X - N_j C_j - C_j^{\mathrm{T}} N_j^{\mathrm{T}} < 0 \tag{28}$$

The final observer gains are given by $L_j = N_j X^{-1}$, for $j = 1, 2, \ldots, r$.

In the case of the disturbed TS fuzzy system (17) a fuzzy observer can be designed with the same structure of (22) and if the the assumptions stated above still hold, the overall error dynamics will be given by

$$\dot{e}(t) = \sum_{j=1}^{r} h_j(z(t)) \{[A_j - L_j\,C]\,e(t) + E_j\,\nu(t)\} \tag{29}$$

An observer design that attenuates the effects of the disturbance input $\nu(t)$ on the error dynamics $e(t)$ to a prescribed level $\rho > 0$ can be achieved by requiring that the error dynamics be internally stable and

$$\sup_{\|\nu(t)\|_2 \neq 0} \frac{\|e(t)\|_2}{\|\nu(t)\|_2} \leq \gamma \tag{30}$$

Following a similar reasoning as above the observer design that satisfy the H_∞ robust criterion (30) can be found by solving the following LMI problem:

Theorem 5 (H_∞ Robust TS Fuzzy Observer Design: Common B and C). *The observer gains L_j for the TS fuzzy observer (22), which make the error dynamics (29) with respect to the perturbed TS fuzzy system (17) internally stable and satisfies the H_∞ disturbance attenuation performance index (30) under the conditions that $B_i = B_j = B$, $C_i = C_j = C$, and $z(t) = \hat{z}(t)$, are given by $L_j = N_j P^{-1}$, where $P = P^T > 0$ and N_j are the solutions to the LMI problem*

$$\text{Minimize } \rho^2 > 0,$$
$$\text{subject to } P = P^T > 0$$
$$\begin{bmatrix} A_j^T P + P\,A_j - C^T N_j^T - N_j\,C + I & P\,E_j \\ E_j^T P & -\rho^2\,I \end{bmatrix} \leq 0 \tag{31}$$
$$\text{for } i - 1, 2, \ldots, r \,.$$

5 Chaos Synchronzation Via Fuzzy Observer Design

The first step in solving the chaos synchronization problem using fuzzy logic is to find an alternative representation for the chaotic system in terms of fuzzy If–Then rules. In particular, letting the consequence part of each rule be a linear subsystem the TS fuzzy model is obtained. There are two basic approaches to construct a TS fuzzy model for a dynamical system, one is to identify the model from input–output data and the other is to derive the fuzzy representation from the system equations.

The identification of a TS fuzzy model is a two part process, first the structure and number of rules are determined and then the parameters adjusting each local linear subsystem are identified. More details of this approach can be found in [32, 34]. When the system equations are available, the TS fuzzy model can be derived using the sector nonlinearity approach.

5.1 Fuzzy Modeling of Chaotic Systems

The basic idea of the sector nonlinearity approach is to find, from the non-linear components of the system equation, a sector in state space where the nonlinearities can be expressed as simple product of the premise variable $z_1(t)$ and a function of the state variables $\eta(x(t))$, in the form

$$\dot{x}(t) = f(x(t)) = \eta(x(t)) z(t) \in [-d, d] \tag{32}$$

To illustrate the sector nonlinearity modeling procedure, consider the following chaotic benchmark systems:

5.1.1 Chua's Circuit

The following ordinary differential equations form the dimensionless version of Chua's circuit [4]:

$$\begin{aligned}
\dot{x}_1(t) &= \alpha\,[x_2(t) - x_1(t) - \Gamma(x_1(t))] \\
\dot{x}_2(t) &= x_1(t) - x_2(t) + x_3(t) \\
\dot{x}_3(t) &= -\beta x_2(t)
\end{aligned} \tag{33}$$

where the so-called Chua's diode $\Gamma(t)$ is a piecewise linear function given by

$$\Gamma(x_1(t)) = m_0 x_1(t) + \frac{1}{2}[m_1 - m_0]\,[|x_1(t) + 1| - |x_1(t) - 1|]$$

As illustrated in Fig. 1, $\Gamma(x_1(t))$ can be divided in two sections from which two fuzzy sets can be defined with the following linguistic descriptions: $M_1 =$ "Close to the origin" and $M_2 =$ "Far from the origin." Then, using the sector nonlinearity approach a TS model can be obtained for (33), defining the extra function

$$g(x_1(t)) = \begin{cases} \frac{\Gamma(x_1(t))}{x_1(t)} & \text{if } x_1(t) \neq 0 \\ m_0 & \text{if } x_1(t) = 0 \end{cases} \tag{34}$$

and chosing the premise variable to be $z_1(t) = x_1(t) \in [-d, \ d]$, with the fuzzy sets described by the membership functions:

$$M_1(z_1(t)) = \frac{1}{2}\left[1 + \frac{g(x_1(t))}{d}\right] \tag{35}$$

$$M_2(z_1(t)) = \frac{1}{2}\left[1 - \frac{g(x_1(t))}{d}\right] \tag{36}$$

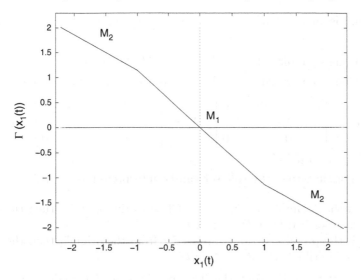

Fig. 1. Chua's diode characteristics

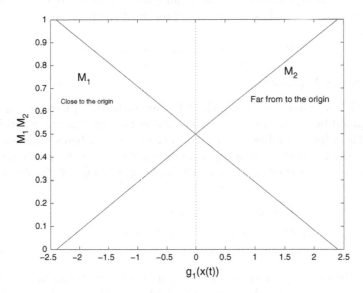

Fig. 2. Membership functions for Chua's circuit

As shown in Fig. 2 the overall value of $x_1(t)$ will by given by

$$x_1(t) = z_1(t) = d\, M_1(z_1(t)) - d\, M_2(z_1(t))$$

From the numerical simulation of (33) for the parameter set $\alpha = 9.0$, $\beta = 100/7$, $m_0 = -5/7$ and $m_1 = -8/7$, the interval of values for the premise variable $z_1(t)$ is found to be $[-3, 3]$.

Then, the chaotic Chua's circuit (33) can be represented by the TS fuzzy model:

$$\text{Chua's circuit rule 1}: \quad \text{If } z_1(t) \text{ is } M_1 \text{ Then } \dot{x}(t) = A_1 x(t) \qquad (37)$$

$$\text{Chua's circuit rule 2}: \quad \text{If } z_1(t) \text{ is } M_2 \text{ Then } \dot{x}(t) = A_2 x(t) \qquad (38)$$

where $A_1 = \begin{bmatrix} \alpha\,(-1+d) & \alpha & 0 \\ 1 & -1 & 1 \\ 0 & -\beta & 0 \end{bmatrix}$, $A_2 = \begin{bmatrix} \alpha\,(-1-d) & \alpha & 0 \\ 1 & -1 & 1 \\ 0 & -\beta & 0 \end{bmatrix}$ and $x(t) = \begin{bmatrix} x_1(t) & x_2(t) & x_3(t) \end{bmatrix}^{\mathrm{T}}$

In linguistic terms, the rule set can be interpreted as:

Rule 1: If the premise variable z_1 is "Close to the origin", then the system is described by $\dot{x}(t) = A_1\, x(t)$.

Rule 2: If the premise variable z_1 is "Far from the origin", then the system is described by $\dot{x}(t) = A_2\, x(t)$.

It is easy to verify that the TS fuzzy model of Chua's circuit is well defined, that is it satisfies the 3C's, and $M_1(z_1) > 0$, $M_2(z_1) > 0$ with $M_1(z_1) + M_2(z_1) = 1$. Then, the overall result of the fuzzy system (37)–(38) is infered by

$$\dot{x}(t) = M_1(z_1)\, A_1\, x(t) + M_2(z_1)\, A_2\, x(t) \qquad (39)$$

5.1.2 Chen's Equation

The chaotic system described in the following ordinary differential equations was proposed by Chen and Ueta in 1999 [38], as the result of the chaotification of a stable solution of Lorenz system [39]. However, Chen's equation has a chaotic attractor that is not a topological equivalent of Lorenz attractor, and it has been shown they are distint elements of a family of chaotic systems [40]

$$\begin{aligned} \dot{x}_1(t) &= a\,[x_2(t) - x_1(t)] \\ \dot{x}_2(t) &= (c-a)\,x_1(t) - x_1(t)\,x_3(t) + c\,x_2(t) \\ \dot{x}_3(t) &= x_1(t)\,x_2(t) - b\,x_3(t) \end{aligned} \qquad (40)$$

Since the only nonlinearities on (40) are quadratic terms and both have $x_1(t)$ in them, choosing the premise variable as $z_1(t) = x_1(t)$ the TS fuzzy model of Chen's equation can be constructed in terms of two fuzzy sets with the linguistic description, $M_1 =$ "Positive value" and $M_2 =$ "Negative value," with the associated membership functions

$$M_1(z_1(t)) = \frac{1}{2}\left[1 + \frac{z_1(t)}{d}\right] \qquad (41)$$

$$M_2(z_1(t)) = \frac{1}{2}\left[1 - \frac{z_1(t)}{d}\right] \qquad (42)$$

Then, the TS fuzzy model of Chen's equation will be given by

$$\text{Chen's equation rule 1}: \quad \text{If } z_1(t) \text{ is } M_1 \quad \text{Then } \dot{x}(t) = A_1 x(t) \qquad (43)$$

$$\text{Chen's equation rule 2}: \quad \text{If } z_1(t) \text{ is } M_2 \quad \text{Then } \dot{x}(t) = A_2 x(t) \qquad (44)$$

where $A_1 = \begin{bmatrix} -a & a & 0 \\ (c-a) & c & -d \\ 0 & d & -b \end{bmatrix}$ and $A_2 = \begin{bmatrix} -a & a & 0 \\ (c-a) & c & d \\ 0 & -d & -b \end{bmatrix}$.

From the numerical simulations of (40) for the parameter set $a = 35$, $b = 3$, and $c = 28$ the premise variable $z_1(t)$ is found to be contained within the interval $[-25, 25]$. While the overall output of the fuzzy model is given by (39)

5.1.3 Duffing Oscillator

This chaotic system is form by a mass-damper-spring system with a forcing term, represented by the following equation [11]:

$$\ddot{x}(t) + p_1 \dot{x}(t) + p_2 x(t) + p_3 x^3(t) = \tau(t) \qquad (45)$$

where p_1, p_2, and p_3 are constant parameters and the forcing term is given by, $\tau(t) = q \cos(\omega t)$ with q a constant gain and ω a constant frequency.

The Duffing oscillator can be rewritten as

$$\begin{aligned} \dot{x}_1(t) &= x_2(t) \\ \dot{x}_2(t) &= -p_2\, x_1(t) - p_3\, x_1^3(t) - p_1\, x_2(t) + \tau(t) \end{aligned} \qquad (46)$$

The cubic term in the Duffing oscillator can be express in terms of the premise variable $z_1(t) = x_1(t)$ and the new function $\phi(t) = x_1^2(t)$, using two fuzzy sets $M_1 =$ "Near zero" and $M_2 =$ "Far from zero" with the associated membership functions

$$M_1(z_1(t)) = \frac{1}{2}\left[1 + \frac{\phi(t)}{d}\right] \qquad (47)$$

$$M_2(z_1(t)) = \frac{1}{2}\left[1 - \frac{\phi(t)}{d}\right] \qquad (48)$$

A TS fuzzy model of the Duffing oscillator will be given by

$$\text{Duffing oscillator rule 1: If } z_1(t) \text{ is } M_1 \text{ Then } \dot{x}(t) = A_1 x(t) + B\,\tau(t) \qquad (49)$$

$$\text{Duffing oscillator rule 2: If } z_1(t) \text{ is } M_2 \text{ Then } \dot{x}(t) = A_2 x(t) + B\tau(t) \qquad (50)$$

where $A_1 = \begin{bmatrix} 0 & 1 \\ [-p_2 - p_3\, d] & -p_1 \end{bmatrix}$, $A_2 = \begin{bmatrix} 0 & 1 \\ [-p_2 + p_3\, d] & -p_1 \end{bmatrix}$, $B = [0, 1]^{\mathrm{T}}$, and $x(t) = [x_1(t), x_2(t)]^{\mathrm{T}}$.

From the numerical simulation of (46) for the parameter set $p_1 = 0.4$, $p_2 = -1.1$, $p_3 = 1.0$, $\omega = 1.8$, and $q = 1.8$, the interval of values for the premise variable $z_1(t)$ is found be $[-2, 2]$.

The *crisp* value of the TS fuzzy model (49)–(50) is infered by

$$\dot{x}(t) = h_1(z(t)) \left[A_1 \, x(t) + B \, \tau(t) \right] + h_2(z(t)) \left[A_2 \, x(t) + B \, \tau(t) \right] \qquad (51)$$

with $h_i(z)$ as describe in (3).

5.2 Fuzzy Chaos Synchronization: Numerical Example

If the output of the chaotic Duffing oscillator (46) is given by

$$y(t) = C \, x(t) = [1, \, 0] \, x(t) = x_1(t) \qquad (52)$$

A *slave* system that synchronizes to the chaotic Duffing oscillator (46) can be constructed as an observer for its TS fuzzy model in the following form:

$$\text{Slave duffing system rule } 1: \quad \text{If } z_1(t) \text{ is } M_1$$
$$\text{Then } \dot{\hat{x}}(t) = A_1\hat{x}(t) + B\tau(t) + L_1 \left[y(t) - \hat{x}_1(t) \right] \qquad (53)$$
$$\text{Slave duffing system rule } 2: \quad \text{If } z_1(t) \text{ is } M_2$$
$$\text{Then } \dot{\hat{x}}(t) = A_2\hat{x}(t) + B\tau(t) + L_2 \left[y(t) - \hat{x}_1(t) \right] \qquad (54)$$

Then, the overall error dynamics will be given by

$$\dot{e}(t) = h_1(z(t)) \, (A_1 - L_1 \, C) \, e(t)$$
$$+ h_2(z(t)) \, (A_2 - L_2 \, C) \, e(t) \qquad (55)$$

using the results presented in Theorem 4, the observer gains can be designed solving the corresponding LMI problem.

In Fig. 3, the results of numerical simulations of the synchronization between (46) and the TS Fuzzy system (53)–(54) are presented. Here the chaotic master and slave systems are on free evolution at the begining, and at $t = 15$ the TS fuzzy observer is activated.

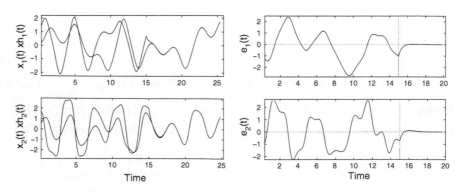

Fig. 3. Fuzzy chaos synchronization results for the Duffing system

6 Digitally Redesigned Takagi–Sugeno Fuzzy Observers

In this section, the case of a hybrid, analog-digital chaos synchronization problem is investigated; the problem originates from the consideration that a digital device is used in the connection between the master and slave systems, resulting on a sampled-data driving signal between two continuous-time chaotic systems.

The proposed solution is the redesign of the TS fuzzy observer designed for continuous-time chaos synchronization. In its original version, digital redesign is a methodology to find a discrete-time controller from a continuous-time controller designed to satisfied a set of control objectives, such that the states of the digitally and the continuous-time controlled closed-loop systems match at least at every sampling instant [26].

6.1 Digital Redesign of Linear Feedback Controllers

Consider a controllable and observable linear system of the form

$$\dot{x}_C(t) = A\,x_C(t) + B\,u_C(t), \qquad x_C(0) = x_0 \qquad (56)$$
$$y_C(t) = C\,x_C(t)$$

where $x_C(t)$, $u_C(t)$, and $y_C(t)$ are the state, input, and output variables defined for (2), with the subindex $(\cdot)_C$ used to indicated that these are the continuous-time variables. A, B, and C are constant matrices of appropriate dimensions.

Let the controller $u_C(t)$ be

$$u_C(t) = -K_C\,x_C(t) + E_C\,r(t) \qquad (57)$$

where the feedback $K_C \in \mathbf{R}^{m \times n}$ and feedforward $E_C \in \mathbf{R}^{m \times m}$ gains are obtained to satisfy a given set of control objectives, and $r(t)$ is an $m \times 1$ reference input.

The continuous-time closed-loop system is given by

$$\dot{x}_C(t) = (A - B\,K_C)\,x_C(t) + B\,E_C\,r(t), \qquad x_C(0) = x_0 \qquad (58)$$

Now, consider a piecewise-constant control law $u_d(t)$ satisfying

$$u_d(t) = u_d(k\,T) \quad \text{for } kT \le t < (k+1)\,T \qquad (59)$$

with $T > 0$ being the sampled-hold period. Here the subindex $(\cdot)_d$ indicates that this variable is digital, that is, it can be consider the output of a zero-order hold device when the input is a discrete-time signal $u_d(k\,T)$.

In particular, assume that $u_d(k\,T)$ has the form

$$u_d(k\,T) = -K_d\,x_d(k\,T) + E_d\,r^*(k\,T) \qquad (60)$$

where $K_d \in \mathbf{R}^{m \times n}$ and $E_d \in \mathbf{R}^{m \times m}$ are the feedback and feedforward digital gains, respectably, and $r^*(k\,T)$ is a piecewise-constant reference determined in terms of $r(k\,T)$ for tracking purpose.

The closed-loop digitally controlled system becomes

$$\dot{x}_d(t) = A\,x_d(t) + B\left[-K_d\,x_d(k\,T) + E_d\,r^*(k\,T)\right], \qquad x_d(0) = x_0 \qquad (61)$$

The digital redesign for the feedback controller (57) consists in finding the digital gains (K_d, E_d) in (60), from the continuous-time controller gains (K_C, E_C) in (57) such that the closed-loop continuous-time controlled states $x_C(t)$ in (58) closely match the closed-loop digitally controlled states $x_d(t)$ in (61), at every sampling instant throughout the whole process.

Different methods to determine appropriate values for (K_d, E_d) from (K_C, E_C) have been proposed in the last couple of decades, mainly by Shieh et al., see for example [26,27]. Of them, a particularly suitable methodology for the digital redesign of observers is the prediction-based method [22, 24], which is derived as follows.

Consider the state solution of (58), $x_C(t)$, at a future time given by $t = t_v = k\,T + v\,T$,

$$x_C(t_v) = \exp^{A(t_v - kT)}x_C(kT) + \int_{kT}^{kT+vT} \exp^{A(kT+vT-\tau)}B\,u_C(\tau)\,\mathrm{d}\tau \qquad (62)$$

where v is a tuning parameter, $0 \le v \le 1$.

Let the controller $u_C(t_v)$ be piecewise constant and the solution (62) can be reduced to

$$x_C(t_v) = G^{(v)}x_C(k\,T) + H^{(v)}u_C(t_v) \qquad (63)$$

where $G^{(v)} = \exp^{vAT}$ and $H^{(v)} = \int_{kT}^{t_v} \exp^{A(t_v-\tau)}B\,\mathrm{d}\tau = \int_0^{vT} \exp^{A\tau}B\,\mathrm{d}\tau = \left(G^{(v)} - I\right)A^{-1}B$.

The shorthand notation $(G^{(v)} - I)A^{-1}B$ of $H^{(v)}$ does not requires the invertiability of A, since the series expansion of $(G^{(v)} - I)$ has a common factor A that cancels A^{-1}.

Under the same assumptions as above, the predicted solution of the digitally controlled system, $x_d(t_v)$ is given by

$$x_d(t_v) = G^{(v)}x_d(k\,T) + H^{(v)}u_d(k\,T) \qquad (64)$$

Thus, it follows that to satisfy the state matching requirement for the predicted states $x_C(t_v)$ and $x_d(t_v)$ under the assumption that $x_C(t_{kT}) = x_d(t_{kT})$, it is necessary to have $u_C(t_v) = u_d(k\,T)$, which leads to the prediction-based digital controller:

$$\begin{aligned} u_d(k\,T) &= -K_C\,x_C(t_v) + E_C\,r(t_v) \\ &= u_C(t_v) = -K_C\,x_d(t_v) + E_d\,r(t_v) \end{aligned} \qquad (65)$$

Using (63) on (65) and solving for $u_d(k\,T)$, one gets the predicted digital controller as

$$u_d(k\,T) = [I + K_C\,H^{(v)}]^{-1}(-K_C\,G^{(v)}x_d(k\,T) + E_C\,r(t_v))$$
$$= K_d^{(v)}x_d(kT) + E_d^{(v)}r^*(k\,T) \qquad (66)$$

A way to ensure the prerequisite $x_C(k\,T) = x_d(k\,T)$ is to set the tunning the parameter to one, $v = 1$. Then, the prediction-based digital redesigned gains are obtained as

$$K_d = [I + K_C\,H]^{-1}K_C\,G \qquad (67)$$
$$E_d = [I + K_C\,H]^{-1}E_C \qquad (68)$$

where $G - \exp^{A\,T}$, $H = (G - I)\,A^{-1}B$, and the discrete-time tracking reference is given by $r^* = r(k\,T + T)$.

6.2 Digitally Redesigned Observer: Sampled Driving Signal

Let a continuous-time observer for system (56) be the system

$$\dot{\hat{x}}_C(t) = A\,\hat{x}_C(t) + B\,u_C(t) + L_C\,[y_C(t) - \hat{y}_C(t)], \quad \hat{x}_C(0) = \hat{x}_0 \qquad (69)$$
$$\hat{y}_C(t) = C\,\hat{x}_C(t)$$

which results in the continuous-time error dynamics

$$\dot{e}_C(t) = A\,e_C(t) - L_C\,[y_C(t) - C\,\hat{x}_C(t)] \qquad (70)$$

From the dual system approach, (70) can be written as

$$\dot{\epsilon}_C(t) = A^{\mathrm{T}}\,\epsilon_C(t) + C^{\mathrm{T}}\left[-L_C^{\mathrm{T}}\epsilon_C(t)\right] \qquad (71)$$

where $\epsilon_C(t) = e_C^{\mathrm{T}}(t)$ and the term $-L_C^{\mathrm{T}}\epsilon_C(t)$ can be consider a static feedback controller for the system (71).

Let the output from (69) $y_C(t)$ be sampled at a constant frequency such that one gets

$$y_d(t) = y_d(k\,T) = C\,x_C(k\,T) \quad \text{for} \quad k\,T \le t < (k+1)\,T \qquad (72)$$

If this sampled output signal is used to construct a continuous-time observer, the resulting hybrid error dynamics $e_d(t)$ can be expressed as the feedback control system

$$\dot{\epsilon}_d(t) = A^{\mathrm{T}}\,\epsilon_d(t) + C^{\mathrm{T}}\left[-L_d\epsilon_d(k\,T)\right] \qquad (73)$$

where $\epsilon_d(k\,T) = x_d^{\mathrm{T}}(k\,T) - \hat{x}_d^{\mathrm{T}}(k\,T)$ represents the sampled error dynamics and L_d can be chosen such that the hybrid error dynamics and the continuous-time error dynamics match at least at every sampling instant $e_C(t)|_{t=kT} \approx$

$e_d(t)|_{t=kT}$. In other words, L_d can be obtained as the digital redesign of the observer gain L_C^T in (71).

Using the prediction-based digital redesign results (67), the digital observer gain L_d that makes the hybrid error dynamics in (73) match the continuous-time error dynamics in (70) is given by

$$L_d = \left[\left(I + L_C^T \bar{H} \right)^{-1} L_C^T \bar{G} \right]^T \tag{74}$$

where $\bar{G} = \exp^{A^T T}$ and $\bar{H} = \left[\bar{G} - I \right] \left(A^T \right)^{-1} C^T$.

6.3 Sampled-Data TS Fuzzy Observer

To solve the chaos synchronization problem when the slave system is driven by a sampled signal $y_d(t)$, the results presented above can be used in order to design a digitally redesigned TS fuzzy observer.

Then a slave system that synchronizes to (2) with a sampled-data driving signal (72) will have the following form:

Sampled Driven Slave System Rule $j: j = 1, 2, \ldots, r$

If $z_1(t)$ is $M_{j,1}$ and $\cdots$ and $z_p(t)$ is $M_{j,p}$ $\qquad$ (75)

Then $\quad \begin{aligned} \dot{\hat{x}}_d(t) &= A_j \hat{x}_d(t) + B_j u_d(t) + L_{d,j} \left[y_d(k\,T) - \hat{y}_d(k\,T) \right] \\ \hat{y}_d(t) &= C\,\hat{x}_d(t) \end{aligned}$

where the observer gains $L_{d,j}$ are the digitally redesigned versions of the continuous-time observer gains $L_{C,j}$, obtained solving the LMI problems presented above, for each rule of the TS fuzzy model.

The block diagram shown in Fig. 4 visualizes the overall fuzzy hybrid chaos synchronization scheme.

6.4 Fuzzy Hybrid Chaos Synchronization: Numerical Examples

6.4.1 Chen's Equation

An observer for synchronization of the chaotic Chen's equation (40) through a sampled driving signal

$$y_d(t) = y_d(k\,T) = x_{d,1}(k\,T) \tag{76}$$

can be designed from the TS fuzzy model (43)–(44) as

Digitally Redesigned

Slave Chen's system rule 1:   If $z_1(t)$ is M_1 Then $\dot{\hat{x}}(t) = A_1 \hat{x}(t)$
$\qquad + L_{d,1} \left[y_d(k\,T) - \hat{x}_{d,1}(k\,T) \right]$ $\qquad$ (77)

Slave Chen's system rule 2:   If $z_1(t)$ is M_2 Then $\dot{\hat{x}}(t) = A_2 \hat{x}(t)$
$\qquad + L_{d,2} \left[y_d(k\,T) - \hat{x}_{d,1}(k\,T) \right]$ $\qquad$ (78)

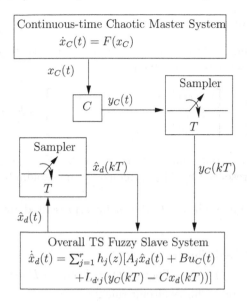

Continuous-time Chaotic Master System
$$\dot{x}_C(t) = F(x_C)$$

$x_C(t)$

C | $y_C(t)$

Sampler
T

Sampler
T

$\hat{x}_d(kT)$

$y_C(kT)$

$\hat{x}_d(t)$

Overall TS Fuzzy Slave System
$$\dot{\hat{x}}_d(t) = \sum_{j=1}^{r} h_j(z)[A_j\hat{x}_d(t) + Bu_C(t)$$
$$+ L_{d,j}(y_C(kT) - Cx_d(kT))]$$

Fig. 4. Block diagram of the sampled-data TS fuzzy synchronization system

with $L_{d,j}$ given by (74) from the values of $L_{C,j}$ obtained from solving the LMIs in (28).

In Fig. 5, the results of numerical simulations of the synchronization between (40) and the digitally redesigned TS fuzzy system (77)–(78) are presented. Here the continuous-time chaotic master is simulated using a Dormand–Prince intergration algorithm with a fixed integration step of $T_f = 0.001$ while driving signal $y(k\,T)$ was sampled with a sampled-hold period of $T_s = 0.025$. Both systems are allow to evolve freely until $t = 5$, then the TS fuzzy observer is activated.

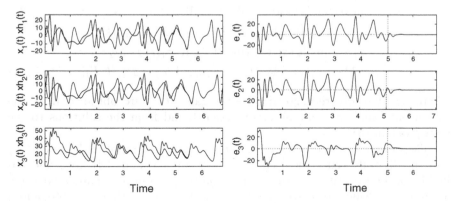

Fig. 5. Fuzzy Hybrid Chaos Synchronization Results for Chen's equation

6.4.2 Chua's Circuit

Consider that the Chua's circuit (33) is affected by a disturbance input $\nu(t)$, such that the disturbed system is given by

$$\begin{bmatrix} \dot{x}_1(t) \\ \dot{x}_2(t) \\ \dot{x}_3(t) \end{bmatrix} = \begin{bmatrix} \alpha \ [x_2(t) - x_1(t) - \Gamma(x_1(t))] \\ x_1(t) - x_2(t) + x_3(t) \\ -\beta x_2(t) \end{bmatrix} + E \ \nu(t) \tag{79}$$

Here the disturbance input matrix E is set to be $E = [1,1,1]^{\mathrm{T}}$ and the disturbances are represented by a high frequency sinuosidal function, $\nu(t) = 0.1 \sin(2\pi \ 500 \ t)$.

The disturbed Chua's circuit (79) can be represented by the TS fuzzy model

Disturbed Chua's circuit rule 1 :

If $z_1(t)$ is M_1 Then $\dot{x}(t) = A_1 x(t) + E \ \nu(t)$ \qquad (80)

Disturbed Chua's circuit rule 2 :

If $z_1(t)$ is M_2 Then $\dot{x}(t) = A_2 x(t) + E \ \nu(t)$ \qquad (81)

where $z_1(t)$, M_1, M_2, A_1, and A_2 are as defined in (37) and (38).

Now, let the disturbed Chua's circuit (79) be a continuous-time master system and let the driving signal be a sampled version of its output given by

$$y_d(t) = y_d(k\ T) = x_{d,1}(k\ T) \tag{82}$$

A *slave* system that synchronizes to the *master* system such that the error dynamics be internally stable and satisfies the performance index $\|T_{e,\nu}(s)\|_\infty$ can be constructed using the results presented above in the form

Digitally redesigned H$_\infty$ robust

Slave chua's system rule 1 :   If $z_1(t)$ *is* M_1 \qquad (83)

Then $\dot{\hat{x}}(t) = A_1 \hat{x}(t) + L_{d,1} \left[y_d(k\ T) - \hat{x}_{d,1}(k\ T) \right]$

Slave Chua's system rule 2 :   If $z_1(t)$ *is* M_2

Then $\dot{\hat{x}}(t) = A_2 \hat{x}(t) + L_{d,2} \left[y_d(k\ T) - \hat{x}_{d,1}(k\ T) \right]$ \qquad (84)

where the digitally redesigned observer gains $L_{d,j}$ are obtained from the continuous-time TS fuzzy observer design found from the solutions to the LMI problem (31).

The simulation results of the robut hybrid synchronization between (79) and the digitally redesigned TS Fuzzy system (83) and (84) are presented in Fig. 6. The continuous-time chaotic *master* system was simulated using a Dormand–Prince fixed step intergration algorithm with $T_f = 0.001$ and a sampled-hold period of $T_s = 0.1$. The observer was activated after $t = 5$.

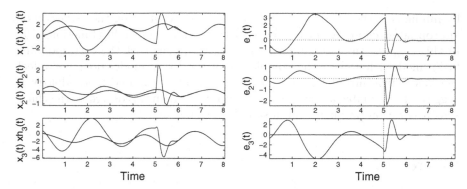

Fig. 6. Robust Fuzzy Hybrid Chaos Synchronization Results for Chua's circuit

7 Concluding Remarks

In this collaboration, the TS fuzzy model is used as an alternative representation of a chaotic system to solve the synchronization problem. In particular, the case of a sampled-data driving signal was investigated. The solution proposed is in the same lines of the basic observer-based solution to the synchronization problem. It is shown that using the fuzzy model representation an H_∞ observer can be designed, using the dual system approach in a PDC format, to synchronize two continuous-time chaotic systems, where the gains are determined from the solution to an LMI minimization problem. To take into consideration the effects of sampling on the driving signal, the design is expressed as hybrid feedback control problem; in this setting the corresponding observer gains are digitally redesigned from the original TS fuzzy observer design. Simulations of three well-known benchmark chaotic systems are used to confirm the effectiveness of the proposed fuzzy hybrid chaos synchronization algorithm.

The use of fuzzy logic controllers to represent nonlinear systems has many advantages, among them, as illustrated in this chapter and the referenced works, is the simplification of the representation and in the case of the TS fuzzy model, the possibility of using well-establish linear techniques to designed controllers and observers for nonlinear/chaotic systems. However, some limitations can be pointed out; the most significant one being the so-called curse of dimensionality. When modeling a nonlinear system, it is important to choose appropriately the structure of the rule set, because if the rule set is poorly chosen a controller may not be found, even using the LMI approach [42]. This is generally because as the number of rules increases the difficulty in finding a feasible solution also increases. In recent years, significant research efforts have been dedicated to develop methods to cope with the issue of dimensionality in fuzzy design. One approach consists simply using relaxed stability conditions on the fuzzy controller design [16]. Other approaches consists on constructing the most efficient model possible by minimizing the number of rules [43–45] or concentrating the attention on appropriately choosing the inference variables and membership functions [19]. In this chapter, the latter approach is taken to construct the corresponding fuzzy models; however, the complexity of the fuzzy model does not affect the basic structure of the results presented. The proposed methodology is suggested as a viable alternative to conventional digital redesign to the chaos synchronization problem from a sampled driving signal.

References

1. H.O. Wang, J. Li, D. Niemann, K. Tanaka: T-S fuzzy model with linear rule consequence and PDC controller: A universal framework for nonlinear control systems. In: *Proc. FUZZ-IEEE 2000*, San Antonio, TX, 2000, pp 549–554
2. L.M. Pecora, T.L. Carroll: Phys Rev Lett. **64**, 821–824 (1990)
3. IEEE Trans. Circuits Syst. I (Special issue) **40**(10) (1993)
4. Syst. Control Lett. (Special issue) **31**(5) (1997)
5. IEEE Trans. Circuits Syst. I (Special issue) **44**(10) (1997)
6. IEEE Trans. Circuits Syst. I (Special issue) **48**(12) (2001)
7. I.I. Blekhman: *Synchronization in Science and Technology* (ASME, New York, 1988)
8. W.W. Chai: *Synchronization in Coupled Chaotic Circuits and Systems* (World Scientific, Singapore 2002)
9. E. Mosekilde, Y. Maistrenko, D. Postov: *Chaotic Synchronization: Applications to Living Systems* (World Scientific, Singapore, 2002)
10. A. Pikovsky, M. Rosenblum, J. Kurths: *Synchronization: A Universal Concept in Nonlinear Science* (Cambridge University Press, UK, 2001)
11. H. Nijmeijer, I.M.Y. Mareels: IEEE Trans. Circ. Syst. I. **44**(10), 882–890 (1997)
12. H. Nijmeijer: Physica D **154**, 219–228 (2001)
13. Ö. Morgül, E. Solak: Int. J. Bifurcation Chaos **7**(6), 1307–1322 (1997)
14. T.L. Liao, N.S. Huang: IEEE Trans. Circ. Syst. I. **46**(19), 1144–11150 (1999)
15. E. Solak, Ö. Morgül, U. Ersoy: Phys. Lett. A 279, 47–55 (2001)
16. K. Tanaka, T. Ikeda, H.O. Wang: IEEE Trans. Fuzzy Syst. **6**(2), 250–265 (1998)
17. K. Tanaka, H.O. Wang: *Fuzzy Control Systems Design and Analysis* (Wiley, New York, 2001)
18. K.Y. Lian, C.S. Chiu, T.S. Chiang, P. Liu: IEEE Trans. Fuzzy Syst. **9**(1), 212–220 (2001)
19. K.Y. Lian, C.S. Chiu, T.S. Chiang, P. Liu: IEEE Trans. Fuzzy Syst. **9**(4), 539–553 (2001)
20. K.Y. Lian, T.S. Chiang, C.S. Chiu, P. Liu: IEEE Trans. Syst. Man Cyber Part B **31**(1), 66–83 (2001)
21. J.G. Barajas-Ramrez, G. Chen, L.S. Shieh: Int. J. Bifurcation Chaos **14**(8), 2721–2733 (2004)
22. S.M. Guo, L.S. Shieh, G. Chen, M. Ortega: Int. J. Bifurcation Chaos **10**(10), 2221–2231 (2000)
23. S.M. Guo, L.S. Shieh, G. Chen, C.F. Lin: IEEE Trans. Circ. Syst. I **47**(11), 1557–1570 (2000)

24. J.G. Barajas-Ramrez, G. Chen, L.S. Shieh: Int. J. Bifurcation Chaos **13**(5), 1197–1216 (2003)
25. B.C. Kuo: *Digital Control Systems* (Holt, Rinehart and Winston, New York, 1980)
26. L.S. Shieh, W.M. Wang, M.K.A. Panicker: ISA Trans. **37**, 201–213 (1998)
27. J.S.H. Tsai, L.S. Shieh, J.L. Zhang: Circuits Syst. Signal Process **12**(1), 37–49 (1993)
28. J. Yen, R. Langari: *Fuzzy Logic, Intelligence, Control and Information* (Prentice-Hall, Upper Saddle River, NJ, 1999)
29. B. Kosko: *Fuzzy Engineering* (Prentice-Hall, Upper Saddle River, NJ, 1997)
30. T. Takagi, M. Sugeno: IEEE Trans. Syst. Man Cyber. **15**, 116–132 (1985)
31. M. Sugeno, G.T. Kang: Fuzzy Sets Syst. **28**, 316–326 (1986)
32. G. Chen, T.P. Trung: *Introduction to Fuzzy Sets, Fuzzy Logic and Fuzzy Control Systems* (CRC Press, Boca Raton, FL, 2000)
33. M. Sugeno, G.T. Kang: Fuzzy Sets Syst. **28**, 329–346 (1986)
34. T. Takagi, M. Sugeno: Fuzzy Sets Syst. **45**, 135–156 (1992)
35. H.O. Wang, K. Tanaka, M.F. Griffin: IEEE-IFESAJ **2**, 531–538 (1995)
36. Y.H. Joo, L.S. Shieh, G. Chen: IEEE Trans. Fuzzy Syst. **7**, 394–408 (1999)
37. H.O. Wang, K. Tanaka, M.F. Griffin: IEEE Trans. Fuzzy Syst. **4**, 14–23 (1996)
38. G. Chen, T. Ueta: Int. J. Bifurcation Chaos **9**, 1465–1466 (1999)
39. T. Ueta, G. Chen: Int. J. Bifurcation Chaos **10**(8), 1917–1931 (2000)
40. J. Lü, G. Chen: Int. J. Bifurcation Chaos, **12**(3), 659–661 (2002)
41. G. Chen, X. Dong: *From Chaos to Order: Methodologies, Perspectives and Applications* (World Scientific, Singapore, 1988)
42. S. Boyd, L. ElGhaoui, E. Feron, V. Balakrishnan: *Linear Matrix Inequalities in Systems and Control Theory* (SIAM, Philadelphia, PA, 1994)
43. B. Kosko: Int. J. Intell. Syst. **10**, 249–255 (1995)
44. K. Mustafa, K.M. Passino: IEEE Trans. Fuzzy Syst. **9**, 194–199 (2001)
45. G. Chen, Y.H. Joo: Introduction to fuzzy control systems. In: *The Handbook of Applied Computational Intelligence*, ed by M.L. Padgett, N.B. Karayiannis, L.A. Zadeh (CRC Press, Boca Raton, FL, 2003)

Bifurcation Phenomena
in Elementary Takagi–Sugeno Fuzzy Systems

Federico Cuesta, Enrique Ponce, and Javier Aracil

Abstract. The relevance of bifurcation analysis in Takagi–Sugeno (T-S) fuzzy systems is emphasized mainly through examples. It is demonstrated that even the most simple cases can show a great variety of behaviors. Several local and global bifurcations (some of them, degenerate) are detected and summarized in the corresponding bifurcation diagrams. It is claimed that by carefully making this kind of analysis it is possible to overcome some criticism raised regarding the blind use of fuzzy systems.

1 Introduction

Fuzzy systems are nonlinear systems and, consequently, they can exhibit multiple equilibria, periodic orbits, and even chaotic attractors. To understand such richness of possible behaviors, the qualitative theory and, in particular, the bifurcation theory are two valuable tools to be known and adequately exploited. Following the ideas already introduced in a preliminary work [1], we aim at this chapter to illustrate by means of elementary examples how these tools can be useful in the analysis of fuzzy systems. Throughout the chapter, some basic bifurcation phenomena will also be introduced.

Dealing with nonlinear systems, one must be aware not only of possible complex dynamical behaviors but, more importantly, also of the influence of changes in parameters or in their structure. Through these changes, even small in magnitude, it is possible to observe drastic qualitative changes in their behavior modes (see [2]). This is the realm of bifurcation theory, which supplies tools to study the points where these changes are produced and the archetypical forms of the state portrait changes in these points. Some comprehensive works for a thorough introduction in the subject are [3–7].

As mentioned before, the bifurcation theory can be a valuable tool for understanding the behavioral richness of nonlinear systems. Roughly speaking, when system parameters are moved and as a result a qualitative change in the system response (to be deduced from a phase portrait, for instance) is observed, it is said that the system undergoes bifurcation phenomena. These phenomena can lead to a change in the number of stationary solutions (equilibrium points), to the appearance of oscillations, or even to more complex behavior (chaos, for instance). Thus, from a certain point of view, possible bifurcations are related to issues of robustness, since the system only displays

F. Cuesta et al.: *Bifurcation Phenomena in Elementary Takagi–Sugeno Fuzzy Systems*,
StudFuzz **187**, 285–315 (2006)
www.springerlink.com

behaviors that are structurally stable [3–5, 8, 9] far from the bifurcation points.

It should be remarked that the bifurcation theory makes it possible to determine not only the values of the parameters for which the qualitative behavior of the system changes but also the kind of behavior that the system will exhibit after such changes.

Even if there are no real parameters in the system to study, it is interesting from the point of view of bifurcation analysis to assume that some parameters are involved and, by studying the possible bifurcations that may arise, it is possible to obtain valuable information about the behavior corresponding to the actual values of parameters. In any case, after a bifurcation analysis it is also possible to split the parameter space on several regions with different asymptotic dynamics, see for instance [10].

These questions are quite relevant for fuzzy systems in which, through the fuzzification–defuzzification process, some "plastic" changes can occur, depending on the specific method used in that process. The changes could be associated to bifurcations, whose consequences should be known by the system user and, maybe, have not attracted enough attention yet (but see [11, 12]). It should be clearly understood that through the fuzzification–defuzzification process a fuzzy system with a component of linguistic rules has been converted into a mathematical object with well-known properties which one should not forget when using such a system. As will be seen in this chapter, some of the qualitative characteristics of these complex behaviors can be captured through a bifurcation analysis.

2 Fuzzy Systems and Bifurcation Theory

Throughout the chapter we will consider affine Takagi–Sugeno (T-S) systems. Such systems are composed, in general, by M rules of the form

$$R_i: \text{ if } x_1 \text{ is } F_i^1 \text{ and } x_2 \text{ is } F_i^2, \ldots, \text{and } x_n \text{ is } F_i^n \text{ then } \dot{\mathbf{x}} = A_i \mathbf{x} + C_i$$

for $i = 1, \ldots, M$, where F_i^j are fuzzy sets, n is the dimension of vector $\mathbf{x}$, and A_i, C_i are constant matrices of adequate dimensions. Taking all the rules in mind, we get the nonlinear dynamical system

$$\dot{\mathbf{x}} = \sum_{i=1}^{M} w_i(\mathbf{x})(A_i \mathbf{x} + C_i) , \tag{1}$$

where

$$w_i(\mathbf{x}) = \frac{\prod_{j=1}^{n} \mu_{F_i^j}(x_j)}{\sum_{i=1}^{M} \prod_{j=1}^{n} \mu_{F_i^j}(x_j)} \tag{2}$$

are nonlinear functions of the state vector $\mathbf{x} \in \mathbb{R}^n$, representing the weight or contribution of each rule R_i to the dynamics of the system. The functions

$\mu_{F_i^j}(x_j)$ represents the membership degree of the variable x_j to the fuzzy set F_i^j.

The mathematical object of (1) belongs to the class of the nonlinear dynamic systems $\dot{\mathbf{x}} = f(\mathbf{x}, p)$, where $p \in P$ stands for the parameters of the system. In a fuzzy dynamic system, the parameter space P is formed, among other things, by the parameters appearing in the membership functions L_i and the consequents.

As indicated in the introduction, the main concern of the bifurcation theory is the analysis of the qualitative changes which take place in the behavior modes of a nonlinear dynamical system as the parameters are varied. Thus, since in fuzzy systems parametric functions are involved, bifurcation theory is well suited to analyze the parametric robustness of fuzzy systems.

The first point to take into account when dealing with nonlinear dynamic systems is that they can have more than one attractor. As each attractor has its own attraction basin, the landscape of such systems can be very complex, with several basins bounded by separatrices. The shape of this landscape can suffer qualitative changes for some critical values of the parameters. These values are called *bifurcation points*. Elementary bifurcations give the simplest ways in which the qualitative structure of the state portrait changes. Fortunately, with a few of these elementary bifurcations many practical situations can be dealt with.

Bifurcation phenomena can be classified into two main classes, namely local and global ones. Roughly speaking, *local bifurcations* are due to changes in the dynamics of a small region of the phase space, typically, a neighborhood of an equilibrium point. When all the eigenvalues of the linearization of the system at one equilibrium point have real parts different from zero, the equilibrium is called *hyperbolic*. As is well known, hyperbolic equilibria whose eigenvalues have negative real parts are asymptotically stable, while if there is some eigenvalue with positive real part, then the equilibrium is unstable. Thus, starting for instance from a stable equilibrium, a crossing of certain eigenvalues through the imaginary axis will lead to a local bifurcation. The most basic bifurcations are those corresponding to a zero eigenvalue (saddle-node, pitchfork, and transcritical bifurcations) or to a pair of pure imaginary eigenvalues (Hopf bifurcation) [3, 5]. For instance, through a supercritical (also called *soft* [5]) Hopf bifurcation a stable limit cycle is born from a point attractor which becomes unstable. Hopf bifurcations can also be detected by means of harmonic balance [13, 14].

Furthermore, there can be *global bifurcations*. In that case, the bifurcation is produced in such a way that it involves phenomena not reducible to locality. That occurs, for instance, when the interaction of a limit cycle with a saddle point is produced. In such a case, an attractor can suddenly appear or disappear. Apart from the quoted situation, which is called a *homoclinic* or *saddle connection*, and its analog involving two equilibria (*heteroclinic connection*) [3, 5], other global bifurcations are, for instance, the saddle-node bifurcation of periodic orbits (two periodic orbits of different stability character collide to disappear or vice versa) and the Hopf bifurcation from infinity, where a periodic orbit of great amplitude comes from (goes to) infinity [15]. Global bifurcations cannot be detected by local analysis and normally need numerical or approximate global methods. Again, the harmonic balance method can help [16–18].

The relevance of qualitative analysis and bifurcations in T-S systems has already been pointed out (see [19], and the references therein). However, their complicated structure from the mathematical point of view explains the lack of theoretical results up to this moment.

To lessen the mentioned difficulties, it is possible to resort to piecewise linear membership functions with local support. These functions clearly induce a partition in the state space of T-S systems (see Fig. 1). Thus, the state space is divided into operating regions X_i, where only one rule and the corresponding affine dynamic is active, and interpolation regions in between them, where several dynamics are present. Nevertheless, the problem remains nontrivial to analyze (bifurcation theory normally assumes a high degree of smoothness for the system).

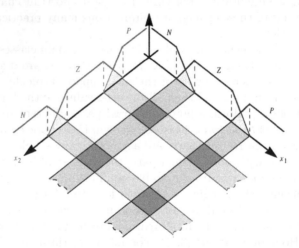

Fig. 1. State space partition induced by piecewise linear membership functions

For the sake of simplicity, in this chapter, the simplest T-S systems are considered, that is, $\mathbf{x} \in \mathbb{R}^2$ and only x_1 with two fuzzy sets will be considered in the antecedents, yielding two rules ($M = 2$ in (2)). As indicated before, in order to facilitate the analysis, normalized piecewise linear membership functions will be assumed; even for this case nontrivial behaviors will be found. To be more precise, all the examples analyzed in this chapter have the following structure:

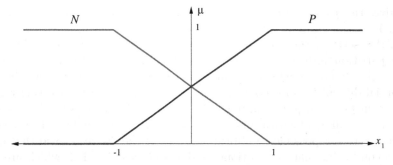

Fig. 2. Membership functions

$$\text{if } x_1 \text{ is N then } \dot{\mathbf{x}} = A_1\mathbf{x} + C_1 \, ,$$

$$\text{if } x_1 \text{ is P then } \dot{\mathbf{x}} = A_2\mathbf{x} + C_2 \, ,$$

(3)

where the linguistic terms N and P are represented by the normalized trape-zoidal membership functions shown in Fig. 2, resulting in the following dynamical system:

$$\dot{\mathbf{x}} = w_1(\mathbf{x})(A_1\mathbf{x} + C_1) + w_2(\mathbf{x})(A_2\mathbf{x} + C_2) \, , \qquad (4)$$

where

$$w_1(\mathbf{x}) = \mu_N(x_1) = \begin{cases} 1 & \text{for} \quad x_1 < -1 \\ \frac{1}{2} - \frac{1}{2}x_1 & \text{for} \quad |x_1| \le 1 \, , \\ 0 & \text{for} \quad x_1 > 1 \end{cases}$$

$$w_2(\mathbf{x}) = \mu_P(x_1) = \begin{cases} 0 & \text{for} \quad x_1 < -1 \\ \frac{1}{2}x_1 + \frac{1}{2} & \text{for} \quad |x_1| \le 1 \, . \\ 1 & \text{for} \quad x_1 > 1 \end{cases}$$

(5)

Note that $\mu_N(x_1) + \mu_P(x_1) = 1$, for all x_1, and that two operating regions and only one interpolating region will appear.

Expressions (3)–(5) define a dynamical system in $\mathbb{R}^2$, which is governed by a piecewise smooth vector field (it is only of class C^0 for $x_1 = 1$ and $x_1 = -1$). Of course, the system is well posed and for every initial condition it is possible to assume the existence and uniqueness of solutions.

Therefore, the dynamic system is equivalent to

$$\dot{\mathbf{x}} = A_1\mathbf{x} + C_1, \quad \text{for } x_1 < -1 \, ,$$

$$\dot{\mathbf{x}} = A_2\mathbf{x} + C_2, \quad \text{for } x_1 > 1 \, ,$$

(6)

and gives rise to a quadratic system in the interpolation region, that is, for $|x_1| < 1$.

In this setting, the bifurcation analysis can take advantage of the state space partition induced by the membership functions. From the point of view of the dynamics which are intrinsic to every operating region, there is no special difficulties in the analysis. In fact, since in every operating region the system becomes purely affine, all that has to be done is to characterize the corresponding linearization, which is given by the matrix $A_i, i = 1, 2$, and to locate the possible equilibrium points. It should be noticed that even *virtual* equilibria, that is, solutions of $A_i \mathbf{x} + C_i = 0, i = 1, 2$, which are not in the corresponding region, do govern the dynamics in the region. In the interpolating region, the analysis of its intrinsic dynamics is more complex, as it is a quadratic system, but no special difficulties are to be expected. The problem arises when the regional dynamics are merged into a unique state space. For instance, the merging of only two affine regions is enough to produce limit cycles (see [20]), which is impossible for each separate region. The interaction of trajectories in different regions can produce interesting global phenomena, in spite of the linear nature of the two systems involved [21].

As will be seen, these elementary T-S systems can display local and global bifurcations, some of them due to the structure of the regions and others originated by the lack of differentiability. In this last case (low differentiability) we speak of degenerate bifurcations. It should be emphasized that these degenerate bifurcations might not persist after some smoothing of the piecewise linear membership functions, even when the perturbations are small. That is not the case for the rest of the bifurcations found in the following examples; the first two have homogeneous consequents and the last one is composed of affine consequents.

Finally, to facilitate the analysis of periodic orbits the Poincaré section method will be used. This technique reduces the problem to the study of discrete dynamics on a space of dimension $n - 1$, where n is the dimension of the original system.

If γ is a periodic orbit of $\dot{x} = f(x)$ with period $T > 0$, and Φ is the flow of the system, then $\Phi_{t+T}(x) = \Phi_t(x)$ for $x \in \gamma$ and $t \in \mathbb{R}$ arbitrary. Moreover, given a local transversal section of $\mathcal{S}$ at $x \in \gamma$, it can be shown that there is an open set $\mathcal{U}$ of x in $\mathcal{S}$ and a unique differentiable function $\rho : \mathcal{U} \to \mathbb{R}$, so that

$$\rho(x) = T, \quad \Phi_{\rho(y)}(y) \in \mathcal{S}, \ \forall y \in \mathcal{U} . \tag{7}$$

where $\rho(y)$ is the time required by the orbit starting at point $y \in \mathcal{U}$ to reach the section $\mathcal{S}$ at a point $P(y)$ (see Fig. 3a).

Thus, there exists an application $P : \mathcal{U} \to \mathcal{S}$, defined by

$$P(y) = \Phi_{\rho(y)}(y), \quad y \in \mathcal{U} . \tag{8}$$

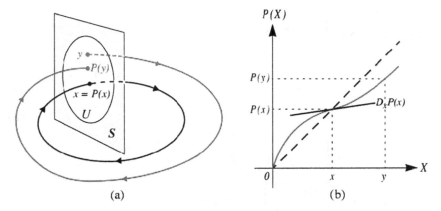

$P(X)$

Fig. 3. (a) Poincaré section in a three-dimensional state space. (b) Resulting Poincaré map in a bi-dimensional space

where P is the Poincaré map or first return map associated to the flow in a neighborhood of the closed periodic orbit γ. Therefore, $P(x) = x$, with $x \in \gamma$, is a fixed point of the Poincaré map.

The linear approximation of P in x is $D_x P(x)$, and it can be shown [22] that the eigenvalues of $D_x P(x)$ are independent on the point $x \in \gamma$ chosen for tracing the section and on the section S itself. Therefore, it is possible to analyze the stability of a periodic orbit from the eigenvalues of $D_x P(x)$, which in case of dimension 2 ($n = 2$) are given by the slope of the curve $P(x)$ at x (see Fig. 3b).

3 Examples

3.1 System with Linear Consequents

As a first example, consider the fuzzy system

$$\text{if } x_1 \text{ is N then } \dot{x} = \begin{bmatrix} 1 & -\frac{1}{2} \\ 1 & 0 \end{bmatrix} x \,,$$

$$\text{if } x_1 \text{ is P then } \dot{x} = \begin{bmatrix} -\beta & -\frac{1}{2} \\ \beta & 0 \end{bmatrix} x \,,$$

(9)

where β is the parameter which can vary (i.e., the bifurcation parameter). The linguistic terms N and P are represented by normalized trapezoidal membership functions (see Fig. 2).

As indicated before, the phase space can be partitioned into two operating regions and one interpolating region according to the membership functions, so that

$$\dot{\mathbf{x}} = \begin{bmatrix} 1 & -\frac{1}{2} \\ 1 & 0 \end{bmatrix} \mathbf{x}, \quad \text{for } x_1 < -1,$$

$$\dot{\mathbf{x}} = \frac{1 - x_1}{2} \begin{bmatrix} 1 & -\frac{1}{2} \\ 1 & 0 \end{bmatrix} \mathbf{x} + \frac{1 + x_1}{2} \begin{bmatrix} -\beta & -\frac{1}{2} \\ \beta & 0 \end{bmatrix}$$

$$\mathbf{x} = \begin{bmatrix} \frac{1-\beta}{2} x_1 - \frac{1+\beta}{2} x_1{}^2 - \frac{1}{2} x_2 \\ \frac{1+\beta}{2} x_1 - \frac{1-\beta}{2} x_1{}^2 \end{bmatrix}, \quad \text{for } |x_1| \leq 1,$$

$$\dot{\mathbf{x}} = \begin{bmatrix} -\beta & -\frac{1}{2} \\ \beta & 0 \end{bmatrix} \mathbf{x}, \quad \text{for } x_1 > 1.$$

(10)

The global dynamics in the phase space is the composition of the three regional dynamics. Therefore, all the trajectories can be determined by gluing together (not only continuously but also with a continuous derivative) trajectories in each region. To start with, it is easily concluded that the dynamics in the left operating region ($x_1 < -1$) is a linear dynamics, independent of the bifurcation parameter. For this region the origin is the only equilibrium governing the dynamics (in fact, one unstable focus), but note that it is out of the region and so it constitutes a virtual equilibrium. Trajectories enter this region from the middle region at $x_1 = -1$ for $x_2 > -2$, and they always return to the middle region at $x_1 = -1$ for $x_2 < -2$.

Now, considering the right operating region, the corresponding dynamics, which is also linear, depends on the value of β. Again, for $\beta \neq 0$, the only equilibrium governing the dynamics is the origin (a virtual equilibrium) and by linear analysis it is possible to make the following assertions:

- for $\beta < 0$ the virtual equilibrium at the origin is a saddle.
- for $\beta = 0$ there appears a continuum of equilibria at the x_1-axis. For $x_1 \geq 1$ these points are actual equilibrium points and the dynamics is rather degenerate as all trajectories are horizontal straight lines (entering the region for $x_2 < 0$ and leaving it for $x_2 > 0$).
- for $0 < \beta < 2$ the dynamics is of a stable focus type. Trajectories enter the region at $x_1 = 1$ for $x_2 < -2\beta$ and always return to the middle region at $x_1 = 1$ for $x_2 > -2\beta$.
- for $\beta = 2$ the dynamics is governed by an improper stable node. Trajectories enter the right region at $x_1 = 1$ for $x_2 < -4$ and always return to the middle region at $x_1 = 1$ for $x_2 \in (-4, -1)$. At $x_1 = 1$ for $x_2 \geq -1$ trajectories leave the region coming from the point at infinity.
- for $\beta > 2$ the dynamics is governed by a stable node with a behavior of trajectories similar to that of the previous case.

The analysis of the middle (interpolating) region dynamics is somehow more complex as it is a quadratic system. First of all, apart from the equilibrium point at the origin (now, an actual equilibrium) if $\beta \neq 1$, there is another equilibrium point at

$$\bar{x}_1 = \frac{1+\beta}{1-\beta}, \qquad \bar{x}_2 = -\frac{4\beta(1+\beta)}{(1-\beta)^2}, \tag{11}$$

which is a virtual equilibrium for $\beta > 0$, and an actual one for $\beta \leq 0$.

Note that for the complete system the value $\beta = 0$ clearly represents a bifurcation value since the dynamics changes at this value. When $\beta > 0$, the system has one equilibrium point. At $\beta = 0$ there appears a half straight line of equilibrium points ($x_1 \geq 1$, with $x_2 = 0$) from which the system inherits one new equilibrium point at $(\bar{x}_1, \bar{x}_2)$ for $\beta < 0$. This bifurcation can be thought of as a degenerate saddle-node bifurcation (DSN).

The linearization of the equilibrium point at the origin is

$$J(0,0) = \begin{bmatrix} \frac{1-\beta}{2} & -\frac{1}{2} \\ \frac{1+\beta}{2} & 0 \end{bmatrix}, \tag{12}$$

so that

$$\text{trace } J(0,0) = \frac{1-\beta}{2}, \qquad \det J(0,0) = \frac{1+\beta}{2}. \tag{13}$$

For $\beta < -1$ the origin is unstable (a saddle point). For $\beta = -1$ it is an unstable nonhyperbolic equilibrium (which is a necessary condition for a bifurcation to occur; the character of this bifurcation will be analyzed later). The origin is also unstable (node or focus) for $-1 < \beta < 1$, becoming stable for $\beta > 1$. Again, when $\beta = 1$, this equilibrium is nonhyperbolic, since its linearization has a pair of pure imaginary eigenvalues, which might be associated to a Hopf bifurcation. To detect the character of this Hopf bifurcation, an additional analysis is required. This can be performed by means of the Poincaré map for different values of β around $\beta = 1$ (see Fig. 6). From this it is concluded that for $\beta = 1$ there is a global nonlinear center (GC) that for $|\beta - 1| \neq 0$, and $|\beta - 1|$ small, does not give rise to any periodic orbit. Notice that this can also be concluded from a mathematical analysis of the global system equations [1] as shown in the appendix. From this appendix it is concluded that for $\beta = 1$ there is a global nonlinear center (GC) that for $|\beta - 1| \neq 0$, and small, does not give rise to any periodic orbit.

To study the character of the second equilibrium point, it suffices to compute from (10) the corresponding linearization, namely,

$$J(\bar{x}_1, \bar{x}_2) = \begin{bmatrix} \frac{(1-\beta)}{2} - (1+\beta)\bar{x}_1 & -\frac{1}{2} \\ \frac{(1+\beta)}{2} - (1-\beta)\bar{x}_1 & 0 \end{bmatrix}$$

$$= \begin{bmatrix} -\frac{1+6\beta+\beta^2}{2(1-\beta)} & -\frac{1}{2} \\ -\frac{1+\beta}{2} & 0 \end{bmatrix}, \tag{14}$$

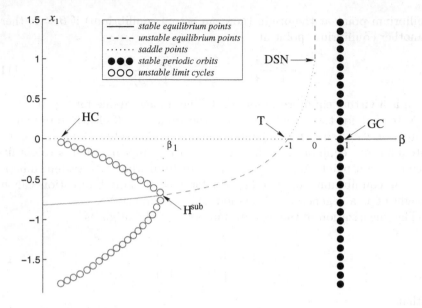

Fig. 4. Bifurcation diagram of Example 1 (HC = homoclinic connection; H^{sub} = subcritical Hopf; T = transcritical; DSN = degenerate saddle node; GC = global center)

and it should be remarked that

$$\text{trace } J(\bar{x}_1, \bar{x}_2) = -\frac{1 + 6\beta + \beta^2}{2(1 - \beta)}, \qquad \det J(\bar{x}_1, \bar{x}_2) = -\frac{1 + \beta}{4}. \qquad (15)$$

Thus, the point $(\bar{x}_1, \bar{x}_2)$ is a saddle point for $\beta > -1$. When $\beta = -1$ the system undergoes a bifurcation, since this equilibrium and the equilibrium at the origin coalesce.

Analyzing the sign of the trace in (15), it can be concluded that it is positive both for $\beta > 1$ and for $\beta \in (\beta_1, \beta_2)$, where β_1, β_2 are the roots of the quadratic $1 + 6\beta + \beta^2 = 0$, that is

$$\beta_1 = -3 - 2\sqrt{2}, \qquad \beta_2 = -3 + 2\sqrt{2}. \qquad (16)$$

With this information it is not difficult to see that at $\beta = -1$ a transcritical bifurcation takes place (two equilibrium points collide, interchanging their stability properties).

Also, another bifurcation arises when $\beta = \beta_1$, since the trace changes its sign with a positive determinant. In fact, the system undergoes a subcritical Hopf bifurcation (to be denoted as H^{sub}), as the point $(\bar{x}_1, \bar{x}_2)$ (an unstable focus for $\beta > \beta_1$) becomes a stable focus for $\beta < \beta_1$, with one unstable limit cycle around it. As predicted by the bifurcation theory, the amplitude of this limit cycle evolves with the bifurcation parameter and it can be approximated by an expression that is $O(|\beta-\beta_1|^{\frac{1}{2}})$ for $|\beta-\beta_1|$ small. Therefore, the unstable limit cycle will remain in the interpolating region for $\beta < \beta_1$ if $|\beta - \beta_1|$ is sufficiently small, stating the attraction basin of the point $(\bar{x}_1, \bar{x}_2)$ (see Fig. 5a).

Letting $|\beta - \beta_1|$ to be larger (with $\beta < \beta_1$), the above unstable limit cycle begins to enter the left operating region, also approaching the origin from its righthand side. Then, for certain value of β, a global bifurcation appears when the limit cycle becomes a loop connecting one branch of the unstable manifold at the origin with one branch of its stable manifold (saddle connection or homoclinic bifurcation HC). After this critical value, the relative positions of these two manifold branches change, giving as a result the disappearance of the limit cycle. Now, the attraction basin of the point $(\bar{x}_1, \bar{x}_2)$ is no longer bounded and, depending on the situation of the initial conditions with respect to the stable manifold of the origin, trajectories go to infinity or to the stable point $(\bar{x}_1, \bar{x}_2)$.

Clearly, for $\beta > 1$ the system exhibits a standard behavior, with no sensitivity to initial conditions and only one equilibrium, which is globally stable. When $\beta < \beta_1$, the system also possesses a stable equilibrium, but it can be said that it is not robust, since its attraction basin is limited. For the intermediate values of β, that is, in the range $[\beta_1, 1]$, the system is unstable, and so it can be considered useless.

Thus, very different system behavior may be found depending on the actual value of β. It must be emphasized that the identification of the adequate value of β turns out to be a critical issue.

The whole analysis for the system (9) can be summarized in the bifurcation diagram of Fig. 4. Also, the corresponding phase portraits for several values of β are sketched in Figs. 5 and 6. In Fig. 6 there are also shown the corresponding Poincaré maps for different values of β with the section S set at $x_1 = 0$. Notice that the slope of the curve in Fig. 6a is greater than 1 and the system is unstable. In Fig. 6b the slope is always 1, which corresponds to a global center. On the other hand, in Fig. 6c the slope is always lower than 1, showing global stability.

3.2 System with Modified Consequent

In this example, we will consider a change in the second rule with respect to the previous one, to use instead

$$\text{if } x_1 \text{ is P then } \dot{\mathbf{x}} = \begin{bmatrix} -1 & -\frac{1}{2} \\ 1 & \beta \end{bmatrix} \mathbf{x}, \tag{17}$$

taking the same normalized trapezoidal membership functions (5) as in Example 1. Now, the system becomes

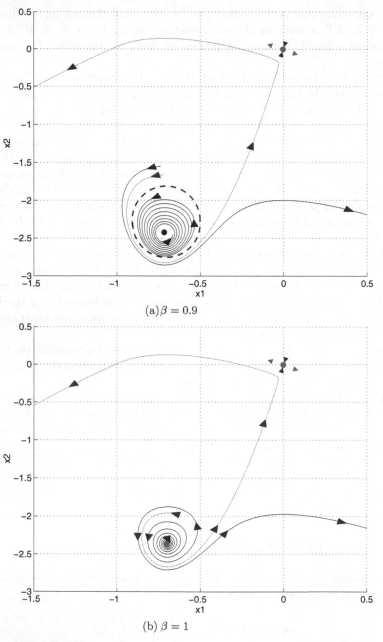

(a)$\beta = 0.9$

(b) $\beta = 1$

Fig. 5. Phase portraits of Example 1 for different values of β. (**a**) For $\beta = -6$, there is an unstable limit cycle stating the attraction basin of a stable equilibrium (*dashed line*). (**b**) For $\beta = -5.6$ there are no stable equilibrium points (at $\beta = \beta_1$ the system undergoes a subcritical Hopf bifurcation within the interpolating region)

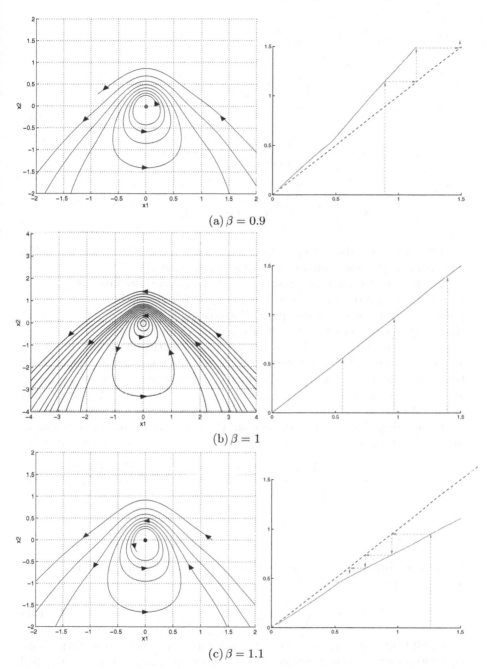

(a) $\beta = 0.9$

(b) $\beta = 1$

(c) $\beta = 1.1$

Fig. 6. Phase portraits of Example 1 for different values of β and their corresponding Poincaré maps. (a) For $\beta = 0.9$ the system has an unstable equilibrium at the origin. (b) For $\beta = 1$ the system has a global center. (c) For $\beta = 1.1$ the origin is the global attractor of the system

$$\dot{\mathbf{x}} = \begin{bmatrix} 1 & -\frac{1}{2} \\ 1 & 0 \end{bmatrix} \mathbf{x}, \quad \text{for } x_1 < -1,$$

$$\dot{\mathbf{x}} = \frac{1-x_1}{2} \begin{bmatrix} 1 & -\frac{1}{2} \\ 1 & 0 \end{bmatrix} \mathbf{x} + \frac{1+x_1}{2} \begin{bmatrix} -1 & -\frac{1}{2} \\ 1 & \beta \end{bmatrix}$$

$$\mathbf{x} = \begin{bmatrix} -x_1^2 - \frac{1}{2}x_2 \\ x_1 + \frac{\beta}{2}x_2 + \frac{\beta}{2}x_1 x_2 \end{bmatrix}, \quad \text{for } |x_1| \le 1,$$

$$\dot{\mathbf{x}} = \begin{bmatrix} -1 & -\frac{1}{2} \\ 1 & \beta \end{bmatrix} \mathbf{x}, \quad \text{for } x_1 > 1.$$

(18)

Repeating the above steps, the first remark is that the dynamics in the left operating region is identical to that of Example 1.

For the right operating region ($x_1 > 1$) the linear dynamics is governed by a virtual equilibrium at the origin for $\beta \ne 0.5$. This equilibrium is a saddle point for $\beta > 0.5$ and a stable point (focus or node) for $\beta < 0.5$. For $\beta = 0.5$ there appears a continuum of equilibria making up half of the straight line $2x_1 + x_2 = 0$ with $x_1 \ge 1$ (for $x_1 < 1$ they constitute virtual equilibria) as shown in Fig. 10. Trajectories enter the middle interpolating region from the right region, coming from the point at infinity, at $x_1 = 1$ for $x_2 \ge -2$, and they return to the right region at $x_1 = 1$ for $x_2 < -2$ approaching asymptotically one of the continuum of equilibria.

Concerning the interpolating region, the origin is always an actual equilibrium, and for $\beta \notin (-4, 0]$ there are two other equilibrium points, namely

$$\bar{x}_1^1 = \frac{-\beta + \sqrt{\beta^2 + 4\beta}}{2\beta}, \qquad \bar{x}_2^1 = \frac{-2 - \beta + \sqrt{\beta^2 + 4\beta}}{\beta}, \qquad (19)$$

and

$$\bar{x}_1^2 = \frac{-\beta - \sqrt{\beta^2 + 4\beta}}{2\beta}, \qquad \bar{x}_2^2 = \frac{-2 - \beta - \sqrt{\beta^2 + 4\beta}}{\beta}, \qquad (20)$$

both coalescing for $\beta = -4$ at the point $(-\frac{1}{2}, -\frac{1}{2})$. The linearization matrix at $(-\frac{1}{2}, -\frac{1}{2})$ is

$$J\left(-\frac{1}{2}, -\frac{1}{2}\right) = \begin{bmatrix} 1 & -\frac{1}{2} \\ 2 & -1 \end{bmatrix}, \qquad (21)$$

with null trace and null determinant. Therefore, this point is a nonhyperbolic equilibrium with two zero eigenvalues. The corresponding bifurcation is called a Bogdanov–Takens bifurcation (see [5]; this bifurcation needs two parameters to make all the possible behaviors nearby visible). Thus the point $(-\frac{1}{2}, -\frac{1}{2})$ is a *cuspidal point*, and by moving β a one-dimensional section of the unfolding of the Bogdanov–Takens bifurcation will become visible.

The equilibrium $(\bar{x}_1^1, \bar{x}_2^1)$ of the interpolation region is an actual equilibrium both for $\beta \leq -4$ and for $\beta \geq 0.5$, and a virtual equilibrium for $0 < \beta < 0.5$. On the other hand, the point $(\bar{x}_1^2, \bar{x}_2^2)$ is an actual equilibrium only for $\beta \leq -4$ (a virtual equilibrium for $\beta > 0$).

By considering the analysis of both the right and the interpolation regions, $\beta = 0.5$ clearly represents a bifurcation value for the whole system. When $\beta < 0.5$ (with $|\beta - 0.5|$ small), the only actual equilibrium is the origin. At $\beta = 0.5$ there appears a halfstraight line of equilibrium points ($x_1 \geq 1$, with $x_2 = -2x_1$) from which the system inherits a new equilibrium point at $(\bar{x}_1^1, \bar{x}_2^1)$ for $\beta > 0.5$, in a (DSN) bifurcation (see Fig. 10).

Returning to the analysis of the interpolation region, we will proceed to analyze the three equilibrium points within the region [namely, the origin, (19), and (20)]. The linearization of the equilibrium point at the origin is

$$J(0,0) = \begin{bmatrix} 0 & -\frac{1}{2} \\ 1 & \frac{\beta}{2} \end{bmatrix}, \tag{22}$$

so that

$$\text{trace } J(0,0) = \frac{\beta}{2}, \qquad \det J(0,0) = \frac{1}{2}. \tag{23}$$

Thus, the origin is stable for $\beta < 0$ (a node for $\beta < -2\sqrt{2}$, and a focus for $-2\sqrt{2} < \beta < 0$). For $\beta = 0$ it is a nonhyperbolic equilibrium. And it is unstable for $\beta > 0$ (a focus for $0 < \beta < 2\sqrt{2}$, and a node for $\beta > 2\sqrt{2}$).

For $\beta = 0$ a Hopf bifurcation could be expected. It is interesting to note that the system for $\beta = 0$ is the same as that of Example 1 for $\beta = 1$. From Lemma A1 of Appendix A, at $\beta = 0$ there is a global nonlinear center. However, for $\beta > 0$, and small, there now appears a stable limit cycle, as a result of a supercritical Hopf bifurcation at the infinity (the existence of this bifurcation can be confirmed by using the techniques introduced in [15]). Notice that this limit cycle did not exist in Example 1. The limit cycle can also be characterized based on the Poincaré map as shown in Fig. 7. Particularly, the slope in Fig. 7c shows that it is a stable limit cycle. The amplitude of the stable limit cycle decreases as the bifurcation parameter grows. Furthermore, it is possible to show that the periodic orbit disappears suddenly at $\beta = 0.5$, in a degenerate global bifurcation, due to the appearance of the continuum of equilibria in the right region (see Fig. 8).

In the case of point $(\bar{x}_1^1, \bar{x}_2^1)$, the linearization process results in

$$J(\bar{x}_1^1, \bar{x}_2^1) = \begin{bmatrix} \frac{\beta - \sqrt{\beta^2 + 4\beta}}{\beta} & -\frac{1}{2} \\ \frac{-\beta + \sqrt{\beta^2 + 4\beta}}{2} & \frac{\beta + \sqrt{\beta^2 + 4\beta}}{4} \end{bmatrix}, \tag{24}$$

and so,

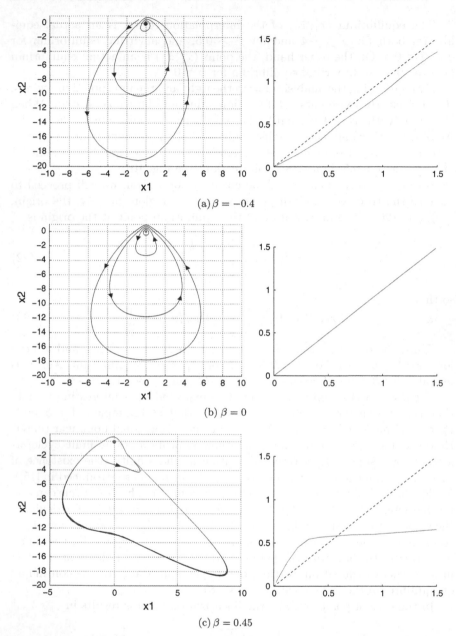

(a) $\beta = -0.4$

(b) $\beta = 0$

(c) $\beta = 0.45$

Fig. 7. Phase portraits of Example 2 for different values of β and their corresponding Poincaré map. (**a**) For $\beta = -0.4$ the origin is the global attractor of the system. (**b**) For $\beta = 0$ the system has a global center. (**c**) For $\beta = 0.45$ there is a stable limit cycle surrounding the origin which is unstable (at $\beta = 0$ the system undergoes a supercritical Hopf bifurcation at infinity)

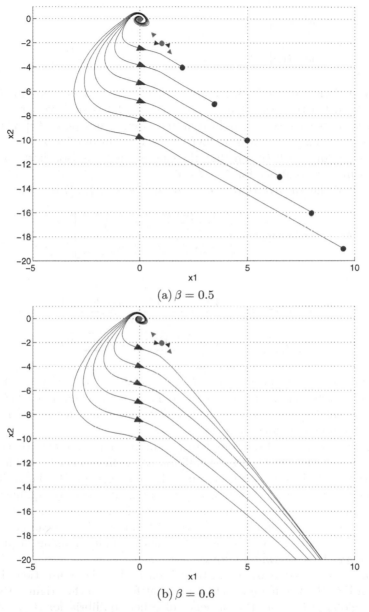

(a) $\beta = 0.5$

(b) $\beta = 0.6$

Fig. 8. Phase portraits of Example 2 for different values of β. (a) For $\beta = 0.5$ the system has a half line of stable equilibrium points at $x_2 = -2x_1$ with $x_1 \geq 1$. Appearance of this continuum of equilibria causes that the periodic orbit in Fig. 7c disappears suddenly. (b) For $\beta = 0.6$ there are no stable equilibrium points

$$\text{trace } J\big(\bar{x}_1^1, \bar{x}_2^1\big) = \frac{4\beta + \beta^2 + (\beta - 4)\sqrt{\beta(\beta + 4)}}{4\beta}, \tag{25}$$

$$\det J\big(\bar{x}_1^1, \bar{x}_2^1\big) = \frac{\sqrt{\beta(\beta + 4)}(\beta - \sqrt{\beta(\beta + 4)})}{4\beta}. \tag{26}$$

Trace (25) is positive both for $\beta < -4$ and $\beta > \frac{4}{3}$ (being zero for $\beta = -4$ and $\beta = \frac{4}{3}$). On the other hand, determinant (26) is positive for $\beta < -4$. Therefore, it is easy to conclude that the equilibrium point $(\bar{x}_1^1, \bar{x}_2^1)$ is an unstable focus for $\beta < -4$, a nonhyperbolic equilibrium for $\beta = -4$, and a saddle point for $\beta > 0$ (recall that the equilibrium does not exist for $-4 < \beta \leq 0$, being a virtual one for $0 < \beta < 0.5$).

Similarly, the linearization at equilibrium point $(\bar{x}_1^2, \bar{x}_2^2)$ is

$$J\big(\bar{x}_1^2, \bar{x}_2^2\big) = \begin{bmatrix} \frac{\beta + \sqrt{\beta^2 + 4\beta}}{\beta} & -\frac{1}{2} \\ \frac{-\beta - \sqrt{\beta^2 + 4\beta}}{2} & \frac{\beta - \sqrt{\beta^2 + 4\beta}}{4} \end{bmatrix}, \tag{27}$$

giving

$$\text{trace } J\big(\bar{x}_1^2, \bar{x}_2^2\big) = \frac{4\beta + \beta^2 + (4 - \beta)\sqrt{\beta(\beta + 4)}}{4\beta}, \tag{28}$$

$$\det J\big(\bar{x}_1^2, \bar{x}_2^2\big) = \frac{\sqrt{\beta(\beta + 4)}(\beta + \sqrt{\beta(\beta + 4)})}{4\beta}. \tag{29}$$

Now, trace (28) is positive for $\beta > 0$ and it is zero for $\beta = -4$. Also, determinant (29) is null only for $\beta = -4$. Thus, equilibrium point $(\bar{x}_1^2, \bar{x}_2^2)$ is a saddle point for $\beta < -4$, a nonhyperbolic equilibrium for $\beta = -4$, and a virtual unstable node for $\beta > 0$.

Therefore, at $\beta = -4$ the complete system undergoes a bifurcation similar to the saddle-node bifurcation of equilibria (it could be called a saddle-focus bifurcation), which is in fact a section of the more general Bogdanov–Takens bifurcation. Thus, for $\beta > -4$, with $|\beta + 4|$ small, the only equilibrium of the system is the origin, which is globally stable. However, for $\beta < -4$ the system exhibits three equilibria (see Fig. 9), due to the appearance of two new equilibria (an unstable focus and a saddle point), making in principle the stability of the origin only local.

Finally, the whole analysis can be summarized in the bifurcation diagram shown in Fig. 10. The main conclusion is that for $\beta < 0$ the origin is the only (quasi-) global attractor (the presence of other equilibria for $\beta \leq -4$ does not preclude this assertion, since excepting these equilibrium points makes all trajectories tend to the origin).

3.3 System with Affine Consequents

The last example deals with an affine (includes an offset term) T-S fuzzy system

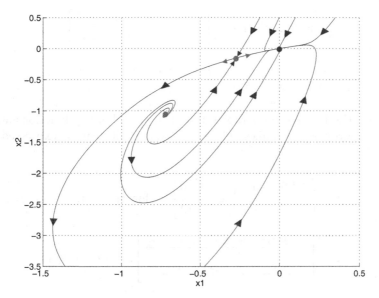

Fig. 9. Phase portrait of Example 2 for $\beta = -5$; the origin is the only stable equilibrium point of the system

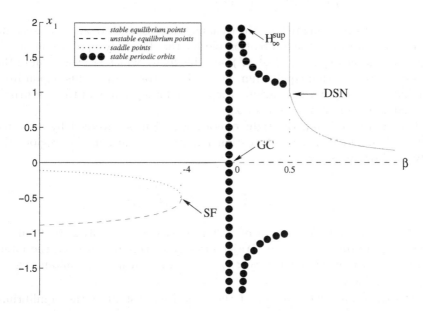

Fig. 10. Bifurcation diagram of Example 2 (SF = saddle focus from a cuspidal point; H_∞^{sup} = supercritical Hopf at infinity; DSN = degenerate saddle node; GC = global center)

$$\text{if } x_1 \text{ is N then } \dot{\mathbf{x}} = \begin{bmatrix} -4.5 & -1.5 \\ 3 & -6 \end{bmatrix} \mathbf{x} + \begin{bmatrix} 0 \\ 12 \end{bmatrix},$$

$$\text{if } x_1 \text{ is P then } \dot{\mathbf{x}} = \begin{bmatrix} \beta & -3 \\ 6 & \beta \end{bmatrix} \mathbf{x} + \begin{bmatrix} 0 \\ -12 \end{bmatrix}, \tag{30}$$

where β is the bifurcation parameter, and the membership functions are the same as in previous examples. Therefore, system (30) can be expressed as

$$\dot{\mathbf{x}} = \begin{bmatrix} -4.5 & -1.5 \\ 3 & -6 \end{bmatrix} \mathbf{x} + \begin{bmatrix} 0 \\ 12 \end{bmatrix}, \quad \text{for } x_1 < -1,$$

$$\dot{\mathbf{x}} = \begin{bmatrix} \frac{(2\beta+9)x_1^2+(2\beta-9)x_1-3x_1x_2-9x_2}{4} \\ \frac{3x_1^2-15x_1+(\beta+6)x_1x_2+(\beta-6)x_2}{2} \end{bmatrix}, \quad \text{for } |x_1| \le 1, \tag{31}$$

$$\dot{\mathbf{x}} = \begin{bmatrix} \beta & -3 \\ 6 & \beta \end{bmatrix} \mathbf{x} + \begin{bmatrix} 0 \\ -12 \end{bmatrix}, \quad \text{for } x_1 > 1.$$

In the left operating region, the dynamics corresponds to an affine system, being independent of the bifurcation parameter. The only equilibrium in this region should be $(x_1, x_2) = (\frac{-4}{7}, \frac{12}{7})$, a stable focus, which is always out of the region, being a virtual equilibrium point. Trajectories enter this region from the middle (interpolating) region at $x_1 = -1$ for $x_2 > 3$, and they return to the middle region at $x_1 = -1$ for $x_2 < 3$.

On the other hand, the right operating region is governed by an affine system depending on the bifurcation parameter. Thus, in this region, the only equilibrium is given by

$$\bar{x}_1 = \frac{36}{18 + \beta^2}, \qquad \bar{x}_2 = \frac{12\beta}{18 + \beta^2}, \tag{32}$$

which is an actual equilibrium of system (31) for $\beta \in [-3\sqrt{2}, 3\sqrt{2}]$, and a virtual equilibrium elsewhere. Trajectories enter this region from the interpolating region at $x_1 = 1$ for $x_2 < \frac{\beta}{3}$, and they return to the middle region at $x_1 = -1$ for $x_2 > \frac{\beta}{3}$.

The trace and determinant of the linearized system at the equilibrium point $(\bar{x}_1, \bar{x}_2)$ are

$$\text{trace } J(\bar{x}_1, \bar{x}_2) = 2\beta, \quad \det J(\bar{x}_1, \bar{x}_2) = 18 + \beta^2. \tag{33}$$

Therefore for $\beta < 0$, the equilibrium is a stable focus; when $\beta = 0$, it is a nonhyperbolic equilibrium and, for $\beta > 0$, it is an unstable focus. At

$\beta = 0$ the trace vanishes, with a positive determinant, which suggests the existence of a Hopf bifurcation as shown in Figs. 12 and 13 (indeed, as affine systems cannot undergo Hopf bifurcations, it is better to speak of a Hopf-like bifurcation). Analyzing this bifurcation, it can be concluded that there appears a linear center around point $(2, 0)$, which is restricted to the right operating region.

The outermost periodic orbit of this center is tangent to the border of the region at $(x_1, x_2) = (1, 0)$, and it is a semistable periodic orbit (stable from its inner side but unstable from its outer side). Trajectories starting near this semistable periodic orbit from its outer side tend to another stable periodic orbit of greater amplitude, which lies partially in the interpolating region. For $\beta < 0$, and small, the center disappears, leaving one unstable limit cycle (which arises from the outermost periodic orbit of the center, in a bifurcation analogous to the one studied in [23]), whose amplitude grows as β decreases, so lying partially in the middle region and coexisting with the stable limit cycle. For $\beta > 0$, and small, the center disappears without giving rise to new limit cycles but the stable limit cycle persists.

With respect to the interpolating region, apart from the origin, another two equilibria could exist for $\beta \notin [\frac{-168-12\sqrt{21}}{25}, \frac{-168+12\sqrt{21}}{25}]$ (assuming also that $\beta \neq -12$), namely,

$$
\begin{aligned}
\bar{x}_1^1 &= -\frac{2\beta^2 - 63 - 3\alpha}{63 + 21\beta + 2\beta^2}, \\
\bar{x}_2^1 &= -\frac{2\beta^3 - (99 - 2\alpha)\beta^2 - (1071 + 3\alpha)\beta - 27\alpha + 2268}{2(\beta + 12)(63 + 21\beta + 2\beta^2)}
\end{aligned}
\tag{34}
$$

and

$$
\begin{aligned}
\bar{x}_1^2 &= -\frac{2\beta^2 - 63 + 3\alpha}{63 + 21\beta + 2\beta^2}, \\
\bar{x}_2^2 &= -\frac{2\beta^3 - (99 + 2\alpha)\beta^2 - (1071 - 3\alpha)\beta + 27\alpha + 2268}{2(\beta + 12)(63 + 21\beta + 2\beta^2)},
\end{aligned}
\tag{35}
$$

with $\alpha = \sqrt{25\beta^2 + 336\beta + 1008}$.

Point $(\bar{x}_1^1, \bar{x}_2^1)$ is an actual equilibrium only for $\beta \geq 3\sqrt{2}$, while the point $(\bar{x}_1^2, \bar{x}_2^2)$ is an actual equilibrium only for $\beta \geq -3\sqrt{2}$.

The linearization of the equilibrium point at the origin is

$$
J(0, 0) = \begin{bmatrix} \frac{2\beta - 9}{4} & -\frac{9}{4} \\ -\frac{15}{2} & \frac{\beta - 6}{2} \end{bmatrix},
\tag{36}
$$

so that

$$
\text{trace } J(0, 0) = \beta - \frac{21}{4}, \qquad \det J(0, 0) = \frac{2\beta^2 - 21\beta - 81}{8}.
\tag{37}
$$

The trace vanishes for $\beta = \frac{21}{4}$, but with a negative determinant. The determinant is zero both for $\beta_1 = -3$ and $\beta_2 = \frac{27}{2}$. Thus, the origin is a stable node for $\beta < \beta_1$. For $\beta = \beta_1$, it is a nonhyperbolic equilibrium. For $\beta_1 < \beta < \beta_2$ the origin is a saddle point. For $\beta = \beta_2$ it is again a nonhyperbolic equilibrium, being always an unstable node for $\beta > \beta_2$.

In fact, both at $\beta = \beta_1$ and at $\beta = \beta_2$, the system undergoes a transcritical bifurcation, due to the collision of the origin with equilibrium $(\bar{x}_1^2, \bar{x}_2^2)$ and $(\bar{x}_1^1, \bar{x}_2^1)$, respectively.

The linearization at equilibrium $(\bar{x}_1^1, \bar{x}_2^1)$ is

$$J(\bar{x}_1^1, \bar{x}_2^1) = \begin{bmatrix} \dfrac{(4\beta+18)\bar{x}_1^1 - 3\bar{x}_2^1 + 2\beta - 9}{4} & -\dfrac{3\bar{x}_1^1 + 9}{4} \\ \dfrac{6\bar{x}_1^1 + (\beta+6)\bar{x}_2^1 - 15}{2} & \dfrac{(\beta+6)\bar{x}_1^1 + \beta - 6}{4} \end{bmatrix}, \tag{38}$$

so that

$$\operatorname{trace} J(\bar{x}_1^1, \bar{x}_2^1) = \frac{(6\beta + 30)\bar{x}_1^1 - 3\bar{x}_2^1 + 4\beta - 21}{4}, \tag{39}$$

$$\det J(\bar{x}_1^1, \bar{x}_2^1) = \frac{\gamma(\bar{x}_1^1)^2 + \zeta\bar{x}_1^1 + (6\beta + 72)\bar{x}_2^1 + \eta}{8}, \tag{40}$$

with $\gamma = 4\beta^2 + 42\beta + 126$, $\zeta = 6\beta^2 - 3\beta - 153$, and $\eta = 2\beta^2 - 21\beta - 81$.

Thus, trace (39) is zero for

$$\beta = \frac{-108404}{5583} \approx -19.417,$$

$$\beta = \frac{-157769}{39304} \approx -4.014, \tag{41}$$

$$\beta = \frac{286644}{9169} \approx -31.262,$$

with a positive determinant only at $\beta \approx -4.014$ (note that although a Hopf bifurcation could be possible, the point $(\bar{x}_1^1, \bar{x}_2^1)$ at this value is a virtual equilibrium). On the other hand, determinant (40) is zero for several values of β, namely,

$$\beta_3 = \frac{-168 - 12\sqrt{21}}{25} \approx -8.92,$$

$$\beta_4 = \frac{-168 + 12\sqrt{21}}{25} \approx -4.52, \quad \text{and} \tag{42}$$

$$\beta_2 = 13.5,$$

with the trace being nonzero at these values, and point $(\bar{x}_1^1, \bar{x}_2^1)$ a nonhyperbolic equilibrium.

In a similar way, analyzing the trace and determinant of the linearized system at point $(\bar{x}_1^2, \bar{x}_2^2)$, it can be concluded that the trace is always negative, and the determinant is zero for

$$\beta = \beta_3, \quad \beta = \beta_4, \quad \beta = \beta_1$$

with point $(\bar{x}_1^2, \bar{x}_2^2)$ being a nonhyperbolic equilibrium at these values.

Note that equilibria $(\bar{x}_1^1, \bar{x}_2^1)$ and $(\bar{x}_1^2, \bar{x}_2^2)$ do not exist for $\beta \in [\beta_3, \beta_4]$. The system undergoes a saddle-node bifurcation of virtual equilibria at $\beta = \beta_4$, where two branches of equilibria appear, corresponding to point $(\bar{x}_1^1, \bar{x}_2^1)$ (a stable node) and point $(\bar{x}_1^2, \bar{x}_2^2)$ (a saddle point). Furthermore, these branches will become actual equilibria of the whole system at $\beta > -3\sqrt{2}$ for $(\bar{x}_1^2, \bar{x}_2^2)$, and $\beta > 3\sqrt{2}$ for $(\bar{x}_1^1, \bar{x}_2^1)$.

Therefore, from the point of view of the whole system, $\beta = -3\sqrt{2}$ represents a bifurcation value; in fact, a nonsmooth saddle-node bifurcation (NSSN) takes place just at the boundary between the interpolating (middle) and the right operating regions. When $\beta < -3\sqrt{2}$, the only actual equilibrium is the origin (a globally stable node). At $\beta = -3\sqrt{2}$ two new equilibria appear starting from point $(x_1 = 1, x_2 = -\sqrt{2})$: equilibrium $(\bar{x}_1^2, \bar{x}_2^2)$, which is a saddle point, and equilibrium $(\bar{x}_1, \bar{x}_2)$ on the right operating region, which is a stable node. Thus, for $\beta \geq -3\sqrt{2}$ the stability or instability of the origin will be only local (see Fig. 11). It should be remarked that this bifurcation (NSSN) occurs at a point where the system is not differentiable. However, such a phenomenon can be detected with this methodology.

On the other hand, at $\beta = \beta_2$, the system undergoes a transcritical bifurcation corresponding to the collision of point $(\bar{x}_1^1, \bar{x}_2^1)$ with the origin. Thus, for $\beta < \beta_2$, point $(\bar{x}_1^1, \bar{x}_2^1)$ is an unstable node, while for $\beta > \beta_2$ it is a saddle point. In a similar way, at $\beta = \beta_1$ there exists a transcritical bifurcation due to the collision of point $(\bar{x}_1^2, \bar{x}_2^2)$ with the origin. Thus, for $\beta < \beta_1$, $(\bar{x}_1^2, \bar{x}_2^2)$ is a saddle point, while for $\beta > \beta_2$ it is a stable node.

Finally, two global bifurcations also appear. First, a saddle-node bifurcation of periodic orbits takes place at $\beta \approx -0.05$, resulting in the appearance of an unstable limit cycle, together with a stable limit cycle around it (see Fig. 12). The amplitude of the stable periodic orbit grows with the value of the bifurcation parameter, while the amplitude of the unstable limit cycle decreases with β. Thus, the unstable periodic orbit will decrease until it connects at $\beta = 0$ with the outermost orbit of the local center (see Fig. 12c).

As mentioned before, for $\beta > 0$, the only periodic orbit is the stable one. This limit cycle will grow in amplitude approaching the saddle point at the origin. Thus, for $\beta \approx 0.028$ the periodic orbit becomes a loop connecting one branch of the unstable manifold of the origin with one branch of its stable manifold, in a global homoclinic bifurcation. For $\beta > 0.028$ the relative positions of these manifold branches change, giving as a result the disappearance of the limit cycle (see Fig. 13).

The whole bifurcation analysis is summarized in Fig. 14, which shows the bifurcation diagram for this example. The main conclusions are that for $\beta \leq -3\sqrt{2}$ the origin is the only global attractor of the system, and for $-3\sqrt{2} < \beta < 0$ the system exhibits two stable attractors. Also it can be concluded that for $\beta > -3$ the origin is not an operating point for the system.

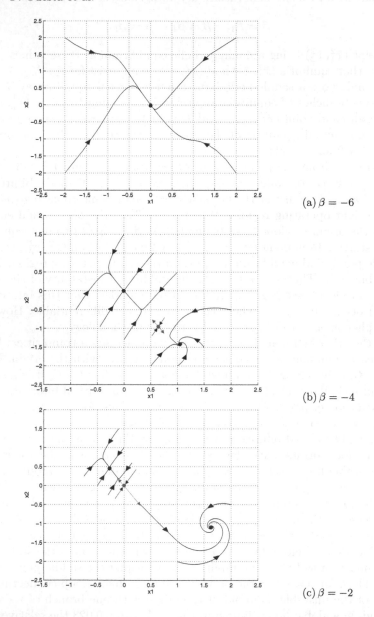

(a) $\beta = -6$

(b) $\beta = -4$

(c) $\beta = -2$

Fig. 11. Phase portraits of Example 3 for different values of β. (**a**) For $\beta = -6$ the origin is the global attractor of the system. (**b**) For $\beta = -4$ the system has two stable equilibria and a saddle point (at $\beta = -3\sqrt{2}$ the system undergoes a degenerate saddle node bifurcation). (**c**) For $\beta = -2$ the system also has two stable equilibria but the origin is a saddle point (at $\beta = -3$ a transcritical bifurcation takes place)

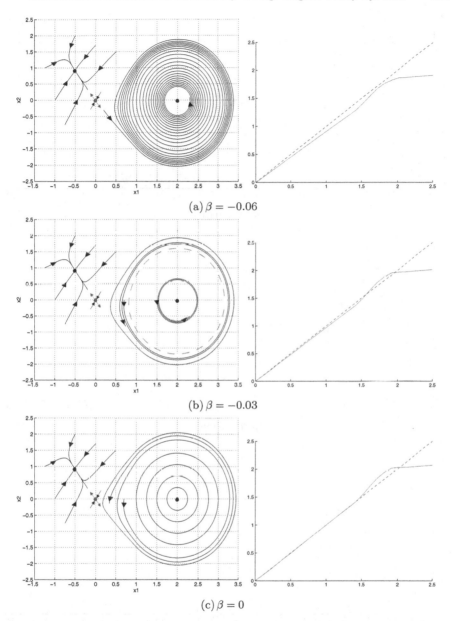

Fig. 12. Phase portraits of Example 3 for different values of β and their corresponding Poincaré maps. (**a**) For $\beta = -0.06$ the system has two stable equilibria and a saddle point. (**b**) For $\beta = -0.03$ there exists an unstable periodic orbit and a stable limit cycle around it (at $\beta = -0.05$ the system undergoes a saddle-node bifurcation of periodic orbits). (**c**) For $\beta = 0$ there is a stable limit cycle and a local center around point $(2,0)$ in the right operating region (the outermost periodic orbit of the center is tangent to the border of the region at point $(1,0)$)

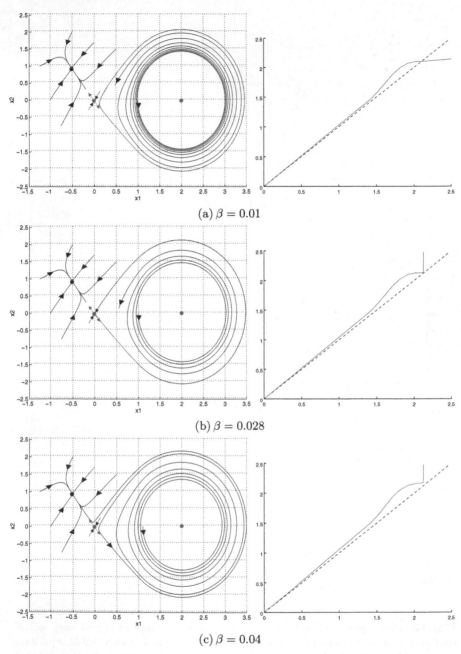

(a) $\beta = 0.01$

(b) $\beta = 0.028$

(c) $\beta = 0.04$

Fig. 13. Phase portraits of Example 3 for different values of β and their corresponding Poincaré maps. (**a**) For $\beta = 0.01$ the system has only one (stable) limit cycle whose amplitude grows with β. (**b**) For $\beta = 0.028$ the limit cycle connects with the saddle point in a homoclinic connection. (**c**) For $\beta = 0.04$ the system has only one stable equilibrium

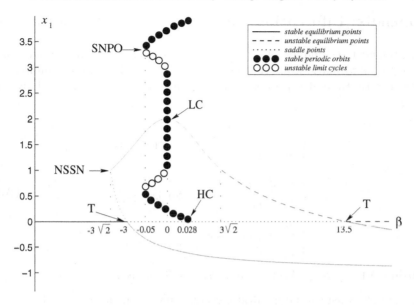

Fig. 14. Bifurcation diagram of Example 3 (NSSN = nonsmooth saddle node; T = transcritical; SNPO = saddle node of periodic orbits; LC = local center; HC = homoclinic connection)

4 Summary

In this chapter, the variety of behaviors of some elementary T-S fuzzy systems has been studied with the tools supplied by the bifurcation theory. Qualitative analysis and bifurcations are of high relevance in every family of nonlinear systems, and so in T-S-systems. It has been shown that even very elementary T-S systems can display local and global bifurcations, some of them due to the parameters in the consequents of the rules and to the lack of differentiability at the boundaries. In the latter case (low differentiability) we speak of degenerate bifurcations. It should be emphasized that these degenerate bifurcations might not persist after some smoothing of the piecewise linear membership functions, even when the perturbations are small ones. It has also been shown that small variations in the values of the parameters associated with the fuzzy rules can give rise to unexpected behavior.

It has been stressed that the task is nontrivial even for the simplest cases, but as seen in the examples analyzed, the information gained is of great value, for instance, in studies of robustness, since the system only displays structurally stable behaviors far from the bifurcation points.

Appendix: Bifurcation Analysis of Example 1 for $\beta = 1$

The standard techniques of Hopf bifurcation analysis do not provide any interesting information, apart from the fact that in this case the bifurcation is degenerate. Here, it is proposed an alternative approach. For $\beta = 1$, system (10) is

$$
\begin{cases}
\dot{x}_1 = x_1 - \frac{1}{2}x_2 \\
\dot{x}_2 = x_1
\end{cases}
, \quad \text{for } x_1 < -1,
$$

$$
\begin{cases}
\dot{x}_1 = -x_1^2 - \frac{1}{2}x_2 \\
\dot{x}_2 = x_1
\end{cases}
, \quad \text{for } |x_1| \leq 1, \tag{A.1}
$$

$$
\begin{cases}
\dot{x}_1 = -x_1 - \frac{1}{2}x_2 \\
\dot{x}_2 = x_1
\end{cases}
, \quad \text{for } x_1 > 1.
$$

Lemma A1. *System* (A.1) *has a global nonlinear center.*

Proof First, consider the quadratic system that coincides with system (A.1) in the middle region. Such quadratic system is not difficult to integrate by rewriting it as follows

$$
\left(x_1^2 + \frac{x_2}{2} \right) \frac{dx_2}{dx_1} + x_1 = 0 , \tag{A.2}
$$

and observing that e^{2x_2} is an integrating factor. Thus, trajectories are implicitly given by

$$
e^{2x_2} \left(x_1^2 + \frac{x_2}{2} - \frac{1}{4} \right) = K \tag{A.3}
$$

for each value of the constant K.

From (A.3) it is easily deduced that all trajectories of the quadratic system are symmetric with respect to the x_2-axis, since they are defined by

$$
x_1 = \pm \sqrt{\frac{1}{4} - \frac{x_2}{2} + Ke^{-2x_2}} , \tag{A.4}
$$

provided that there exists a range of values of x_2 where the expression in the radical is positive. In fact, that range always appears for $K \geq -\frac{1}{4}$. For instance, when $K = 0$, the radical is positive for $x_2 \leq \frac{1}{2}$ and the trajectory is given by the parabola $x_2 = \frac{1}{2} - 2x_1^2$.

For $K = -\frac{1}{4}$, the corresponding trajectory degenerates in a point (the origin), while for $K \in (-\frac{1}{4}, 0)$ trajectories form closed curves (see Fig. A.1).

Obviously, from these trajectories only the portion in the interval $-1 \leq x_1 \leq 1$ represents actual trajectories for system (A.1). However, the complete system (A.1) is invariant under the following transformations:

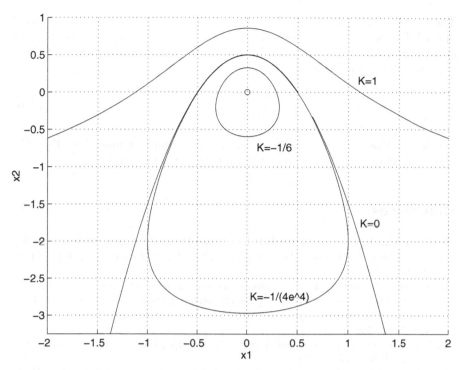

Fig. A.1. Trajectories of the quadratic system (A.2) for different values of K. When $K \in (-\frac{1}{4}, 0)$ trajectories are closed curves; these closed curves belong completely to the middle region only for $K \in (-\frac{1}{4}, -\frac{1}{4e^4})$

$$(x_1, x_2, t) \to (-x_1, x_2, -t), \quad \text{for } |x_1| \leq 1 \,,$$
$$(x_1, x_2, t) \to (-x_1, -x_2, t), \quad \text{for } |x_1| \geq 1 \,,$$

which indicates that apart from the aforementioned symmetry of trajectories with respect to the x_2-axis in the middle region, trajectories are symmetric with respect to the origin in the outer regions. In these regions the dynamics are of focus type (one unstable and one stable) and so all the trajectories are closed curves. The conclusion follows.

Lemma A2. *For $|\beta - 1| \neq 0$ and small, the system (10) has no periodic orbits.*

Proof. For sake of brevity, technical details will be omitted. After an elemental scaling of time, system (10) can be written in Liénard form, that is

$$\dot{x}_1 = F(x_1) - x_2$$
$$\dot{x}_2 = g(x_2) \tag{A.5}$$

for $|\beta - 1|$ being small, system (10) has only one equilibrium at the origin.

Then a necessary condition for existence of periodic orbits, which is based upon Filippov transformations, is given in Theorem 5 of Cherkas [24] (see also the related remark therein). Basically, the condition needed is that there exist $u < 0 < v$ such that the system equation

$$F(u) = F(v)$$
$$G(u) = G(v) \qquad \qquad (A.6)$$

is fulfilled, where $G(u) = \int_0^u g(x) \, dx$. Detailed computations show that the above system cannot be compatible, and the conclusion follows.

References

1. F. Cuesta, E. Ponce, J. Aracil: IEEE Trans. Fuzzy Syst. **9**(2), 355–368 (2001)
2. H.O. Wang, K. Tanaka, M.F. Griffin: IEEE Trans. Fuzzy Syst. **4**(1), 14–23 (1996)
3. J. Guckenheimer, P. Holmes: *Nonlinear Oscillations, Dynamical Systems and Bifurcations of Vector Fields* (Springer, Berlin Heidelberg New York, 1983)
4. J.K. Hale, H. Koçak: *Dynamics and Bifurcations* (Springer, Berlin Heidelberg New York, 1991)
5. Y.A. Kuznetsov: *Elements of Applied Bifurcation Theory* (Springer, Berlin Heidelberg New York, 1995)
6. A.H. Nayfeh, B. Balachandran: *Applied Nonlinear Dynamics* (Wiley, New York, 1995)
7. S.H. Strogatz: *Nonlinear Dynamics and Chaos* (Addison-Wesley, Reading, MA, 1995)
8. E.H. Abed, H.O. Wang, A. Tesi: Control of bifurcations and chaos. In: *The Control Handbook*, ed by W.S. Levine (IEEE Press, Boca Raton, FL, 1996)
9. P. Glendinning: *Stability, Instability and Chaos* (Cambridge University Press, Cambridge, 1994)
10. I. Mareels, S. Van Gils, J.W. Polderman, A. Ilchmann: Asymptotic dynamics in adaptive gain control. In: *Advances in Control (Highlights of ECC'99)*, ed by P.M. Frank (Springer, London, 1999), pp 29–63
11. J. Aracil, A. Ollero, A. Garcia-Cerezo: IEEE Trans. Syst. Man, Cybern. **19** 988–1007 (May 1989)
12. J. Aracil, F. Gordillo, T. Álamo: "Global stability analysis of second-order fuzzy control systems. In: *Advances in Fuzzy Control*, ed by R. Palm, D. Driankov, H. Hellendorn (Springer, Berlin Heidelberg New York, 1997), pp 11–31
13. F. Gordillo, J. Aracil, T. Álamo: Determining limit cycles in fuzzy control systems. In: *FUZZ-IEEE'97*, Barcelona, 1997, pp 193–198
14. J. Moiola, G. Chen: *Hopf Bifurcation Analysis: A Frequency Domain Approach* (World Scientific, Singapore, 1996)
15. J. Llibre, E. Ponce: Nonlinear Anal. **36**, 623–653 (1999)
16. M. Basso, R. Genesio, A. Tesi: Nonlinear Dyn. **13**, 339–360 (1997)
17. F. Cuesta, F. Gordillo, J. Aracil, A. Ollero: IEEE Trans. Fuzzy Syst. **7**(5), 508–520 (1999)
18. J. Llibre, E. Ponce: Dyn. Stabil. Syst. **11**(1), 49–88 (1996)

19. M. Sugeno: IEEE Trans. Fuzzy Syst. **7**(2) (April 1999)
20. E. Freire, E. Ponce, F. Rodrigo, F. Torres: Int. J. Bifurc. Chaos **8**(11), 2073–2097 (1998)
21. V. Carmona, E. Freire, E. Ponce, F. Torres: Instability in the simplest class of continuous switched linear systems with stable component. In: *Proc. 16th IFAC World Congr.* Prague, July 4–8, 2005
22. R. Abraham, J. Marsden: *Foundations of Mechanics* (Benjamin, New York, 1978)
23. E. Freire, E. Ponce, J. Ros: Int. J. Bifurc. Chaos **9**(5), 895–907 (1999)
24. L.A. Cherkas: Differential Eq. **13**, 529–547 (1977)

Self-Reference, Chaos, and Fuzzy Logic*

Patrick Grim

Abstract. Self-reference and paradox introduce a spectrum of nonlinear phenomena in fuzzy logic. Working from the example of the Liar paradox, and using iterated functions to model self-reference, sentences can be constructed with the dynamical semantics of fixed-point attractors, fixed-point repellors, and full chaos on the [0,1] interval. The paper also extends the analysis to pairs and triples of mutually referential sentences, which generate strange attractors and semantic fractals in two and three dimensions.

1 Introduction

Chaos theory and fuzzy logics form two of the most intriguing and promising areas of recent mathematical research. In what follows I want to explore a region of fuzzy logic which exhibits a wide range of the phenomena of chaos theory. The route to that region is via a consideration of some of the dynamics of paradox.

The basics of the fuzzy logic used here are outlined in Sect. 1. In Sect. 2 the classical example of the Liar paradox is used to introduce iterated functions as a way of modeling self-reference. Section 3 shows how self-referential sentences in fuzzy logic display the dynamical semantics of fixed-point attractors and fixed-point repellers. Section 4 is devoted to the Chaotic Liar, a fuzzy self-referential sentence with a dynamical semantics that is chaotic in the full mathematical sense. Section 5 traces similar results with regard to pairs of mutually referential sentences, introducing strange attractors and fractals in two dimensions. In Sect. 6 the exploration is extended to triples of mutually referential sentences and to dynamical phenomena in three dimensions.

What I want to outline are some of the dynamical phenomena of self-referential fuzzy logic, proceeding for the most part by example. In a somewhat different context, Buckley [1] expressed a hope that research in fuzzy logic would reveal fuzzy systems with chaotic behavior, "and then we can define, and study, fuzzy fractals" (p. 20)[1]. The work outlined here fulfills that hope, here using self-reference in fuzzy logic.

---

* This is an adaptation of "Self-Reference and Chaos in Fuzzy Logic." *IEEE Trans. Fuzzy Syst.* **1** 237–253 (1993)

[1] For related work see also [2, 3]

P. Grim: *Self-Reference, Chaos, and Fuzzy Logic*, StudFuzz **187**, 317–359 (2006)
www.springerlink.com
© Springer-Verlag Berlin Heidelberg 2006

It should perhaps be emphasized that the attempt here is not to *solve* the paradoxes. Two thousand years of attempted solutions can hardly be said to have met with conspicuous success. Three-valued, multi-valued, infinite-valued, gapped, and antifoundational logics all prove vulnerable to strengthened versions of the Liar, and there seems little reason to think that fuzzy logics should be any exception in that regard (for a critical survey see, for example, Chap. 1 in [4]). But a *solution* to the phenomena of self-reference may be the wrong thing to look for. In something of the spirit of Herzberger [5], Gupta [6], and Gupta and Belnap [7], the attempt here is rather to open for investigation the semantical dynamics of self-reference and self-referential reasoning for study in their own right.

Although I attempt to draw some speculative conclusions, much remains to be done in terms of generalization and interpretation. Intriguing and beautiful formal phenomena appear in the semantics of self-referential sentences within fuzzy logic, but it is not always clear why they appear, how they generalize, or what the formal phenomena really mean. In a different context the approach outlined is used to extend Gödel-like limitative results to both chaos theory and fuzzy logic [8, 9]. Here, however, the concentration is on the chaotic phenomena themselves.

2 A Simple Fuzzy Logic

The philosophical premise of all fuzzy logic is a denial of the assumption in classical logic that all sentences or propositions are either fully true or fully false. Fuzzy logic is built on the premise that truth comes rather in degrees: that a sentence of proposition could take any truth-value between 0 (for full falsity) and 1 (for full truth). For a proposition p, the numerical truth value $v(p)$ of that proposition might be any real value in the $[0, 1]$ interval.

The semantics for classical logic is outlined in terms of the familiar Boolean truth tables; fuzzy logic requires an analog in which the truth value of compound sentences can be calculated from the truth value of their components. The fuzzy logic that will be used here is built on the Łukasiewicz system $Ł_{\aleph 1}$[2].

Given the value $v(p)$ for a proposition, its negation is taken to be 1 minus that value:

$$\nu(\sim p) = 1 - \nu(p)$$

The value of a conjunction is the value of its least conjunct. The value of a disjunction, on the other hand, is the value of its greatest disjunct:

$$\nu(p \wedge q) = \min[\nu(p), \nu(q)]$$
$$\nu(p \vee q) = \max[\nu(p), \nu(q)]$$

---

[2] See for example [10, 11]

We follow Łukasiewicz in calculating the value of "if ... then" as 1 if the antecedent is less true than the consequent, and the absolute distance between the two otherwise:

$$\nu(p \rightarrow q) = \begin{cases} 1 & \text{if } \nu(p) < \nu(q) \\ 1 - \text{abs}(\nu(p) - \nu(q)) & \text{otherwise} \end{cases}$$

or equivalently

$$\nu(p \rightarrow q) = \min[1, 1 - \nu(p) + \nu(q)].$$

The biconditional, then, can be treated as a conjunction of conditionals in each direction or as 1 minus the absolute difference between the values of its components:

$$\begin{aligned} \nu(p \leftrightarrow q) &= \nu((p \rightarrow q) \vee (q \rightarrow p)) \\ &= \min[\min[1, 1 - \nu(p) + \nu(q)], \min[1, 1 - \nu(q) + \nu(p)]] \\ &= 1 - \text{abs}(\nu(p) - \nu(q)). \end{aligned}$$

Were we to restrict propositional values to 0 and 1, as in classical logic, the formulas above would give us the classical connectives. But of course the formulas above are not the only ones that will generalize classical logic in this way. Formal considerations cast a strong presumption in favor of this treatment of conjunction and disjunction in terms of min and max [12], and an only slightly weaker presumption in favor of this treatment of negation. The same cannot be said for Łukasiewicz implication, however. With regard to implication it must simply be admitted that there are a number of alternatives.[3]

The distinguishing mark of true fuzzy logics, as opposed to mere infinite-valued logics, is the use of a denumerable set of "linguistic" truth values beyond the numerical truth values $\nu(p)$ outlined above.[4] Linguistic truth values are themselves represented by fuzzy sets, standardly generated from a fuzzy set "true" and its converse "false" through recursive application of algorithmic hedges such as "very," "more or less," "slightly," and the like [11, 15–19].

Nothing in the basic Łukasiewicz logic, however, dictates what shape a basic fuzzy set for "true" is to take. Here as in the case of implication there are clear alternatives. For present purposes I will use what may be the simplest and most clearly justified fuzzy set for "true," generated by importing the Tarski convention T directly into $Ł_{\aleph 1}$.

---

[3] In [8], for example, we use an implication definable as ($\sim p \vee q$), which gives us Rescher's system $S_{\aleph}^{\supset}$ [10], first developed in [13]. As Gaines [14] notes, this approach offers a direct fuzzification of predicate calculus.

It should also be noted that a stronger intuitive argument might be made for the Łukasiewicz biconditional than for the conditional itself. It is the biconditional rather than the conditional that is most directly relied on in what follows

[4] See, however, [14] on different senses of "fuzzy logic"

Following Tarski [20], and using $T(p)$ for the claim that "p" is true, Tarski's convention T specifies that p is true if and only if

$$T(p) \leftrightarrow p \, .$$

Using the outline for the Łukasiewicz biconditional,

$$\nu(T(p) \leftrightarrow p) = \nu((T(p) \rightarrow p) \wedge (p \rightarrow T(p))$$
$$= 1 - \mathrm{abs}(\nu(T(p)) - \nu(p)) \, .$$

if we assume that the Tarskian schema itself takes the value of "1", for absolute truth, it must then be the case that

$$\nu(T(p)) = \nu(p) \, .$$

What direct importation of the Tarskian T-schema into $L_{\aleph 1}$ gives us is thus the basic fuzzy set for "true" outlined in [17, 18] (see also [21, 22]). With "false" as the complement of "true" and modeling "very" and "fairly" in terms of squares and square roots, respectively—a treatment fairly consistent across the literature—we get the basic set of linguistic truth values indicated in Fig. 1.[5]

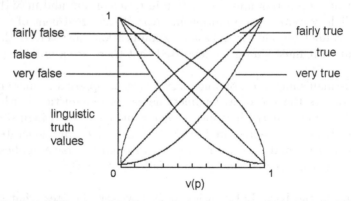

Fig. 1. Baldwin's fuzzy sets for linguistic truth-values, using a direct importation of the Tarski T-schema

As noted, there are alternaives to this treatment of "true." Zadeh [11] characterizes a fuzzy set for "true" as

---

[5] For the incorporation of Tarski via the Łukasiewicz biconditional I am obliged to Gary Mar and Paul St. Denis. Here I do not mean to suggest that the standard modeling for "very" and "fairly" is any way beyond question, of course. On this see also the discussion of Sect. 4 and note 17

$$\mu_{\text{true}}(\nu) = 0 \quad \text{for } 0 \le \nu \le \alpha$$

$$= 2\left(\frac{\nu - \alpha}{1 - \alpha}\right)^2 \quad \text{for } \alpha \le \nu \le \frac{\alpha + 1}{2}$$

$$= 2\left(\frac{\nu - 1}{1 - \alpha}\right)^2 \quad \text{for } \frac{\alpha + 1}{2} \le \nu \le \alpha$$

Here α is a parameter "which indicates the subjective judgment about the minimum value of ν in order to consider a statement as 'true' at all" (see p. 124 in [19]). A sketch of Zadeh's basic fuzzy sets for "true" and "false" [where $\mu_{\text{false}}(\nu) = \mu_{\text{true}}(1 - \nu)$], using an α of .6, as shown in Fig. 2. Corresponding sets for "very" and "fairly" are also indicated.

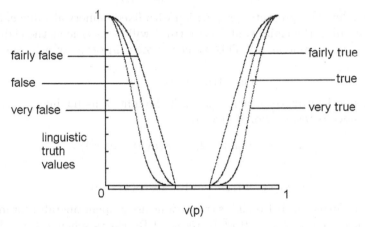

Fig. 2. Zadeh's fuzzy sets for linguistic truth-values

The results that follow are built on Baldwin's fuzzy logic rather than Zadeh's for reasons of simplicity. Baldwin's outline is simpler not only in algorithmic and graphic terms but also in terms of its justification as a direct importation of the Tarski T-schema into Łukasiewicz $L_{\aleph 1}$. A range of results similar to those below, however, would emerge using Zadeh's or more complicated fuzzy logics as well. At several points results are indicated which will hold for any fuzzy logic with $L_{\aleph 1}$ as a base.

It is important to note that although the underlying semantic model of our logic is expressed in terms of numerical truth values, the propositions admissible in the language of the logic itself can use only linguistic truth values. "p is very true" is thus a type of sentence for which this fuzzy logic provides a valuation scheme; "p is .75 true" is not. The perceived artificiality of sentences of the latter sort seems to be have been part of Zadeh's initial motivation for moving to linguistic truth values (see, for example, [11]), and much of the common suspicion of the "artificial numbers" of fuzzy logics can be dealt with by viewing those numbers merely as artifacts of the semantic model. The language for which that model is designed need not contain numerical truth values at all.[6]

Consider also a language, however, in which we *are* allowed to claim numerical truth values for particular propositions—a language in which "p is .75 true," for example, *is* allowed as a sentence of the language. Suppose, moreover, that we have a claim q that we know to be .75 true. In that case, the claim that "p is .75 true" will intuitively be as true as the claim that p has the same value as q. For this particular case, and using the biconditional quite naturally to represent "has the same value as," we can express that basic intuition as

$$p \text{ is .75 true } \leftrightarrow (p \leftrightarrow q) \,.$$

In general, let $[v]$ represent a proposition with fixed numerical value v. By the same reasoning, the claim that "p is v true" will be as true as the claim that p has the same value as $[v]$. This we can express in terms of a biconditional as

$$p \text{ is } v \text{ true } \leftrightarrow (p \leftrightarrow [v]) \,.$$

If this general intuitive principle is itself thought of as having a value of 1, the Łukasiewicz biconditional gives us

$$\nu(p \text{ is } v \text{ true}) = 1 - \text{abs}(\nu(p) - \nu([v])) \,,$$

or simply

$$\nu(p \text{ is } v \text{ true }) = 1 - \text{abs}(\nu(p) - v) \,.$$

This, it turns out, is Rescher's truth-value assignment operator for infinite-valued logic (see p. 89 in [10]).[7] In terms of Rescher's schema, the value for a proposition Vvp to the effect that a proposition p has a numerical truth value v is given by

$$\nu(Vvp) = 1 - \text{abs}(v - \nu(p)) \,.$$

This relation between Baldwin's fuzzy sets for linguistic truth values and Rescher's Vvp schema for attributions of numerical truth values is perhaps even clearer when envisaged graphically. With numerical values on axes, the graph for a proposition "$V1p$" to the effect that a proposition p has value 1, given Rescher's treatment, is precisely Baldwin's graph for "p is true" in Fig. 1. The Rescher graph for "$V0p$" corresponds to Baldwin's graph for "p is false."

The infinite-valued Rescher scheme is central to that of Mar and Grim [8], Grim et al. [9, 23]. The emphasis in what follows, however, is on dynamical semantic phenomena within the more strictly fuzzy logic outlined above.

---

[6] The case of set-theoretical semantics is perfectly parallel: the fact that a semantic model is written in terms of sets need not commit us to thinking that the sentences modeled are themselves about sets

[7] Here as at many other points I am obliged to Gary Mar

3 Self-Reference as Iteration: The Example of the Liar

The oldest and most famous of the paradoxes is the Liar paradox: a sentence that says of itself that it is false:

$$\text{This sentence is false.} \tag{1}$$

Consider for a moment the Liar sentence in the context of a classical logic. If (1) is true, it must be false. If (1) is false—since it *says* it is false—it must be true. When one first approaches the Liar, one is forced into an intuitive pattern of reasoning that seems to oscillate in conclusion between "true" and "false." If true, the Liar must be false...but then if false, it must be true...but then if true, it must be false:

$$\text{T, F, T, F, T, F,}\ldots\,{}^{8}$$

In intuitive terms, this alternation is a temporal one. One is first drawn to the conclusion that the Liar must be true, later forced to the conclusion that it must be false, and so forth. It is also the case that finite reasoners such as ourselves are generally smart enough—or logically unprincipled enough, or both—to break out of such a series rather than to continue it indefinitely. In what follows, I want to abstract from both of these points. The dynamical semantics of the Liar can be thought of as a series of revised semantic values forced by something like the standard Liar argument. As such, the points of oscillation represent not so much discrete times as discrete abstract steps in

---

[8] Semantic paradox has had of course a long and distinguished career in philosophical and mathematical logic. It lies at the core of Cantor's diagonal argument and the paradise of transfinite infinities it offers. Russell's paradox, discovered as a simplification of Cantor's argument, was historically instrumental in motivating axiomatic set theory. Gödel himself [24] explicitly uses the Liar paradox (and a relative known as the Richard paradox) to motivate his incompleteness theorems, and the limitative theorems of Tarski [20], Church [25], and Turing [26] can all be seen as exploiting the basic reasoning of the Liar. In the mid-1960s, Gregory Chaitin [27] developed an interpretation of Gödel's theorem in terms of the notion of algorithmic randomness by formalizing the Berry paradox, itself a simplification of the Richard paradox.

Philosophers have repeatedly attempted to find solutions to the semantic paradoxes by seeking patterns of semantic stability—hence the proliferation of "truth-value gap solutions" of the 1960s and 1970s (see [28–30]). Efforts in the direction of finding patterns of stability within the paradoxes continued in the 1980s with the works of Hans Herzberger [5] and Anil Gupta [6]. Later work includes that of Jon Barwise and John Etchemendy [31], using Aczel set theory with an antifoundation axiom to characterize Liar-like cycles.

The work of this chapter, in contrast, can be seen as an attempt to study complex patterns of *in*stability in the general domain of self-reference and paradox

a pattern of protracted reasoning.[9] If we think of these as values arrived at by a reasoner, that reasoner should be thought of as an idealized reasoner acting purely on logical principle and without time constraints.[10]

Using 1 for truth and 0 for falsity, we can then model the classical semantic behavior of the Liar in terms of a sequence of values x_n, where x_0 is an initial or "seed" value and

$$x_{n+1} = 1 - x_n .$$

With a seed value of 1, the dynamic semantics of the Liar within a classical logic can thus be graphed as in Fig. 3.

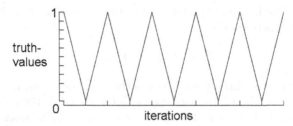

Fig. 3. 'This sentence is false.' The dynamical semantics of the classic Liar

A seed value of 0, of course, shows an identical oscillation shifted one iteration to the left. A central idea in the work that follows is to use iterated

---

[9] In p. 218 of [32], dynamical systems are spoken of intuitively as the description of the time behavior of a point moving about on some sort of surface according to a rule that describes how one point is to follow another. Here as in other contexts, I think that restriction to time cannot be taken too literally. Within ecological studies iterative "times" may in fact be generations; within epidemiological studies "times" may be (variable) periods of vulnerability to infection, and of course the pure mathematics of iterated functions need not be thought of in terms of literal time at all.

Many attempts to *solve* the Liar and similar paradoxes, of course, rely on denying that there *is* a genuine oscillation here, by insisting that at each step one is using a distinct truth predicate, say, or has ascended to a higher metalanguage. That move itself, however, is quite clearly counterintuitive. (As Barwise and Etchemendy note, "When we think about the Liar on an intuitive level, there is an inclination to claim that the truth value "flips back and forth." First we see that it is false, then that it is true, then that it is false, and so forth." [31, p. 136]). Here and throughout, the attempt (like that of Herzberger [5], Gupta [6], and Gupta and Belnap [7]) is to track the intuitive dynamical semantics of self-referential sentences rather than to sacrifice semantic intuitions in a too-quick search for a "solution" to the paradoxes

[10] The idealizations and abstractions at issue clearly make our semantic model "objectivist" in spirit. For a critique of objectivist approaches in general see esp. 205 ff. in [33]

algorithms of this type to model the dynamical semantics of self-referential sentences in general.

As a first example, consider the Liar sentence within the context of the fuzzy logic outlined above—a variation we might term the Fuzzy Liar. Given the Baldwin fuzzy set for "false," our algorithm remains the same, though now of course we need to consider numerical truth-values in the full [0, 1] interval.

For a seed value of .3, the Fuzzy Liar gives an oscillation between .3 and .7 (Fig. 4).

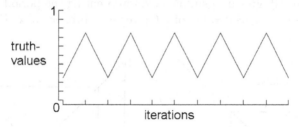

Fig. 4. 'This sentence is false' in a fuzzy logic. The dynamical semantics of the Fuzzy Liar with an initial value of .3

For any seed value x, in fact, the dynamical semantics of the Fuzzy Liar will be a simple oscillation between x and $(1 - x)$. The one fixed point is .5.

Often the dynamical semantics of self-referential sentences is better illustrated using a web diagram. In a graph such as that in Fig. 5 we start with a plotted function for, say, "false." An initial seed value a (.3, in this case) is plotted as $(a, 0)$, and a line drawn from that point vertically to the

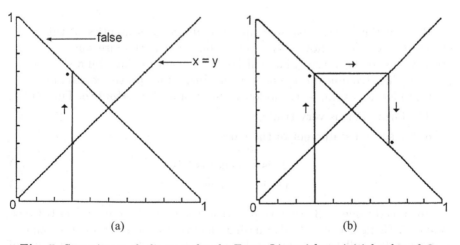

Fig. 5. Steps in a web diagram for the Fuzzy Liar with an initial value of .3

326   P. Grim

function $f(x)$ (Fig. 5a). We read off our y value here as indicating that for $x = .3$, $f(x)$ is .7. In order to represent the iteration of that function, we now draw a horizontal line to a point $(f(a), f(a))$ on the diagonal $x = y$, thereby converting our previous y value to a new x value. We then draw a vertical line again from that point to our function at $(f(a), f(f(a))$ (Fig. 5b). The y value of this intersection point indicates that $f(x)$ for $x = .7$ is .3 (Fig. 5b). We graph the results of repeated iteration by continuing the process, at each step converting our y value to an x value by reflecting off the $x = y$ diagonal, giving us a new point of intersection with our plotted function.

Within a web diagram of this sort it is clear that a seed value of .3 for the Fuzzy Liar will give us a simple box, representing the period 2 oscillation between .3 and .7. A seed value of .86 gives us a broader box (Fig. 6).

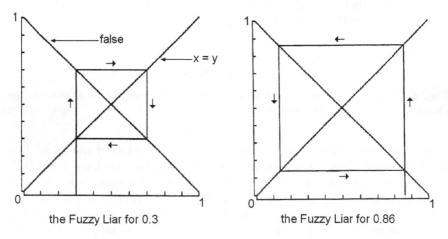

the Fuzzy Liar for 0.3   the Fuzzy Liar for 0.86

Fig. 6. The Fuzzy Liar for values of 0.3 and 0.86

Because the Fuzzy Liar sets up an oscillational semantics of this type, statements which are not self-referential but which attribute some linguistic truth value to the Fuzzy Liar will have a semantics that follows the same pattern. The dynamical semantics of the Fuzzy Liar proves contagious.

Consider, for example, the following fuzzy statement *about* the Fuzzy Liar:

The Fuzzy Liar is very true

or the second statement of the pair:

$$\text{This sentence is false.} \tag{1}$$
$$\text{(1) is very true.} \tag{2}$$

For a seed value of .3, we have seen that, the Fuzzy Liar alternates between .3 and .7. Using the standard squaring function for "very true," the value of statement (2) will then alternate between .09 and .49. For a seed value of

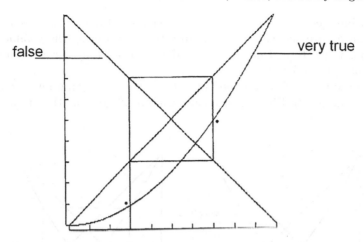

false   very true

Fig. 7. Derivative semantics for the statement that the Fuzzy Liar is very true

.54, to use another example, the Fuzzy Liar alternates between .54 and .46. In that case, statement (2) will correspondingly alternate between .2916 and .2116.[11]

The derivative semantic behavior of statement (2) can be thought of in terms of a web diagram as shown in Fig. 7. Let us start with an initial seed value of .3 for the Fuzzy Liar, represented on the x-axis as (.3, 0). If sentence (1) thus has a value of .3, sentence (2) will have a value of .09, reflected by the fact that a line drawn vertically from (.3, 0) intersects our function for "very true" at (.3, .09). Given a seed value of .3, however, we will also be forced to a revised value for the Fuzzy Liar of .7, reflected by the fact that our vertical line intersects the function for the Fuzzy Liar at (.3, .7).

Here we are interested in successive values for (2). In order to obtain the next value for (2), however, we cannot simply reflect (.3, .09) off the $x = y$ diagonal as before. We must instead work from the revised value .7 for the Fuzzy Liar, reflect *that* off the $x = y$ diagonal, and then drop a line vertically

---

[11] On the pattern of the Liar, the intuitive reasoning here will proceed something as follows. If we assume that (1) has a value of .3, then (2) will be fairly false, with a value of .09. On the assumption that (1) has a value of .3, however, since (1) *says* it is false, (1) will be fairly true—it will have a value of .7. But then (2) will not be so far off after all, receiving a value of .49. But if (1) has a value of .7.... Despite this alternation, the spirit of Zadeh's extension principle (see 416 ff. in [11]) appears to be preserved at each step.

The intuition that we should nonetheless be able to say *some*thing constant about the truth value of (1), despite its oscillation, can perhaps be addressed only in a language in which we explicitly introduce predicates such as "is not consistently true" or "has no constant truth value". In this chapter I have concentrated on chaotic relatives of the Liar; in that stronger language, I believe, we should expect chaotic relatives of the Strengthened Liar

to again intersect our "very true" function for (2). As it turns out, w can also graph progressive values for (2) directly by reflecting the y value of our earlier point (.3, .09) off not the $x = y$ diagonal but the mirror image left to right of our graph for (2).

An initial value of .41 for the Fuzzy Liar gives us a graph for sentence (2) shown on the left in Fig. 8. An initial value of .8 gives us the graph on the right.

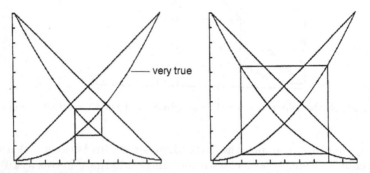

Fig. 8. Derivative semantics for the claim that the Fuzzy Liar is very true, using initial values of .41 and .8 in the Fuzzy Liar

In each case our graph again shows a box, though here in a different position, reflecting the fact that semantic values for sentences which attribute truth values to the Fuzzy Liar will oscillate in the same way that semantic values for the Fuzzy Liar itself do.

4 Attractor and Repeller Fixed Points in the Phenomena of Self-Reference

To this point we have concentrated on the simple Liar. The same basic techniques also allow us to model a wider spectrum of self-referential sentences with a wider class of dynamic semantic behaviors.

Consider, for example, a sentence we might call the Modest Liar:

$$\text{This sentence is fairly false.} \tag{3}$$

In terms of our basic logic, the dynamical semantics of (3) can be modeled using the following algorithm:

$$x_{n+1} = \sqrt{1 - x_n} \ .$$

In a simple bounce diagram, for a seed value of .314, this gives us the behavior shown in Fig. 9.

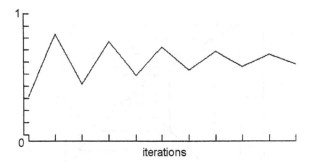

Fig. 9. 'This sentence is fairly false.' The dynamical semantics of the Modest Liar, for a seed value of .314

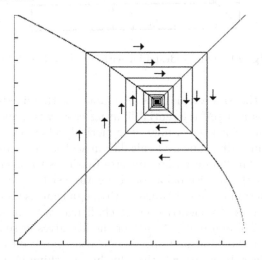

Fig. 10. 'This sentence is fairly false.' A web diagram of the Modest Liar, for a seed value of .3

The semantical behavior of the Modest Liar is still clearer, however, in a web diagram (Fig. 10).

For any seed value, the Modest Liar converges inexorably on a fixed-point attractor of $\frac{-1\pm\sqrt{5}}{2}$.[12] Figure 11, for example, shows the Modest Liar started with a seed of .99.

---

[12] The solution to $\sqrt{1-x}$ is $\frac{-1\pm\sqrt{5}}{2}$. Only $\frac{-1+\sqrt{5}}{2}$ appears within our semantic interval $[0,1]$, however. Similar comments apply with respect to $x = (1-x)^2$ and the semantic fixed point $\frac{3-\sqrt{5}}{2}$ for the Emphatic Liar below. In an entertaining knights-and-knaves exploration of some of these ideas, Nathaniel Hellerstein (*Isle of Paradox and Other Logic Adventures*, unpublished manuscript) refers to the Modest Liar as the Golden Liar, pointing out that its attractor fixed point is $1/\phi$, where ϕ is the golden ratio. The repeller fixed point for the Emphatic Liar is similarly $1 - 1/\phi$. The golden ratio ϕ itself turns up in a number of surprising

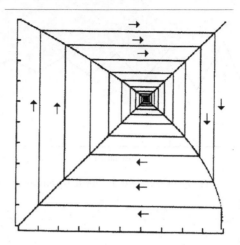

Fig. 11. The Modest Liar with a seed value of .99

The fact that the Modest Liar has a semantics with a fixed-point attractor of this sort makes it a self-referential limiting case with regard to fuzziness. Whatever element of a fuzzy range of numerical values we might assign to such a sentence initially, repeated calculation will force us to a very precise and single value. The Modest Liar converts fuzziness to precision.

It is clear that the precise behavior of the Modest Liar relies on the use of a square root function to model "fairly." Although that use is fairly consistent in the literature, it is also clearly open to challenge: why insist on the square root in particular? The use of $\sqrt[3]{}$, $\sqrt[4]{}$, or the like gives us semantic behavior similar to that of the Modest Liar, though converging on a different fixed point. It is not clear, however, whether this has anything important to tell us about the appropriateness of using different roots in modeling hedges such as "fairly".

Consider also an Emphatic Liar:

$$\text{This sentence is very false.} \tag{4}$$

Semantic values for the Emphatic Liar will be determined by

$$x_{n+1} = (1 - x_n)^2 .$$

For a seed value of .3 the Emphatic Liar forces a series of revised values which eventually converge on the familiar oscillation between 0 and 1, characteristic of the classical Liar (Fig. 12): With one exception, the Emphatic Liar will force any numerical value in the [0, 1] interval to the oscillation of

places: ϕ is the limit of the Fibonacci series $1/1, 2/1, 3/2, 5/3, 8/5, \ldots; \phi - 1 = 1/\phi; \phi = \sqrt{1 + \sqrt{1 + \sqrt{1+}}} \ldots;$ etc. Here see pp 203–206 in [34]

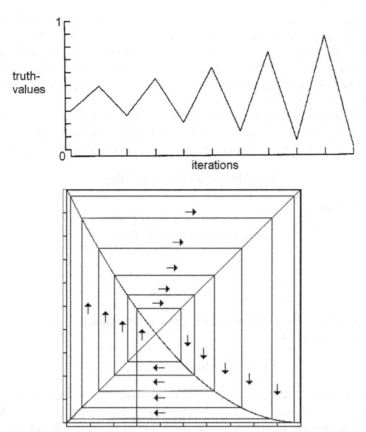

Fig. 12. 'This sentence is very false.' The Emphatic Liar converges on semantical dynamics characteristic of the classic Liar

the classical Liar. The semantical dynamics of the Emphatic Liar is that of a fixed repeller point at $\frac{3-\sqrt{5}}{2}$. The one point that is not forced out to the behavior of the classical Liar is the point $\frac{3-\sqrt{5}}{2}$ itself. The Emphatic Liar, like the Modest Liar, can thus be seen as a limiting case to fuzziness. The Modest Liar, however, forces convergence to a precise nonclassical value. With one exception, the Emphatic Liar forces revised values to a Liar-like oscillation between the two classical values of 0 and 1.[13]

---

[13] Nathaniel Hellerstein has also suggested an Equivocal Liar:

This sentence is not very true, with an algorithm

$$x_{n+1} = 1 - x_n^2 .$$

Here $\frac{-1+\sqrt{5}}{2}$ is a *repelling* fixed point, forcing values out to an oscillation between 0 and 1

332   P. Grim

The use of squaring to model "very" is of course open to question in precisely the same way that the use of square roots to model 'fairly' is. An Emphatic Liar using $(1 - x_n)^3$ or $(1 - x_n)^4$ would still converge to a Liar-like oscillation, though from a different repeller point. Here again it remains unclear whether this has anything to tell us about the proper modeling of hedges such as "very" or "fairly".

What of statements which attribute fuzzy truth values *to* sentences such as the Modest and Emphatic Liars? Here as in the case of the Fuzzy Liar we will have sentences with a dependent semantics. Consider, for example,

$$\text{The Modest Liar is very false} \tag{5}$$

or sentence (6):

$$\text{This sentence is fairly false} \tag{3}$$

$$\text{(3) is very false} \tag{6}$$

For a seed value of .3, the Modest Liar gives us the following series, converging as we have seen on $\frac{1+\sqrt{5}}{2}$:

$$.3, .83667, .40415, .77191, \ldots {}^{14}$$

If the Modest Liar has a value of .3, however, the claim that it is very false takes a value of $(1-.3)^2$ or .49. Given a revised value of .83667 for the Modest Liar, we are forced to revise the value of (6) accordingly, to .02668. Thus, the progressive dynamics of the Modest Liar forces a corresponding dynamics for sentence (6):

$$.49, .02668, .35503, .05202, \ldots$$

This pattern can also be illustrated in a web diagram. Figure 13 graphs functions for both the Modest Liar and "very false". For a given seed value x, we draw a line vertically from $(x, 0)$ to intersect the Modest Liar function. The y value of this point of intersection (x, y) is our revised numerical truth value for the Modest Liar, and reflection off the $x = y$ line will give us the next value for the Modest Liar.

At each step, the value of (6) depends on that of the Modest Liar, and is in fact the value at which our line representing the x value intersects the function line for (6). Thus we can think of progressive values for (6) as intersection points on the "very false" curve directly below the progressive points on the graph for the Modest Liar.

In this case we can also graph the dynamics of the dependent sentence (6) more directly, however, by reflecting its values off the x^4 curve. Progressive values for (6) starting with two different seed values for (3) thus appear as

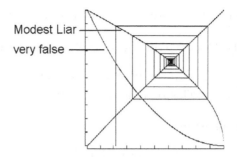

Fig. 13. Derivative semantics for the claim that the Modest Liar is very false

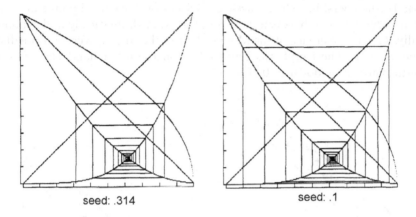

seed: .314   seed: .1

Fig. 14. Attribution of fuzzy truth-values to the Modest Liar also converge to a fixed point

in Fig. 14. Not surprisingly, attribution of fuzzy linguistic truth-values to the Modest Liar shows convergence to a single fixed point as well.

To this point we have concentrated on fuzzy relatives of the Liar. We should however also mention fuzzy relatives of the Truth-teller. Within classical logic, sentence (7) proves troublesome in a manner different from but related to the troubles of the Liar:

$$\text{This sentence is true.} \qquad (7)$$

Unlike the Liar, the problem with (7) is not that it cannot consistently be assigned a value of either true or false. On the contrary, (7) can consistently be assigned *either* value, with no apparent basis on which to prefer one rather than the other. In a fuzzy logic, using the algorithm

$$x_{n+1} = x_n \; ,$$

---

[14] Only approximate values are shown, rounded off for the sake of simplicity

the Truth-teller can still consistently be assigned any value in the [0, 1] interval.

Consider also the Modest Truth-teller and the Emphatic Truth-teller:

$$This\ sentence\ is\ fairly\ true \tag{8}$$

$$This\ sentence\ is\ very\ true \tag{9}$$

with corresponding algorithms

$$x_{n+1} = \sqrt{x_n}$$
$$x_{n+1} = x_n{}^2$$

Both Truth-tellers, like their classical predecessor, have fixed points at 0 and 1. For fuzzy values in between, however, they vary dramatically and symmetrically. The self-reference of the Modest Truth-teller drives every intermediate value up to 1. The Emphatic Truth-teller, on the other hand, drives every intermediate value down to 0 (Fig. 15).

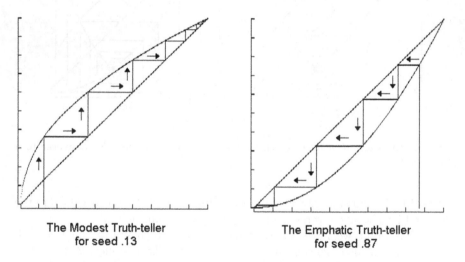

The Modest Truth-teller
for seed .13

The Emphatic Truth-teller
for seed .87

Fig. 15. The Modest Truth-teller: 'This sentence is fairly true.' The Emphatic Truthteller: 'This sentence is very true'

The Truth-tellers, like the Modest Liar, thus force fuzziness to precision through the iteration of semantic self-reference. In the case of the Truth-tellers as in the case of the Emphatic Liar, that inexorable self-precision is the more remarkable since the values one is driven to in each case are *classical* values. In the case of the Truth-tellers, moreover, one is driven to *stable* classical values.

One lesson of sentences such as the Modest Truth-teller and the Emphatic Truth-teller is that dynamical semantics may introduce the need for logical categories beyond the traditional "tautology" and "contradiction." A classical tautology is one which (instantly, as it were) takes a value of "1" for any value assigned to its components. A dynamic tautology, we might propose, is one which converges through iteration on a value of "1" for any initial value. The Modest Truth-teller might thus be proposed as a dynamic tautology; the Emphatic Truth-teller as a dynamic contradiction.[15]

From this first sampling of examples it is clear that fuzzy self-reference opens up a realm of dynamical semantics far richer than anything dreamt of in classical logic. It is also clear that there are some surprises in the transition from classical to fuzzy logic. The Fuzzy Liar, for example, is a quite direct fuzzification of the standard Liar. But it is not the Fuzzy Liar but the quite different Emphatic Liar that converges on the familiar oscillation between 0 and 1.

As outlined in the next section, it is also the case that fuzzy self-reference is capable of giving us full semantic chaos in the strict mathematical sense.

5 Fuzzy Chaos

Consider the following sentence, which I will call the Chaotic Liar:

This sentence is true if and only if it is false.   (10)

Using the Łukasiewicz biconditional and modeling self-reference in terms of iterated algorithms, semantic values for the Chaotic Liar will be given by

$$x_{n+1} = 1 - \text{abs}((1 - x_n) - x_n) \ .$$

By expanding our language slightly we can also express the Chaotic Liar in other ways. If "p is as true as q" or "the value of p is the same as the value of q" is treated fairly naturally as taking the value

$$1 - \text{abs}(\nu(p) - \nu(q)) \ ,$$

for example, then the Chaotic Liar can alternatively be expressed as

This sentence is as true as it is false   (11)

or

The value of this sentence is the same as its negation.   (12)

A seed value of .314 for any of these gives us a series of values that begins as in Fig. 16.

---

[15] The very different behavior of the Modest Truth-teller and the Emphatic Truth-teller is inevitable given a modeling of "very" and "fairly" by squares and square roots—or, for that matter, by cubes and cube roots or the like. It has been suggested, however, that this is itself a mark against such a modeling: are "very" and "fairly" so different that "this is fairly true" should converge on pure truth and "this is very true" should converge on pure falsity?

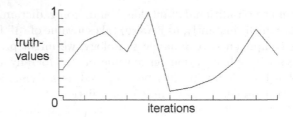

Fig. 16. 'This sentence is true if and only if it is false.' The dynamical semantics of the Chaotic Liar with a seed value of .314

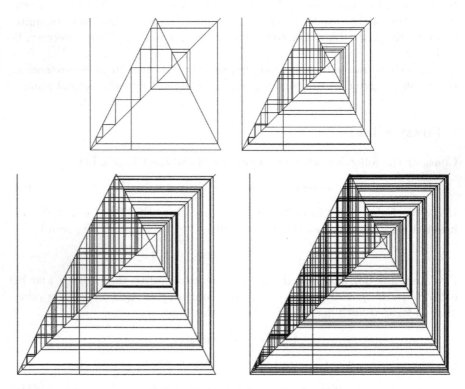

Fig. 17. 'This sentence is true if and only if it is false.' A web diagram for the Chaotic Liar with a seed value of .314

The semantical behavior of the Chaotic Liar is perhaps best exhibited by its web diagram, shown in four progressive stages of development in Fig. 17.[16]

---

[16] One peculiarity of this function is that standard rounding off within the binary arithmetic of computers in fact disguises its chaoticity: although it *is* provably chaotic on the [0, 1] interval, it does not show up as such on the computer screen. In order to graph something closer to the function's true behavior it is thus standard to "cancel out" the effect of rounding-off by introducing a small

The dynamical semantics of the Chaotic Liar qualifies as chaotic in the precise mathematical sense of, for example, Devaney [36].[17] What this amounts to is a range of surprising semantic features:

(1) The Fuzzy Liar, we have seen, has a simple period of 2 for almost all values. For sentences such as the Chaotic Liar, on the other hand, any repeating period we might care to choose, however high, will be generated by some initial value. For such sentences there is no upper limit to semantic periodicity.

(2) It is also the case that there will be numerical truth values for such sentences which eventually move from any arbitrarily small semantic region to any other. There is thus no range of degrees of truth, however small, such that values within that range assigned to the Chaotic Liar will safely stay there on iteration. For *any* such region, some semantic values will eventually escape to any other semantic region we might name.

(3) Finally, no matter what numerical truth value x we might start off with as an estimate for such sentences, there will be numerical truth values arbitrarily close to our initial value which, upon iterative recalculation through our sentences, eventually move as far from corresponding iterations of x as we might choose to specify. There is thus no initial range "close enough" to a starting estimate x that differences within that range will not make a significant difference. Within any distance from x, how-

---

element of randomness, and that has been done for the illustrations here. On this point I am obliged to John Milnor for discussion

[17] Given a set $J, f : J \to J$ is chaotic on J if

a. f shows sensitive dependence on initial conditions
b. f is topologically transitive
c. the set of periodic points is dense in J

Here let us use the notation $f^n(x)$ to stand for the composition or iteration of the function $f(x)$ n times, i.e.,

$$f^n(x) = \underbrace{\ldots f(f(f(x)))\ldots n \text{ times}}$$

a. $f : J \to J$ shows sensitive dependence on initial conditions if there exist points arbitrarily close to any $x \in J$, which eventually separate from x by any chosen distance δ or more under iteration of f, i.e., there exist $\delta > 0$ such that, for any $x \in J$ and any neighborhood N of x, there exist $y \in N$ and $n \geq 0$ such that $\mathrm{abs}(f^n(x) - f^n(y)) > \delta$.

b. $f : J \to J$ is topologically transitive if it has points which eventually move under iteration from one arbitrarily small neighborhood to any other, i.e., for any pair of open sets $U, V \subseteq J$ there exists some $n > 0$ such that $f^n(U) \cap V$ is nonempty.

c. The set of periodic points of J, $\mathrm{PER}(J)$, is the set of all $x \in J$ such that $f^n(x) = x$ for some natural number n, i.e., $\mathrm{PER}(J) = \{x \in J : \exists n f^n(x) = x\}$. $\mathrm{PER}(J)$ is dense in J if $\mathrm{PER}(J)$ together with all its limit points is equal to J

ever small, is another value iterations of which will eventually diverge from iterations of x enormously—as enormously, within the limits of our semantic range, as we might care to specify.

Although stronger and weaker characterizations of chaos appear in the literature, the central element in all versions is this last feature, sensitive dependence on initial conditions. The graph in Fig. 18 shows the basic idea of sensitive dependence for the Chaotic Liar. Here iteration graphs are superimposed for seed values starting with .314 and increasing by .001.

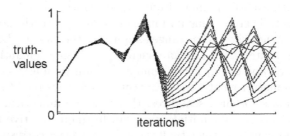

Fig. 18. Sensitivity to initial conditions: Dynamics of the Chaotic Liar with initial values of .314 increasing by .001

The basic algorithm for the Chaotic Liar is in fact a very simple and paradigmatically chaotic function, known as a tent map because of the shape of its graph and more familiar in the mathematical guise $x_{n+1} = 1 - \text{abs}(2x_n - 1)$ or

$$x_{x+1} = \begin{cases} 2x_n & \text{for } 0 \le x \le 0.5 \\ 2(1 - x_n) & \text{for } 0.5 \le x \le 1 \end{cases}$$

(see 171 ff. in [36]). Though this function appears in Robert May's seminal paper on chaos theory and ecology [37], its role within self-referential fuzzy logic comes as something of a surprise.[18]

Here let me also offer a second simple sentence which shows chaotic behavior within fuzzy logic—a sentence I will call the Fuzzy Logistic:

> It is very false that this sentence is true if it is false .   (13)

Here semantic values will be given by

$$x_{n+1} = (1 - (1 - \text{abs}((1 - x_n) - x_n)))^2 .$$

Note that the algorithm for the Chaotic Liar is embedded within that of the Fuzzy Logistic. We can in fact obtain the Fuzzy Logistic from the Chaotic Liar simply by adding the prefix "It is very false that...."

---

[18] The similar role of that algorithm in a Rescher multi-valued logic appears in [8, 9]

Here as before there are of course alternative phrasings. If we take "differs in value from" as the negation of "is as true as," the Fuzzy Logistic can be phrased as

It is very true that this sentence differs in value from its negation   (14)

or

The value of this sentence is very different from that of its negation.   (15)

For values in the $[0, 1]$ interval our algorithm amounts in each case to

$$\begin{aligned} x_{n+1} &= ((1 - x_n) - x_n)^2 \\ &= (1 - 2x_n)^2 \\ &= 1 - 4x_n(1 - x_n) . \end{aligned}$$

This, it turns out, is an inverted form of the logistic or quadratic equation, perhaps the most familiar and thoroughly studied sample of chaos.[19] For an intitial value of .314 the web digram of the Fuzzy Logistic develops as shown in Fig. 19

Chaos can also be expected to appear within other fuzzy logics by way of other self-referential sentences. Although our tour of fuzzy chaos has been confined in general to the simple Baldwin fuzzy logic outlined in Sect. 2, it is perhaps worth noting a route by which chaos will appear within *any* fuzzy logic with the standard Łukasiewicz base, regardless of the fuzzy set it introduces for "true." Within the Zadeh fuzzy logic or any other based on $L_{\aleph 1}$, consider the prospect of a sentence p which amounts to a biconditional between itself and its negation:

$$p = (p \leftrightarrow \sim p) .$$

Given simply the basic Łukasiewicz biconditional, the algorithm for such a sentence will be

$$x_{n+1} = 1 - \text{abs}((1 - x_n) - x_n) ,$$

and any such sentence will thus amount to the Chaotic Liar.[20]

---

[19] An additional negation would of course give us the Logistic without inversion:

It is not very false that this sentence is true if it is false

or

It is not very true that this sentence differs in value from its negation

or

The value of this sentence is not very different from that of its negation

I am obliged to Nathaniel Hellerstein for his seminal work on the Fuzzy Logistic, communicated in private correspondence

[20] For Zadeh fuzzy logic in particular, Paul St. Denis has suggested the following chaotic sentence:

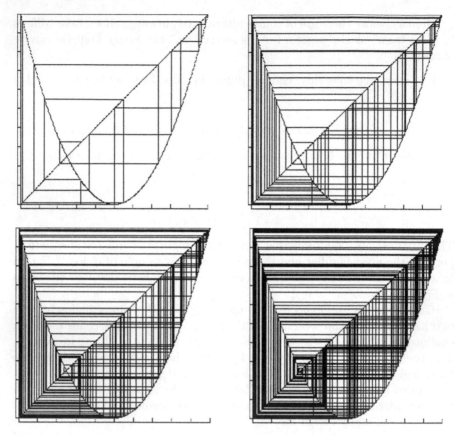

Fig. 19. 'It is very false that this sentence is true iff it is false.' Dynamics for the Fuzzy Logistic

6 Fuzzy Self-Reference in Two Dimensions

Beyond the traditional Liar lies an infinite series of Liar cycles, the simplest of which is perhaps the Dualist. In medieval form it appears as an exchange between Socrates and Plato:

> Socrates: What Plato is about to say is true.
>
> Plato: Socrates speaks falsely.

It is not the case that this sentence is fairly not true, or it is not the case that the negation of this sentence is fairly not true.

Here our algorithm is

$$x_{x+1} - \max[1 - \sqrt{1 - Z(x_n)}, 1 - \sqrt{1 - Z(1 - x_n)}],$$

where $Z(x_n)$ indicates the degree of membership in Zadeh's fuzzy set "true" (using .5 as α) of a sentence with numerical truth value x_n

More simply, we have two sentences each of which is about the truth value of the other:

X: Y is true

Y: X is false

With the tools of our basic fuzzy logic we can also introduce fuzzy variations on the Dualist. Consider for example,

X: X $\leftrightarrow$ Y

Y: Y $\leftrightarrow$ it is very false that X

or equivalently

X: X is true if and only if Y is

Y: Y is true if and only if X is very false.

The focus in previous sections was on single sentences which force a series of revised values. Here the situation is somewhat more complex: we have a pair of sentences which, for any initial seed values (x_0, y_0), forces a series of pairs of revised values. For the X and Y above, revised values can be calculated in terms of the following algorithms:

$$x_{n+1} = 1 - \mathrm{abs}(x_n - y_n)$$
$$y_{n+1} = 1 - \mathrm{abs}(y_n - (1 - x_n)^2)$$

If we start with seed values of .25 and .25, for example, we are forced to the following series of revised values:[21]

$$(1, .6875), (.6875, .3125), (.625, .7852), (.8398, .3555), (.5156, .6702), \ldots$$

If we plot these pairs as Cartesian coordinates, the pentagonal attractor as shown in Fig. 20 appears.

The persistence of such an attractor, for various seed values, is clear from an overlay diagram for seed values (x, y) where x and y range from 0 to 1 in intervals of .05 (Fig. 21). Throughout the $[0, 1]$ interval values are attracted to and trapped within the same clearly defined region.

Consider also a second fuzzy Dualist variation:

X: It is very false that (X $\leftrightarrow$ Y)

Y: It is fairly false that (Y $\leftrightarrow\sim$ X)

or more colloquially

X: It is very false that: X is true if and only if Y is.

Y: It is fairly false that: Y is true if and only if X is false.

---

[21] Here again numbers are rounded off for pres entational simplicity

342   P. Grim

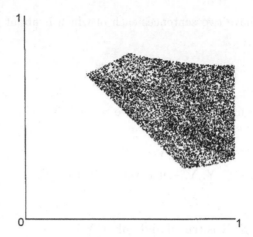

Fig. 20. The two-dimensional attractor for a Dualist pair of mutually referential sentences, with seed values of .25 and .25 (simultaneous calculation).
       X: X is true if and only if Y is
       Y: Y is true if and only if X is very false

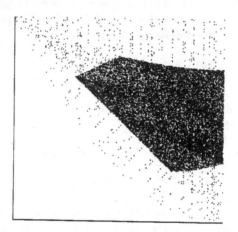

Fig. 21. Persistence of the attractor: Overlay diagram for seed values for x and y from 0 to 1 in intervals of .05 (simultaneous calculation)

Here successive values can be calculated using the following algorithms:

$$x_{n+1} = (1 - (1 - \text{abs}(x_n - y_n)))^2$$
$$Y_{n+1} = \sqrt{(1 - (1 - \text{abs}(y_n - (1 - x_n))))}$$

or more simply

$$x_{n+1} = (x_n - y_n)^2$$
$$x_{y+1} = \sqrt{\text{abs}(y_n - (1 - x_n))}.$$

In this case our attractor is shown in Fig. 22 as an overlay diagram.

Fig. 22. Overlay diagram for a second Dualist pair
    X: It is very false that $(X \leftrightarrow Y)$
    Y: It is fairly false that $(Y \leftrightarrow \sim X)$ (simultaneous calculation)

Here let me finally offer one further fuzzy Dualist:

    X: It is very false that $(X \leftrightarrow Y)$
    Y: It is very false that X is false if and only if Y is very true,

with successive values

$$x_{n+1} = (x_n - y_n)^2$$
$$y_{n+1} = ((1 - x_n) - y_n^2)^2 .$$

The attractor for this final variation, once again in overlay form, appears in Fig. 23.

In our calculation of revised values for the Fuzzy Dualists above, it should be noted, we have assumed a simultaneous calculation of numerical truth values for sentences X and Y. Given a pair of seed values (x, y), we have calculated a new value for X in terms of those values and have simultaneously calculated a new value for Y in terms of those same values.

Evaluation of the sentences of a Fuzzy Dualist pair might also proceed sequentially. In at least some contexts, it might be argued, a more natural way to approach such a pair of sentences would be to begin with seed values (x, y), to calculate the value of X in terms of those seed values, but then to calculate the value of Y using the newer or most recent value computed for X.

This second pattern of reasoning with regard to the same pairs of sentences above can be represented with a slight change in our algorithms: in each case x_n is replaced in the second algorithm with x_{n+1}.

Our algorithms for the first variation on the Dualist above, for example,

$$X: X \leftrightarrow Y$$
$$Y: Y \leftrightarrow \text{it is very false that } X$$

will now be

$$x_{n+1} = 1 - \text{abs}(x_n - y_n)$$
$$y_{n+1} = 1 - \text{abs}(y_n - (1 - x_{n+1})^2) \ .$$

Using this alternative form of calculation the same pair of sentences give us the attractor shown in Fig. 24.

A similar change from a simultaneous to sequential pattern of reasoning in the case of our other fuzzy Dualist variations gives us other attractors. Figure 25 shows the sequential attractor for

$$X: \text{It is very false that } (X \leftrightarrow Y)$$
$$Y: \text{It is fairly false that } (Y \leftrightarrow\sim X)$$

$$x_{n+1} = (x - y)^2$$
$$y_{n+1} = \sqrt{\text{abs}(y_n - (1 - x_{n+1}))}$$

Figure 26 shows the sequential attractor for

$$X: \text{It is very false that } (X \leftrightarrow Y)$$
$$Y: \text{It is very false that } X \text{ is false if and only if } Y \text{ is very true}$$

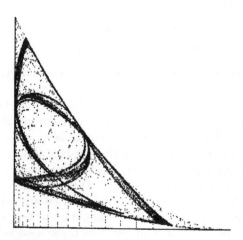

Fig. 23. Overlay diagram for a third Dualist pair
X: It is very false that $(X \leftrightarrow Y)$
Y: It is very false that X is false if and only if Y is very true
(simultaneous calculation)

$$x_{n+1} = (x_n - y_n)^2$$
$$y_{n+1} = ((1 - x_{n+1}) - y_n^2)^2$$

Although these figures illustrate a clear difference between simultaneous and sequential updating, it must be confessed that their interpretation remains much less clear. Simultaneous updating might be said to be more appropriate to a God's eye view of the informational dynamics of the fuzzy Dualist, in which all information is received and processed simultaneously at each step. Sequential updating, on the other hand, might be thought to be more appropriate to beings capable of processing only the information of a single sentence at a time. In more realistic applications to repeated sequences of mutually referential sentences, the difference might be appropriate to contexts in which two sources provide information about each other, and in which the second source of information may or may not be aware of reports coming from the first source. (For further work on epistemic chaos of this type, and the attempt to control it, see Chap. 2 in [9]).

Certain aspects of the fuzzy dynamics of Dualist variations can also be graphed using what are known as escape-time diagrams. For each pair of points (x, y) on the Cartesian plane, we can graph in terms of color the number of iterations required for the series of values to reach a certain threshold. We might choose a threshold as a certain distance from the origin, for example, with the origin itself representing "double falsity" for our two sentences.

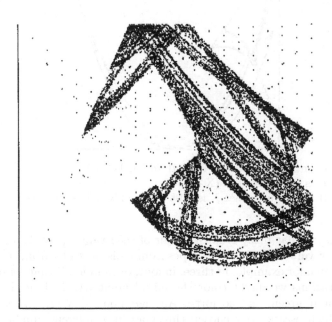

Fig. 24. Sequential calculation for the first Dualist pair:
X: X is true if and only if Y is
Y: Y is true if and only if X is very false

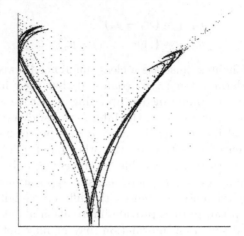

Fig. 25. Sequential calculation for the second Dualist pair:
X: It is very false that (X ↔ Y)
Y: It is fairly false that (Y ↔∼ X)

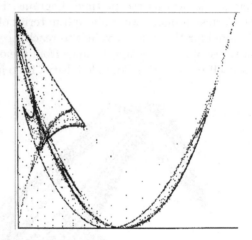

Fig. 26. Sequential calculation for the third Dualist pair:
X: It is very false that (X ↔ Y)
Y: It is very false that X is false if and only if Y is very true

Within a particular fuzzy Dualist a pair of seed values (say .1, .5) may give
us a series of values which first escapes from a distance of 1 from the origin in
two iterations, for example, in three, in four, or in more. If that series escapes
in two iterations we might color the initial point (.1, .5) blue; if three, we
might color it green, and so forth. Another point (say .2, .4) may give us
a series which escapes our chosen threshold in a different number of iterations, and will correspondingly be given a different color. The general idea of

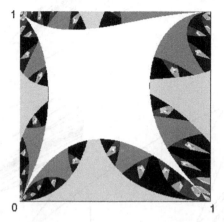

Fig. 27. Escape-time diagram for a Fuzzy Dualist with simultaneous calculation
X: X is true if and only if Y is
Y: Y is true if and only if X is very false

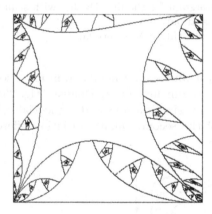

Fig. 28. Tracery of escape-time diagram, showing points at which number of iterations changes

escape-time diagrams also appears in the graphing of the familiar Mandelbrot set.

Figure 27 shows an escape-time diagram of this type for the first Fuzzy Dualist above.

The fractal character of such a graph is clearer if we emphasize merely the interfaces of different colored areas: points at which the number of iterations required to pass the chosen threshold changes. The escape-time diagram now appears as the tracery shown in Fig. 28.

Fig. 29. Escape-time diagram for the first Dualist with sequential calculation.
X: X is true if and only if Y is
Y: Y is true if and only if X is very false

Consider in contrast an escape-time diagram for a sequential pattern of reasoning with regard to our first Fuzzy Dualist (Fig. 29). Figure 30 shows escape-time diagrams for simultaneous and sequential calculations (left and right, respectively) of the second Dualist variation offered above, using an escape threshold of .8.

Fig. 30. Escape-time diagrams for the second Dualist, using simultaneous (*left*) and sequential (*right*) calculations and an escape threshold of .8

 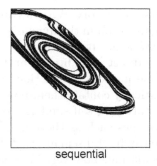

simultaneous   sequential

Fig. 31. Escape-time diagrams for the third Dualist, using simultaneous and sequential calculations, plotted for values between −1.4 and 2.4 with a threshold of .85.

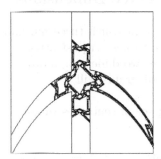 

Fig. 32. Variant escape-time diagrams measuring iterations to a point at which x and y values are separated by at least .5

In all of the escape time diagrams considered to this point we have confined our values (x, y) to the unit interval, reflecting the fact that the semantics of our fuzzy logic limits numerical truth values to the $[0, 1]$ interval. In some cases, however, it is easy to see that the characteristics of points within the $[0, 1]$ interval are merely part of a larger pattern. Figure 31 shows our third Fuzzy Dualist variation for values between −1.4 and 2.4 and with a threshold of .85.[22]

Other escape-time diagrams are possible using other parameters. Figure 32 shows an escape-time diagram in which what we measure is the number of iterations required before a series starting from a pair of seed values (x, y) reaches a value (x', y') such that x and y are separated by a distance of at least .5.

In many of these images a deep fractal character—self-affinity at descending scales—is clearly evident. This is a familiar companion to chaos within

---

[22] The range of Fuzzy Dualist variations is so immense as to be intimidating. For a few other small samples the reader is referred to [8, 9, 33]

dynamical systems theory. What its appearance here indicates is the presence of fractal organization in the dynamical semantics of self-referential sentences within a fuzzy logic. It is tempting to speculate that different varieties of self-reference, direct or indirect, can themselves be thought of as abstractly fractal in some intuitive sense: that self-referential sentences or sets of sentences semantically contain themselves, or images of themselves, in much the way that fractals show self-similarity on different scales. It might thus be proposed that images such as those above give more explicit visual expression to the inherently fractal semantics of different patterns of self-reference. This remains speculation, however. Here as elsewhere it proves easier to graph certain semantic characteristics than to fully understand them.

7 Fuzzy Triplists Modeled in Three Dimensions

Beyond the Dualist lie Triplist variations, in which three mutually referential sentences speak of each other's truth values. Triplist attractors must be graphed as three-dimensional rather than two-dimensional objects. The correlates to two-dimensional escape-time diagrams will be three-dimensional escape-time solids.

Consider, for example, the following set of sentences (a colon is used to avoid ambiguity):

X: It is very false that: $X \leftrightarrow \sim (Y \leftrightarrow Z)$

Y: It is very false that: $Y \leftrightarrow \sim Z$

Z: It is very false that: $Z \leftrightarrow \sim (X \leftrightarrow Y)$

In the fuzzy Dualists, our sentences forced us through a series of revisions for initial seed values (x, y) for sentences X and Y. In the case of this fuzzy Triplist, our sentences will force us through a similar series of revisions for seed values (x, y, z). For the sentences above, these revised values can be calculated as

$$x_{n+1} = (\text{abs}(y_n - z_n) - x_n)^2$$
$$y_{n+1} = ((1 - z_n) - y_n)^2$$
$$z_{n+1} = (\text{abs}(x_n - y_n) - z_n)^2 \ .$$

If we plot revised values for these sentences starting with seed values of .23, .34, .45, the attractor in Fig. 33 appears. In Fig. 34 the attractor is rotated in three dimensions. Despite its apparent depth when viewed "full face," it is clear that the attractor for this first fuzzy Triplist is still confined to a plane.

Here as before we can also compute revised values sequentially rather than simultaneously, with the following changes in our formulas:

$$x_{n+1} = (\mathrm{abs}(y_n - z_n) - x_n)^2$$
$$y_{n+1} = ((1 - z_n) - y_n)^2$$
$$z_{n+1} = (\mathrm{abs}(x_{n+1} - y_{n+1}) - z_n)^2 .$$

With that sequential calculation, the looping attractor of Figs. 35 and 36 appears.

Consider also a second fuzzy Triplist variation:

$$X: \sim (X \leftrightarrow \text{it is very true that } (Y \leftrightarrow Z))$$
$$Y: \sim (Y \leftrightarrow \text{it is very true that } (X \leftrightarrow Z))$$
$$Z: \sim (Z \leftrightarrow \text{it is very true that } (X \leftrightarrow Y)),$$

with the following algorithms for revised values:

Fig. 33. X: It is very false that: X $\leftrightarrow \sim$ (Y $\leftrightarrow$ Z)
   Y: It is very false that: Y $\leftrightarrow \sim$Z
   Z: It is very false that: Z $\leftrightarrow \sim$(X $\leftrightarrow$ Y)
Triplist attractor in two dimensions (simultaneous calculation)

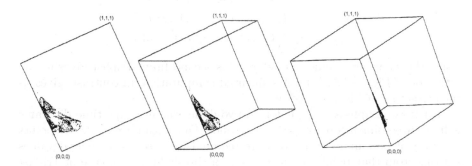

Fig. 34. The same Triplist attractor in three dimensions (simultaneous calculation)

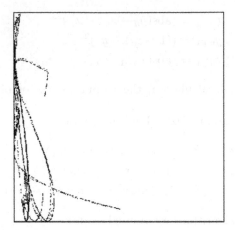

Fig. 35. Sequential calculation for the Triplist attractor in Figs. 33 and 34, shown in two dimensions

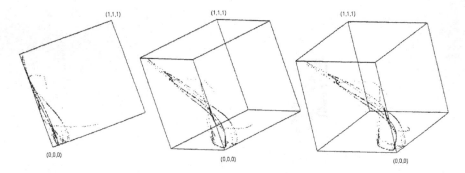

Fig. 36. Sequential calculation for the Triplist attractor in Figs. 33 and 34, shown in three dimensions

$$x_{n+1} = \text{abs}((1 - \text{abs}(y_n - z_n))^2 - x_n)$$
$$y_{n+1} = \text{abs}((1 - \text{abs}(x_n - z_n))^2 - y_n)$$
$$z_{n+1} = \text{abs}((1 - \text{abs}(x_n - y_n))^2 - z_n) .$$

Using the same seed values as before, this second fuzzy Triplist gives us the attractor of Figs. 37 and 38. A sequential computation, in contrast, gives us the attractors of Figs. 39 and 40.

For fuzzy Triplists, the analog to two-dimensional escape-time diagrams will be three-dimensional escape-time solids. Once again we can color points in terms of how many iterations are required for a series of revised values starting from that point to reach a certain threshold. In the case of Triplist variations, however, we will be coloring points (x, y, z) in a three-dimensional space.

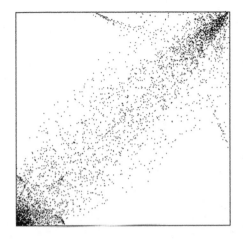

Fig. 37. X: $\sim$(X $\leftrightarrow$ it is very true that (Y $\leftrightarrow$ Z))
  Y: $\sim$(Y $\leftrightarrow$ it is very true that (X $\leftrightarrow$ Z))
  Z: $\sim$(Z $\leftrightarrow$ it is very true that (X $\leftrightarrow$ Y))
Triplist attractor in two dimensions (simultaneous calculation)

In Figs. 41 and 42 we use sequential calculation for the first Triplist variation and simultaneous calculation for the second, with a chosen threshold in each case of $\sqrt{x^2 + y^2 + z^2} = 1$. Both escape-time solids are shown from two angles, in a space extending roughly from -2.5 to $+5$ for each of our three values.

There is no upper limit to the size of sets of mutually self-referential sentences that might be considered, of course, nor any upper limit to the number of dimensions appropriate for modeling their semantical dynamics. Beyond the three-dimensional semantic phenomena of Triplist variations lie the four-dimensional semantic phenomena of the Quadruplists, the five-dimensional semantic phenomena of Quintuplists, and so on.

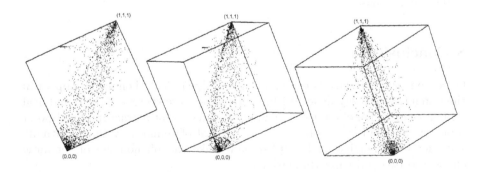

Fig. 38. The same Triplist attractor in three dimensions (simultaneous calculation)

Fig. 39. Sequential calculation for the Triplist attractor in Figs. 37 and 38, shown in two dimensions

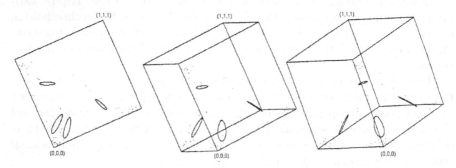

Fig. 40. Sequential calculation for the Triplist attractor in Figs. 37 and 38, shown in three dimensions

8 Conclusion

Once we introduce self-reference, a range of dynamical phenomena appear in the semantics of fuzzy logic. Within such a logic are sentences the dynamical semantics of which exhibit the behavior of fixed-point attractors, fixed-point repellers, and full chaos on the [0, 1] interval of semantic values. Mutually self-referential Dualist and Triplist pairs take the phenomena of semantic chaos into two and three dimensions.

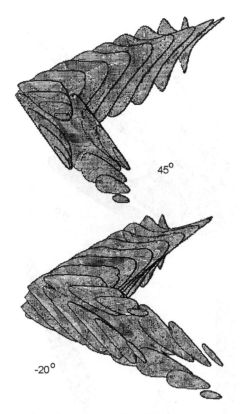

Fig. 41. X: It is very false that: X $\leftrightarrow\sim$ (Y$\leftrightarrow$ Z)
        Y: It is very false that: Y $\leftrightarrow\sim$Z
        Z: It is very false that: Z $\leftrightarrow\sim$(X $\leftrightarrow$Y)
Escape-time solid using sequential calculation

A great deal of further formal exploration, generalization, and application clearly remains to be done.[23] Perhaps it is not out of place, however, to close with some admittedly philosophical speculations.

Logical systems have typically been introduced with certain semantical expectations. Self-reference has a tendency to violate those expectations. Classical logic is a prime example. Within such a logic the expectation is that every proposition will be simply true or false. With the introduction of semantical self-reference, however, we are confronted with the classical Liar:

---

[23] One meta-mathematical application is mentioned in the introduction: References [8] and [9] each contain a sketch of Gödel-like limitative results for chaos theory in the context of formal systems for real arithmetic, motivated by a close relative of the sentence that appears here as the Chaotic Liar. It is clear that one class of extensions would take these into the context of fuzzy logics

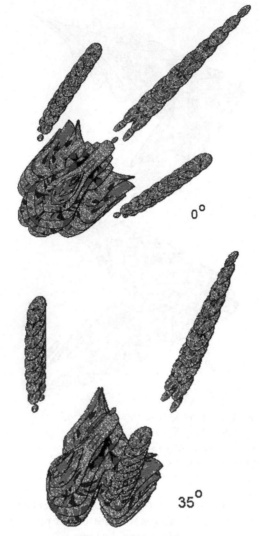

Fig. 42. X: ∼(X ↔ it is very true that (Y ↔ Z))
Y: ∼(Y ↔ it is very true that (X ↔ Z))
Z: ∼(Z ↔ it is very true that (X ↔ Y))
Escape-time solid using simultaneous calculation

$$\text{This sentence is false.} \tag{1}$$

The dynamical semantics of that sentence seems to be an oscillation, and the attempt to assign either of our supposedly exhaustive semantical categories results in simple contradiction. A similar story, relying on strengthened versions of the Liar, can be told for multi-valued, infinite-valued, gapped, and antifoundational logic (see Chap. 1 in [1]). In each case self-reference seems to violate initial expectations by forcing us to recognize new categories of semantical behavior.

What the work above shows is that something similar also happens when self-reference is introduced into fuzzy logic. Fuzzy logic was constructed to incorporate an important range of intuitive phenomena not provided for in classical logics, facilitating a range of applications. One assumption carried over from its classical predecessors, however, was that semantic values, however fuzzy, would nonetheless be tolerably well behaved and manageably stable. Here again the introduction of self-reference seems to violate semantical expectations. In the context of fuzzy logic, self-reference introduces a range of patterns of semantic *in*stability as diverse and complex as the phenomena of chaos theory generally.

Acknowledgments

The work presented here is an expansion of collaborative work on infinite-valued logics and chaos which appears in [8, 9, 23]. I am indebted to Paul St. Denis for programming assistance and for repeatedly bringing my attention back to the Łukasiewicz biconditional. Matt Neiger developed the programming required for three-dimensional escape-time solids in Sect. 5. I am indebted to Gary Mar for fruitful collaboration and for central good ideas.

References

1. J.J. Buckley: Fuzzy dynamical systems I. In: *Proc. IFSA '91 (Mathematics Section)*, Brussels, pp 16–20.
2. G.-Y. Wang, J.-P. Ou, P.-Z. Wang: Dynamic fuzzy sets and fuzzy processes. In: *Proc. 3rd IFSA Congr.*, ed by J. Bezdek, Seattle, 1989, pp 276–279
3. P. Diamond: Chaos and fuzzy representations of dynamical systems. In: *Int. Symp. on Fuzzy Systems*, Iizuka, Japan, July 1992, pp 51–58
4. P. Grim: *The Incomplete Universe* (MIT Press, Cambridge, MA, 1991)
5. H. Herzberger: J. Philosophical Logic **11**, 61–102 (1982) [Reprinted in *Recent Essays on Truth and the Liar Paradox*, ed by R.L. Martin (Oxford University Press, New York, 1984), pp 133–174]
6. A. Gupta: J. Philosophical Logic **11**, 1–60 (1982) [Reprinted in *Recent Essays on Truth and the Liar Paradox*, ed by R.L. Martin (Oxford University Press, New York, 1984), pp 175–235]

7. A. Gupta, N. Belnap: *The Revision Theory of Truth* (MIT Press, Cambridge, MA, 1993)
8. G. Mar, P. Grim: Noûs **25**, 659–694 (1991)
9. P. Grim, G. Mar, P.St. Denis: *The Philosophical Computer: Exploratory Essays in Philosophcial Compuiter Modeling* (MIT Press, Cambridge, MA, 1998)
10. N. Rescher: *Many-Valued Logic* (McGraw-Hill, New York, 1969)
11. L.A. Zadeh: Synthese **30**, 407–428 (1975)
12. R.E. Bellman, M. Giertz: Inf. Sci. **5**, 149–156 1973
13. Z.P. Dienes: J. Symbolic Logic **14**, 95–97 (1949)
14. B.R. Gaines: Int. J. Man-Machine Studies **8**, 623–668 (1976)
15. L.A. Zadeh: J. Cybern. **2**, 4–34 (1972)
16. R.E. Bellman, L.A. Zadeh: Local and fuzzy logics. In: *Modern Uses of Multiple-Valued Logic*, ed by J. Michael Dunn, G. Epstein (D. Reidel, Dordrecht, 1977)
17. J.F. Baldwin: *Fuzzy Sets Syst.* **2**, 309–325 (1979)
18. J.F. Baldwin: Fuzzy logic and fuzzy reasoning. In: *Fuzzy Reasoning and Its Applications*, ed by E.H. Mamdani, B.R. Gaines (Academic Press, New York, 1981)
19. H.-J. Zimmerman: *Fuzzy Set Theory and Its Applications* (Kluwer-Nijhoff, Dordrecht, 1985)
20. A. Tarski: Studien Philosophica I, 261–405 (1935) [Trans. by J.H. Woodger as "The concept of truth in formalized languages" in *Logic, Semantics, Meta-Mathematics*, 2nd edn, ed by J. Corcoran (Hackett Pub. Co., Indianapolis, 1983, pp 152–278]
21. Y. Tsukamoto: An approach to fuzzy reasoning method. In: *Advances in Fuzzy Set Theory and Applications*, ed by M.M. Gupta et al. (North-Holland, Amsterdam, 1979, pp 137–149)
22. L. Ding, Z. Shen, M. Mukaidono: A new method for approximate reasoning. In: *Nineteenth Int. Symp. on Multiple-Valued Logic*, IEEE Computer Society Press, Washington, 1989, pp 179–189
23. P. Grim, G. Mar, M. Neiger, P.St. Denis: Comput. Graphics **17**, 609–612 (1993)
24. K. Gödel: Monatschefte für Mathematik und Physik **38**, 173–198 (1931) [Translated by J. van Heijenoort as "On formally undecidable propositions of *Principia Mathematica* and related systems I" in *Kurt Gödel: Collected Works*, vol. I, ed by S. Feferman et al. (Oxford University Press, New York, 1986, pp 144–195]
25. A. Church: J. Symbolic Logic **1**, 40–41, 101–102 (1936)
26. A. Turing: *Proc. Lond. Math. Soc.* **42**, 230–265 (1936) [Reprinted in *The Undecidable*, ed by Martin Davis (Raven Press, Hewlett, New York, 1965, pp 116–303]
27. G. Chaitin: *Information, Randomness, and Incompleteness—Papers on Algorithmic Information Theory*, 2nd edn (World Scientific, Singapore, 1990)
28. B. van Fraassen: J. Philosophy **65**, 136–152 (1968)
29. R.L. Martin: A category solution to the liar. In: *The Paradox of the Liar*, ed by R.L. Martin (Ridgeview Press, Reseda, CA, 1970)
30. S. Kripke: J. Philosophy **72**, 690–716 (1975) [Reprinted in *Recent Essays on Truth and the Liar Paradox*, ed by R.L. Martin (Oxford University Press, New York, 1984), pp 53–81]
31. J. Barwise, J. Etchemendy: *The Liar* (Oxford University Press, New York, 1987)

32. J.L. Casti: *Alternate Realities* (Wiley, New York, 1989)
33. G. Lakoff: *Women, Fire, and Dangerous Things* University of Chicago Press, Chicago, 1987)
34. C. Pickover: *Computers and the Imagination* (St. Martin's Press, New York, 1991)
35. R.L. Devaney: *An Introduction to Chaotic Dynamical Systems*, 2nd edn (Addison-Wesley, Menlo Park, CA, 1989)
36. K. Falconer: *Fractal Geometry: Mathematical Foundations and Applications* (Wiley, New York, 1990)
37. R. May: Nature **261**, 459–467 (1976)

Chaotic Behavior
in Recurrent Takagi–Sugeno Models

Alexander Sokolov and Michael Wagenknecht

Abstract. We investigate dynamic systems which are modeled by recurrent fuzzy rule bases widely used in applications. The main question to be answered is "Under which conditions recurrent rule bases show chaotic behavior in the sense of Li–Yorke?" We determine the minimal number of rules of zero-order and first-order Takagi–Sugeno models with chaotic orbits. We also consider the case of an arbitrary number of rules in such models and high-order time delay case. This chapter is the first from a series of papers where we will consider arbitrary types of consequent functions, noncomplete or contradictory rule bases, vectors in the rule antecedents, Mamdani model, and methods of chaos identification by backward mapping.

1 Introduction

One of the main features in modeling dynamic systems is prediction (respectively simulation). On the other hand, there do exist very simple dynamic systems which are unpredictable in principle and show chaotic behavior [1, 2]. If some dynamic model is chaotic, we cannot perform long-time predictions due to its extreme sensitivity with respect to initial conditions. There is an extensive literature for classical (unfuzzy) chaotic dynamic systems.

We can distinguish several approaches of how to investigate dynamic systems based on different classes of mathematical models that can be used for their description. The kind of the model applied heavily depends on our idea about the necessary accuracy of description. Generally, increasing accuracy implies increasing complexity.

During the last two decades, Takagi–Sugeno (TS) fuzzy rule bases have successfully been applied as an excellent tool to describe many complex dynamic processes in engineering recurrent. In the simplest case of one-time delay, TS model has the form

$$
\begin{aligned}
R_1 &: \text{If } x_k = L_1 \quad \text{then } x_{k+1} = f_1(x_k) \\
R_2 &: \text{If } x_k = L_2 \quad \text{then } x_{k+1} = f_2(x_k) \\
&\;\;\vdots \\
R_N &: \text{If } x_k = L_N \quad \text{then } x_{k+1} = f_N(x_k),
\end{aligned}
\tag{1}
$$

where L_i are linguistic variables (terms) and $f_i(x)$ are real-valued functions.

A. Sokolov and M. Wagenknecht: *Chaotic Behavior in Recurrent Takagi–Sugeno Models*, StudFuzz **187**, 361–389 (2006)
www.springerlink.com

If f_i is constant, we speak of zero-order TS models; if f_i is linear, we call it first-order TS model. We can represent the recurrent system as closed-loop dynamic system with interior structure (1), as shown in Fig. 1.

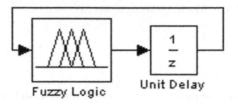

Fuzzy Logic   Unit Delay

Fig. 1. Recurrent fuzzy rule base

In this chapter we consider TS models and their properties for modeling of chaos, and in the process we shall answer the following questions:

1. Can dynamic TS rule bases be chaotic?
2. Which conditions are necessary for chaos?
3. How can chaotic behavior be recognized by analyzing the TS rule base structure and parameter values?

First of all we need to give necessary terms, definitions of chaos, and its properties for future investigations.

2 On the Nature of Chaos

The basic operation of dynamical systems is iteration. Iteration is the repetition of a process. Iteration begins with some initial value or input and proceeds with the output of one application of the process as the input to the next application of the process. While iteration may be applied to any process, in dynamics it is usually applied to a mathematical function. For example, consider the positive square root function $F(x) = \sqrt{x}$ and the initial input $x_0 = 256$. Iteration yields the following:

$$\sqrt{256} = 16 \,,$$
$$\sqrt{16} = 4 \,,$$
$$\sqrt{4} = 2 \,,$$
$$\sqrt{2} = 1.414214\ldots \,,$$
$$\sqrt{1.414214\ldots} = 1.189207\ldots \,,$$
$$\sqrt{1.189207\ldots} = 1.090508\ldots \,,$$
$$\sqrt{1.090508\ldots} = 1.044274\ldots \,,$$
$$\vdots$$
$$\sqrt{1} = 1 \,.$$

For convenience, $F^i(x)$ is defined as the ith application of $F(x)$. Thus,

$$F^1(256) = 16 ,$$
$$F^2(256) = 4 ,$$
$$F^3(256) = 2 ,$$
$$\vdots$$

The sequence of successive iterates of a point is called the *orbit* of that point. The orbit analysis of $F(x) = \sqrt{x}$ reveals for any positive x_0

$$\lim_{n \to \infty} F^n(x_0) = 1 .$$

Obviously, 1 is a *fixed point* of $F(x)$.

Not all functions obey such a simple behavior as above. For example, consider the square function $S(x) = x^2$. Obviously, $S(x)$ has a fixed point at 1; however, it also has orbits tending to 0 and ∞:

$$\lim_{n \to \infty} S^n(x_0) = \begin{cases} 0, & \text{if} \quad 0 < x_0 < 1 \\ 1, & \text{if} \quad x_0 = 1 \\ \infty, & \text{if} \quad x_0 > 1 \end{cases} .$$

In the case of square root function, 1 is an *attracting* fixed point of $F(x)$. For the square function, 1 is a *repelling* fixed point of $S(x)$. As a rule, it is easier to find an attracting fixed point than a repelling one.

Another type of orbit is the so-called *periodic orbit* or *cycle*. A periodic orbit eventually returns to where it began. That is, the orbit x_0 is periodic, if there exists an integer n such that $F^n(x_0) = x_0$. In this case, x_0 is a *periodic point* of period n. The smallest such n is the *prime period* of the orbit. For example, consider the reciprocal function $R(x) = 1/x$. Both 1 and –1 are fixed points of $R(x)$. Any other initial point x_0 generates a cycle of period 2, oscillating between x_0 and $1/x_0$.

However, many realistic systems are rather complex and nonregular. It may happen that an orbit has many different periodic points, and its behavior is "strange" (e.g., very sensitive with respect to initial conditions, etc.). This is the realm of chaotic dynamic systems [3]. Chaos entails the concept of *sensitivity to initial conditions*, namely

- neighbored trajectories in the phase space diverge exponentially.
- chaotic systems are inherently unpredictable, i.e.
  1. one can never specify system's initial state to arbitrary precision and
  2. negligible differences in initial conditions amplify the system's evolution in time.
- geometric interpretation
  1. under the action of the equations of motion, volumes of the phase space are stretched (trajectories separation) and folded (boundedness of solutions, i.e., trajectories do not go off to infinity).

2. produces "mixing" or "homogenization" of initial states, which are eventually spread over the entire attractor according to some probability distribution called *invariant measure*.

- mathematical definition: existence of invariant sets containing
    1. countable sets of periodic motions,
    2. uncountable sets of aperiodic motions, and
    3. "dense" orbits which come arbitrarily close to any point in the set.

There are a number of definitions of chaos [4, 5]:

1. The Li–Yorke definition
2. Devaney's, Kloeden's, and Wiggin's definition
3. Definition based on topological mixing
4. Definition based on Smales's horeshoe
5. Definition based on transversal homoclinic points
6. Definition based on symbolic dynamics

In our investigation we will mainly use notions and theorems given by Li–Yorke, and Kloeden. Historically, Li and Yorke gave the first definition of chaos [1]. They considered a mapping $f : I \to I$ (I is the unit interval) into itself

$$x_{n+1} = f(x_n) \, . \tag{2}$$

Definition 1. (Li and Yorke [1, 6]) *A mapping $f : I \to I$ is chaotic if*

1. *there exists a positive integer K ($K = 1$ in [7]) such that the iterative scheme (2) has a cycle of period k for each $k \geq K$.*
2. *the iterative scheme (2) has a scrambled set, i.e., an uncountable set $S \subset I$ containing no cyclic points of f such that*
    a. *$f(S) \subset S$;*
    b. *for every $x_0, y_0 \in S$ with $x_0 \neq y_0$,*

$$\lim_{n \to \infty} \sup \left| f^n(x_0) - f^n(y_0) \right| > 0 \, ; \tag{3}$$

   c. *for every $x_0 \in S$ and any cyclic point y_0 of f,*

$$\lim_{n \to \infty} \sup \left| f^n(x_0) - f^n(y_0) \right| > 0 \, .$$

3. *there exists an uncountable subset $S_0 \subset S$ such that for all $x_0, y_0 \in S$,*

$$\lim_{n \to \infty} \inf \left| f^n(x_0) - f^n(y_0) \right| > 0 \, . \tag{4}$$

Analogous definitions apply for more general metric spaces replacing absolute values by distances.

Theorem 1. (Li and Yorke [1, 8]) *If the function $f : I \to I$ is continuous on a compact interval I and there exists a point, $a \in I$, for which $f^3(a) \leq a < f(a) < f^2(a)$ (or $f^3(a) \geq a > f(a) > f^2(a)$), then f has a cycle of length 3 and is chaotic (in the sense of Definition 1).*

The following theorem gives sufficient conditions for the existence of chaos for Banach space mappings and is particularly suited for TS models because of its flexibility.

Theorem 2. (Kloeden [6]) *Let f be a continuous mapping of a Banach space B into itself and suppose that there exists nonempty compact subsets A, B of B and integers $n_1, n_2 \geq 1$ such that*

(i) A is homeomorphic to a convex subset of B,
(ii) $A \subseteq f(A)$,
(iii) f is expanding on A, i.e., there exists a constant $\lambda > 1$ such that

$$\lambda \|x - y\| \leq \|f(x) - f(y)\| \quad \text{for all } x, y \in A \,,$$

(iv) $B \subset A$,
(v) $f^{n_1}(B) \cap A = \varnothing$,
(vi) $A \subseteq f^{n_1+n_2}(B)$, and
(vii) $f^{n_1+n_2}$ is injective on B (onc to one).

Then mapping f is chaotic in the sense of Definition 1.

There are many examples of mathematical models with chaotic behavior with respect to the definitions given above. The most well-known one-dimensional ones are the logistic equation and tent mapping. The *logistic equation* is of the form

$$P_{n+1} = cP_n(1 - P_n) \,, \tag{5}$$

where c is a constant. Given an initial value P_0, constant c, and the logistic equation, future values can be determined. Note that this is simply the iteration of the function $F(x) = cx(1 - x)$. Another example of chaotic behavior can be illustrated with the so-called *tent mapping* [6, 7]:

$$x_{n+1} = \begin{cases} 2x_n, & \text{if} \quad 0 \leq x_n \leq 1/2 \\ 2(1 - x_n), & \text{if} \quad 1/2 \leq x_n \leq 1 \end{cases} \,. \tag{6}$$

3 Modeling Chaos by Takagi–Sugeno Rule Bases with One-Time Delay Case

In this chapter we consider the TS models (1) with the following generally accepted restrictions [8]:

1. Complete rule base.
2. Noncontradictory rule set.

The membership functions $\mu_i(x)$ for the linguistic terms $L_i, i = \overline{1, N}$ have the following properties.

1. Delimitation: $\mu_i(x) \in [0, 1], x \in X$, i.e., domain of state variable x.

2. Convexity : $\begin{cases} \mu_i(x) \text{ monotonically increases for } x < a_i \\ \mu_i(x) \text{ monotonically decreases for } x > a_i \end{cases}$,

where a_i is a core position of appropriated linguistic variable L_i.
3. Partition: $\sum_i \mu_i(x) > 0$ for all $x \in X$.
4. Feedback correspondence: $\mu_i(a_j) = 0, i \neq j$.

If the range of f_i is a finite set of linguistic variables, the recurrent fuzzy system (1) can be regarded as a *linguistic automata* and under certain conditions chaotic behavior can be proved [8]. But in general case it is important to find the chaotic properties for any case of output function $f(x)$ and not just for the case that this function maps into sets of linguistic variable.

For TS fuzzy model we will answer the following questions:

1. What is the minimal number of rules for different order TS models implying chaos?
2. What are the relationships between the parameters of a TS model implying chaos?
3. Which methods for identification of chaos in TS models do exist?

3.1 Modeling Chaos
with Scalar Zero-Order Takagi–Sugeno Model

In the zero-order TS model the consequents of rules (1) are constants. In this case let $f_i(x_k) = A_i$. The transition function of any TS mapping for (1) is given by

$$f(x) = \frac{\sum_i \mu_i(x) A_i}{\sum_i \mu_i(x)} \quad \forall x \in X ,$$

where μ_i is defined on X and means the membership function describing the corresponding linguistic variable L_i. We suppose $X = I = [0, 1]$, if not stated otherwise. Let the membership functions (MF) μ_i satisfy the following additional conditions:

1. They belong to the triangular class of MF.
2. $\mu_i(a_i) = 1$ and $\mu_i(a_j) = 0$, if $j \neq i$, where a_i is the core position (mean value);
3. They are globally normalized, i.e.,

$$\sum_i \mu_i(x) = 1 \quad \forall x \in I . \tag{7}$$

Remark 1. The MFs need not be symmetrical.

In this case the transition function of (1) reads as

$$f(x) = \sum_i \mu_i(x) A_i \quad \forall x \in I .$$

We now investigate how many rules are necessary to imply chaotic systems. According to [2] it is sufficient to find a mapping fulfilling the assumptions of Theorem 1.

Lemma 1. *The minimal number of rules of zero-order TS model for occurrence of chaos (according to the above definitions) is three.*

Proof. If we have one rule, the transition function $f : x_k \leftarrow x_{k+1}$ of model (1) according to (8) is

$$f(x) = \mu_1(x)A_1 \quad \forall x \in I .$$

According to (7), $\mu_1(x) = 1 \quad \forall x \in I$. Hence, $f(x) = A_1 \ \forall x \in I$ and $x_1 = f(x_0) = x_0$ and there is no chaos. For two rules we get

$$f(x) = \mu_1(x)A_1 + \mu_2(x)A_2 \quad \forall x \in I . \tag{8}$$

Let the core positions fulfill. Then for $x \in [0, a_1]$, $f(x) = A_1$; for $x \in [a_2, 1]$, $f(x) = A_2$; and for $x \in [a_1, a_2]$, $f(x) = \frac{a_2 - x}{a_2 - a_1}A_1 + \frac{x - a_1}{a_2 - a_1} - A_2$.

Obviously, mapping $f(x)$ in (8) is a monotonic function again excluding chaos. If we have three rules then rule base is

$$\begin{aligned} R_1\colon &\text{ If } x_k = L_1 \text{ then } x_{k+1} = A_1 , \\ R_2\colon &\text{ If } x_k = L_2 \text{ then } x_{k+1} = A_2 , \\ R_3\colon &\text{ If } x_k = L_3 \text{ then } x_{k+1} = A_3 . \end{aligned} \tag{9}$$

Then

$$\begin{aligned} &\text{for } x \in [0, a_1]\colon \quad f(x) = A_1; \\ &\text{for } x \in [a_3, 1]\colon \quad f(x) = A_3; \\ &\text{for } x \in [a_1, a_2]\colon \quad f(x) = \frac{a_2 - x}{a_2 - a_1}A_1 + \frac{x - a_1}{a_2 - a_1}A_2 ; \\ &\text{for } x \in [a_2, a_3]\colon \quad f(x) = \frac{a_3 - x}{a_3 - a_2}A_2 + \frac{x - a_2}{a_3 - a_2}A_3 . \end{aligned}$$

Hence, $f(x)$ is a piecewise linear function. For the special choice $A_1 = A_3 = a_1$ and $A_2 = a_3$ with $a_1 = 0, a_2 = 0.5$, and $a_3 = 1$, it is a tent mapping (5) and therefore chaotic [9].

Now we consider which value could be set for consequents A_1, A_2, A_3. Of course, in general case each of these can take on any value from unit interval $I = [0, 1]$. But taking into account that

$$\begin{aligned} &\text{for } x \in [0, a_1]\colon \quad f(x) = A_1 , \\ &\text{for } x \in [a_3, 1]\colon \quad f(x) = A_3 , \end{aligned}$$

the transition function is a constant outside of $[a_1, a_3]$ and that is why we will find chaotic mapping in this interval and consider the mapping $f : [a_1, a_3] \rightarrow [a_1, a_3]$.

This means that first of all we consider the case $A_i \in [a_1, a_3], i = 1, 2, 3$, (See Fig. 2). Letting function $f(x)$ be convex on $[a_1, a_3]$, more general cases, including the case of concave mapping, will be considered below.

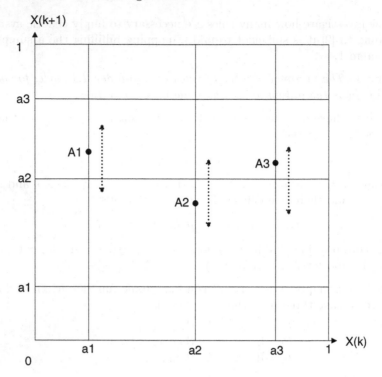

Fig. 2. Moving of consequents positions

Theorem 3. *A rule base (9) with mapping* $f : [a_1, a_3] \rightarrow [a_1, a_3]$ *is chaotic in the sense of Li and Yorke on the interval* $[a_1, a_3] \subseteq I$ *if the following conditions are satisfied:*

(a) $A_1 \in [a_1, f_2^{-1}(a_2))$,
(b) $A_2 = a_3$, *and*
(c) $A_3 = a_1$.

Proof. Because we consider convex mapping, conditions (b) $A_2 = a_3$ and (c) $A_3 = a_1$ are necessary for providing mapping $f : [a_1, a_3] \rightarrow [a_1, a_3]$. Let us apply Theorem 2 for mapping $f : [a_1, a_3] \rightarrow [a_1, a_3]$. Then it is sufficient to find appropriate compact subsets A and B of I. Let $A = [\xi, \Psi] \subset [a_2, a_3]$ and $B = [\theta, \Psi] \subset A$, with ξ, θ, and Ψ to be determined.

Note that $\xi > a_2, \Psi < a_3$, and $\xi < \theta$; the left increasing part of tent mapping is denoted as $f_1(x)$ and the right decreasing part as $f_2(x)$ (in the case of zero-order TS model these functions are straight lines). Then $f(A) = [f_2(\Psi), f_2(\xi)]$. Let $f_2(\xi) = \Psi$. Then when $f_2(\Psi) \leq a_2$, we have $A \subseteq f(A)$.

We shall now prove that f is expanding on A, i.e., it is sufficient (and necessary) to have

$$\left\|\frac{A_3 - A_2}{a_3 - a_2}\right\| > 1 \ .$$

When $A_2 = a_3$ (condition (b)), it is clear that it comes to agreement about $A_3 \in [a_1, a_2)$ (including of course condition (c)). Now it is necessary to check the conditions

$$f(B) \cap A = \varnothing \ ,$$
$$A \subseteq f^2(B) \ .$$

Let us find $f(B)$. We have $f(B) = [f_2(\Psi), f_2(\theta)]$. To satisfy the condition of $f(B) \cap A = \varnothing$ let us assume that $f_2(\theta) = a_2$. It is easy to see that because $\Psi > \theta$ and so $f_2(\Psi) < a_2$. According to above assumption we can write $f^2(B) = [f_1(f_2(\Psi)), a_3]$. Depending of the value of A_1 we have several cases.

Case 1: $a_1 \leq A_1 < a_2$

In this case we can find that if $\Psi > \theta$, then $f_1(f_2(\Psi)) \leq \xi$ and the condition $A \subseteq f^2(B)$ is satisfied assuredly. As a result we can write that

$$\theta = f_2^{-1}(a_2), \quad f_1(f_2(\Psi)) \leq \xi, \quad \xi = f_2^{-1}(\Psi) \ .$$

It is easy to see that when the conditions of Theorem 3 hold, the values ξ, θ and Ψ can be found. It is clear that for this we must have the following: $\exists x \in [a_1, a_2]$, when $f_1(x) = a_2$. So if we choose Ψ such that $f_1(f_2(\Psi)) = a_2$ holds, we guarantee providing the condition $A \subseteq f^2(B)$. It is always possible when $a_1 \leq A_1 \leq a_2$.

Case 2: $a_2 < A_1 \leq z$, where $z = f_2(z)$ is a fixed point of the mapping $f : [a_1, a_3] \to [a_1, a_3]$.

In this case providing the conditions of Kloeden theorem is not so easy because $f_1(x) > a_2$. Moreover, we need to investigate the possibility of fundamental of Kloeden conditions in this case, namely the property of mapping of *compact set into itself*, on which we are finding chaos.

Let us check this now that we have the mapping $f : [a_1, a_3] \to [a_1, a_3]$. Because $[a_1, a_3] = [a_1, a_2] \cup [a_2, a_3]$ we find each part of mapping separately. It is easy to see that $f : [a_2, a_3] \to [a_1, a_3]$. Now find $f : [a_1, a_2] \to [A_1, a_3]$. The next step is mapping $f : [A_1, a_3] \to [a_1, f_2(A_1)]$. Because $a_2 < A_1$ we have inequality $f(A_1) > A_1$ and we can rewrite the mapping $f : [A_1, a_3] \to [a_1, f_2(A_1)]$ as $f : [A_1, a_3] \to [a_1, a_2] \cup [a_2, f_2(A_1)]$.

The next step of iteration gives $f : [a_1, f_2(A_1)] \to [A_1, a_3] \cup [f_2^2(A_1), a_3]$. Because of the expanded properties of function $f_2(x)$ we have $f_2(A_1) < A_1$ and can rewrite the above expression as $f : [a_1, f_2(A_1)] \to [f_2^2(A_1), a_3]$. If $f_2^2(A_1) \leq a_2$ then the next step is $f : [f_2^2(A_1), a_3] \to [a_1, a_3]$ and we have mapping $f : [a_1, a_3] \to [a_1, a_3]$. If $f_2^2(A_1) > a_2$, we continue the action of mapping and due to expanded properties of $f_2(x)$ we find, at last, power $2k$ when $f_2^{2k}(A_1) \leq a_2$ holds.

Now we need to find appropriate values for $\xi < \theta$ and $\Psi > \theta$. Because $A_1 \leq z$ and it can take on arbitrary values and taking into account that

$f_2(x)$ is expanded on $[a_2, a_3]$ let us assume that $\xi = f^k(A_1) < \theta$ and $\Psi = f^{k+1}(A_1) > \theta$, where integer $k \geq 0$. Here $f^k(A_1)$ means use the function $f(x) = \{f_1(x), f_2(x)\}$ k times, i.e., $f^k(A_1) = \underbrace{f(f(\dots f_2(A_1)))}_{k}$. It is easy to see that such value k can be found because $f_2(x)$ is expanded. Then subsets A and B will be as follows:

$$A = [f^k(A_1), f^{k+1}(A_1)], \quad B = [\theta, f^{k+1}(A_1)] \, .$$

Then $f(A) = [f_2^{k+2}(A_1), f_2^{k+2}(A_1)], \quad f_2^{k+2}(A_1) < a_2$. It follows that $A \subseteq f(A)$.

It is clear that $A \cap f(B) = \emptyset$ because $f(B) = [f_2^{k+2}(A_1), a_2]$. Let us find such $f^n(B), n > 1$, that $A \subseteq f^n(B)$:

$$f^2(B) = \left[f_1\left(f_2^{k+2}(A_1)\right), a_3\right] \, ,$$
$$f^3(B) = \left[a_1, f_2\left(f_1\left(f_2^{k+2}(A_1)\right)\right)\right] \, ,$$
$$f^4(B) = \left[A_1, f_1\left(f_2\left(f_1\left(f_2^{k+2}(A_1)\right)\right)\right)\right] \, .$$

Denoting $\Phi = f_1\left(f_2\left(f_1\left(f_2^{k+2}(A_1)\right)\right)\right)$ we can have two cases: $\Phi \geq \theta$ or $\Phi < \theta$. When $\Phi \geq \theta$ we can rewrite

$$f^4(B) = [A_1, \theta] \cup [\theta, \Phi] \, .$$

Then $f^5(B) = [a_2, f_2(A_1)] \cup [f_2(\Phi), a_2]$. It is easy to see that for $f^5(B)$ the condition $A \subseteq f^5(B)$ holds when $k = 0$. If $k > 0$, we find

$$f^6(B) = \left[f_2^2(A_1), a_3\right] \cup \left[f_1(f_2(\Phi)), a_3\right] \, .$$

Drop the set $[f_1(f_2(\Phi)), a_3]$ and we could continue to find

$$f^7(B) = \left[a_1, f_2^3(A_1)\right] \, ,$$
$$f^8(B) = \left[f_2^4(A_1), A_1\right] \, .$$

Continuing in such way it is clear that for expanded mapping we find appropriate iteration for satisfying condition $A \subseteq f^n(B)$.

If $\Phi < \theta$ we can find such number N. Because of the expanded mapping $f_2(x)$ we find such iteration

$$f^N(B) = \left[f_2^{N-1}(A_1), \theta\right] \cup \left[\theta, f_2^{N-1}(\Phi)\right]$$

satisfying the condition $A \subseteq f^n(B)$. So, we have chaotic behavior if $a_2 < A_1 \leq z$.

Case 3: $z < A_1 < \theta$.

Let us investigate the property of mapping of compact set into itself, on which we are finding chaos. Let us check this as now we do not deal with the mapping $f : [a_1, a_3] \to [a_1, a_3]$. It is easy to see that $f : [a_2, a_3] \to [a_1, a_3]$.

Now find $f : [a_1, a_2] \rightarrow [A_1, a_3]$. The next step is mapping $f : [A_1, a_3] \rightarrow [a_1, f_2(A_1)]$.

Because $\theta > A_1 > z$ we have the inequality $a_2 < f(A_1) < z < A_1$ and we can rewrite the mapping $f : [A_1, a_3] \rightarrow [a_1, f_2(A_1)]$ as $f : [A_1, a_3] \rightarrow [a_1, a_2)] \cup [a_2, f_2(A_1)]$.

The next step of iteration gives $f : [a_1, f_2(A_1)] \rightarrow [A_1, a_3] \cup [f_2^2(A_1), a_3)]$. Because of the expanded properties of function $f_2(x)$ we have $f_2^2(A_1) > A_1$ and can rewrite the above expression as $f : [a_1, f_2(A_1)] \rightarrow [A_1, a_3]$.

Hence we do not have chaotic mapping because it is based on the non compact set $[a_1, f_2(A_1)] \cup [A_1, a_3]$, where $f(A_1) < A_1$:

$$f\colon [A_1, a_3] \rightarrow [a_1, f_2(A_1)] \ ,$$
$$f\colon [a_1, f_2(A_1)] \rightarrow [A_1, a_3] \ .$$

In this case we cannot use the Kloeden theorem directly for this mapping, but can use this theorem for f^2 mapping and find appropriate sets.

Let us consider new mapping $g = f^2 : [A_1, a_3] \rightarrow [A_1, a_3]$. Now we need to find appropriate values for $\xi < \theta$ and $\Psi > \theta$. Let $A = [\xi, \Psi] \subset [A_1, a_3]$ and $B = [\xi, \theta] \subset A$. Find the necessary mappings for Kloeden theorem:

$$g(A) = f\left[f[\xi, \Psi]\right] = f\left[f_2(\xi), f_2(\Psi)\right] = f\left[[f_2(\Psi), a_2] \cup [a_2, f_2(\xi)]\right]$$
$$= [f_1(f_2(\Psi)), a_3] \cup [f_2(f_2(\xi)), a_3]$$

and

$$g(B) = f\left[f[\xi, \theta]\right] = f\left[a_2, f_2(\xi)\right] = f_2\left[f_2(f_2(\xi)), a_3\right] \ .$$

It is easy to see that to provide properties $A \subseteq f(A)$ and $A \cap f(B) = \emptyset$ let $f_1(f_2(\Psi)) \leq \xi$ and $f_2(f_2(\xi)) > \Psi$. We can find such higher power of $g(B)$ to satisfy the condition $A \subseteq f^k(B)$ because $g(x)$ is expanded on $[A_1, a_3]$. Actually we have

$$g([\theta, a_3]) = f(f_2([\theta, a_3])) = f_1([a_1, a_2]) = [A_1, a_3] \ .$$

That is, $z < A_1 < \theta$ gives the chaotic orbits.

Case 4: $\theta \leq A, < a_3$

In this case we deal with not expanding mapping $g = f^2$. As in the case above we can write

$$f : [a_1, a_2] \rightarrow [A_1, a_3] \ ,$$
$$f : [A_1, a_3] \rightarrow [a_1, f_2(A_1)] \ .$$

From the condition $\theta \leq A_1 < a_3$ we have $a_2 \geq f_2(A_1) > a_1$. It means that new iteration gives

$$f\colon [a_1, f_2(A_1)] \rightarrow [A_1, f_1(f_2(A_1))] \subseteq [A_1, a_3] \ .$$

That is why we do not have chaotic behavior in this case.

Finally, it is easy to show that $f^2(B) = f_1(f_2(B))$ is injective on B (one to one). Thus, f is chaotic in the sense of Definition 1.

Figure 3 illustrates Theorem 3.

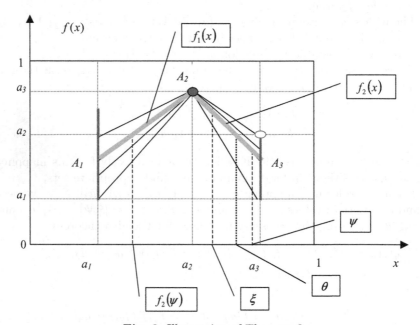

Fig. 3. Illustration of Theorem 3

Remark 2. When $a_1 < A_3 < a_2$, it is easy to see that we deal with mapping $f^* : [A_3, a_3] \rightarrow [A_3, a_3]$ instead of previous mapping $f : [a_1, a_3] \rightarrow [a_1, a_3]$. In this case we can consider f^* as a renewed mapping, namely, $g : [b_1, b_3] \rightarrow [b_1, b_3]$, where $b_1 = A_3, b_2 = a_2, b_3 = a_3, g(b_1) = f(A_3), g(b_2) = a_3$ and $g(b_3) = b_1$. Then Theorem 3 can be used for mapping $g : [b_1, b_3] \rightarrow [b_1, b_3]$.

Remark 3. When $A_2 \neq a_3$ it is easy to see that we deal with mapping $f^* : [a_1, A_2] \rightarrow [a_1, A_2]$ instead of the previous mapping $f : [a_1, a_3] \rightarrow [a_1, a_3]$. In this case we can consider f^* as a renewed mapping, namely, $g : [b_1, b_3] \rightarrow [b_1, b_3]$, where $b_1 = a_1, b_3 = A_2, g(a_2) = b_3$, and $b_2 = a_2$. It is clear that $A_2 > a_2$. Then there are two possible cases.

The first case

If $a_2 < A_2 \leq a_3$, we will observe chaos when conditions of Theorem 3 hold for mapping g. (For this we substitute all symbols in Fig. 3, namely $f \to g, a \to b$ and we obtain strict equality again). If $A_2 \in (a_2, a_3]$ we consider the mapping $g : [b_1, b_3] \to [b_1, b_3]$, where $b_1 = a_1, b_3 = A_2, g(a_2) = b_3$ and $b_2 = a_2$ and check the conditions of Theorem 3 for mapping g.

The second case

If $a_3 < A_2 \leq 1$ the transition function is not triangular and we have horizontal interval with length γ (Fig. 4). This mapping is chaotic on classic Cantor set. In our case chaos is presented not on the set $[a_1, a_3]$, but on set C^* pertaining to Cantor set C. The set C^* is constructed on $[a_1, a_3]$ and includes as well $(\lambda_{\min}, \lambda_{\max}) \not\subset C^*$ and so on. So, in this case we can have chaos, but not in interval $[a_1, a_3]$ as in the previous cases. That is why we never obtain chaotic orbits on whole interval. However, in Cantor set we can find different closely set points that produce different orbits. But it is true only in the limit.

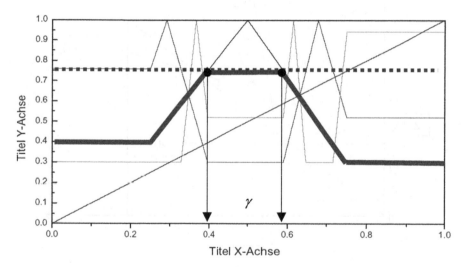

Titel Y-Achse

Titel X-Achse

Fig. 4. Interval $\gamma \subset (a_1, a_3)$

We could summarize the above remark as follows:

1. If $A_2 = a_3$, we have the original conditions of Theorem 3 regarding A_1, A_3 for chaos on interval $[a_1, a_3]$.

2. If $a_2 < A_2 < a_3$, we say about chaos on interval $[a_1, A_2] \subset [a_1, a_3]$ when the remaining conditions of Theorem 3 are fulfilled with following substitutions: We consider the mapping $g : [b_1, b_3] \to [b_1, b_3]$, where

$b_1 = a_1, b_3 = A_2, g(a_2) = b_3$ and $b_2 = a_2$ and check the conditions of Theorem 3 for mapping g.

3. If $a_3 < A_2 \leq 1$, we say about chaos pertaining to Cantor set constructed on interval $[a_1, a_3]$ and using interval λ as excluded one. The conditions for A_1, A_3 remain effective.

3.2 More General Cases of Consequents

3.2.1 Case $0 \leq A_1 < a_1$

It is easy to see (Fig. 5) that we have repelling fixed point Z. In this case for $x_0 \in [a_1, Z)$ we have orbit $x_{k+1} = f_1(x_k)$ with $\lim x_k = a_1$. That is why we have an evident condition $A_3 = Z$. We now refine the conditions of Theorem 3.

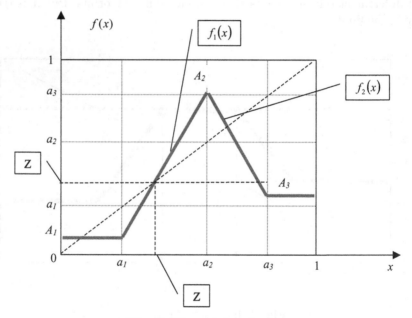

Fig. 5. The case $0 \leq A_1 < a_1$

Theorem 4. *A rule base (9) with mapping* $f : I \to I$ *is chaotic in the sense of Li and Yorke on the interval* $[A_3, a_3] \subseteq I$ *if the following conditions are satisfied:*

(a) $A_1 \in [0, f_2^{-1}(a_2))$,
(b) $A_2 = a_3$, *and*

(c) $A_3 = \begin{cases} a_1, & \text{if} \quad A_1 \geq a_1, \\ Z, & \text{if} \quad A_1 < a_1, \text{ where } f_1(Z) = Z. \end{cases}$

3.2.2 Case of Concave Mapping

We could use the same approach for investigation of concave mapping. Here we show one case when $A_1 = a_3, A_2 = a_1$, and $A_3 = a_3$ (Fig. 6). The remaining cases are similar to convex mapping investigations and are therefore omitted.

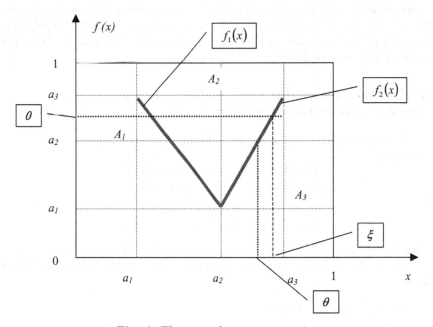

Fig. 6. The case of concave mapping

We have the following theorem that is opposite to Theorem 3 for convex mapping.

Theorem 5. *A rule base (7) with mapping $f : I \to I$ is chaotic in the sense of Li and Yorke on the interval $[a_1, a_3] \subseteq I$ if the following conditions are satisfied:*

(a) $A_1 = a_3$,
(b) $A_2 = a_1$, and
(c) $A_3 \in (a_2, a_3]$.

Proof. Because we consider concave mapping, condition (b), $A_2 = a_1$, is necessary for providing mapping $f : [a_1, a_3] \to [a_1, a_3]$. Let us apply Theorem

2 for mapping $f : [a_1, a_3] \to [a_1, a_3]$ and find n_1, n_2. Then it is sufficient to find appropriate compact subsets A and B of I. Let $A = [\theta, a_3] \subset [a_2, a_3]$ and $B = [\theta, \xi] \subset A$, with ξ, θ to be determined.

Note that $\theta < \xi < a_3$; left part of tent mapping is denoted as $f_1(x)$, and right part as $f_2(x)$ (see Fig. 6). Then $f(A) = [f_2(\theta), a_3]$. Let $f_2(\theta) = a_2$. Then we have $A \subseteq (A)$.

Now it is necessary to check the condition $f(B) \cap A = \emptyset$. Let us find $f(B)$. We have $f(B) = [a_2, f_2(\xi)]$. To satisfy the condition of $f(B) \cap A = \emptyset$ let us assume that $f_2(\xi) = \theta$. That is why condition (c) is necessary.

According to the above assumption we can write $f^2(B) = [a_1, a_2]$ and $f^3(B) = [a_1, a_3]$. Thus for $n_1 = 1, n_2 = 2$ we have $A \subseteq f^{n_1+n_2}(B)$. The remaining of the assumption are the same as with Theorem 3. Thus f is chaotic in the sense of Definition 1.

3.3 The Case of an Arbitrary Number of Rules in the Zero-Order Takagi–Sugeno Model

Let us consider the case of zero-order TS model, when rule base consists of several rules:

$$\begin{aligned}
&R_1: \text{ If } x_k = L_1 \text{ then } x_{k+1} = A_1 , \\
&R_2: \text{ If } x_k = L_2 \text{ then } x_{k+1} = A_2 , \\
&\quad\vdots \\
&R_N: \text{ If } x_k = L_N \text{ then } x_{k+1} = A_N .
\end{aligned} \tag{10}$$

Under the above conditions with respect to the MF and assuming $0 \leq a_1 < a_2 < \cdots < a_n \leq 1$ for the core positions, $f(x)$ is again piecewise linear, namely,

for $x \in [0, a_1]$: $f(x) = A_1$;

for $x \in [a_N, 1]$: $f(x) = A_N$;

for $x \in [a_1, a_2]$: $f(x) = \dfrac{a_2 - x}{a_2 - a_1} A_1 + \dfrac{x - a_1}{a_2 - a_1} A_2$;

$\vdots$

for $x \in [a_{N-1}, a_N]$: $f(x) = \dfrac{a_N - x}{a_N - a_{N-1}} A_{N-1} + \dfrac{x - a_{N-1}}{a_N - a_{N-1}} A_N$.

Therefore, if we found a triple (a_i, a_j, a_k) with $a_i < a_j < a_k$ and monotonous functions between appropriated core positions for which the conditions of Theorems 3–5 are satisfied, we would have a chaotic behavior in $[a_i, a_k]$. Such an interval $[a_i, a_k] \subseteq I$ is known has *cluster of chaos*. The example below illustrates two clusters in the mapping (Fig. 7).

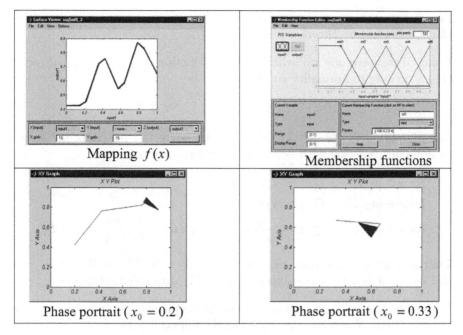

Fig. 7. Mapping with two clusters of chaos

The rule base is as follows:

$$R_1\colon \text{If } x_k \text{ is } L_1 \text{ then } x_{k+1} = 0.425 \ ;$$
$$R_2\colon \text{If } x_k \text{ is } L_2 \text{ then } x_{k+1} = 0.8 \ ;$$
$$R_3\colon \text{If } x_k \text{ is } L_3 \text{ then } x_{k+1} = 0.5 \ ;$$
$$R_4\colon \text{If } x_k \text{ is } L_4 \text{ then } x_{k+1} = 0.9 \ ;$$
$$R_5\colon \text{If } x_k \text{ is } L_5 \text{ then } x_{k+1} = 0.65 \ .$$

3.4 Modeling of Chaos with Scalar First-Order Takagi–Sugeno Model

The first-order TS model is given by

$$
\begin{aligned}
R_1\colon & \text{If } x_k = L_1 \text{ then } x_{k+1} = A_1 x_k + B_1 \ , \\
R_2\colon & \text{If } x_k = L_2 \text{ then } x_{k+1} = A_2 x_k + B_2 \ , \\
& \vdots \\
R_N\colon & \text{If } x_k = L_N \text{ then } x_{k+1} = A_N x_k + B_N \ ,
\end{aligned}
\tag{11}
$$

where L_i are the linguistic variables and A_i and B_i are numerical coefficients. The MFs and core positions are the same as above.

First of all we want to clarify how many rules are necessary to have chaotic behavior.

Lemma 2. *The minimal number of rules of first-order TS model for modeling chaos according to Theorem 1 is two.*

Proof. If we have one rule the transition function $f : x_k \rightarrow x_{k+1}$ of model (11) according to normality conditions will be

$$f(x) = \mu_1(x)(A_1 x + B_1) \quad \forall x \in X .$$

Again, $\mu_1(x) = 1 \quad \forall x \in I$. Then $f(x) = A_1 x + B_1 \quad \forall x \in I$. Hence, $f(x)$ is a monotonous function with no chaos. If we have two rules then

for $x \in [0, a_1]$: $f(x) = A_1 x + B_1$;

for $x \in [a_2, 1]$: $f(x) = A_2 x + B_2$;

for $x \in [a_1, a_2]$: $f(x) = \dfrac{a_2 - x}{a_2 - a_1}(A_1 x + B_1) + \dfrac{x - a_1}{a_2 - a_1}(A_2 x + B_2).$

That is, $f(x)$ is a parabola and for $A_1 = 4, A_2 = 0, B_1 = 0, B_2 = 0, a_1 = 0,$ and $a_2 = 1$ it is a logistic mapping of type (5) with chaotic features [9].

Now let us derive general conditions for coefficients A_i and B_i to provide chaotic f. For $N = 2$ the rule base has the following form:

$$\begin{aligned} &R_1\text{: If } x_k = L_1 \text{ then } x_{k+1} = A_1 x_k + B_1 , \\ &R_2\text{: If } x_k = L_2 \text{ then } x_{k+1} = A_2 x_k + B_2 . \end{aligned} \tag{12}$$

In Figs. 8 and 9 the case $a_1 = 0.25, a_2 = 0.75, A_1 = 2, B_1 = 0, A_2 = -2,$ and $B_2 = 2$ is depicted. As for the zero-order TS model we can formulate an analogous theorem for the first-order TS model.

Theorem 6. *A rule base (12) is chaotic in the sense of Li and Yorke on the interval $[a_1, a_2] \subset I$ if the coefficients A_1, A_2, B_1 and B_2 are the solutions of the following equations:*

$$\begin{cases} A_1 a_1 + B_1 = \phi , \\ A_2 a_2 + B_2 = \phi , \\ A_1(a_1 + a_2)A_2(a_1 + a_2) + 2B_1 + 2B_2 = 4a_2 , \end{cases}$$

with certain $\phi \in [a_1, \frac{a_1 + a_2}{2})$.

Proof. According to Theorem 2 it is sufficient to find the appropriate compact convex sets A and B in $[a_1, a_2]$ for the transition function $f : [a_1, a_2] \rightarrow [a_1, a_2]$. Let $A = [\xi, \Psi] \subset (\frac{a_1 + a_2}{2}, a_2]$ and $B = [\theta, \Psi] \subset A$, with $\xi, \theta,$ and Ψ to be determined. Note that $\xi > \frac{a_1 + a_2}{2}, \Psi < a_2,$ and $\xi < \theta$; the left branch of a parabola is denoted as $f_1(x)$ and the right branch as $f_2(x)$ (see Fig. 10). Then $f(A) = [f_2(\Psi), f_2(\xi)]$. Let $f_2(\xi) = \Psi$. Now we need $f_2(\Psi) \le \frac{a_1 + a_2}{2}$ to satisfy $A \subseteq f(A)$.

Let us find $f(B)$. We have $f(B) = [f_2(\Psi), f_2(\theta)]$. To satisfy the condition of $f(B) \cap A = \phi$ let us assume that $f_2(\theta)] = \frac{a_1 + a_2}{2}$. Because we have $\Psi > \theta$ therefore $f_2(\Psi) < \frac{a_1 + a_2}{2}$. That is why condition $\phi \in [a_1, \frac{a_1 + a_2}{2})$ is necessary. Note that conditions (13) provide the following properties:

$$f(a_1) = \phi, \quad f(a_2) = \phi, \quad f\left(\frac{a_1 + a_2}{2}\right) = a_2 .$$

According to the above assumption we can write $f^2(B) = [f_1(f_2(\Psi)), a_2]$. When $f_1(f_2(\Psi)) = \frac{a_1 + a_2}{2}$, the condition $A \subseteq f^2(B)$ is satisfied. It is clear that $\exists x \in [a_1, \frac{a_1 + a_2}{2}]$, when $f_1(x) = \frac{a_1 + a_2}{2}$.

As a result we can write that

$$\theta = f_2^{-1}\left(\frac{a_1 + a_2}{2}\right), \quad \Psi = f_2^{-1}\left(f_2^{-1}\left(\frac{a_1 + a_2}{2}\right)\right), \quad \xi = f_2^{-1}(\Psi) .$$

It is clear that in this case condition $\xi < \theta$ holds. Now we can rewrite that $f_2(\xi) > f_2(\theta)$ and at last we have $\Psi > \frac{a_1 + a_2}{2}$.

We shall now prove that f is expanding on A. This means that $|f'(x)| > 1$, for $x \in [\xi, \Psi]$. Obviously, it is enough to prove this statement only for point $x = \xi$. Let us find $x^* \in \left[\frac{a_1 + a_2}{2} + a_2\right]$ such that $|f'(x^*)| = 1$. According to conditions (13) we can write the derivative of transition function after evident substitutions as

$$f'(x) = \frac{4(\phi - a_2)}{(a_2 - a_1)^2}(2x - (a_1 + a_2))$$

for which we have

$$x^* = \frac{1}{2}\left(a_1 + a_2 - \frac{(a_1 - a_2)^2}{4(\phi - a_2)}\right) .$$

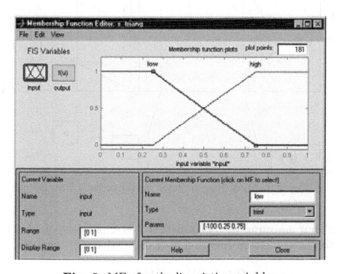

Fig. 8. MFs for the linguistic variables

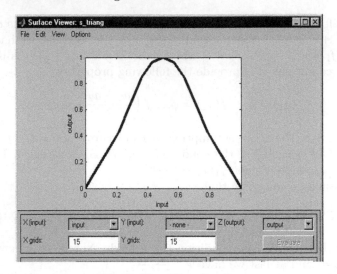

Fig. 9. Transition function

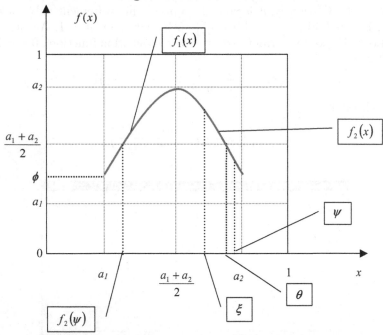

Fig. 10. Modelling of chaos by 1st order TS model

If $\xi > x^*$ our assumption that function f is expanding on A is true. Then according to the transition function we can rewrite this inequality as

$$f(f(f(\xi))) < f(f(f(x^*))) .$$

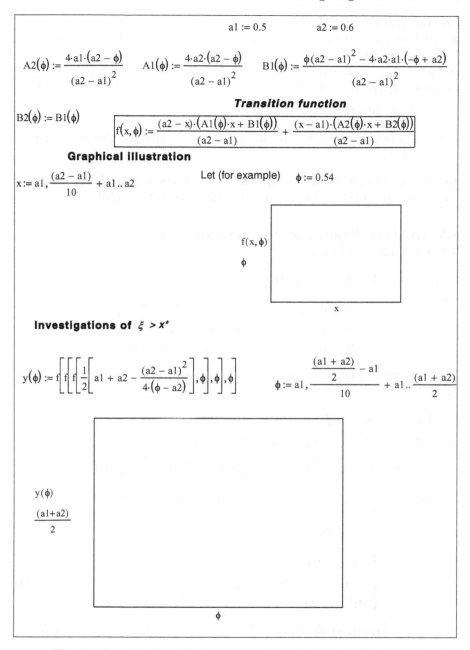

$$a1 := 0.5 \qquad a2 := 0.6$$

$$A2(\phi) := \frac{4 \cdot a1 \cdot (a2 - \phi)}{(a2 - a1)^2} \qquad A1(\phi) := \frac{4 \cdot a2 \cdot (a2 - \phi)}{(a2 - a1)^2} \qquad B1(\phi) := \frac{\phi (a2 - a1)^2 - 4 \cdot a2 \cdot a1 \cdot (-\phi + a2)}{(a2 - a1)^2}$$

Transition function

$$B2(\phi) := B1(\phi)$$

$$f(x, \phi) := \frac{(a2 - x) \cdot (A1(\phi) \cdot x + B1(\phi))}{(a2 - a1)} + \frac{(x - a1) \cdot (A2(\phi) \cdot x + B2(\phi))}{(a2 - a1)}$$

Graphical illustration

$$x := a1, \frac{(a2 - a1)}{10} + a1 .. a2 \qquad \text{Let (for example)} \quad \phi := 0.54$$

$f(x, \phi)$

ϕ

x

Investigations of $\xi > x^*$

$$y(\phi) := f\left[\left[f\left[f\left[\frac{1}{2}\left[a1 + a2 - \frac{(a2 - a1)^2}{4 \cdot (\phi - a2)}\right], \phi\right], \phi\right], \phi\right]\right] \qquad \phi := a1, \frac{\frac{(a1 + a2)}{2} - a1}{10} + a1 .. \frac{(a1 + a2)}{2}$$

$y(\phi)$

$\frac{(a1+a2)}{2}$

ϕ

Fig. 11. Investigation of convex properties of set A with MathCad

Because $f(f(f(\xi))) = \frac{a_1+a_2}{2}$, we need to prove that $f(f(f(x^*))) > \frac{a_1+a_2}{2}$ to see that we made the numerical experiment (unfortunately, analytical transformation is not enough evidently) with different values of $\phi \in [a_1, \frac{a_1+a_2}{2})$ and any values of $a_1, a_2 \in I$. The result of numerical modeling in MathCad is shown in Fig. 11. (Let $B_1 = B_2$). Finally, the injectivity on B follows trivially.

Figure 10 illustrates the situation.

Remark 4. Besides we have another degree of freedom to choose the coefficients in (13).

Remark 5. We can design positive or negative convex parabolas for chaos modeling. For negative one we need to change conditions (13).

3.5 Arbitrary Number of Rules of First-Order Takagi–Sugeno Model

First, we consider the case of three rules:

$$
\begin{aligned}
R_1&: \text{ If } x_k = L_1 \text{ then } x_{k+1} = A_1 x_k + B_1 \;, \\
R_2&: \text{ If } x_k = L_2 \text{ then } x_{k+1} = A_2 x_k + B_2 \;, \\
R_3&: \text{ If } x_k = L_3 \text{ then } x_{k+1} = A_3 x_k + B_3 \;.
\end{aligned} \tag{13}
$$

Now we have to solve equations (13) twice: for ranges $[a_1, a_2]$ and $[a_2, a_3]$. Moreover,

$$
\begin{aligned}
&\text{for } x \in [0, a_1]: \quad f(x) = A_1 x + B_1 \;; \\
&\text{for } x \in [a_3, 1]: \quad f(x) = A_3 x + B_3 \;; \\
&\text{for } x \in [a_1, a_2]: \quad f(x) = \frac{a_2 - x}{a_2 - a_1}(A_1 x + B_1) + \frac{x - a_1}{a_2 - a_1}(A_2 x + B_2) \;; \\
&\text{for } x \in [a_2, a_3]: \quad f(x) = \frac{a_3 - x}{a_3 - a_2}(A_2 x + B_2) + \frac{x - a_2}{a_3 - a_2}(A_3 x + B_3) \;.
\end{aligned}
$$

For $x \in [a_1, a_2]$ we derive

$$
\begin{cases}
A_1 a_1 + B_1 = \Psi \;, \\
A_2 a_2 + B_2 = \Psi \;, \\
A_1(a_1 + a_2) + A_2(a_1 + a_2) + 2B_1 + 2B_2 = 4a_2 \;,
\end{cases} \tag{14}
$$

and for $x \in [a_2, a_3]$ we derive

$$
\begin{cases}
A_3 a_3 + B_3 = \Psi \;, \\
A_2(a_2 + a_3) + A_3(a_2 + a_3) + 2B_2 + 2B_3 = 4a_3 \;.
\end{cases} \tag{15}
$$

In (15) coefficients A_2 and B_2 are already known from (14). Hence, for an arbitrary number of rules we have to solve equation sets of type (15), where part of coefficients is know from previous calculations. Moreover, $f(a_i) = f(a_{i+1})$ leading to $f(a_i) = \Psi, i = 1, \ldots, N$. Therefore the following obvious statement is true.

Statement 1. *If the first-order TS model is constructed as directly connected parabolas according to Theorem 8 there will be only one cluster of chaos for any number of rules that will appear in the first parabola (orientation from left to right).*

4 Modeling of Chaos by Takagi–Sugeno Rule Bases with High-Order Time Delay Case

For description of many complex dynamic processes we often use the 0th recurrent TS fuzzy rule bases with high-order mapping. These rules have the following form:

$$
\begin{aligned}
R_1 : \quad & \text{If } x_k \text{ is } L_{1j_k} \text{ and } x_{k+1} \text{ is } L_{1j_{k+1}} \text{ and} \cdots \text{ and } x_{k+n} \text{ is} \\
& L_{1j_{k+n}} \text{ then } x_{k+n+1} = A_1 , \\
R_2 : \quad & \text{If } x_k \text{ is } L_{2j_k} \text{ and } x_{k+1} \text{ is } L_{2j_{k+1}} \text{ and} \cdots \text{ and } x_{k+n} \text{ is} \\
& L_{2j_{k+n}} \text{ then } x_{k+n+1} = A_2 , \\
& \vdots \\
R_M : \quad & \text{If } x_k \text{ is } L_{Mj_k} \text{ and } x_{k+1} \text{ is } L_{Mj_{k+1}} \text{ and} \cdots \text{ and } x_{k+n} \text{ is} \\
& L_{Mj_{k+n}} \text{ then } x_{k+n+1} = A_M ,
\end{aligned}
\tag{16}
$$

where L_{2j_q} are linguistic variables (terms) and A_i are numerical constants.
The transition function of such TS mapping for (16) is given by

$$
f(x) = \frac{\sum_{j=1}^{M} \prod_{k=1}^{n} \mu_{jk}(x) A_i}{\sum_{j=1}^{M} \prod_{k=1}^{n} \mu_{jk}(x)} \quad \forall x \in X .
\tag{17}
$$

We suppose that assumptions about membership function agree with those for scalar case (see Sect. 3).

Such rule bases can demonstrate the chaotic behavior in sense of Li and Yorke. So it is important to find the properties of the consequents that could help to recognize chaos. Earlier in this chapter we considered the one-time delay case of system (16) for $n = 0$ and found the conditions for coefficients A_i that deliver chaotic behavior of recurrent rule base. We proved the statement that the minimum number of rules that are necessary for chaos in TS models is three for triangular membership functions. Moreover, we obtained the necessary and sufficient conditions for coefficients A_i in such rule base.

Much more interesting problem is large-scale time delay TS models that are really important in task of time series analysis and simulation modeling. First we will consider the case of two-time delay model, then three-time delay, and at last the general case. Herein we consider the mapping $f : I \times I \times \ldots \times I \to I$, in which $I = [0, 1]$.

4.1 Chaos in Two-Time Delay Takagi–Sugeno Model

In this case model (14) can be rewritten as

$$
\begin{aligned}
R_1: &\text{ If } x_k \text{ is } L_{1j_k} \text{ and } x_{k+1} \text{ is } L_{1j_{k+1}} \text{ then } x_{k+2} = A_1 \,, \\
R_2: &\text{ If } x_k \text{ is } L_{2j_k} \text{ and } x_{k+1} \text{ is } L_{2j_{k+1}} \text{ then } x_{k+2} = A_2 \,, \\
&\vdots \\
R_M: &\text{ If } x_k \text{ is } L_{Mj_k} \text{ and } x_{k+1} \text{ is } L_{Mj_{k+1}} \text{ then } x_{k+2} = A_M \,.
\end{aligned}
\tag{18}
$$

It is easy to see that the following propositions hold. Three rules in two-time delay TS rule base are necessary and sufficient for producing chaos. Let us consider the following rule base:

$$
\begin{aligned}
R_1: &\text{ If } x_k \text{ is } L_1 \text{ then } x_{k+2} = A_1 \,, \\
R_2: &\text{ If } x_k \text{ is } L_2 \text{ then } x_{k+2} = A_2 \,, \\
R_3: &\text{ If } x_k \text{ is } L_3 \text{ then } x_{k+2} = A_3 \,.
\end{aligned}
\tag{19}
$$

If the linguistic variables L_i and consequents A_i in (22) satisfy the conditions of Theorem 3 for one-time delay case, we have the chaotic orbit $X = \{x_0, x_2, \ldots, x_{2n}, \ldots\}$. In a general case of two-time delay TS model the rule base is as follows:

$$
\begin{aligned}
R_1: &\text{ If } x_k \text{ is } L_1 \text{ and } x_{k+1} \text{ is } L_1 \text{ then } x_{k+2} = A_{11} \,, \\
R_2: &\text{ If } x_k \text{ is } L_1 \text{ and } x_{k+1} \text{ is } L_2 \text{ then } x_{k+2} = A_{12} \,, \\
R_3: &\text{ If } x_k \text{ is } L_1 \text{ and } x_{k+1} \text{ is } L_3 \text{ then } x_{k+2} = A_{13} \,, \\
R_4: &\text{ If } x_k \text{ is } L_2 \text{ and } x_{k+1} \text{ is } L_1 \text{ then } x_{k+2} = A_{21} \,, \\
R_5: &\text{ If } x_k \text{ is } L_2 \text{ and } x_{k+1} \text{ is } L_2 \text{ then } x_{k+2} = A_{22} \,, \\
R_6: &\text{ If } x_k \text{ is } L_2 \text{ and } x_{k+1} \text{ is } L_3 \text{ then } x_{k+2} = A_{23} \,, \\
R_7: &\text{ If } x_k \text{ is } L_3 \text{ and } x_{k+1} \text{ is } L_1 \text{ then } x_{k+2} = A_{31} \,, \\
R_8: &\text{ If } x_k \text{ is } L_3 \text{ and } x_{k+1} \text{ is } L_2 \text{ then } x_{k+2} = A_{32} \,, \\
R_9: &\text{ If } x_k \text{ is } L_3 \text{ and } x_{k+1} \text{ is } L_3 \text{ then } x_{k+2} = A_{33} \,.
\end{aligned}
\tag{20}
$$

Let a_1, a_2, and a_3 be the core positions of appropriate linguistic variables L_1, L_2, and L_3. Then rule base (20) can be considered as lattice with coordinates a_1, a_2, and a_3 on each axis (see Fig. 12). The nodes of the lattice contains the appropriate values $A_{11}, A_{12}, \ldots, A_{33}$.

We use Theorem 2 for the general case of two-time delay TS model (20). Let us try to generalize the Theorem 2 on the vector case $f : x, x \to x$. Then we have the following theorem.

Theorem 7. *A rule base (20) with mapping $f : [a_1, a_3] \times [a_1, a_3] \to [a_1, a_3]$ is chaotic on $x \in [a_1, a_3] \subset I$ in the sense of Li and Yorke if the following conditions are satisfied:*

(a) $A_{22} = a_3$,
(b) $min(A_{11}, A_{12}, A_{13}) \in [a_1, a_2]$, and

(c) $min(A_{31}, A_{32}, A_3) = a_1$.

Proof. According to Theorem 2 it is necessary to find the appropriate sets A and B for the transition function $f : [a_1, a_3] \times [a_1, a_3] \rightarrow [a_1, a_3]$. Let us extend the conditions of Theorem 2 on vector case. Assume that $A = [[\xi, \Psi] \times [\theta, \Psi]]$ and $B = [[\xi, \Psi] \times [\theta, a_2]] \subset A$ (see Fig. 13), with $\xi, \Psi,$ and θ to be determined. Let $\theta < a_2$ and $\Psi > \xi > a_2$ as well.

We can rewrite the general mapping $f : x_k, x_{k+1} \rightarrow x_{k+2}$ in the state space $X = (x_1, x_2)$, where $x_1 = x_k, x_2 = x_{k+1}$. Then we can write

$$\begin{pmatrix} x_1 \\ x_2 \end{pmatrix}_{k+1} = \begin{pmatrix} x_2 \\ f(x_1, x_2) \end{pmatrix}_k \tag{21}$$

It is clear that according to (21) we obtain $X_{k+1} = F(X_k)$, where $F = \begin{pmatrix} x_2 \\ f(x_1, x_2) \end{pmatrix}$. Then we can consider Theorem 2 in a vector case. Let $A =$

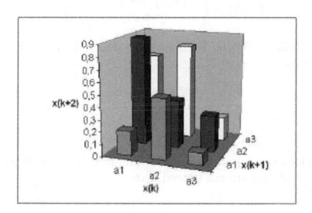

Fig. 12. Example of rule base (25)

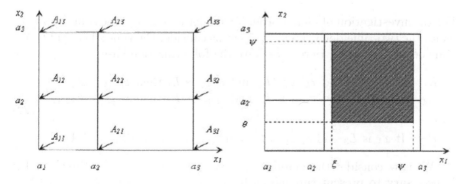

Fig. 13. Rule base representation and illustration of sets A, B

$\left(\begin{bmatrix} \xi, \Psi \\ \theta, \Psi \end{bmatrix} \right)$. Then $F(A) = \left(\begin{smallmatrix} [\theta, \Psi] \\ [f(\Psi, \circ), f(\xi, a_2)] \end{smallmatrix} \right)$. Here $f(\xi, a_2)$ is a value of func-tion F on the straight line $x_1 = \xi$ and $f(\Psi, \circ)$ means some value of F on the straight line $x_1 = \Psi$. Let us assume that $f(\xi, a_2) = \Psi$. To provide condition (ii) of Theorem 2 in a vector case $A \subseteq F(A)$ we have the following inequality:

$$f(\Psi, \circ) \leq \theta < a_2 .$$

Because

$$B = \left(\begin{bmatrix} \xi, & \Psi] \\ [\theta, & a_2] \end{bmatrix} \right)$$

we can find

$$F(B) = \left(\begin{smallmatrix} [\theta, a_2] \\ [f(\Psi, \circ), f(\xi, a_2) = \Psi] \end{smallmatrix} \right) .$$

It is clear that $F(B) \cap A = \emptyset$ (condition (v) of Theorem 2 for $n_1 = 1$). Let us find $F^2(B)$. Here we have

$$F^2(B) = \left(\begin{smallmatrix} [f(\Psi, \circ), f(\xi, a_2) = \Psi] \\ [f(\theta, \circ), a_3] \end{smallmatrix} \right) .$$

To provide condition (vi) of Theorem 2 we demand that $f(\theta, \circ) \leq \theta \leq a_2$. Then conditions

$$f(\theta, \circ) \leq \theta \leq a_2$$
$$f(\xi, a_2) = \Psi$$
$$f(\Psi, \circ) < a_2$$

define conditions (a)–(c) of Theorem 9. The remaining conditions of Theorem 2 are evident and are not proved here.

4.2 Chaos in Three and Higher Time Delay Takagi–Sugeno Model

For the investigation of higher order TS model we propose the same approach as for the two-time delay TS model. Let us consider the general case of three-time delay TS model. Herein we have the following rule base:

R_1: If x_k is L_1 and x_{k+1} is L_1 and x_{k+2} is L_1 then $x_{k+3} = A_{11}$,
R_2: If x_k is L_1 and x_{k+1} is L_1 and x_{k+2} is L_2 then $x_{k+3} = A_{12}$,
$\vdots$ (22)

R_{27}: If x_k is L_3 and x_{k+1} is L_3 and x_{k+2} is L_3 then $x_{k+3} = A_{33}$.

We now consider the mapping $f : x_k, x_{k+1}, x_{k+2} \rightarrow x_{k+3}$. First of all it is necessary to present this model in the state space form $X_{k+1} = F(X_k)$, where $X = (x_1, x_2, x_3)$. We have

$$\begin{pmatrix} x_1 \\ x_2 \\ x_3 \end{pmatrix}_{k+1} = \begin{pmatrix} x_2 \\ x_3 \\ f(x_1, x_2, x_3) \end{pmatrix}_k \tag{23}$$

According to the proposed approach we have the following theorem.

Theorem 8. *A rule base* (22) *with mapping* $f : [a_1, a_3] \times [a_1, a_3] \times [a_1, a_3] \to [a_1, a_3]$ *is chaotic on* $x \in [a_1, a_3] \subset I$ *in the sense of Li and Yorke if the following conditions are satisfied:*

(a) $A_{222} = a_3$,
(b) $min(A_{111}, A_{112}, \dots A_{133}) \in [a_1, a_3]$, *and*
(c) $min(A_{311}, A_{312}, A_{333}) = a_1$.

Proof. According to Theorem 2 it is necessary to find the appropriate sets A and B for the transition function $f : [a_1, a_3] \times [a_1, a_3] \times [a_1, a_3] \to [a_1, a_3]$. Assume that $A = [[\xi, \Psi] \times [\theta, \Psi] \times [\theta, \Psi]]$ and $B = [[\xi, \Psi] \times [\theta, a_2] \times [\xi, \Psi]] \subset A$ (see Fig. 14), with ξ, Ψ and θ to be determined. Let $\theta < a_2$ and $\Psi > \xi > a_2$ as well.

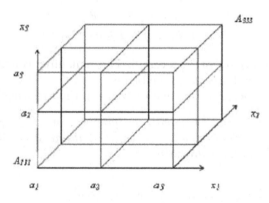

Fig. 14. Rule base representation for three time delay model

According to (23) we can write

$$F = \begin{pmatrix} x_2 \\ x_3 \\ f(x_1, x_2, x_3) \end{pmatrix}.$$

Let $A = \left(\begin{bmatrix} \xi, \Psi \\ \theta, \Psi \\ \theta, \Psi \end{bmatrix} \right)$; then $F(A) = \left(\begin{bmatrix} \theta, \Psi \\ \theta, \Psi \\ f(\Psi, \circ, \circ), \Psi \end{bmatrix} \right)$. To provide condition (ii) of Theorem 2 in a vector case $A \subseteq F(A)$ we have the following inequality:

$$f(\Psi, \circ, \circ) \le \theta < a_2 .$$

Because

$$B = \begin{pmatrix} [\xi, \Psi] \\ [\theta, a_2] \\ [\xi, \Psi] \end{pmatrix}$$

we can find

$$F(B) = \begin{pmatrix} [\theta, a_2] \\ [\xi, \Psi] \\ [f(\Psi, \circ, \circ), \Psi] \end{pmatrix}.$$

It is clear that $F(B) \cap A = \emptyset$ (condition (v) of Theorem 2 for $n_1 = 1$). Let us find $F^2(B)$. Herein we have

$$F^2(B) = \begin{pmatrix} [\xi, \Psi] \\ [f(\Psi, \circ, \circ), \Psi] \\ [f(\theta, \circ, \circ), a_3] \end{pmatrix}.$$

To provide condition (vi) of Theorem 2 we demand that $f(\theta, \circ, \circ) \le \theta \subseteq a_2$. Then conditions

$$f(\theta, \circ) \le \theta \le a_2$$
$$f(\xi, a_2, a_2) = \Psi$$
$$f(\Psi, \circ, \circ) \le \theta \le a_2$$

define conditions (a)–(c) of Theorem 9. The remaining conditions of Theorem 2 are evident and are not proved here.

Thus we can generalize our investigations in the following Theorem.

Theorem 9. *A rule base* (16) *with mapping* $f : [a_1, a_3] \times [a_1, a_3] \times \cdots \times [a_1, a_3] \to [a_1, a_3]$ *is chaotic on* $x \in [a_1, a_3] \subset I$ *in the sense of Li and Yorke if the following conditions are satisfied:*

(a) $A_{22...2} = a_3$,
(b) $min(A_{11...1}, A_{11...2}, \ldots, A_{13...3}) \in [a_1, a_2]$, *and*
(c) $min(A_{311...1}, A_{31...2}, \ldots, A_{33...3}) = a_1$.

Remark 6. The restriction $A_{22...2} = a_3$ can be weakened according to Remark 3 with the same reservations.

Remark 7. Conditions (b) of Theorems 7–9 can be weakened according to Theorem 2 as $min(A_{11...1}, A_{11...2}, \ldots, A_{13...3}) \in [a_1, f^{-1}(a_2, a_2, \ldots, a_2)]$. The proof is the same as in scalar case.

Remark 8. Conditions (c) of Theorems 7–9 can be weakened according to Theorem 2 as it was done in Remark 2.

5 Summary

This chapter continues the investigation of recurrent fuzzy rule bases. We analyzed important properties of TS recurrent models with respect to their transition functions. In future we plan to investigate the chaotic behavior of Mamdani models responding to real tasks in time series analysis and TS models without MF limitations like globally normalized.

References

1. T.Y. Li, J.A. Yorke: Amer. Math. Monthly **82**, 985–992 (1975)
2. M. Yamaguti, S. Ushiki: Physica D 618–626 (1981)
3. Nonlinear dynamics and chaos in the science [Ecol/Math 195b]. In: Freshman Colloquium, Fall 2002. Available at: http://bill.srnr.arisona.edu/classes/195b.html
4. X.C. Fu, W. Lu, P. Ashwin, J. Duan: Symbolic representations of iterated maps. 2002. Available at: http://citeseer.nj.nec.com/fu00symbolic.html
5. B. Kieninger: Analyse dreier Chaosdefinitionen für stetige Abbildungen auf metrischen Räumen (Analysis of three definitions of chaos for continuous mapping in metric spaces). Diploma Thesis, University of Augsburg (1998)
6. P.E. Kloeden: Fuzzy Sets Syst. **42**, 37–42 (1991)
7. P.E. Kloeden: Cycles and chaos in higher dimensional difference equations. In: Proc. 9th Int. Conf. Nonlinear Oscillations, vol 2, (Naukova Dumka, Kiev, 1984) pp 184–187
8. R. Kempf, J. Adamy: Fuzzy Sets Syst. **140**, 259–284 (2003)
9. M. Yamaguti: Comput. Math. Applic. **28**, 263–267 (1994)
10. L. Corcoran, R.L. Wainwright: The performance of a genetic algorithm on an objective function. In: Proc. 7th Oklahoma Conf. Artificial Intelligence (1993) pp 200–206. Available at: http://citeseer.nj.nec.com/cachedpage/487609/

Theory of Fuzzy Chaos for the Simulation and Control of Nonlinear Dynamical Systems

Oscar Castillo and Patricia Melin

Abstract. This chapter introduces the basic concepts of dynamical systems theory and several basic mathematical methods for controlling chaos. The main goal of this chapter is to provide an introduction to and a summary of the theory of dynamical systems, with particular emphasis on fractal theory, chaos theory, and chaos control. We first define what is meant by a dynamical system, then we define an attractor, and then the concept of the fractal dimension of a geometrical object. We also define the Lyapunov exponents as a measure of the chaotic behavior of a dynamical system. On the other hand, the fractal dimension can be used to classify geometrical objects because it measures the complexity of an object. The chapter also describes mathematical methods for controlling chaos in dynamic systems. These methods can be used to control a real dynamic system; however, due to efficiency and accuracy requirements we were forced to use fuzzy logic to model the uncertainty, which is present when numerical simulations are performed. We also describe in this chapter a new theory of chaos using fuzzy logic techniques. Chaotic behavior in nonlinear dynamical systems is very difficult to detect and control. Part of the problem is that mathematical results for chaos are difficult to use in many cases, and even if one could use them there is an underlying uncertainty in the accuracy of the numerical simulations of the dynamical systems. For this reason, we can model the uncertainty of detecting the range of values where chaos occurs, using fuzzy set theory. Using fuzzy sets, we can build a theory of fuzzy chaos, where we can use fuzzy sets to describe the behaviors of a system. We illustrate our approach with two cases: Chua's circuit and Duffing's oscillator.

1 Basic Concepts of Dynamical Systems

In this section we present a brief overview of the field of nonlinear dynamical systems and fractal theory. Recently research has shown that many simple nonlinear deterministic systems can behave in an apparently unpredictable and "chaotic" manner [1]. The existence of complicated dynamics has been discussed in the mathematical literature for many decades, with important contributions by Poincaré, Birkhoft, Smale, and Kolmogorov and his students, among others. Nevertheless, it is only recently that the wide-ranging impact of "chaos" has been recognized. Consequently, the field is now undergoing explosive growth, and many applications have been made across a broad spectrum of scientific disciplines—robotics, engineering, physics, chemistry,

O. Castillo and P. Melin: *Theory of Fuzzy Chaos for the Simulation and Control of Nonlinear Dynamical Systems*, StudFuzz **187**, 391–414 (2006)
www.springerlink.com   © Springer-Verlag Berlin Heidelberg 2006

fluid mechanics, and economics, to name several. We start with some basic definitions of concepts used in this chapter.

Dynamic System: This is a set of mathematical equations that allows one, in principle, to predict the future behavior of the system given the past. One example is a system of first-order ordinary differential equations in time:

$$\frac{dx}{dt} = G(x,t) \ , \tag{1}$$

where $x(t)$ is a D-dimensional vector and G is a D-dimensional vector function of x and t. Another example is a map.

Map: A map is an equation of the following form:

$$x_{t+1} = F(x_t) \ , \tag{2}$$

where the "time" t is discrete and integer valued. Thus, given x_0, the maps gives x_1. Given x_1, the map gives x_2, and so on.

Dissipative System: In Hamiltonian (conservative) systems such as the ones arising in Newtonian mechanics of particles (without friction), phase space volumes are preserved by time evolution (the phase space is the space of variables that specify the state of the system). Consider, for example, a two-dimensional phase space (q, p), where q denotes a position variable and p a momentum variable. Hamilton's equations of motion take the set of initial conditions at time $t = t_0$ and evolve them in time to the set at time $t = t_1$. Although the shapes of the sets are different, their areas are the same. By a dissipative system we mean one that does not have this property. Areas should typically decrease (dissipate) in time so that the area of the final set would be less than the area of the initial set. As a consequence of this, dissipative systems typically are characterized by the presence of attractors.

Attractor: If one considers a system and its phase space, then the initial conditions may be attracted to some subset of the phase space (the attractor) as time $t \to \infty$. For example, for a damped harmonic oscillator the attractor is the point at rest. For a periodically driven oscillator in its limit cycle, the limit set is a closed curve in the phase space.

Strange Attractor: In the above two examples, the attractors were a point, which is a set of dimension zero, and closed curve, which is a set of dimension one. For many other attractors the attracting set can be much more irregular (some would say pathological) and, in fact, can have a dimension that is not an integer. Such sets have been called "fractal," and when they are attractors, they are called strange attractors. The existence of a strange attractor in a physically interesting model was first demonstrated by Lorenz.

Chaotic Attractor: By this term we mean that if we take two typical points on the attractor that are separated from each other by a small distance $\Delta(0)$ at $t = 0$, then for increasing time t they move apart exponentially fast. That is, in some average sense

$$\Delta(t) \sim \Delta(0) \exp(\lambda t) \ , \tag{3}$$

with $\lambda > 0$ (where λ is called the Lyapunov exponent). Thus a small uncertainty in the initial state of the system rapidly leads to inability to forecast its future. It is typically the case that strange attractors are also chaotic.

One of the most prominent, chaotic, continuous-time dynamical systems is the "Lorenz attractor," named after the meteorologist E.N. Lorenz who investigated the three-dimensional, continuous-time system

$$
\begin{aligned}
x' &= s(-x + y) \\
y' &= rx - y - xz \quad s, r, b > 0 \\
z' &= -bz + xy
\end{aligned}
\tag{4}
$$

emerging in the study of turbulence in fluids. For r above the critical value of $r = 28.0$, trajectories of (4) evolve in a rather unexpected way. Suppose that a trajectory starts at an initial value near the origin. For some time the trajectory regularly spirals outward from one fixed point, then the trajectory jumps to a region near another fixed point and does the same thing. As trajectories starting at different initial values all converge to and remain in the same region near the two fixed points, the region is considered an "attractor." It is a "strange attractor" because it is neither a point nor a closed curve. In general, this chaotic behavior can only occur for systems of at least three simultaneous nonlinear differential equations or for systems of at least a one-dimensional nonlinear map [2].

Fractal geometry is a mathematical tool for dealing with complex systems that have no characteristic length scale. A well-known example is the shape of a coastline. When we see two pictures of a coastline on two different scales, we cannot tell which scale belongs to which picture: both look the same. This means that the coastline is scale invariant or, equivalently, has no characteristic length scale. Other examples in nature are rivers, cracks, mountains, and clouds. Scale-invariant systems are usually characterized by noninteger ("fractal") dimensions.

The dimension tells us how some property of an object or space changes as we view it at increased detail. There are several different types of dimension. The fractal dimension d_f describes the space filling properties of an object. Three examples of the fractal dimension are the self-similarity dimension, the capacity dimension, and the Hausdorff–Besicovitch dimension. The topological dimension d_t describes how points within an object are connected together. The embedding dimension d_e describes the space in which the object is contained.

The fractal dimensions d_f are useful and important tools to quantify self-similarity and scaling. Essentially, the dimension tells us how many new pieces are resolved as the resolution is increased. The self-similarity dimension can be applied only to geometrical self-similar objects, where the small pieces are exact copies of the whole object. However, the capacity dimension can be used to analyze irregularly shaped objects that are statistically self-similar. On the other hand, the Hausdorff–Besicovitch dimension requires more complex mathematical tools. For this reason, we will limit our discussion here to the capacity dimension.

A ball is the set of points within radius r of a given point. We determine $N(r)$ the minimum number of balls required so that each point in the object is contained within at least one ball of radius r. In order to cover all the points of the object, the balls may need to overlap. The capacity dimension is defined by the following equation:

$$dc = \lim_{r \to 0} \frac{\log N(r)}{\log(1/r)} \,. \tag{5}$$

The capacity dimension defined as above is a measure of the space filling properties of an object because it gives us an idea of how much work is needed to cover the object with balls of changing size.

A useful method to determine the capacity dimension is to choose balls that are the nonoverlapping boxes of a rectangular coordinate grid. $N(r)$ is then the number of boxes with side of length r that contain at least one point of the object. Efficient algorithms have been developed to perform this "box counting" for different values of r, and thus determine the box counting dimension as the best fit of log $N(r)$ versus $\log(1/r)$.

The fractal dimension d_f characterizes the space-filling properties of an object. The topological dimension d_t characterizes how the points that make up the object are connected together. It can have only integer values. Consider a line that is so long and wiggly that it touches every point in a plane and thus covers an area. Because it covers a plane, its space-filling fractal dimension $d_f = 2$. However, no matter how wiggly it is, it is still a line and thus has topological dimension $d_t = 1$. Thus, the essence of a fractal is that its space-filling properties are larger than one anticipates from its topological dimension. Thus we can now present a formal definition of a fractal [3], namely, that an object is a fractal if and only if

$$d_f > d_t \,.$$

However, there is no one definition that includes all the objects or processes that have fractal properties.

Despite the identification of fractals in nearly every branch of science, too frequently the recognition of fractal structure is not accompanied with any additional insight as to its cause. Often we do not even have the foggiest idea as to the underlying dynamics leading to the fractal structure. The chaotic dynamics of nonlinear systems, on the other hand, is one area where considerable progress has been made in understanding the connection with fractal geometry. Indeed, chaotic dynamics and fractal geometry have such a close relationship that one of the hallmarks of chaotic behavior has been the manifestation of fractal geometry, particularly for strange attractors in dissipative systems [4]. For a practical definition we take a "strange attractor," for a dynamic system, to be an attracting set with fractal dimension. For example,

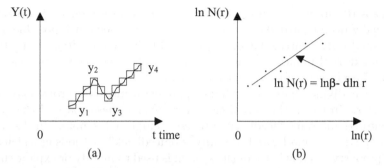

Fig. 1. Fractal dimension of a time series: (a) curve and the boxes covering it, and (b) the logarithmic regression to find d

the famous Lorenz strange attractor has a fractal dimension of about 2.06. Also, we think that beyond only this relationship between strange attractors and the fractal dimension of the set, there is a deeper relationship between the underlying dynamics of a system and the fractal nature of its behavior. We will explore this relationship in more detail in the following sections of this chapter.

Let us consider as an example the use of the fractal dimension as a mathematical model of the time series in the following form:

$$d = \frac{\log(N)}{\log(1/r)} , \qquad (6)$$

where d is the fractal dimension for an object of N parts, each scaled down by a ratio r. For an estimation of this dimension we can use the following equation:

$$N(r) = \beta[1/r^d] , \qquad (7)$$

where $N(r) = $ number of boxes contained in a geometrical object and $r = $ size of the box. We can obtain the box dimension of a geometrical object [3] counting the number of boxes for different sizes and performing a logarithmic regression on this data. For our particular case the geometrical object consists of the curve constructed using the set of points from the time series. Fig. 1a shows the curve and the boxes used to cover it, and in Fig. 1b the corresponding logarithmic regression is illustrated.

2 Controlling Chaos

More than two decades of intensive studies on nonlinear dynamics have posed the question on the practical applications of chaos [5]. One of the possible answers is to control chaotic behavior in such a way as to make it predictable. Indeed, nowadays the idea of controlling chaos is an appealing one.

Chaos occurs widely in engineering and natural systems; historically, it has usually been regarded as a nuisance and is designed out if possible. It has been noted only as irregular or unpredictable behavior, often attributed to random external influences. More recently, there have been examples of the potential usefulness of chaotic behavior [5].

We can divide chaos controlling approaches into two broad categories: firstly those in which the actual trajectory in the phase space of the system is monitored and some "feedback" process is employed to maintain the trajectory in the desired mode, and secondly "nonfeedback" methods in which some other property or knowledge of the system is used to modify or exploit chaotic behavior. Feedback methods do not change the controlled systems and stabilize unstable periodic orbits or strange chaotic attractors, while nonfeedback methods slightly change the controlled system, mainly by a small permanent shift of control parameter, changing the system behavior from chaotic attractor to periodic orbit which is close to the initial attractor.

We describe in this section several methods by which chaotic behavior in a dynamical system may be modified, displaced in parameter space, or removed. The Ott–Grebogi–Yorke (OGY) method [6] is extremely general, relying only on the universal property of chaotic attractors, i.e, they have embedded within them infinitely many unstable periodic orbits. On the other hand, the method requires following the trajectory and employing a feedback control system, which must be highly flexible and responsive; such a system in some experimental configurations may be large and expensive. It has the additional disadvantage that small amounts of noise may cause occasional large departures from the desired operating trajectory.

The nonfeedback approach is inevitable much less flexible and requires more prior knowledge of equations of motion. On the other hand, to apply such a method, we do not have to follow the trajectory. The control procedures can be applied at any time and we can switch from one periodic orbit to another without returning to the chaotic behavior, although after each switch, transient chaos may be observed. The lifetime of this transient chaos strongly depends on initial conditions. Moreover, in a nonfeedback method we do not have to wait until the trajectory is close to an appropriate unstable orbit; in some cases this time can be quite long. The dynamic approach can be very useful in mechanical systems, where feedback controllers are often very large. In contrast, a dynamical absorber having a mass of the order of 1% of that of the control system is able, as we will see later, to convert chaotic behavior to periodic one over a substantial region of parameter space. Indeed, the simplicity by which chaotic behavior may be changed in this way may actually motivate the search for, and exploitation of, chaotic behavior in practical systems.

The essential property of a chaotic trajectory is that it is not asymptotically stable. Closely correlated initial conditions have trajectories, which quickly become uncorrelated. Despite this obvious disadvantage, it has been

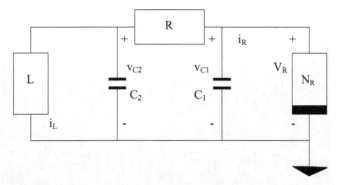

Fig. 2. Chua's circuit

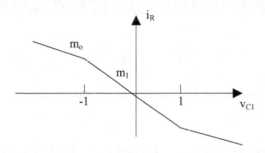

Fig. 3. i_R–v_{C1} characteristic of the non-linear resistor

established that control leading to the synchronization of two chaotic systems is possible.

The methods described in this section are illustrated by the example of Chua's circuit [7] shown in Fig. 2. Chua's circuit contains three linear energy storage elements (an inductor and two capacitors), a linear resistor, and a single nonlinear resistor N_R, namely Chua's diode with a three-segment piecewise linear v–i characteristic defined by

$$f(v_{c1}) = m_0 v_{c1} + \frac{1}{2}(m_1 - m_0)(|v_{c1} + 1| - |v_{c1} - 1|) , \qquad (8)$$

where the slopes in the inner and outer regions are m_0 and m_1, respectively (see Fig. 3).

In this case the state equations for the dynamics of Chua's circuit are as follows:

$$C_1 \frac{dv_{c1}}{dt} = G(v_{c2} - v_{c1}) - f(v_{c1})$$

$$C_2 \frac{dv_{c2}}{dt} = G(v_{c1} - v_{c2}) - i_L$$

$$L \frac{di_L}{dt} = v_{c2} , \qquad (9)$$

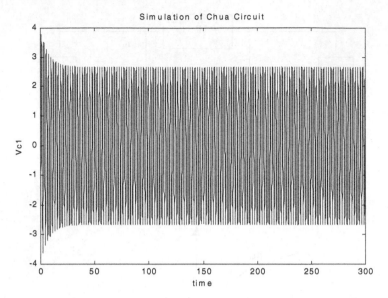

Fig. 4. Plot of variable v_{C1} of Chua's circuit

where $G = 1/R$.

It is well known that for $R = 1.64\,\mathrm{k\Omega}, C_1 = 10\,\mathrm{nF}, C_2 = 99.34\,\mathrm{nF}, m_1 = -0.76\,\mathrm{mS}, m_0 = 0.41\,\mathrm{mS}$, and $L = 18.46\,\mathrm{mH}$, Chua's circuit operate on the chaotic double-scroll Chua's attractor. We show in the following figures the simulation of Chua's circuit for initial conditions $(-3, -3, -10)$. Figure 4 shows the plot of variable v_{C1} in time. In this figure, we can appreciate the erratic behavior of this variable. Figure 5 shows the plot of variable v_{C2} across time, which is similar to the behavior of v_{C1}. Figure 6 shows a two-dimensional view of the double-scroll Chua's attractor. Finally, in Fig. 7 we can appreciate a three-dimensional view of Chua's attractor. The chaotic dynamics of Chua's circuit have been widely investigated (e.g., see [8]). One of the main advantages of this system is the very good accuracy between numerical simulations of the model and experiments on real electronic devices. Experiments with this circuit are very easy to perform, even for nonspecialists.

2.1 Controlling Chaos Through Feedback

2.1.1 Ott–Grebogi–Yorke Method

Ott et al. [6] have proposed and developed a method by which chaos can always be suppressed by shadowing one of the infinitely many unstable periodic orbits embedded in the chaotic attractor. The basic assumptions of this method are as follows:

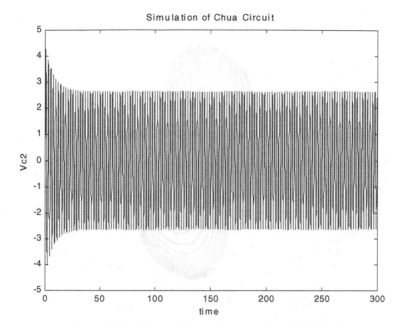

Fig. 5. Plot of variable v_{C1}Vc2 for Chua's circuit

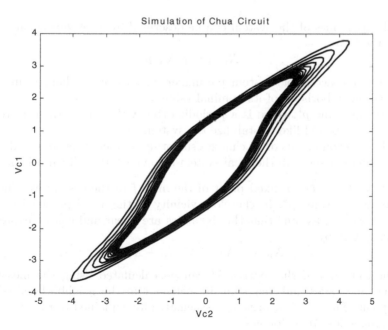

Fig. 6. Bi Two-dimensional view of Chua's attractor

Simulation of Chua Circuit

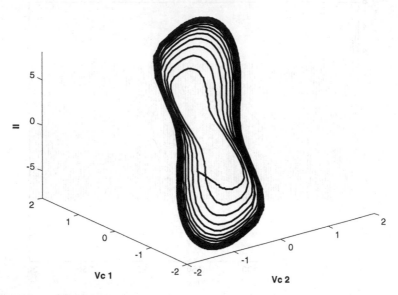

Fig. 7. Three-dimensional view of Chua's attractor

(a) The dynamics of the system can be described by an n-dimensional map of the form
$$X_{n+1} = f(X_n, p) . \tag{10}$$
(b) p is some accessible system parameter which can be changed in some small neighborhood of its nominal value p^*.
(c) For this value p^*, there is a periodic orbit within the attractor around which we would like to stabilize the system.
(d) The position of this orbit changes smoothly with changes in p, and there are small changes in the local system behavior for small variations of p.

Let X_F be a chosen fixed point of the map f of the system existing for the parameter value p^*. In the close vicinity of this fixed point with good accuracy we can assume that the dynamics are linear and can be expressed approximately by
$$X_{n+1} - X_F = M(X_n - X_F) . \tag{11}$$

The elements of the matrix M can be calculated using the measured chaotic time series and analyzing its behavior in the neighborhood of the fixed point. The OGY algorithm is schematically explained in Fig. 8 and its main properties are as follows:

(a) No model of dynamics is required. One can use either full information from the process or a delay coordinate embedding technique using single variable experimental time series.

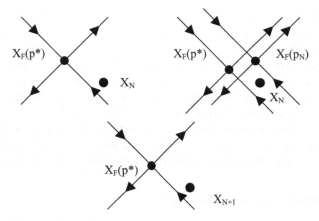

Fig. 8. General idea of the Ott–Grebogi–Yorke method

(b) Any accessible variable (controllable) system parameter can be used as the control parameter.
(c) In the absence of noise and error, the amplitude of applied control signal must be large enough (exceed a threshold) to achieve control.
(d) Inevitable noise can destabilize the controlled orbit, resulting in occasional chaotic bursts.
(e) Before settling into the desired periodic mode, the trajectory exhibits chaotic transients, the length of which depends on the actual starting point.

In [9] the OGY method was applied to control chaos in Chua's circuit. Using a specific software package, unstable periodic orbits were found embedded in the attractor which could serve as goals of control. The controlling method was implemented in the way shown in Fig. 9. The computer was used for data acquisition, identification of the chaotic system in terms of unstable periodic orbits, and calculation of the control signal. When applying the OGY method to control chaos in a real electronic circuit the main problem encountered was the noise introduced due to inevitable noise of the circuit elements.

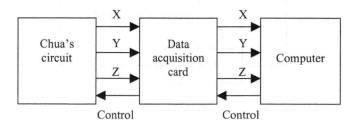

Fig. 9. Practical implementation of the Ott–Grebogi–Yorke OGY method

The method was found to be very sensitive to the noise level—very small signals sometimes are hidden within the noise, and control is impossible.

Generally, the experimental application of the OGY method requires a permanent computer analysis of the state of the system. The changes of the parameters, however, are discrete in time and this leads to some serious limitations. The method can stabilize only those periodic orbits in which maximal Lyapunov exponent is small compared to the reciprocal of the time interval between parameter changes. Since the corrections of the parameter are rare and small, the fluctuation noise leads to occasional bursts of the system into regions far from the desired periodic orbit, especially in the presence of noise.

2.1.2 Pyragas's Control Methods

A different approach to feedback control, which helps to avoid the above-mentioned problems, was proposed by Pyragas [10]. This method is based on the construction of a special form of a time continuous perturbation, which does not change the form of the desired unstable periodic orbit, but under certain constraints can stabilize it. Two feedback-controlling loops, shown in Fig. 10, have been proposed.

A combination of feedback and periodic external force is used in the first method (Fig. 10a). The second method (Fig. 10b) does not require any external source of energy and it is based on self-controlling delayed feedback.

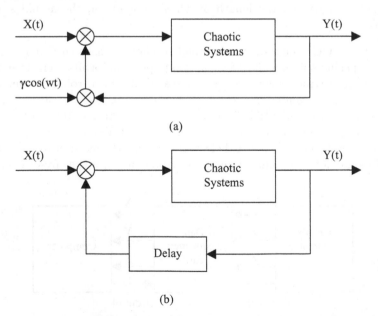

(a)

(b)

Fig. 10. Feedback controlling loops; (**a**) control by periodic external perturbation, and (**b**) control by time delay

If the period of external force or a time delay is equal to the period of one of the unstable periodic orbits embedded in the chaotic attractor, it is possible to find a constant K, which allows stabilization of the unstable periodic orbit. This approach, being noise resistant, can easily be used in experimental systems. The first of Pyragas's methods can be considered as the special case of the direct application of classical controlling methods to the problem of controlling chaos.

The dynamical system

$$X' = f(X) \,, \tag{12}$$

where $X \in R^n$, is controllable if there exists a control function $u(t)$, such that

$$X' = f(X) + u(t) \tag{13}$$

allows to move trajectory $X(t)$ from point X_0 at time t_0 to the desired point X in finite time t.

The controllability concept can be applied to the chaos controlling problems. For example, for Chua's circuit the equations for the controlled circuit are

$$\begin{aligned} X' &= a(Y - X - f(X)) \\ Y' &= X - Y + Z - K(Y - Y^*) \\ Z' &= -bY \,. \end{aligned} \tag{14}$$

This approach is illustrated in Fig. 11. The main advantages of this method are as follows:

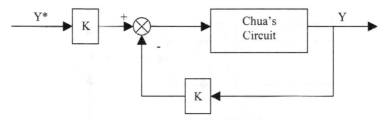

Fig. 11. Closed loop feedback control configuration

(a) Any solution of the original system can be a goal of the control (fixed point, unstable periodic orbit, etc.)
(b) The controller has a very simple structure.
(c) Access to system parameters is not required.
(d) It is not affected by small parameter variations.

2.1.3 Controlling Chaos by Chaos

In this section, we show that the chaotic behavior of one system can be controlled by coupling it with another one which can also be chaotic [5]. Thus we consider two chaotic systems, which we call A and B respectively,

$$X' = f(X)$$
$$Y' = g(X)\,, \tag{15}$$

where $X, Y \in R^n$, and we use the controlling strategy, which is illustrated in Fig. 12; the two systems are coupled through the operators λ, μ, which have a very simple linear form. We assume that some or all state variables of both systems A and B can be measured, so that we can measure signal $X(t)$ from system A and signal $Y(t)$ from B, and that the systems are coupled in such a way that the differences D_1 and D_2 between the signals $X(t)$ and $Y(t)$ are

$$F_1(t) = \lambda[X(t) - Y(t)] = \lambda D_1(t)$$
$$F_2(t) = \mu[Y(t) - Y(t)] = \mu D_2(t) \tag{16}$$

used as control signals introduced, respectively, into each of the chaotic systems A and B as negative feedback. We take $\lambda, \mu > 0$ to be experimentally adjustable weights of the perturbation.

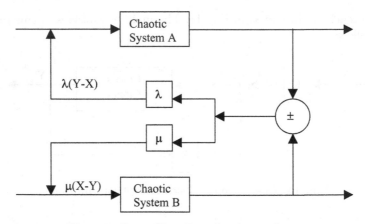

Fig. 12. Controlling chaos-by-chaos scheme

Using the coupling given in Fig. 12, it has been shown that one chaotic system coupled with the other one can significantly change the behavior of one of them (unidirectional coupling, i.e., λ or $\mu = 0$) or of both systems (mutual coupling, i.e., λ, $\mu \neq 0$). This property allows us to describe the above procedure as the "controlling chaos by chaos" method. In [11], rigorous conditions are given, under which chaotic attractors of systems A and B are equivalent, or the evolution of one of them is forced to take place on the attractor of the other one. Kapitaniak [5] shows an example of coupling two Lorenz chaotic attractors, which results in chaos control and increase of the predictability.

2.2 Controlling Chaos Without Feedback

2.2.1 Control Through Operating Conditions

Virtually all engineering and most natural systems are subjected during operation to external forcing. This forcing will contain (and hopefully be dominated by) planned and intentional components; it will also almost invariably contain unintentional "noise." Smart design and control of this forcing is often able to annihilate, or shift to a harmless region of parameter space, an unwanted chaotic behavior.

In this case, the method consists of finding the chaotic region in parameter space by analytical and numerical methods [5]. Then based on this region change the parameters to control the dynamical system. The procedure described in this section is based on the direct change of one of the system parameters to shift system behavior from chaotic to periodic, close to the chaotic attractor. It cannot be called a control method in the sense of the methods described before, but it illustrates that having a system designed as chaotic, we obtain easy access to different types of periodic behavior.

2.2.2 Control by System Design

In this section, we explore the idea of modifying or removing chaotic behavior by appropriate system design. It is clear that, to a certain extent, chaos may be "designed out" of a system by appropriate modification of parameters, perhaps corresponding to modification of mass or inertia of moving parts. Equally clearly, there exist strict limits beyond which such modifications cannot go without seriously affecting the efficiency of the system itself.

In this section, we describe a method for controlling chaos in which the chaos effect is achieved by coupling the chaotic main system to a simpler autonomous system (controller), usually linear, as shown in Fig. 13. This method [5] is developed for chaotic systems in which for some reason it is difficult, if not impossible, to change any parameter of the main system. In particular consider the coupling of the chaotic system

$$X' = f(X, \mu) , \tag{17}$$

where $X \in R^n$, $n \geq 3$, and $\mu \in R$ is a system parameter, to another (simpler) asymptotically stable system (controller) described by

$$Y' = g(Y, e) , \tag{18}$$

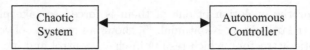

Fig. 13. Coupling scheme

where $Y \in R^n$ and e is a vector denoting the controller's parameters, and where at least one of the parameters e_i can be easily changed. For practical reasons, the dimension m of the controller system (18) should be chosen as low as possible. Since the method was mainly designed for controlling chaos in mechanical systems, we choose $m = 2$, i.e., a one degree of freedom controller (the simplest mechanical system). The equations for the extended system are

$$X' = f(X, \mu) + AY$$
$$Y' = g(Y, e) + BX , \tag{19}$$

where A and B are the coupling matrices. Since the Y subsystem is asymptotically stable, the role of the controller is to change the behavior of the system from chaotic to some desired periodic, possibly constant, operating regime.

The idea of this method is similar to that of the so-called dynamical vibration absorber. A dynamical vibration absorber is a one-degree of freedom system, usually mass on a spring, which is connected to the main system as shown in Fig. 14.

Although such a dynamical absorber can change the overall dynamics substantially, it usually need only be physically small in comparison with the main system, and does not require an increase of excitation force. It can be easily added to the existing system without major changes of design or

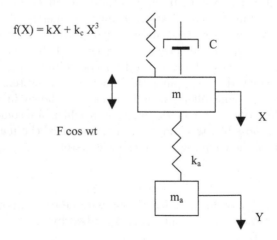

Fig. 14. Dynamical damper as chaos controller

construction. This contrasts with devices based on feedback control, which can be large and costly.

To explain the role of dynamical absorbers in controlling chaotic behavior let us consider the Duffing oscillator, coupled with an additional linear system:

$$X'' + aX' + bX + cX^3 + d(X - Y) = B_0 + B_1 \cos wt \,, \tag{20a}$$

$$Y'' = e(Y - X) = 0 \,, \tag{20b}$$

where a, b, c, d, e, B_0, B_1, and w are constants. Here d and e are the characteristic parameters for the absorber, and we take e as the control parameter.

It is well known that the Duffing's oscillator shows chaotic behavior for certain parameter regions. As has been mentioned in the previous section, in many cases the route to chaos proceeds via a sequence of period doubling bifurcations, and in such cases this method provides an easy way of switching between chaotic and periodic behavior. Figure 15 shows a two-dimensional view of the chaotic behavior in Duffing's oscillator. Figure 16 shows a plot of variable X across time $[0, 350]$. Figure 17 shows a plot of variable X' across time. And finally, Fig. 18 shows a three-dimensional view of the strange attractor.

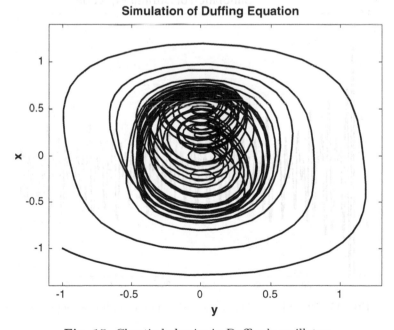

Fig. 15. Chaotic behavior in Duffing's oscillator

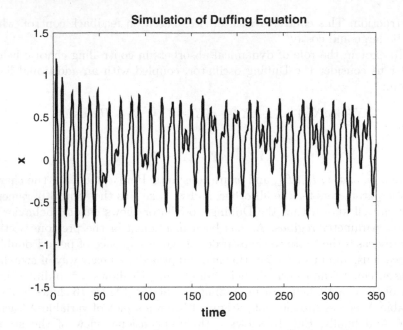

Fig. 16. Plot of variable X across time for Duffing's oscillator

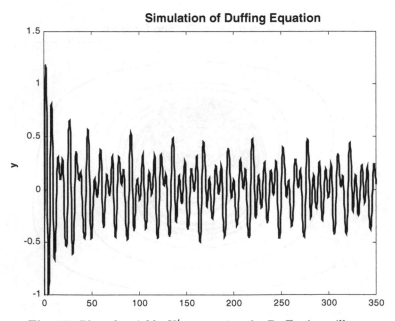

Fig. 17. Plot of variable X' across time for Duffing's oscillator

Simulation of Duffing's Oscillator in Three Dimensional Space

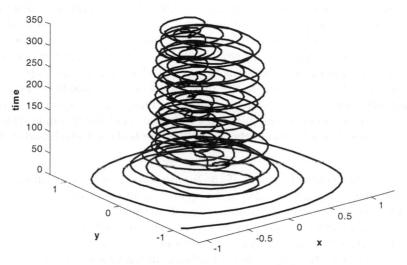

Fig. 18. Simulation of Duffing's oscillator (20) for $e = 0.01$

Let us consider the parameters of (20a) to be fixed at values $a = 0.077, b = 0, c = 1.0, B_0 = 0.045, B_1 = 0.16$, and $w = 1.0$, then we can find [5] that we have chaos for $e \in [0, 0.10]$, and we can control this chaos by increasing e above 0.10. As this method is designed mainly for experimental applications, we shall now briefly suggest some guidelines for applying it:

(1) The coupled system has to be as simple as possible.
(2) The coupling e should be chosen as small as possible.
(3) If it is possible one should couple the controller in such a way that the locations of the fixed points of the original system are not changed.

2.3 Method Selection

Although the methods described in the previous sections have been developed mainly by physicists and mathematicians, generally most of them can be applied to control engineering systems.

In particular, the nonfeedback methods can practically always be used. Their applications are straightforward and do not require special complicated controllers to be used. The main disadvantage of these methods is that the goal of controlling has to be determined by trial and error method.

The motivations for using feedback systems to control chaos are the following: feedback controllers are easy to implement, especially in electrical systems, they can perform the job automatically, and stabilize the overall control system efficiently. On the other hand, conventional feedback controllers are designed for nonchaotic systems. A chaotic system's sensitivity to initial conditions may lead to the impression that in chaotic systems their sensitivity to small errors makes them very difficult. Such an impression may lead to the argument that once the control is initiated there is no need for further monitoring of the system's dynamics, nor feeding back this information in order to sustain the control. Indeed, it turns out that conventional feedback control of chaotic systems is generally difficult, but not impossible. Recently, Castillo and Melin [12–14] used neural networks and fuzzy logic for identification and control of chaotic systems. In many cases, a specially implemented feedback method can guarantee stabilization of the dynamic system. To summarize, the selection of the controlling method has to be based on the following:

(1) The goal of controlling (e.g., if the suppression of chaos is the main goal, then nonfeedback methods can be applied in an easier way).
(2) The level of noise in the system (e.g., if the level of noise is large, then Pyragas's methods can be more effective than the OGY approach).
(3) The particular characteristics of the system. (Generally, in electrical systems one can try to use both feedback and nonfeedback methods. In mechanical systems where the suppression of chaos is the main goal of controlling, nonfeedback methods are recommended.)

3 Towards a Theory of Fuzzy Chaos

For a given dynamical system expressed as a nonlinear differential equation:

$$\frac{d\mathbf{y}}{dt} = f(t, \mathbf{y}) \qquad \mathbf{y}(0) = \mathbf{y}_0 \,. \tag{21}$$

or as nonlinear difference equation:

$$\mathbf{y}_{t+1} = f(\mathbf{y}_t, \ldots) \qquad \mathbf{y}(0) = \mathbf{y}_0 \,. \tag{22}$$

We can have many different types of dynamic behaviors, for the above equations, depending on the parameter values, and also depending on the analytical properties of the function f. Also, there exists a fundamental difference between (1) and (2), namely, that differential equations can only exhibit chaos when they are at least three-dimensional. However, difference equations can exhibit chaos even for the one-dimensional case.

In particular, we can have "chaotic behavior," defined formally as sensitive dependence on initial conditions for many real dynamical systems [2]. However, in numerical simulations we usually have uncertainty related to numerical errors in the methods and also in the initial values. For this reason, it is very difficult to identify precisely real chaotic behavior [15–17]. We can relax the traditional mathematical definition of "chaos" by using the theory of fuzzy logic [18], in this way obtaining a new more realistic definition of chaotic behavior. We assume that we have a dynamical system in the real line given as

$$y_t = f(y_{t-1}, \theta) \,. \tag{23}$$

In this case, we can associate chaotic behavior with the number of period doublings (or bifurcations) that occur when the parameter θ is varied. According to this fact, we can state the following definition.

Definition 1. (Chaotic behavior according to period doublings) *A one-dimensional dynamical system shows fuzzy chaos when the number of period doublings is considered to be large:*

IF number of period doublings is large THEN behavior is fuzzy chaos

We can also state a definition of fuzzy chaos based on the fact that complex behavior is related to the fractal dimension of a strange attractor.

Definition 2. (Fuzzy chaos by the fractal dimension) *A one-dimensional dynamical system shows fuzzy chaos, when the value of the fractal dimension is large (close to a numeric value of 2 for the plane):*

IF the fractal dimension is large THEN behavior is fuzzy chaos.

Also, the value of the fractal dimension has to be calculated from the time series using the box counting algorithm.

4 Controlling Chaotic Behavior Using Fuzzy Chaos

In any of the above-mentioned methods for controlling chaos we have that a specific parameter is used to change the dynamics of the system from a chaotic to a stable behavior. For example, for the specific case of Duffing's oscillator the parameter e of (20b) can be used for controlling the chaotic behavior of the oscillator. However, the crisp interval [0, 0.10] for parameter e in which chaotic behavior occurs is not really an accurate range of values. Of course, for $e = 0$ we can expect chaotic behavior, but as e increases in value, real chaotic behavior is more difficult to find. In the crisp boundary of $e = 0.10$, things are more dramatic, one can either find cyclic stable behavior or unstable behavior depending on the conditions of the experiment or simulation. For this reason, it is more appropriate to use the proposed concept of "fuzzy chaos," which will allow us to model the uncertainty in identifying this chaotic behavior. In this case, a membership function could be defined to represent this uncertainty in

O. Castillo and P. Melin

Table 1. Comparison between the methods for controlling chaos

	Traditional Chaos Definition (%)	New Fuzzy Chaos Definition (%)
Chua's circuit	98.50	99.50
Duffing's oscillator	96.00	98.50

finding chaotic behavior, and also this is really helpful in controlling chaotic behavior as we can take action even before completely chaotic behavior is present. For the case of the Duffing's oscillator we can define fuzzy rules for identifying specific dynamic behaviors. For example, chaotic behavior can be given by the rule

IF e is Small THEN behavior is fuzzy chaos.

In the above fuzzy rule, the linguistic term "small" has to be defined by the appropriate membership function. Other similar rules can be established for identifying different dynamic behavior for the system. One obvious advantage of this approach is that we are able to have relative evidence of chaotic behavior before there is complete instability. As a consequence of this fact we can take action in controlling this chaotic behavior sooner than with traditional methods. A sample fuzzy rule for controlling chaos is as follows:

IF behavior is fuzzy chaos THEN increase is small positive.

This fuzzy rule simply states that when fuzzy chaos is present then we must increase slightly the value of e. Of course, linguistic terms, like "small positive," need to be defined properly. Table 1 shows the comparison between the methods for controlling chaos for the two dynamic systems considered in this chapter. The table shows the efficiency and accuracy for controlling chaotic behavior for the two cases described before. We did consider a sample of 200 different experimental conditions for both dynamical systems, and compare the relative number of times that a particular method was able to really control chaotic behavior. The implementation of the fuzzy chaos approach for behavior identification was done in MATLAB. Figure 19 shows the membership functions of the linguistic variable corresponding to the parameter e, in which there are three linguistic values.

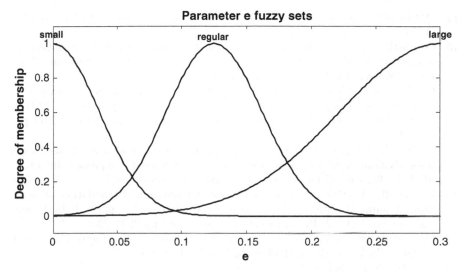

Fig. 19. Membership functions for parameter "e"

5 Conclusions

In this chapter we have presented a new theory of fuzzy chaos for nonlinear dynamical systems. We can apply this theory for behavior identification. We also presented a new method for controlling nonlinear dynamical systems. This method is based on a hybrid fuzzy–chaos approach to achieve the control of a particular dynamical system given its mathematical model.

References

1. C. Grebogi, E. Ott, J.A. Yorke: Science **238**, 632–637 (1987)
2. R. Devaney: *An Introduction to Chaotic Dynamical Systems* (Addison-Wesley, Reading, MA, 1989)
3. B. Mandelbrot: *The Fractal Geometry of Nature* (Freeman, San Francisco, CA, 1987)
4. S.N. Rasband: *Chaotic Dynamics of Non-Linear Systems* (Wiley Interscience, New York, 1990).
5. T. Kapitaniak: *Controlling Chaos: Theoretical and Practical Methods in Non-Linear Dynamics* (Academic, New York, 1996)
6. E. Ott, C. Grebogi, J.A. Yorke: Phys. Rev. Lett. **64**, 1196–1199 (1990)
7. L.O. Chua: IEICE Trans. Fund. 704–734 (1993)
8. R. Madan: *Chua's Circuit: Paradigm for Chaos* (World Scientific, Singapore, 1993)
9. M.J. Ogorzalek: IEEE Trans. Circuit Syst. **40**, 700–721 (1993)
10. K. Pyragas: Phys. Lett. 421–428 (1992).
11. L. Kocarev, T. Kapitaniak: J. Phys. A. **28**, 249–254 (1995)

12. O. Castillo, P. Melin: J. Math. Model. Simulation Syst. Anal. **18/19**, 767–770 (1995)
13. O. Castillo, P. Melin:Automated mathematical modelling and simulation of dynamical engineering systems using artificial intelligence techniques. In: *Proc. CESA'96*, Gerf EC Lille, 1996, pp 682–687
14. O. Castillo, P. Melin: Mathematical modelling and simulation of robotic dynamic systems using fuzzy logic techniques and fractal theory. In: *Proc. IMACS World Congress'97*, vol 5 (Wissenschaft & Technik Verlag, Berlin, 1997) pp 343–348
15. O. Castillo, P. Melin: Modelling, simulation and behavior identification of nonlinear dynamical systems with a new fuzzy-fractal-genetic approach. In: *Proc. of IPMU'98*, vol 1 (EDK Publishers, Paris, 1998) pp 467–474.
16. O. Castillo, P. Melin: A general method for automated simulation of nonlinear dynamical systems using a new fuzzy-fractal-genetic approach. In: P*roc.s CEC'99*, vol 3 (IEEE Press, Piscataway, NJ, 1999) pp 2333–2340
17. O. Castillo, P. Melin: *Soft Computing for Control of Non-linear Dynamical Systems* (Springer, Berlin Heidelberg New York, 2001)
18. L.A. Zadeh: Inf. Sci. **8**, 43–80 (1975)

Complex Fuzzy Systems
and Their Collective Behavior

Maide Bucolo, Luigi Fortuna, and Manuela La Rosa

Abstract. This work aims at being a contribution for the characterization of a new class of complex systems built as arrays of coupled fuzzy logic based chaotic oscillators and an investigation on their collective dynamical features. Different experiments were carried out varying the parameters related to the single-unit dynamics, as Lyapunov exponent, and to the macrosystem structure, as the number of connections. Four types of global behaviors have been identified and characterized distinguishing their patterns as follows: the *spatiotemporal chaos*, the *regular synchronized behavior*, the *transition phase*, and the *chaotic synchronized behavior*. These collective behaviors and the synchronization capability have been highlighted by defining a mathematical indicator which weights the slight difference among a wide number of spatiotemporal patterns. To investigate the effects due to the network architecture on the synchronization characteristics, complex fuzzy systems have been reproduced using fuzzy chaotic cells connected through different topologies: regular, "small worlds," and random.

1 Introduction

This chapter discusses the study of the dynamical behavior of spatiotemporal complex systems built by coupling fuzzy logic chaotic units. The artificial reproduction of phenomena like spatiotemporal chaos and synchronization constitutes an interesting approach to the study and the understanding of collective dynamics that are commonly observed in nature [1]. In particular the synchronization of oscillators is a classical topic related to different research fields: from circuit theory to biological and social systems. In the last 10 years literature ever more detailed on synchronization is continuously dealt [2]; circuits and systems based on synchronization principles are today usually adopted in different applications. The theme of synchronization has catalyzed an ever increasing scientific attention in the fascinating field of complex systems and networks.

A review study [3] explores in details the behavior of arrays of a large number of oscillators by considering both the effects due to the dynamics of the single unit and those related to the network topology with the question "how does the global system reach the synchronization?" in mind. Moreover the results shown in [4] represents the key reference in the area of the "Small World Theory," and Chaps. 8 and 9 discuss the role played by the

M. Bucolo et al.: *Complex Fuzzy Systems and Their Collective Behavior*,
StudFuzz **187**, 415–437 (2006)
www.springerlink.com

network topology to tame the global dynamics of coupled nonchaotic oscillators. Using the "small-worlds" topology [5], a new complex architecture is obtained starting from a regular lattices and rewiring some connections with probabilistic rules. The introduction of these structures in complex system displays an enhancement in the speed of signal propagation, synchronizability, and computational power compared to regular networks with the same number of elements.

Finally, [6] exploits the possibility of using the network topology to synchronize arrays of chaotic circuits, underlining the value of the results obtained in the case of arrays with nonchaotic oscillators.

Starting from these remarks the role of this chapter is twofold: The first one is to illustrate a new class of complex systems, the arrays of coupled fuzzy chaotic oscillators, and the second one, is to give a contribution to characterize the synchronization of these systems, both in a qualitative and a quantitative way.

This study of complex systems has focused on the relationships among the features of the fundamental cell, the architecture of the network under consideration, and the spatiotemporal dynamics achieved [3]; therefore, their characterization can be seen from two different points of view as a *single-unit* and as a *macrosystem*. At the single-unit level, the attention is pointed toward the effects of nonlinear dynamics of each unit due to its own parameters, without taking into account the neighborhood interferences on the collective behavior. At the macrosystem level, the aim is to correlate the global dynamics with the topology of the complex system through the number of the connections and their spatial distribution.

A generalization of the synchronization principles for an array of fuzzy logic chaotic based dynamical systems is suggested and evaluated as alternative approach to build locally connected fuzzy complex systems by varying both the rules driving the cells and the network architecture. A discrete fuzzy oscillator that shows a chaotic behavior has been chosen as fundamental unit. This fuzzy system is obtained through a linguistic description of the stretching and folding features for an assigned value of the Lyapunov exponent [7]. At macro-level, initially, the considered fuzzy cells have been connected with a regular topology. Different structures have been implemented using different parameter values as the Lyapunov exponent and the number of connections, and a qualitative comparison of their dynamic features have been performed. The definition of a synchronization index gives the opportunity to quantify the synchronization properties of the considered collective behavior. The introduction of this index speeds up the process of pattern comparison, and therefore it is possible to consider a wide number of topologies and architectures for a more exhaustive and detailed study. In the second phase, the investigation at the macro-level has been carried on comparing the global dynamics obtained through different topologies: regular, small worlds, and random [5].

In Sect. 2 the features of a regular fuzzy chain have been described choosing the fuzzy chaotic cell and the topology scheme. In the first step a chaotic fuzzy oscillator with Lyapunov exponent equal to 0.1 has been considered as fundamental element of the array and the spatiotemporal dynamics have been opportunely characterized pointing out the synchronization through both a qualitative pattern analysis and a frequency analysis. The same study has been carried on varying the features of the chaotic fuzzy oscillator; the Lyapunov exponent has been set equal to 0.9. In Sect. 3 the definition of a spatiotemporal synchronization index is explained in details and, above all, the capability of becoming synchronization quantitative indicator is shown clearly in the different implemented cases. In Sect. 4 the analysis aims at investigating on the relationship between the synchronization features and the complex fuzzy dynamics obtained varying the topology under consideration, i.e., the regular, the small-world, and the random.

The reported results underline the possibility to extend the principles of synchronization, as well as auto-organization, to this new class of complex fuzzy system, as it is evident in the pattern formation. This investigation opens a new way to see the complex real-world phenomena that sometimes are not easy to describe using mathematical structures.

2 Complex Fuzzy System

To well characterize the spatiotemporal dynamics of an extended nonlinear fuzzy one-dimensional array, the fundamental choices are to set the dynamic of the elementary cell and the network structure. A discrete chaotic fuzzy oscillator has been chosen as basic brick of this network structure, and its chaotic behavior is defined varying the Lyapunov exponent in a suitable range. The macrosystem has been obtained by connecting of a large number of identical cells in a regular configuration. The study of the global dynamics of this fuzzy system starts from the characterization of the spatiotemporal patterns through the visual inspection and the frequency analysis.

2.1 Single Unit: Chaotic Fuzzy Oscillator

Using a linguistic approach it is possible to describe simple object based on concepts and laws of behavior and interaction. A chaotic fuzzy oscillator has been characterized using two variables x and d,respectively the nominal value of the state and the uncertainty on the center value, that generate the desired evolution and perform the stretching and folding features. The discrete dynamic of this fuzzy oscillator could be considered as a two-dimensional chaotic map with the following structure:

$$x\,(k+1) = \Psi\,(x(k), d(k))$$
$$d\,(k+1) = \Phi\,(x(k), d(k))$$

$$(1)$$

Table 1. Fuzzy inference system: The rules

$x(k)/d(k)$	Zero	Small	Large	Verylarge
x.large. left	x.small.left/ zero	x.small.left/ medium	x.small.left/ verylarge	x.small.right/ large
x.small. left	x.large.left/ zero	x.large.left/ medium	x.large.left/ verylarge	x.large.left/ small
x.small. right	x.large.right/ zero	x.large.right/ medium	x.large.right/ verylarge	x.large.right/ small
x.large. right	x.small.right/ zero	x.small.right/ medium	x.small.right/ verylarge	x.small.left/ large

where Ψ and Φ are the fuzzy inference functions described through a set of fuzzy rules for each variable as reported in Table 1.

The range of the variables has been chosen according to a suitable embedding zone performing a main oscillatory dynamics summed to an uncertainly evolution. The designed fuzzy sets are shown in Fig. 1; the relation between a specific chaotic dynamics and the fuzzy set has been formalized through the following equations (2) where the center of each membership function is evaluated starting from a value of the Lyapunov exponent (l).

$$
\frac{C_{\text{medium}}}{C_{\text{small}}} = e^l
$$
$$
\frac{C_{\text{verylarge}}}{C_{\text{large}}} = e^l \, .
\tag{2}
$$

Figure 2 shows the chaotic time series generated by a fuzzy oscillator with $l = 0.1$.

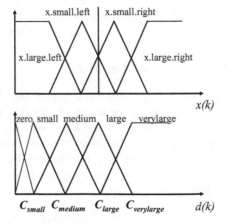

Fig. 1. Fuzzy inference system: the fuzzy sets

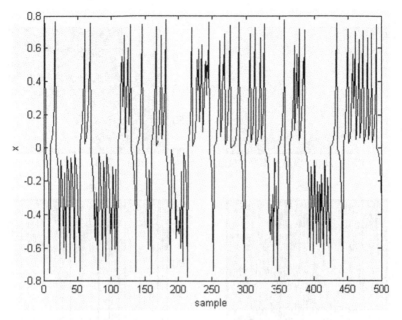

Fig. 2. Time series of the fuzzy chaotic oscillator with $l = 0.1$

2.2 Macrosystem: Chain of Fuzzy Oscillators

The macrosystem configuration has been set with a regular distribution of the connections (C), it means that each fundamental fuzzy unit has $2C$ neighbors, half on its right (C) and half on its left (C). The equations of the single element of the array have been rewritten according to the number of oscillators (N) and the diffusion coefficient (D) that weights the information exchange, as reported in (3):

$$x_i\,(k+1) = \Psi\left(x_i(k) + D\left(-2Cx_i(k) + \sum_{\substack{j=-C \\ j \neq 0}}^{C} x_{i+j}\,(k)\right), d_i(k)\right) \quad (3)$$

$$d_i\,(k+1) = \Phi\,(x_i(k), d_i(k)) \quad i = 1 \dots N$$

The state value of each fuzzy oscillator depends on its own state value in the previous sample time and on the contributions coming from the state values of the other fuzzy oscillators connected through a bi-directional information exchange. In this phase the one-dimensional array has been designed coupling $N = 200$ discrete fuzzy chaotic oscillators with $l = 0.1$ through a constant diffusion coefficient $D = 0.005$, and the single-unit starts its evolution from random initial conditions.

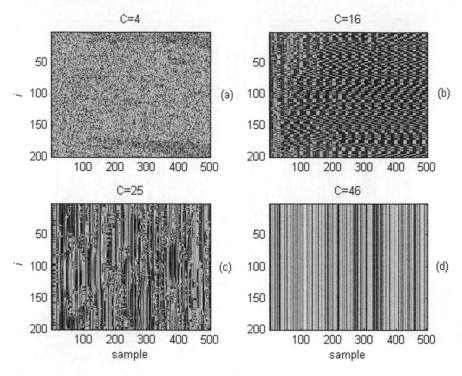

Fig. 3. Spatiotemporal patterns of the fuzzy chains ($l = 0.1$): (**a**) $C = 4$, (**b**) $C = 16$, (**c**) $C = 25$, and (**d**) $C = 46$

Varying in the fuzzy chain the number of connections for each unit ($2C$), four different global dynamics have been observed; the pattern formation related with these four different collective behaviors are displayed in Fig. 3. Each spatiotemporal map is built for a particular value of C where the index associated to each cell (i) is reported versus time in sample. Meanwhile, the state value of each cell (x_i) is represented through a 64-color scale: blue is for low value and red for high values.

In Fig. 3, particularly, four behaviors can be distinguished with increasing the numbers of connections (C):

- for $C = 4$ *spatiotemporal chaos* (Fig. 3a);
- for $C = 16$ *regular synchronized behavior* (Fig. 3b);
- for $C = 25$ *transition phase* (Fig. 3c);
- for $C = 46$ *chaotic synchronized behavior* (Fig. 3d).

Summarizing the four types of collective behaviors: an initial spatiotemporal chaos, two types of synchronization, regular and chaotic, and a transition phase. The behaviors of the nominal value of two cells (x_i) are plotted against the time series trends in Fig. 4 and their Fourier spectra in Fig. 5, where solid line is for x_{100} and dotted line is for x_{101}, this glance both in time and

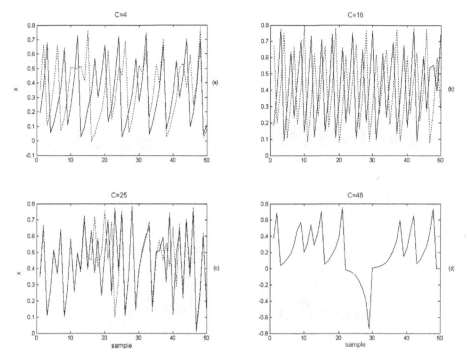

Fig. 4. Trend of two cells (x_{100}, x_{101}) in the fuzzy chains ($l = 0.1$): (a) $C = 4$, (b) $C = 16$, (c) $C = 25$, and (d) $C = 46$

frequency explains more in details the different spatiotemporal dynamics of each map.

Considering the *spatiotemporal chaos* obtained with $C = 4$, the following considerations can raise:

- Each chaotic oscillator evolves with its own dynamics according to the different initial conditions (Fig. 4a).
- The absence of a regular oscillatory behavior is identifiable in the spread Fourier spectra (Fig. 5a).

Considering the *regular synchronized behavior* obtained with $C = 16$, the results underline that

- for both cells regular oscillations are visible (Fig. 4b).
- for both cells the value of picks in a particular time frequency can be seen distinctly (Fig. 6b).

Considering the *transition phase* obtained with $C = 25$, the results are that

- the two cells time evolutions become almost the same in long time intervals (Fig. 4c).

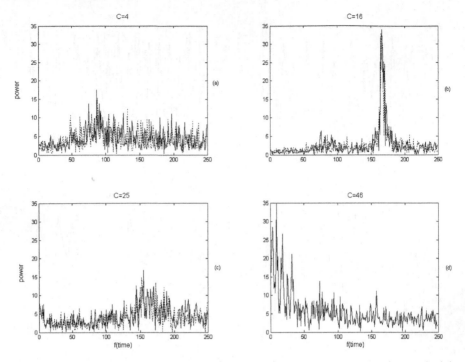

Fig. 5. Fourier spectrum of two cells (x_{100}, x_{101}) in the fuzzy chains ($l = 0.1$): (**a**) $C = 4$, (**b**) $C = 16$, (**c**) $C = 25$, and (**d**) $C = 46$

• the spread Fourier spectra show the disappearance of the regular oscillatory behavior identifiable in the narrow pick (Fig. 5c).

Considering the *chaotic synchronized behavior* obtained with $C = 46$, the results clearly highlight a perfect synchronization of the two cells and the chaotic feature of the signals:

• The two time series are identical and chaotic showing a precise synchronization in time (Fig. 4d).
• Both cells have a broad band spectra, and these are perfectly superimposed (Fig. 5d).

The same types of collective behaviors, previously characterized for $l = 0.1$, have been obtained by maintaining the same structure of chain and varying the chaotic dynamic of the fuzzy oscillator setting $l = 0.9$. Also in this experiment, the previously defined spatiotemporal patterns are recognized: the spatiotemporal chaos (Fig. 6a), the regular synchronization (Fig. 6b), the transition phase (Fig. 6c), and the chaotic synchronization (Fig. 6d).

Moreover, the trends of the two cells (x_{100}, x_{101}) and their Fourier spectra are reported in Figs. 7 and 8. The spatiotemporal chaos is recognizable in the independent chaotic evolution of the cells and in their spread frequency power

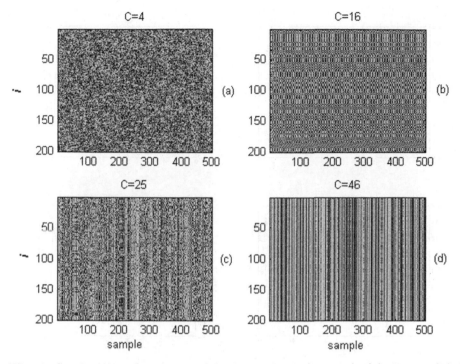

Fig. 6. Spatiotemporal patterns in the fuzzy chains ($l = 0.9$): (**a**) $C = 4$, (**b**) $C = 16$, (**c**) $C = 25$, and (**d**) $C = 46$

distribution as reported in Figs. 7a–8a. The regular synchronized behavior is characterized by a sharp sequence of picks at specific frequency values (Figs. 7b and 8b). The characteristics of the chaotic synchronized behavior are shown in Figs. 7d and 8d. These results can be qualitatively compared with the one obtained for $l = 0.1$ (Figs. 3–5); as it can be seen the synchronization reached in the second experiment ($l = 0.9$, $C = 46$) is not as well defined as the one of the first experiment ($l = 0.1$), it is possible to have the same result by increasing the number of connections.

3 The Collective Dynamics Through the Syncronization Index

The exploited collective dynamics underline the necessity to introduce a mathematical indicator to weight easily the slight difference among a wide number of spatiotemporal patterns. A synchronization index that suits with this requirement is presented in this section; its capability is widely investigated by the renewed analysis of the previous experiments and the presentation of new ones. The spatiotemporal analyses dealt previously have been

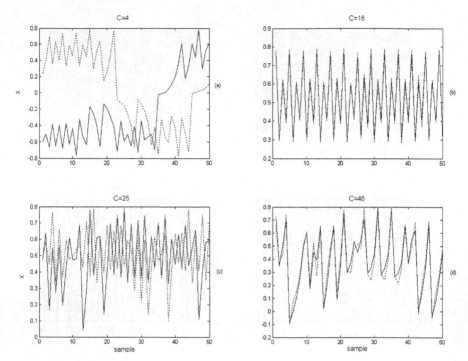

Fig. 7. Trend of two cells (x_{100}, x_{101}) in the fuzzy chains $(l = 0.9)$: **(a)** $C = 4$, **(b)** $C = 16$, **(c)** $C = 25$, and **(d)** $C = 46$

extended through this new indicator by tuning in suitable range both the Lyapunov exponent and the number of connections for a more complete and detailed study.

This evaluation index has enhanced the procedure for the comparison and the identification of the complex network features versus the adopted system parameters, otherwise a direct inspection of all the spatiotemporal maps should be necessary for each parameter variation, and therefore heavy computational strength.

3.1 Synchronization Index

The synchronization index has been introduced to evaluate the synchronization degree of a system made up of coupled units [8]. Let us collect in the rows of the matrix A all the N signals generated by the N subunits of the system and evaluate the covariance matrix as follows: $W^{(N x N)} = AA^{\mathrm{T}}$. The synchronization index takes into account the eigenvalues of W (the square of the singular values of the matrix A). If all the signals are uncorrelated, all the singular values will be no zero. If all the signals are identical, there will be only one no zero singular value and the rank of the matrix W is equal to

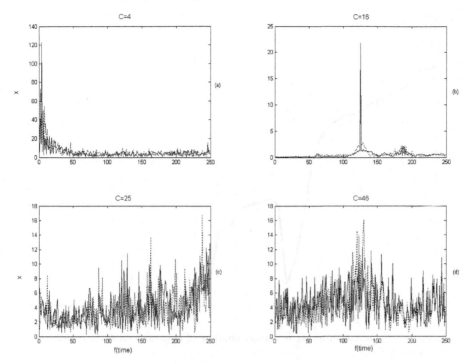

Fig. 8. Fourier spectrum of two cells (x_{100}, x_{101}) in the fuzzy chains ($l = 0.9$): (**a**) $C = 4$, (**b**) $C = 16$, (**c**) $C = 25$, and (**d**) $C = 46$

1. Of course if signals are similar but not identical, very small but not null singular values can be found. The synchronization index is thus defined as the minimum number m of eigenvalues, whose sum is greater than a percentage ξ of the trace of W as in (4):

$$\sigma = \min \; m \left| \sum_{i=1}^{m} \lambda_i^* \right| > \xi \cdot \mathrm{Tr}\,(W) \tag{4}$$

where $\lambda_i^{\mathrm{sort}}$ is one of the ith largest eigenvalue of the covariance matrix W. Starting from a N coupled systems (N signals), this index is in the range from 1 (total synchronization) to $[\xi N] + 1$ (no synchronization) and gives information about the total number of different dynamics in the system. In the following experiments, the percentage ξ has been assumed to be the 98%, and in relation with the number of array elements $N = 200$ the index range is $1 \leq \sigma(0.98) \leq 196$.

3.2 Synchronization Index Versus Network Parameters

Assuming the Lyapunov exponent of cell to be 0.1 and the fuzzy array structure to be regular, in the first analysis this synchronization index has been

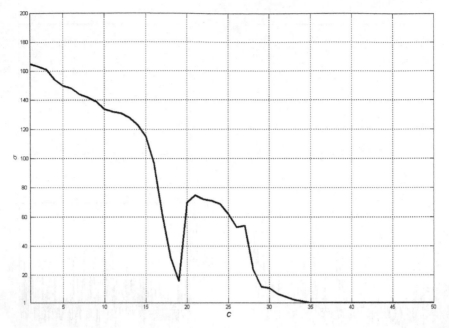

Fig. 9. Synchronization index σ versus number of connections C in regular fuzzy chains with $l = 0.1$

evaluated to compare the collective dynamics varying the value of the connections c from 1 to 50.

In Fig. 9 the values of the index σ are plotted versus the number of connections. This graph shows a specific association between the different ranges of connections and the previously defined spatiotemporal dynamics:

- For $1 < C < 15$, high values of σ characterize the spatiotemporal chaos.
- For $15 < C < 20$, the system exhibits the increasing of the synchronization (regular synchronized behavior) quantified by the decreasing of the synchronization index.
- For $20 < C < 30$, the transition phase takes place and the values of σ become higher.
- For $C > 30$, the index decreases to 1 and the chaotic synchronized behavior is presented.

These results have been compared to the 2D Fourier transform to prove the feasibility of this index for the analysis of the transitions state in complex system varying the network parameters.

A characterization of all the spatiotemporal dynamic features of the array that have been performed through the evaluation of the 2D Fourier transforms is shown in Fig. 10. Looking at the three-dimensional graphs obtained in relation to the four C values used previously, it is possible to recognize

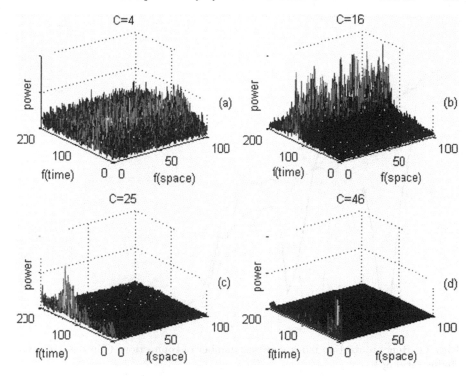

Fig. 10. Fourier 2D transform of regular fuzzy chain with $l = 0.1$. (a) $C = 4$, (b) $C = 16$, (c) $C = 25$, and (d) $C = 46$

the initial chaotic state, the transition phase, and two main synchronized behaviors, the regular one and the chaotic one. Figure 10 can be commented as follows:

- for $C = 4$, the spread number of picks both in space frequency and time frequency is visible (Fig. 10a).
- for $C = 16$, the value of picks at a particular time frequency becomes bigger (Fig. 10b).
- for $C = 25$, the network dynamics performs a transition from the regular behavior to the chaotic synchronized one (Fig. 10c).
- for $C = 46$, the distribution of picks, only in the time–frequency axis, proves the perfect space synchronization of the cells with a chaotic dynamics (Fig. 10d).

This correspondence between the synchronization index and the 2D Fourier transform opens the possibility to characterize a large series of experiments quantifying the influence of the parameters and the network topologies on complex dynamics.

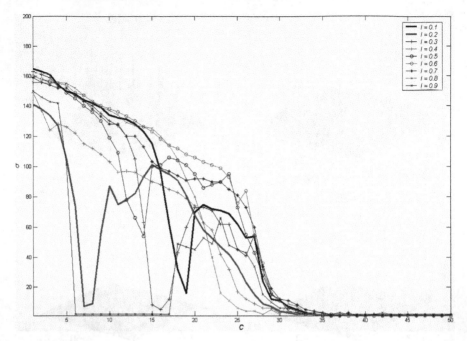

Fig. 11. Synchronization index σ versus number of connections C in regular fuzzy chains varying l

In order to enhance the self-synchronization properties of lattices in relation to the chaotic dynamics of the fundamental cell, the Lyapunov exponent l has been varied in a range from 0.1 to 0.9. In Fig. 11 the behavior of the synchronization index in the regular chains for different Lyapunov exponents versus the number of connections (C) is described; different collective behaviors are clearly distinguished. As the value of the Lyapunov exponent increases, the transition from regularization to chaotic synchronization occurs with different trends underling nonlinear effects affecting chain behaviors, and the following considerations can arise:

- The spatiotemporal chaos occurs with a relatively small number of connections for all values of the Lyapunov exponent.
- The regular synchronized behavior takes place at different degree for different ranges of the number of connections, although the perfect synchronization is for $l = 0.9$.
- The transition phase is not always present, sometimes varying the number of connections the spatiotemporal chaos becomes the synchronized dynamics following a smooth trend.
- The chaotic synchronization is typically quantified for all values of the Lyapunov exponent when the number of connections is $C > 30$.

Regular   Small-world   Random

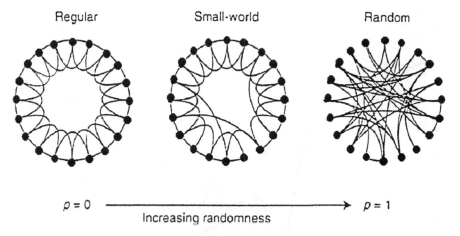

$p = 0$ $\longrightarrow$ $p = 1$

Increasing randomness

Fig. 12. Small world rewiring scheme [5]

This index does not give a direct distinction between the chaotic synchronization and the regular one, but by the frequecy analysis of the time series it is possible to have this information quickly.

4 The Collective Behavior Versus the Network Topology

The evolution of high-order complex fuzzy systems has been studied characterizing the effects of the topology on the collective dynamics. The structure of the regular fuzzy chain has been modified varying the structure of the network connections continuously starting from the regular topology and ending to the random one, passing through different small worlds topologies.

Regular arrays are highly clustered and their path length grows linearly with the number of elements N, whereas networks with random connections are poorly clustered but the mean distance between two vertices grows logarithmically with N. The introduction of randomness in a regular chain, obtained through the small worlds rewiring procedure, interpolates from a regular network to a random one with the same number of vertices and edges; an example of a ring changing is shown in Fig. 12.

The initial configuration is the previous regular array with N vertices, each connected to the same number $2C$ of neighbors. The rewiring is performed choosing a vertex and an edge and reconnecting them with probability p, and the end of the bond to a vertex is chosen uniformly random. All the duplicate edges are forbidden and all the vertices are processed in clockwise direction. All small worlds chains have many vertices with sparse connections, but the graphs never become disconnected guaranteeing the relations (5), as follows:

$$N \gg 2C \gg \ln(N) \gg 1 \tag{5}$$

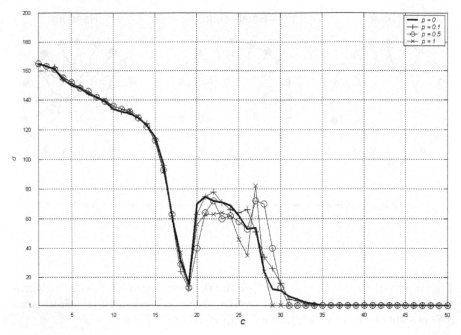

Fig. 13. Synchronization index σ versus number of connections C in fuzzy chains with $l = 0.1$ varying the topology: regular ($p = 0$), small worlds ($p = 0.1, p = 0.5$), and random ($p = 1$)

This procedure moves the graph structure between regularity ($p = 0$) and disorder ($p = 1$) obtaining new structural properties. An extensive description of the small world theory is reported in the Appendix A.

Small values of the small-worlds rewiring probability p generate few rewired connections and give birth to interesting nonlinear effects: the mean distance of each pair of vertices decreases, but the small number of removed edges gives clustering characteristics at local level as regular networks. The random topology with probability $p = 1$ modifies all the connections.

The analysis of the synchronization performances have been carried out building chains of $N = 200$ fuzzy units and with the Lyapunov exponents l from 0.1 to 0.9 through different topologies. The probability of rewiring p has been investigated in the entire range from the regular topology value ($p = 0$) to the random topology one ($p = 1$).

In Figs. 13–16 the synchronization index of these fuzzy chains with several values of the Lyapunov exponent is reported in the case of four rewiring probabilities ($p = 0$, $p = 0.1$, $p = 0.5$, and $p = 1$).

The trends of the synchronization index versus the number of connections for $l = 0.1$, $l = 0.3$, $l = 0.6$, and $l = 0.9$ are reported in Figs. 13–16, respectively, and the following considerations may arise:

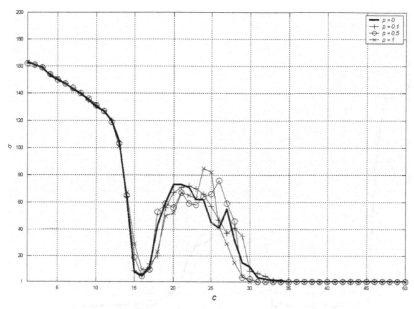

Fig. 14. Synchronization index σ versus number of connections C in fuzzy chains with $l = 0.3$ varying the topology: regular ($p = 0$), small worlds ($p = 0.1$, $p = 0.5$), and random ($p = 1$)

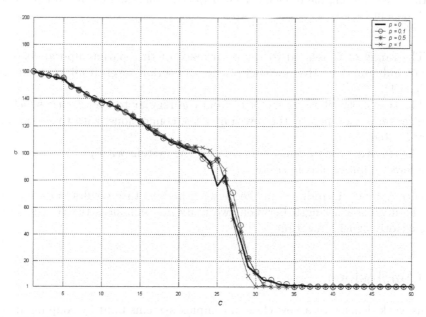

Fig. 15. Synchronization index σ versus number of connections C in fuzzy chains with $l = 0.6$ varying the topology: regular ($p = 0$), small worlds ($p = 0.1, p = 0.5$), and random ($p = 1$)

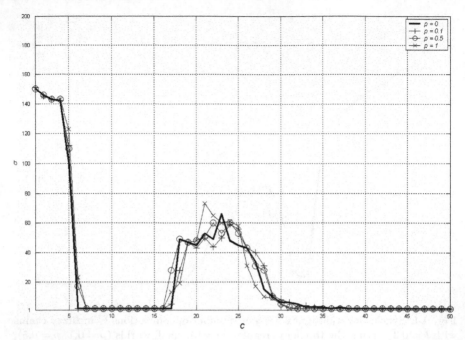

Fig. 16. Synchronization index σ versus number of connections C in fuzzy chains with $l = 0.9$ varying the topology: regular ($p = 0$), small worlds ($p = 0.1, p = 0.5$), and random ($p = 1$)

- The range of C related to the occurrence of the spatiotemporal chaos is not influenced by the rewiring of the connections even if all are modified ($p = 1$).
- The transition phase range of C from the regular to the chaotic synchronization is modified by the rewiring of the connections, with the exception of $l = 0.6$.
- The chaotic synchronization occurs for the system with $p = 1$ considering a fewer number of connections.

The complexity introduced in the network architecture results in decreasing of the mean path length between two units and enhances the transition to the chaotic synchronization.

5 Conclusions

This work deals with a new class of complex systems built by coupling fuzzy logic chaotic units and the study of the their collective spatiotemporal behavior, focusing the attention on the synchronization features. Both qualitative and quantitative analyses of generated patterns have been carried out through the introduction and characterization of a mathematical indicator. A regular one-dimensional array has been built by fixing the Lyapunov exponent value of the fuzzy chaotic oscillators and connecting them. Through the visual inspection and the frequency analysis four types of global behaviors have been identified and characterized distinguishing their patterns as follows: the spatiotemporal chaos, the regular synchronized behavior, the transition phase, and the chaotic synchronized behavior.

A synchronization index that weights easily the slight difference among a wide number spatiotemporal patterns is defined and validated. By means of this quantitative indicator the spatiotemporal analyses have been extended by varying in a suitable range both the Lyapunov exponent and the number of connections giving a more complete and detailed study. This synchronization index speeds the procedure for the comparison and the characterization of the complex network features versus the adopted system parameters, otherwise a direct inspection of all the spatiotemporal maps should be necessary for each parameter variation with a heavy computational strength.

By increasing the number of connections it is possible to distinguish specific ranges for each of the four types of collective behaviors. In order to enhance the self-synchronization properties of lattices in relation to the chaotic dynamics of the fundamental cell the Lyapunov exponent l has been varied in a range from 0.1 to 0.9.

As the value of the Lyapunov exponent increases, the transition from regularization to chaotic synchronization occurs with different trends underling nonlinear effects affecting chain behaviors. The regular synchronized behavior takes place for different ranges of the number of connections; the transition phase is not always present; the chaotic synchronization is typically exhibited for all values of the Lyapunov exponent when the number of connections is $C > 30$.

In the last phase, the structure of the regular fuzzy chain has been modified varying the structure of the network connections continuously starting from the regular topology and ending to the random one, passing through the different small worlds topologies.

The trends of the synchronization index versus the number of connections underline as the complexity introduced in the network architecture results in decreasing of the mean path length between two units and enhances the transition to the chaotic synchronization. Moreover, the transition phase range of the connection number from regular to chaotic synchronization is affected by the rewiring of the connections.

This study extends the principles of synchronization, auto-organization, and pattern formation to this new class of complex fuzzy systems suitable to describe the complex real-world phenomena not easy to describe using mathematical structures.

Appendix

Watts and Strogatz [3] have proposed, in 1998, a new complex network architecture characterized by a regular lattices rewired in order to introduce increasing amounts of disorder: the small-worlds theory. Many real networks are characterized by topologies between order and randomness. A famous manifestation of small world features is the so-called "six-degrees of separation" principle by the psychologist Milgram [9]. This concept is based on the notion that everyone in the world is connected to everybody else through a chain of six mutual connections. These complex systems are realized to exhibit both high clustered structures and reduced path lengths.

Two main parameters characterize the anatomy of a generic network and both are fundamental to investigate the dynamic behavior of the whole system:

- average path length and
- clustering coefficient.

Average path length: The distance between two cells in a generic network is the number of edges along the shortest path between them. The average path length L is defined as the mean of all the distances between two nodes and it gives information about the size of the whole network. The higher the average path length, the higher is the separation between every pair of cells.

Clustering coefficient: The clustering coefficient c is defined as the fraction of pairs of neighbors of a node which are neighbors to each others. It is evident that in a large network the cells connected to a particular cell can be connected with each others too. This is evident if we consider networks formed by people with a friendship connection. The friend of a friend is usually a friend. Opposite values of the clustering coefficient can be calculated for massive and random networks.

The structural characteristics of the networks are fundamental to investigate the collective behaviors in high dimensional systems constituted by nonlinear units communicating to reach a common macrobehavior. In recent studies networks topologies are investigated focusing, in particular, on four architectures that can be described by using a mathematical model:

- regular coupled networks
- random graphs
- small worlds

Regular coupled networks: Networks like chains, grids, and fully connected graphs give typical examples of regular structures. The regular configuration is often adopted to model high-dimensional system to give an order structure to a complex behavior. This hypothesis focuses the attention on the dynamic features of the single node when it is both isolated or connected to the others. The node could be a generic dynamical system with a stable point, a limit cycle, or a chaotic behavior. Regular lattices are highly clustered and their path length grows linearly with the number of edges N.

Random graphs: Erdos and Renyi (ER) [10] have studied random graphs by varying the number of nodes and links. An ER random graph is obtained imaging a set of buttons on the floor. With probability p every pair of buttons is tied with a thread. According to the value of probability p the graph properties change. The average path length of a random graph is $L = p(N - 1)$. For small values of p, graphs are constituted by separated components. As p, the number of links, increases the network becomes more compact. When $p \cong \ln N/N$ the system can be considered as a unique entity. The clustering coefficient of a random graph is equal to $p = L/N \ll 1$. This means that in a friendship network the probability that two friends are friends themselves is not greater than the probability that two randomly chosen persons are connected.

The ER networks are homogeneous systems whose connectivity follows a Poisson distribution. Each node in the graph is connected to each other node through short paths and the maximum "degree of separation" grows with $\log N$. Random graphs are an ideal model but due to their building simplicity these are used to model gene networks, ecosystems, and spread of diseases [11, 12].

Small worlds: The introduction of randomness in regular lattices, obtained through a rewiring procedure, interpolates from a regular network to a random one with the same number of vertices and edges. This is the case of small worlds systems in which in a regular topology are identifiable as small amounts of irregular connections. Watts and Strogatz (WS) [5] have introduced a new class of networks based on small worlds features that tune a graph between a regular lattice and a random network. Small world networks have two fundamental features: short paths and high clustering.

The starting configuration is a ring configuration with N vertices, each connected to the same number $2C$ of neighbors. The rewiring is realized by choosing a vertex and an edge and reconnecting, with probability p, the end of the bond to a vertex chosen uniformly random over the ring. All the duplicate edges are forbidden and all the vertices are processed in a clockwise direction. This procedure tunes the graph between regularity ($p = 0$) and disorder ($p = 1$) obtaining interesting structural properties (see Fig. 12).

In small worlds networks, small values of the rewiring probability p allow the introduction of few far-connections that give birth to interesting nonlinear effects. The mean distance of each pair of vertices decreases, but the number of removed edges is small enough to maintain the same regular network clustering characteristics at local level.

Average path length and clustering coefficient of small worlds versus rewiring probability is shown in Fig. 17. The functional study of small-worlds connectivity for dynamics systems shows an enhancement of network speed and exchange of information.

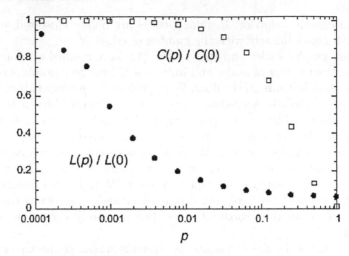

Fig. 17. Average path length L and clustering coefficient c versus rewiring probability in small worlds networks. The values are normalized by the values of a regular topology $L(0)$ and $c(0)$ [5]

Acknowledgments

The authors are grateful to Prof. Salvo Baglio for his helpful suggestions and encouragement to develop this study. The activity is partially supported by the Italian "Ministero dell'Istruzione, dell'Università e della Ricerca" (MIUR) under the FIRB Project RBNE01CW3M_001.

References

1. Y. Bar-Yam: *Dynamics of Complex Systems* (Addison-Wesley, Reading, MA, 1997)
2. A. Pikosky, M. Rosenblum, J. Kurths: *Synchronization. A Universal Concept in Nonlinera Sciences* (Cambridge University Press, Cambridge, UK, 2001)
3. S.H. Strogatz: Nature **410** (2001)
4. D.J. Watts: *Small Worlds* (Princeton University Press, Princeton, NJ, 1999)
5. J. Watts, S.H. Strogatz: Nature **393** (1998)
6. P. Arena, R. Caponetto, L. Fortuna, A. Rizzo, M. La Rosa: Int. J. Bifurcation Chaos **5** (2000)
7. G. Manganaro, S. Baglio, L. Fortuna: Electron. Lett. **32** (1996)
8. L. Fortuna, M.La Rosa, D. Nicolosi, G. Sicurella: spatiotemporal dynamics towards self-synchronization index. In: *Proc. 12th Int. IEEE Workshop on Nonlinear Dynamics of Electronic Systems*, Evora, Portugal, 2004
9. S. Milgram, *Psychol. Today* **2** (1967)
10. P. Erdos, A. Rényi: Publ. Math. Inst. Hung. Acad. Sci. **5** (1960)

11. R. May: *Stability and Complexity in Model Ecosystems* (Princeton University Press, Princeton, NJ, 1973)
12. S.A. Kauffman: J. Theor. Biol. **22** (1969)

Real-Time Identification and Forecasting of Chaotic Time Series Using Hybrid Systems of Computational Intelligence

Yevgeniy Bodyanskiy and Vitaliy Kolodyazhniy

Abstract. In this chapter, the problems of identification, modeling, and forecasting of chaotic signals are discussed. These problems are solved with the use of the conventional techniques of computational intelligence as radial basis neural networks and learning neuro-fuzzy architectures, as well as novel hybrid structures based on the Kolmogorov's superposition theorem and using the neo-fuzzy neurons as elementary processing units. The need for the solution of the forecasting problem in real time poses higher requirements to the processing speed, so the considered hybrid structures can be trained with the proposed algorithms having high convergence rate and providing a compromise between the smoothing and tracking properties during the processing of nonstationary noisy signals.

1 Introduction

In many fields of physics, biology, chemistry, economics, medicine, etc. there is a wide class of deterministic nonlinear systems whose behavior seems to be random, but is not in fact. Moreover, statistical analysis of signals generated by such systems (second moments, autocorrelation functions, spectra) indicates that they are broadband random processes generated by a deterministic object. This phenomenon is quite paradoxical in itself.

Such systems are referred to as chaotic, and are the subject of close attention for both the theoreticians and experts in many quite different fields [1–8].

Although a chaotic process generated by a nonlinear deterministic system looks like a stochastic one, it is actually not stochastic. Its most important feature is the extreme sensitivity to the initial conditions, i.e., if the same system is started from the initial conditions $x(0)$ and $x(0)+\varepsilon$, where ε is a very small value, then its trajectories will diverge exponentially with time, tending to totally different domains of attraction referred to as "strange attractors." Using a more rigorous definition, we can say that a strange attractor is an attracting set in the phase space in which the chaotic trajectories are moving, but is neither an equilibrium point nor a boundary cycle.

In principle, future behavior of a chaotic system is completely determined by its past because of its determinism, but in practice any uncertainty or imprecision in the selection of initial conditions dramatically complicate the problem of analysis. Recently, artificial neural networks and fuzzy inference

Y. Bodyanskiy and V. Kolodyazhniy: *Real-Time Identification and Forecasting of Chaotic Time Series Using Hybrid Systems of Computational Intelligence*, StudFuzz **187**, 439–480 (2006)
www.springerlink.com

systems due to their universal approximation properties and learning capabilities have been increasingly popular for the analysis and modeling of chaotic systems [9–14].

Aside from the chaotic motion itself, some other types of behavior similar to chaos are closely related to nonlinear dynamic systems, first of all:

- transient chaos, which is a motion that looks purely chaotic on a finite time interval, i.e., the trajectory first develops by the strange attractor, but then transforms into a periodic or quasiperiodic motion;
- quasiperiodic oscillations, which are the oscillations with two or more aliquant frequencies;
- bifurcations, which are abrupt changes of character of motion with a slight change of one or several system parameters;
- mixtures like "chaos + quasiperiodic oscillations", and many more.

What all the considered types of behavior have in common is their fractal structure, i.e., the self-similarity of the analyzed processes in different spatial and temporal scales. In principle, any substantially nonlinear dynamic system can demonstrate a chaotic behavior with a certain combination of its parameters. However, in practice, quite a limited number of such structures were studied and are used. Among them, the most popular are as follows:

- Logistic equation describing the growth of a biological population,

$$x(k + 1) = wx(k)\left(1 - x(k)\right), \quad 0 \le w \le 4, \quad 0 \le x(0) \le 1, \qquad (1)$$

where $k = 0, 1, 2, \ldots$ is the current discrete time;
- A modification of (1) [2, 4]

$$\begin{cases} x(k+1) = w_1 x(k) - w_2 x^2(k)\,, \\ x(k+1) = x^2(k) + \theta\,, \\ x(k+1) = 1 - w_1 x^2(k) + w_2 x(k-1)\,, \\ x(k+1) = wx^3(k) + (1-w)x(k), \quad \text{and so on}\,, \end{cases} \qquad (2)$$

where w_1, w_2, θ are some scalar parameters;
- The Mackey–Glass model

$$x(k + 1) = w_1 x(k) + \frac{w_2 x(k - \tau)}{1 + x^{10}(k - \tau)}, \quad \tau \ge 17\,, \qquad (3)$$

where τ is a time delay;
- Two-dimensional Henon mapping

$$\begin{cases} x(k+1) = 1 - w_x x^2(k) + y(k)\,, \\ y(k+1) = w_y x(k)\,; \end{cases} \qquad (4)$$

- Mandelbrot equations

$$\begin{cases} x(k+1) = x^2(k) - y^2(k) + \theta_x\,, \\ y(k+1) = 2x(k)y(k) + \theta_y\,; \end{cases} \qquad (5)$$

- May model
$$x(k+1) = x(k) \exp(w(1 - x(K))) \ ; \tag{6}$$

- Holmes model
$$\begin{cases} x(k+1) = y(k) \ , \\ y(k+1) = -w_x x(k) + w_y y(k) - y^3(k) \ ; \end{cases} \tag{7}$$

- Carrey–Yorke model
$$\begin{cases} x(k+1) = wx(k) \ , \\ y(k+1) = y(k) + x^2(k) \end{cases} \tag{8}$$

and some other using trigonometric functions [13]
$$\begin{cases} x(k+1) = \sin(k + \sin k^2) \ , \\ x(k+1) = x(k) + w \sin 2\pi x(k), \ \text{and so on} \ . \end{cases} \tag{9}$$

Benoit Mandelbrot [5] introduced the set of the so-called fractal equations
$$\begin{cases} x(k+1) = wx(k)(1 - x(k)) \ , \\ x(k+1) = x^2(k) + \theta \ , \\ x(k+1) = w(x(k) + \frac{1}{x(k)}) \ , \\ x(k+1) = wx^2(k) + \theta \ , \\ x(k+1) = wx^3(k) + \theta \ , \end{cases} \tag{10}$$

in which $x(k)$ is the complex variable generating some quite complex fractal geometrical structures.

As we already mentioned, chaos superficially resembles a random process, but is not random. Thus the problem of modeling and identification of signals appeared, which consists in the determination of where the nature of the signals is random or deterministically chaotic. In practice, several methods are used: from the simplest like the power spectrum and Poincaré mapping to the more sophisticated, related to Lyapunov numbers, fractal dimension, and the Hurst exponent.

One of the characteristic features of a chaotic system is the occurrence of a wide spectrum of frequencies at its output when a harmonic or a constant signal is fed to the input. Although this spectrum superficially resembles that of the white noise, the autocorrelation function of a chaotic process, in contrast to the delta function of the white noise, does not look like a single spike.

The Poincaré mapping is the phase-plane portrait of the system on the plane $x(k)$ and $\Delta x(k) = x(k+1) - x(k)$. If this mapping is "spread" all over the phase plane, then we have a stochastic process. If we see a deterministic curve, than we have chaos.

The Lyapunov exponents are used to check the sensitivity of the system to the variations of the initial conditions. If the initial conditions in the phase space are specified as a vector $x(0)$ defined in the hypersphere with a small radius ε, then this hypersphere will evolve with time into an ellipsoid with the maximum semiaxis

$$\varepsilon(k) = \varepsilon 2^{\lambda k} , \qquad (11)$$

where λ is the Lyapunov exponent determined by the expression

$$\lambda(x(0), \varepsilon) = \lim_{N \to \infty} \frac{1}{N} \lg \frac{\|x(N)\|}{\|x(0)\|} . \qquad (12)$$

A positive Lyapunov exponent indicates that we have a chaotic signal.

The fractal dimension index characterizes the geometrical structure of the strange attractor, and is a special measure of filling of the phase space by the phase portrait of the identified signal. Note that a fractional fractal dimension is the main indication of the presence of chaos.

Following the results [13], consider an arbitrary nonlinear dynamic system, and, having outlined some state y in the neighborhood of its attractor, describe a small hypersphere of radius ε around this point. Then define the distribution function of the observations with respect to y as

$$\rho(y) = \lim_{N \to \infty} \frac{1}{N} \sum_{k=1}^{N} \delta(y - x(k)) , \qquad (13)$$

where $\delta(y - x(k))$ is the n-dimensional delta function and N is the number of observations.

It is also worthy to note that the function $\rho(y)$ of the strange attractor is in a way similar to the distribution function of a random variable.

Let us introduce some function $f(y)$ such that

$$\int_{-\infty}^{\infty} f(y)\rho(y) \, dy = f < \infty , \qquad (14)$$

being a measure of change in the number of points in the hypersphere while its radius ε tends to zero.

For the points inside the hypersphere, the following condition holds true

$$\|y - x(k)\| < \varepsilon \qquad (15)$$

or, what is the same,

$$\varepsilon - \|y - x(k)\| > 0 . \qquad (16)$$

At the same time, the function $f(\cdot)$ can be defined as

$$f(x) = \left(\frac{1}{N-1} \sum_{\substack{l=1 \\ l \neq k}}^{N} h(\varepsilon - \|y - x(l)\|) \right)^{q-1} \qquad (17)$$

where q is a nonnegative integer number and $h(x)$ is the Heaviside function

$$h(x) = \begin{cases} 1 & \text{if } x \geq 0 , \\ 0 & \text{otherwise .} \end{cases} \tag{18}$$

Substituting then (13) and (17) into (14), we obtain the expression

$$R(q, \varepsilon) = \int_{-\infty}^{\infty} \left(\frac{1}{N-1} \sum_{\substack{l=1 \\ l \neq k}}^{N} h(\varepsilon - ||y - x(l)||) \right)^{q-1} \left(\frac{1}{N} \sum_{k=1}^{N} \delta(y - x(k)) \right) dy , \tag{19}$$

which, taking into account the obvious relation for the delta function

$$\int_{-\infty}^{\infty} f(y)\delta(y - x(k))dy = f(x(k)) , \tag{20}$$

can be rewritten as

$$R(q, \varepsilon) = \frac{1}{N} \sum_{k=1}^{1} \left(\frac{1}{N-1} \sum_{\substack{l=1 \\ l \neq k}}^{N} h(\varepsilon - ||x(k) - x(l)||) \right)^{q-1} . \tag{21}$$

The function (21) is referred to as correlation, and has a meaning of probability that two points $x(k)$ and $x(l)$ in the neighborhood of the attractor are situated at the distance of ε from each other. The limit behavior of this function when $\varepsilon \to 0$ is described by the relation

$$R(q, \varepsilon) = \varepsilon^{(q-1)D_q} , \tag{22}$$

where the exponent D_q is referred to as the fractal dimension of the attractor. Taking the logarithm of (22), we can formally define the fractal dimension as the expression

$$D_q = \lim_{\varepsilon \to 0} \frac{\log R(q, \varepsilon)}{(q-1) \log \varepsilon} , \tag{23}$$

which for $q = 223$ assumes a simple form often used for calculations and called in this case the correlation dimension of the attractor D_2.

2 Identification of Chaotic Signals in Real Time Using the Hurst Exponent

The methods for the identification of chaos considered above have one common disadvantage: they cannot be used in real time. As a consequence, it is difficult to determine the moments when the properties of the observed signals are changing. This difficulty can be overcome with the index popular in the analysis of fractal time series, which is referred to as Hurst exponent H [3]. This index is related to the correlation function and is used for the estimation of the chaotic or stochastic character of the identified time series.

The Hurst exponent for an arbitrary time series can be computed as

$$\frac{S(k)}{\sigma(k)} = (\alpha k)^H , \tag{24}$$

where $S(k)$ is the range of the sequence of accumulated values $y(l, k)$, which is calculated according to the expressions

$$\begin{cases} S(k) = \max_{1 \leq l \leq k} y(l, k) - \min_{1 \leq l \leq k} y(l, k) , \\ y(l, k) = \sum_{l=1}^{k} (x(l) - \bar{x}(l)) , \\ \bar{x}(l) = \frac{1}{l} \sum_{p=1}^{l} x(p) , \\ \bar{x}(k) = \frac{1}{k} \sum_{l=1}^{k} x(l); \end{cases} \tag{25}$$

$\sigma(k)$ is the standard deviation

$$\sigma(k) = \sqrt{\frac{1}{k} \sum_{l=1}^{k} (x(l) - \bar{x}(k))^2} ; \tag{26}$$

and α is a nonnegative parameter, chosen empirically in the general case.

As one can see, the Hurst exponent can be calculated from a sample of observations. The estimates obtained in such a way are averaged. For real-time calculation, the expressions (25) and (26) should be rewritten in a recursive form. It can be readily seen that

$$\bar{x}(k+1) = \bar{x}(k) + \frac{1}{k+1}(x(k+1) - \bar{x}(k)) , \tag{27}$$

$$\sigma^2(k+1) = \sigma^2(k) + \frac{1}{k+1}\left((x(k+1) - \bar{x}(k+1))^2 - \sigma^2(k)\right) , \tag{28}$$

$$y(l, k+1) = y(l, k) + (x(k+1) - \bar{x}(k+1)) , \tag{29}$$

$$\begin{aligned} y_{\max}(k+1) &= \max\left\{y_{\max}(k), y(l, k+1)\right\} \\ &= y_{\max}(k) - 0,5(1 - \text{sign}(y_{\max}(k) - y(l, k+1) \\ &\quad \times (y_{\max}(k) - y(l, k+1))) , \end{aligned} \tag{30}$$

$$\begin{aligned} y_{\min}(k+1) &= \min\left\{y_{\min}(k), y(l, k+1)\right\} \\ &= y_{\min}(k) - 0,5(1 - \text{sign}(y_{\min}(k) - y(l, k+1) \\ &\quad \times (y_{\min}(k) - y(l, k+1))), \end{aligned} \tag{31}$$

$$S(k+1) = y_{\max}(k+1) - y_{\min}(k+1) , \tag{32}$$

whence

$$H(k+1) = \frac{\log S(k+1) - \log \sigma(k+1)}{\log \alpha + \log(k+1)}. \tag{33}$$

Analysis of the expression (33) shows that the result of calculations strongly depends on the undetermined parameter α and the size of the available sample. This circumstance can lead to obtaining qualitatively opposite results for the same system.

This problem can be easily solved with neural network techniques. Let us rewrite (24) as

$$\log \frac{S(k)}{\sigma(k)} = H \log \alpha + H \log k, \tag{34}$$

and introduce the training signal

$$d(k) = \log \frac{S(k)}{\sigma(k)}, \tag{35}$$

and an adjustable structure like an adaptive linear associator

$$y(k) = \theta + H \log k \tag{36}$$

and use the Widrow–Hoff learning algorithm for artificial neural networks to find the estimates of the searched parameters:

$$\begin{pmatrix} \theta(k+1) \\ H(k+1) \end{pmatrix} = \begin{pmatrix} \theta(k) \\ H(k) \end{pmatrix} + \frac{d(k) - \theta(k) - H(k) \log k}{1 + (\log k)^2} \cdot \begin{pmatrix} 1 \\ \log k \end{pmatrix}, \tag{37}$$

$$\log \alpha(k+1) = \frac{\theta(k+1)}{H(k+1)}. \tag{38}$$

An architecture for the calculation of the Hurst exponent H and the parameter α is shown in Fig. 1.

Observing in real time the variations of the Hurst exponent, we can make conclusions about the nature of the analyzed signal. Oscillations of $H(k)$ about the level of 0.5 indicate that the time series is of stochastic nature; sharp deviations from this value are a sure sign of the occurrence of chaotic motion in the system.

3 Dynamic Reconstruction of Chaotic Signals with Known Structure

A more complex problem in the analysis of chaotic signals is the dynamic reconstruction, which consists in the recovery of the model generating the analyzed time series on the basis of the sample $(x(0), x(1), x(2), \ldots, x(k), \ldots)$. Here, similarly to the classical identification problem [15, 16], this problem can be considered in two aspects: parametric reconstruction, when the structure of the model is known and only the parameters must be recovered, and

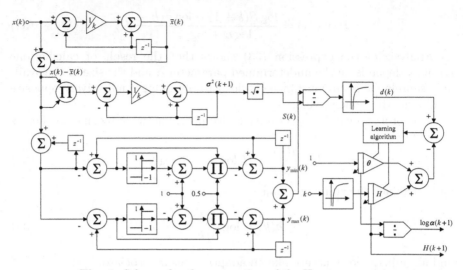

Fig. 1. Scheme for the computing of the Hurst exponent

structural reconstruction, when neither the structure nor the parameters are known a priori.

First consider the situation when the reconstructed signal is generated by the simplest Mandelbrot model (10)

$$x_c(k+1) = x_c^2(k) + \theta_c \,, \tag{39}$$

where

$$\begin{cases} x_c(k) = x_1(k) + ix_2(k) \,, \\ \theta_c = \theta_1 + i\theta_2, \quad i = \sqrt{-1} \,. \end{cases} \tag{40}$$

It is supposed that the parameters θ_1 and θ_2 are unknown.

Such a time series can be obtained using the elementary scheme shown in Fig. 2.

Having rewritten (39) as

$$x(k+1) = \begin{pmatrix} x_1(k+1) \\ x_2(k+1) \end{pmatrix} = \begin{pmatrix} x_1^2(k) - x_2^2(k) \\ 2x_1(k)x_2(k) \end{pmatrix} + \begin{pmatrix} \theta_1 \\ \theta_2 \end{pmatrix} \tag{41}$$

$$= \begin{pmatrix} \Psi_1(x_c(k)) \\ \Psi_2(x_c(k)) \end{pmatrix} + \begin{pmatrix} \theta_1 \\ \theta_2 \end{pmatrix} = \Psi(x_c(k)) + \theta,$$

then introducing the error vector

$$e(k) = \begin{pmatrix} e_1(k) \\ e_2(k) \end{pmatrix} = x_c(k)I_2 - x(k) = x_c(k)I_2 - \psi\left(x_c(k-1)\right) - \theta \tag{42}$$

and the learning criterion in which $I_2 = [1,1]^T$

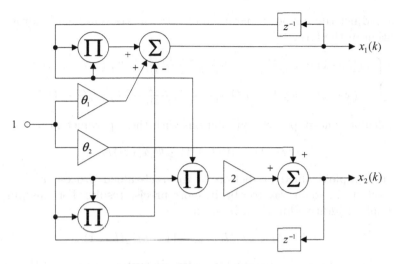

Fig. 2. Chaos generator

$$E(k) = \frac{1}{2}\|e(k)\|^2 = \frac{1}{2}\|\tilde{x}(k) - \theta\|^2 , \qquad (43)$$

we can write the recursive procedure for its minimization

$$\theta(k+1) = \theta(k) + \eta(k)\left(\tilde{x}(k) - \theta(k)\right) , \qquad (44)$$

where

$$\tilde{x}(k) = x(k)I_2 - \Psi(x_c(k-1)), \quad I_2 = (1,1)^{\mathrm{T}} . \qquad (45)$$

It can be readily seen that (45) coincides with the Kohonen learning rule [17]; at the same time the recovered parameters can be used to obtain a pair of predicted values of the observed time series as

$$\hat{x}(k+1) = \Psi(x_c(k)) + \theta(k+1) . \qquad (46)$$

Consider a more complicated structure of a complex-valued chaotic process

$$
x(k+1) = \begin{pmatrix} x_1(k+1) \\ x_2(k+1) \end{pmatrix} = \begin{pmatrix} w_{11}\left(x_1^2(k) - x_2^2(k)\right) + \theta_1 \\ w_{21}x_1(k)x_2(k) + \theta_2 \end{pmatrix}
$$

$$
= \begin{pmatrix} w_{11}\Psi_1(x_c(k)) + \theta_1 \\ w_{21}\Psi_2(x_c(k)) + \theta_2 \end{pmatrix} = \begin{pmatrix} w_{11} & \theta_1 \\ w_{21} & \theta_2 \end{pmatrix} \odot \begin{pmatrix} \Psi_1(x_c(k)) & 1 \\ \Psi_2(x_c(k)) & 1 \end{pmatrix} \quad (47)
$$

$$
\times I_2 = W \odot \Psi(x_c(k))I_2 ,
$$

where $\odot$ is the symbol of Scott (component-wise) product.

Having rewritten (47) component-wise

$$
\begin{cases} x_1(k+1) = (w_{11}, \theta_1)(\Psi_1(x_c(k)), \ 1)^{\mathrm{T}} = w_1\Psi_1(X_c(k)) , \\ x_2(k+1) = (w_{21}, \theta_2)(\Psi_2(x_c(k)), \ 1)^{\mathrm{T}} = w_2\Psi_2(X_c(k)) , \end{cases} \qquad (48)
$$

we can adjust the unknown parameters with the conventional Widrow–Hoff algorithm in the form

$$\begin{cases} w_1(k+1) = w_1(k) + \eta \dfrac{x(k) - w_1(k)\Psi_1(x_c(k-1))}{1 + \Psi_1^2(x_c(k-1))} \Psi_1^{\mathrm{T}}(x_c(k-1)) , \\ w_2(k+1) = w_2(k) + \eta \dfrac{x(k) - w_2(k)\Psi_2(x_c(k-1))}{1 + \Psi_1^2(x_c(k-1))} \Psi_2^{\mathrm{T}}(x_c(k-1)) \end{cases} \tag{49}$$

and calculate one-step ahead predictions with these parameters

$$\hat{x}(k+1) = W(k+1) \odot \Psi(x_c(k))I_2 . \tag{50}$$

The complex-valued prediction of the real-valued process $x(k)$ can be "contracted" in some way to obtain more precise results. For this purpose, the simplest additive form can be used

$$\hat{x}(k+1) = c\hat{x}_1(k+1) + (1-c)\hat{x}_2(k+1) , \tag{51}$$

where c is a nonnegative parameter, determining the accuracy of the prediction.

Introducing $((k+1) \times 1)$ vectors of signals and errors

$$\begin{cases} X(k) = (x(1), x(2), \dots, x(k))^{\mathrm{T}} , \\ \hat{X}(k) = (\hat{x}(1), \hat{x}(2), \dots, \hat{x}(k))^{\mathrm{T}} , \\ \hat{X}_i(k) = (\hat{x}_i(1), \hat{x}_i(2), \dots, \hat{x}(k))^{\mathrm{T}} , \quad i = 1, 2 , \\ V(k) = X(k) - \hat{X}(k) , \\ V_i(k) = X(k) - \hat{X}_i(k), \qquad\qquad i = 1, 2 , \\ V(k) = cV_i(k) + (1-c)V^2(k) \end{cases} \tag{52}$$

and solving the differential equation

$$\frac{\partial \|V(k)\|^2}{\partial c} = 0 , \tag{53}$$

we obtain the following two expressions:

$$\begin{cases} c(k) = V_2^{\mathrm{T}}(k) \dfrac{V_2(k) - V_1(k)}{\|V_2(k) - V_1(k)\|^2} , \\ 1 - c(k) = V_1^{\mathrm{T}}(k) \dfrac{V_1(k) - V_2(k)}{\|V_1(k) - V_2(k)\|^2} . \end{cases} \tag{54}$$

Substituting (54) into the last equation of (52), we obtain

$$\begin{aligned} V(k) &= V_2^{\mathrm{T}}(k) \frac{V_2(k) - V_1(k)}{\|V_2(k) - V_1(k)\|^2} V_1(k) + V_1^{\mathrm{T}}(k) \frac{V_1(k) - V_2(k)}{\|V_1(k) - V_2(k)\|^2} V_2(k) \\ &= \left(V_1(k)V_2^{\mathrm{T}}(k) - V_2(k)V_1^{\mathrm{T}}(k) \right) \frac{V_1(k) - V_2(k)}{\|V_1(k) - V_2(k)\|^2} , \end{aligned} \tag{55}$$

whence

$$\|V(k)\|^2 = \left(\left\|\left(\|V_2(k)\|^2 - V_1^{\mathrm{T}}(k)V_2(K)\right)V_1(k)\right.\right.$$
$$\left.\left. + \left(\|V_1(k)\|^2 - V_1^{\mathrm{T}}(k)V_2(k)\right)V_2(k)\right\|^2\right) \times \|V_1(k) - V_2(k)\|^{-4} . \quad (56)$$

Using (56), it is easy to obtain the system of inequalities

$$\begin{cases} \|V(k)\|^2 - \|V_1(k)\|^2 = -\frac{(\|V_1(k)\|^2 - V_1^{\mathrm{T}}(k)V_2(k))^2}{\|V_1(k) - V_2(k)\|^2} \leq 0 , \\[2mm] \|V(k)\|^2 - \|V_2(k)\|^2 = -\frac{(\|V_2(k)\|^2 - V_1^{\mathrm{T}}(k)V_2(k))^2}{\|V_1(k) - V_2(k)\|^2} \leq 0 , \end{cases} \quad (57)$$

indicating that the accuracy of the prediction (51) cannot be worse than the accuracy of any of the components of (50).

For real-time operation, a recursive form is required, which can be obtained by introducing new variables

$$\begin{cases} V_{21}(k) = V_2(k) - V_1(k) , \\ e_1(k+1) = x(k+1) - \hat{x}_1(k+1) , \\ e_2(k+1) = x(k+1) - \hat{x}_2(k+1) , \\ e_{21}(k+1) = e_2(k+1) - e_1(k+1) \end{cases} \quad (58)$$

and rewriting (54) as

$$\begin{cases} c(k+1) = \frac{\eta(k)}{\eta(k+1)}c(k) + \frac{e_2(k+1)e_{21}(k+1)}{\eta(k+1)} , \\[2mm] \eta(k+1) = \eta(k) + e_{21}^2(k+1). \end{cases} \quad (59)$$

Taking into account the following obvious expressions,

$$\begin{cases} V_{21}(k) = \hat{X}_1(k) - \hat{X}_2(k) , \\ e_{21}(k+1) = \hat{x}_1(k+1) - \hat{x}_2(k+1) \end{cases} \quad (60)$$

we finally obtain

$$\begin{cases} c(k+1) = \frac{\eta(k)}{\eta(k+1)}c(k) + \frac{e_2(k+1)\left(\hat{x}_1(k+1) - \hat{x}_2(k+1)\right)}{\eta(k+1)} , \\[2mm] \eta(k+1) = \eta(k) + \left(\hat{x}_1(k+1) - \hat{x}_2(k+1)\right)^2 \end{cases} \quad (61)$$

Figure 3 shows the scheme of the neuron for the dynamic reconstruction of the chaotic process according to (49), (50), and (61).

Similar schemes can be constructed for any of the chaotic models (1–10), assuming that the internal structure of the process is known.

4 Dynamic Reconstruction and Forecasting of Chaotic Signals with Radial Basis Function Networks

Consider the situation when there is no information available on the structure of the system that generates the chaotic signal, and there is only a sample of

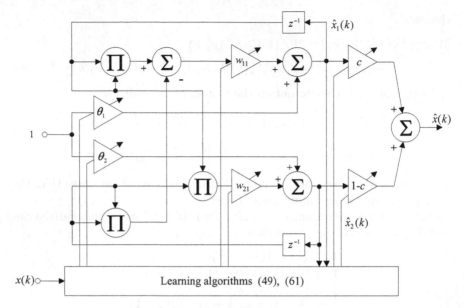

Fig. 3. Neuron for dynamical reconstruction of chaos

observations $x(k)$, distorted in the general case by an additive random noise $\xi(k)$ with the bounded second moment. In such a situation, there is a problem of structural reconstruction whose possibility of solution is determined by the so-called Taken's geometric delay embedding theorem [13]. This theorem states that the behavior of a nonlinear dynamic system can be approximated with sufficient accuracy by some nonlinear transformation $f(x(k))$ of the vector of observations $X(k) = (x(k-1), x(k-2), \ldots, x(k-d))^{\mathrm{T}} \equiv (x_1(k), x_2(k), \ldots, x_d(k))^{\mathrm{T}}$, where d is a positive integer number satisfying the condition

$$d \geq 2n + 1 \,, \tag{62}$$

where n is the dimensionality of the state vector of the system $x(k-i) \equiv x_i(k)$, $i = 1, 2, \ldots, d$.

To solve the problem of the dynamic structural reconstruction, it is natural to use artificial neural networks whose efficiency is explained first of all by their universal approximation properties in combination with a relatively compact representation of the modeled nonlinear system. In [14] it was proposed to use radial basis function (RBF) networks [18, 19] with one hidden layer with nonmonotonic (most often Gaussian) activation functions.

The most important advantage of the RBF networks consists in shorter training time as compared to the conventional multilayer networks trained by means of the error back-propagation technique. At the same time, the construction and training of an RBF network requires the use of at least two procedures. The first one is for the setting of the basis function parameters, and the second one is for the training of the output layer weights. Besides that, when the value of delay d is increased in order to improve the approximation properties, the so-called curse of dimensionality is encountered, which leads to exponential growth of the computational complexity and impossibility of processing of the incoming observations in real time.

To alleviate the above-mentioned problems, it is advisable to develop high-performance learning procedures for the RBF networks that would be able to tune not only the output layer weights in real time, but also the basis functions that is important for efficient identification of complex chaotic systems.

The architecture of the network under study is shown in Fig. 4 and corresponds to that of the generic RBF networks. We consider, without loss of generality, a one-output network.

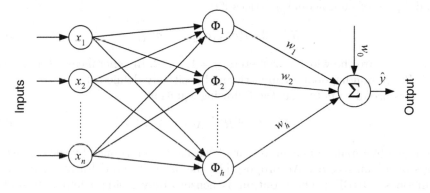

Fig. 4. Radial basis function network

The network output is calculated according to the following equation:

$$\hat{y}(k) = w_0 + \sum_{i=1}^{h} w_i \Phi_i \left(\left\| X(k) - C_i \right\|_{R_i^{-1}}^2 \right)$$

$$= \sum_{i=0}^{h} w_i \varphi_i(X(k)) = w^{\mathrm{T}} \varphi(X(k)),$$

$$\varphi_0(X(k)) \equiv 1, \quad \varphi_i(X(k)) = \Phi_i(\left\| X(k) - C_i \right\|_{R_i^{-1}}^2), \quad i = 1, \ldots, h,$$

$$X(k) = (x_1(k), x_2(k), \ldots, x_d(k))^{\mathrm{T}}, \quad w = (w_0, w_1, \ldots, w_h)^{\mathrm{T}},$$

$$\varphi(X(k)) = (1, \varphi_1(X(k)), \ldots, \varphi_h(X(k)))^{\mathrm{T}}, \tag{63}$$

where $\hat{y}(k)$ is the network output at time instant k, $X(k)$ is the d-dimensional input vector, h is the number of hidden layer units (basis functions), $\Phi_i(\cdot)$ is the ith basis function, C_i is the d-dimensional prototype vector (center) of the ith basis function, R_i^{-1} is the $(d \times d)$ matrix that determines the region of influence of the ith basis function, $\varphi_i(\cdot)$ is the ith generalized basis function, w is the vector of tunable synaptic weights, and $\varphi_i(\cdot)$ is the nonlinear regression vector.

The symbol $\|\cdot\|^2_{R_i^{-1}}$ stands for vector norm with respect to the matrix R_i^{-1}. This norm is calculated as follows:

$$\|X - C_i\|^2_{R_i^{-1}} = (x - C_i)^{\mathrm{T}} R_i^{-1}(x - C) \,. \tag{64}$$

The most widely used RBFs are the multidimensional Gaussian functions, in general form defined as

$$\Phi_i\left(\|X - C_i\|^2_{R_i^{-1}}\right) = \exp\left(-\frac{1}{2}\|X - C_i\|^2_{R_i^{-1}}\right), \tag{65}$$

and R_i are often diagonal matrices [20]:

$$R_i = \mathrm{diag}(\sigma^2_{i,1}, \sigma^2_{i,2}, \ldots, \sigma^2_{i,d}) \,, \tag{66}$$

where $\sigma_{i,j}$ are the width parameters, $j = 1, \ldots, d$. In general case, the matrices R_i are nondiagonal, symmetric, and positively defined [13].

It is useful to note that the following equation

$$(X - C_i)^{\mathrm{T}} R_i^{-1}(X - C_i) = 1 \tag{67}$$

defines a hyperellipsoid centered at C_i with axes determined by the eigenvectors of the matrix R_i. We propose [21] the use of the quadratic radial basis functions (QRBFs) whose output is nonzero only inside a hyperellipsoidal support (receptive field), i.e.,

$$\Phi_i\left(\left\|X - C_i\right\|^2_{R_i^{-1}}\right) = \max\left\{0, 1 - \left\|X - C_i\right\|^2_{R_i^{-1}}\right\}. \tag{68}$$

Note that in [22], the QRBFs are also used, but their receptive fields are hyperspherical, i.e., each basis function has equal radii in all dimensions. The use of hyperellipsoidal supports of arbitrary orientation, determined by the matrices R_i, will improve the approximation accuracy of the network. These receptive field matrices R_i, $i = 1, \ldots, h$, can be viewed as the covariance matrices of the corresponding clusters.

One-dimensional Gaussian function and a QRBF are shown in Fig. 5. The QRBF is localized, requires less computational efforts as compared with the Gaussian RBF, and allows us to derive efficient and yet simple learning procedures.

The task is to choose at every iteration k such parameters $w(k)$, $C_i(k), R_i^{-1}(k)$ that would minimize the local error function

$$E(k) = \frac{1}{2}(d(k) - \hat{y}(k))^2 = \frac{1}{2}e(k)^2 \,, \tag{69}$$

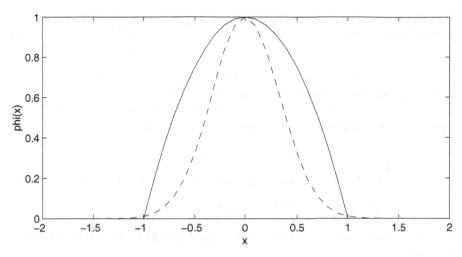

Fig. 5. One-dimensional QRBF with $R = 1$ (solid line) and a Gaussian function with $\sigma = 1/3$ (dashed line), defined on the same universe of discourse

where $d(k)$ is the target output and $e(k)$ is the modeling error at time instant k.

Consider the gradient-descent learning procedures

$$
\begin{cases}
w(k+1) = w(k) - \eta_w \nabla_w E(k) \,, \\
C_i(k+1) = C_i(k) - \eta_{C_i} \nabla_{c_i} E(k), & i = 1, \ldots, h \,, \\
R_i^{-1}(k+1) = R_i^{-1}(k) - \eta_{R_i^{-1}} \partial E(k)/\partial R_i^{-1}, & i = 1, \ldots, h \,,
\end{cases}
\tag{70}
$$

where $\nabla_w E, \nabla_{c_i} E$, are the gradients of the error function (69) with respect to the corresponding parameter vectors; $\partial E/\partial R_i^{-1}$ is the derivative of the error function with the respect to the matrix R_i^{-1}; and η_w, η_C, η_R are the learning rates.

For an arbitrary type of the basis functions $\Phi_i(\cdot)$, the gradients $\nabla_w E$, $\nabla_{c_i} E$ and the derivatives $\partial E/\partial R_i^{-1}$ are

$$
\begin{cases}
\nabla_w E(k) = -e(k)\varphi(x(k)) \,, \\[2mm]
\nabla_{c_i} E(k) = 2e(k)w_i \Phi_i'\left(\left\|X(k) - C_i(k)\right\|_{R_i^{-1}(k)}^2\right) \\
\qquad\qquad \times R_i^{-1}(k)(X(k) - C_i(k)) \,, \\[2mm]
\partial E(k)/\partial R_i^{-1} = -e(k)w_i \Phi_i'\left(\left\|X(k) - C_i(k)\right\|_{R_i^{-1}(k)}^2\right) \\
\qquad\qquad \times (X(k) - C_i(k))(X(k) - C_i(k))^{\mathrm{T}} \,,
\end{cases}
\tag{71}
$$

where $\Phi_i'(\cdot)$ is the derivative of the ith basis function with respect to its argument, which is given by the vector norm (64).

For the basis function (68), the derivative $\Phi_i'(\cdot)$ is as follows:

$$
\Phi_i'\left(\left\|X(k) - C_i(k)\right\|_{R_i^{-1}(k)}^2\right) = \begin{cases} -1, & \left\|X(k) - C_i(k)\right\|_{R_i^{-1}(k)}^2 < 1\,, \\ 0, & \left\|X(k) - C_i(k)\right\|_{R_i^{-1}(k)}^2 \geq 1. \end{cases} \quad (72)
$$

That is, the derivative $\Phi_i'(\cdot)$ of a QRBF equals -1 when the input vector $X(k)$ is inside the receptive field, and 0 otherwise.

Using the expressions (70) and (71), we obtain the complete set of learning procedures for all the parameters of a QRBF network:

$$
\begin{cases} w(k+1) = w(k) + \Delta w(k+1)\,, \\ C_i(k+1) = C_i(k) + \Delta C_i(k+1), & i = 1,\ldots,h\,, \\ R_i^{-1}(k+1) = R_i^{-1}(k) + \Delta R_i^{-1}(k+1), & i = 1,\ldots,h\,, \end{cases} \quad (73)
$$

where the update of the output layer weights $\Delta w(k)$ is calculated according to the equation

$$
\Delta w(k+1) = \eta_w e(k)\varphi(X(k))\,, \quad (74)
$$

and the updates for the hidden layer parameters $\Delta C_i(k+1)$ and $\Delta R_i^{-1}(k+1)$ are calculated as

$$
\Delta C_i(k+1) = -2\eta_C e(k)w_i(k)\Psi_i(k)(k)R_i^{-1}(k)(X(k) - C_i(k))\,, \quad (75)
$$
$$
\Delta R_i^{-1}(k+1) = -\eta_R e(k)w_i(k)\Psi_i(k)(X(k) - C_i(k))(X(k) - C_i(k))^{\mathrm{T}}\,, \quad (76)
$$

where $\Psi_i(k) = \Phi_i'(\left\|X(k) - C_i(k)\right\|_{R_i^{-1}(k)}^2), i = 1,\ldots,h.$

It can be readily seen from (73)–(76) that the parameters of the ith QRBF are updated only when its output is nonzero. This happens when the input vector is inside the corresponding receptive field. Thus, at every time instant only the QRBFs close to the current input are updated. So the learning procedures (73)–(76) resemble the competitive learning rule of the selforganizing maps [17].

It is also very important that the proposed learning procedures do not involve any calculations of nonlinear functions, and are completely based on linear operations.

The convergence properties of the introduced procedures depend on the learning rates η_w, η_c, and η_R. These learning rates can be selected heuristically by the trial and error method and depend on many factors, such as the size of the network and the variance of the inputs. Too small learning rates result in slow learning, and too big learning rates may result in oscillatory behavior or divergence of the learning procedures.

In [23], a fast adaptive learning procedure for a neuro-fuzzy network is proposed. The procedure is characterized by higher learning rate as compared with the conventional gradient descent learning, and is quite simple computationally at the same time. Since the RBF networks and neuro-fuzzy networks are functionally equivalent [24], similar learning methods can be used in both.

Using the procedure, described in [23], we can write the update for the output layer parameters:

$$
\begin{cases}
\Delta w(k+1) = r_w^{-1}(k)e(k)\varphi(X(k)) \, , \\
r_w(k) = \alpha_w r_w(k-1) + \left\| \varphi(X(k)) \right\|^2 \, ,
\end{cases}
\tag{77}
$$

where $0 \le \alpha_w \le 1$ is the forgetting factor for the weights and $r_w^{-1}(k)$ is the learning rate which is recursively computed at every iteration.

Introducing vectors

$$
g_i(k) = -2w_i(k)\Psi_i(k)(k)R_i^{-1}(k)(X(k) - C_i(k)) \, ,
\tag{78}
$$

we can use a similar procedure for the tuning of each QRBF center:

$$
\begin{cases}
\Delta C_i(k+1) = r_{C_i}^{-1}(k)e(k)g_i(k), \qquad i = 1, \ldots, h \, , \\
r_{C_i}(k) = \alpha_C r_{C_i}(k-1) + \left\| g_i(k) \right\|^2 \, ,
\end{cases}
\tag{79}
$$

where $0 \le \alpha_C \le 1$ is the forgetting factor for the centers and $r_{C_i}^{-1}(k)$ is the learning rate of the ith center.

For the tuning of the receptive field matrices R_i^{-1}, we propose the following modification of the procedure (79):

$$
\begin{cases}
\Delta R_i^{-1}(k+1) = r_{R_i}^{-1}(k)e(k)G_i(k), \quad i = 1, \ldots, h \, , \\
r_{R_i}(k) = \alpha_R r_{R_i}(k-1) + \mathrm{Tr}(G_i^{\mathrm{T}}(k)G_i(k)) \, ,
\end{cases}
\tag{80}
$$

where $0 \le \alpha_R \le 1$ is the forgetting factor for all the matrices R_i^{-1}, $r_{R_i}^{-1}(k)$ is the learning rate of the matrix R_i^{-1}, and the matrices G_i are computed as follows:

$$
G_i(k) = w_i(k)\Psi_i(k)(x(k) - C_i(k))(x(k) - C_i(k))^{\mathrm{T}} \, .
\tag{81}
$$

Substituting (78) and (81) into (79) and (80) respectively, we finally obtain

$$
\begin{cases}
\Delta C_i(k+1) = -2r_{C_i}^{-1}(k)e(k)w_i(k)\Psi_i(k)R_i^{-1}(k)(X(k) - C_i(k)), \ i = 1, \ldots, h \, , \\
r_{C_i}(k) = \alpha_C r_{C_i}(k-1) + 4w_i^2(k)\Psi_i^2(k)\|X(k) - C_i(k)\|^2_{R_i^{-2}(k)} \, ,
\end{cases}
\tag{82}
$$

where $\|X(k) - C_i(k)\|^2_{R_i^{-2}(k)} = (R_i^{-1}(k)(X(k) - C_i(k)))^{\mathrm{T}}(R_i^{-1}(k)(X(k) - C_i(k)))$, and

$$
\begin{cases}
\Delta R_i^{-1}(k+1) = r_{R_i}^{-1}(k)e(k)w_i(k)\Psi_i(k)(x(k) - C_i(k)) \\
\qquad\qquad\qquad \times (x(k) - C_i(k))^{\mathrm{T}}, \quad i = 1, \ldots, h \, , \\
r_{R_i}(k) = \alpha_R r_{R_i}(k-1) + w_i^2(k)\Psi_i^2(k)\|X(k) - C_i(k)\|^4 \, .
\end{cases}
\tag{83}
$$

In the learning procedures (82) and (83), each basis function has two individual learning rates for the center and the receptive field matrix. The learning rates are computed recursively at every time step and depend on the forgetting factors α_C and α_R. The forgetting factors can be selected in the same manner as for the linear identification procedures, usually between 0.95 and 1.0.

To demonstrate the applicability of the proposed algorithms, a QRBF network was trained online to predict a chaotic time series, generated by the well-known Mackey–Glass time-delay differential equation [1]:

$$\dot{x}(t) = \frac{0.2\,x(t-\tau)}{1+x^{10}(t-\tau)} - 0.1\,x(t) . \qquad (84)$$

The values of the time series at each integer point were obtained by means of the fourth-order Runge-Kutta method. The time step used in the method was 0.1, initial condition $x(0) = 1.2$, delay $\tau = 17$, and $x(k)$ was derived for $k = 0, \ldots, 51,000$. The values $x(k-18)$, $x(k-12)$, $x(k-6)$, and $x(k)$ were used as inputs ($d = 4$) to predict the value of $x(k+6)$.

The basis functions were created with the help of subtractive clustering [25] of 500 data points for $k = 118, \ldots, 617$. The clustering procedure discovered 10 clusters, whose parameters were used for the initialization of 10 hidden layer units ($h = 10$). We selected initial receptive fields as hyperspheres with the radius of $3\sigma = 0.4689$, where the parameter σ was found by the clustering procedure. Initial output layer weights were set to zero.

The results of the initialization of the hidden layer units are presented in Fig. 6, where the projections of the receptive fields and the chaotic process trajectory are shown on the phase plane $(x(k), x(k-6))$. The numbers from 1 to 10 indicate the centers of the corresponding basis functions.

We tested both the procedures with constant learning rate (74)–(76) and with adaptive learning rate (77), (82) and (83). At first, only the output layer weights were trained with corresponding procedures, while the hidden layer parameters remained unchanged after the initialization. Then we trained the output layer weights and the basis function centers, while the receptive field matrices remained unchanged. Finally, the weights, centers, and receptive field matrices were trained simultaneously. For comparison, we also used the exponentially weighted recursive least-squares method (RLS) for the adjustment of the output layer weights in combination with procedures (75), (76), (82)and (83).

In all the experiments, we trained the QRBF network with the same initial parameters online for $k = 118, \ldots, 50,117$. Then we used the trained network to predict the data points from the checking data set for $k = 50,118, \ldots, 50,617$. To estimate the forecast accuracy, we calculated the root mean squared error on the checking data (RMSE$_{\text{CHK}}$). The results are shown in Table 1.

In procedures (74)–(76), we used the following learning rates, found by the trial and error method: $\eta_w = 0.05, \eta_C = 0.05, \eta_R = 1$. In procedures (77),

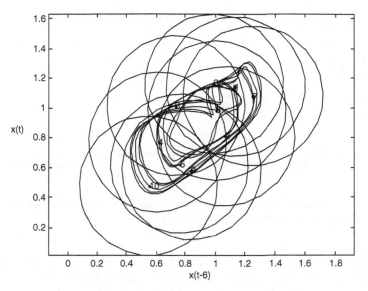

Fig. 6. Two-dimensional projections of the chaotic trajectory and initial receptive fields

(82), and (83) the forgetting factors $\alpha_w, \alpha_c, \alpha_R$ were equal to 0.99. The same value was used for the forgetting factor in the RLS method.

The best forecast accuracy was achieved with the procedures (77), (82), and (83) in the sixth experiment. The forecast of the data from the checking set is shown in Fig. 7. The solid line represents the real chaotic process, and the dashed line represents the forecast. The error is shown in the same figure under the real process and forecast plots, and is very small.

Table 1. Summary of experiments

	Adaptation Method			
No.	Weights	Centers	Receptive Fields	RMSE_{CHK}
1	(74)	None	None	0.0735
2	(74)	(75)	None	0.0196
3	(74)	(75)	(76)	0.0162
4	(77)	None	None	0.0547
5	(77)	(82)	None	0.0193
6	(77)	(82)	(83)	0.0089
7	RLS	None	None	0.0559
8	RLS	(75)	None	0.0261
9	RLS	(75)	(76)	0.0169
10	RLS	(82)	None	0.0175
11	RLS	(82)	(83)	0.0097

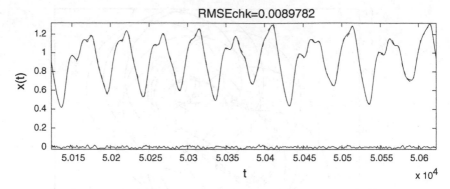

Fig. 7. Forecast of the Mackey–Glass time series after 50,000 time steps of online learning using the procedures (77), (82), and (83)

Interesting enough is that the use of the procedure (77) in the sixth experiment provided better result than the use of much more computationally expensive RLS method in experiment 11. The resulting projections of the receptive fields on the phase plane $(x(k), x(k-6))$ are presented in Fig. 8. Note that the QRBF centers have moved, and the shapes of the projections of the receptive fields transformed from circles into ellipses of arbitrary orientation.

Thus, experimental results confirm high forecast accuracy, provided by the considered nonconventional QRBF network. High performance of the network is achieved due to the use of efficient learning algorithms.

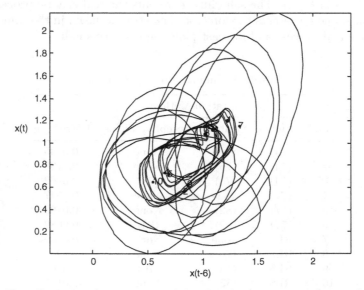

Fig. 8. Two-dimensional projections of the chaotic trajectory and receptive fields after 50,000 iterations of online learning using the procedures (77), (82), and (83)

5 Forecasting of Chaotic Sequences Using Neuro-Fuzzy Networks

Artificial neural networks are essentially a kind of learning "black box," whose properties are not revealed even during the learning process. Non-interpretability of the neural networks is the factor that limits their real-world applications. It also stimulates the emergence and development of the so-called neuro-fuzzy systems [26–28] within the soft computing paradigm.

They combine the linguistic interpretability of fuzzy systems with the learning and universal approximation capabilities of neural networks. Hence, they are an effective tool for the approximation of arbitrary functional relations.

Although a large number of advanced learning procedures for neuro-fuzzy systems have been proposed till now, many of them [26, 29–31] rely on the error back-propagation [32, 33] technique based on the gradient descent algorithm for parameter learning.

Along with its simplicity and satisfactory performance in solving many problems, it has some essential drawbacks that complicate its use for real-time information processing, such as slow convergence, sensitivity to noise, and dependence of its performance on the heuristically selected learning rate.

In [34–36] it is shown that fuzzy systems possess universal approximation capabilities, i.e., they can be used to model various nonlinear systems, described by the following equation:

$$y(k) = f(X(k)) + \xi(k) \,, \tag{85}$$

where $y(k)$ is the output of the system, $X(k) = (x_1(k), \dots, x_d(k))^{\mathrm{T}}$ is the input vector, $k = 0, 1, 2, \dots$ is the discrete time index, d is the number of inputs, $f(\cdot)$ is an arbitrary function, unknown in the general case, and $\xi(k)$ is a stochastic disturbance with bounded second moment. We consider here, without loss of generality, a single-output system.

It is also assumed that the function $f(x)$ is defined on a hyperbox

$$x_j(k) \in [x_j^{\min}, x_j^{\max}], \quad j = 1, \dots, d \,, \tag{86}$$

where $x_j^{\min}$ and $x_j^{\max}$ are the known lower and upper bounds of the jth input.

Let the system (62) be approximated by a model

$$\hat{y}(k) = \hat{f}(x(k)) \,, \tag{87}$$

where $\hat{f}(\cdot)$ is the function approximating $f(\cdot)$.

Below we will consider the implementation of the model (87) as a Takagi–Sugeno fuzzy model [37] with linear rule consequents. The universal approximation properties of such models were proven in [38, 39]. The Takagi–Sugeno models can also be trained with the learning algorithms similar to those of artificial neural networks [29]. Such fuzzy systems with neural network-like learning capabilities are called neuro-fuzzy systems [27, 31]. They can be used to model unknown nonlinear functional relations in the similar way as neural networks, retaining at the same time the structure of fuzzy rules, which is much better understandable than the knowledge in a trained neural network.

Let a neuro-fuzzy model contain n_R rules

$$\text{IF } x_1 \text{ IS } X_{1,i}^{(R)} \text{ AND} \ldots \text{ AND } x_d \text{ IS } X_{d,i}^{(R)} \text{ THEN } \hat{y} = f_i(X) \,, \qquad (88)$$

$$X_{1,i}^{(R)} = X_{j,R_{i,j}}, \quad i = 1,\ldots,n_R, \quad j = 1,\ldots,d \,,$$

where x_j is the jth input, $\hat{y}$ is the output of the model, $X_{j,i}^{(R)}$ is the linguistic label (fuzzy set [40]) of the jth input in the antecedent of the ith rule, R is the matrix $(n_R \times d)$ that determines the structure of the rule base, and $f_i(x)$ is the linear function in the consequent of the ith rule:

$$f_i(x) = \theta_{0,i} + \sum_{j=1}^{d} \theta_{j,i} x_j = \Theta_i^Y \tilde{X} \,, \qquad (89)$$

where $\Theta_i = (\theta_{0,i}, \theta_{1,i}, \ldots, \theta_{d,i})^{\mathrm{T}}$ is the vector of the consequent parameters of the ith rule, $\theta_{0,i}, \theta_{1,i}, \ldots, \theta_{d,i}$ are scalars, and $\tilde{X} = (1, X^{\mathrm{T}})^{\mathrm{T}}$ is the extended input vector. The discrete time index k is left out for simplicity in the description of the neuro-fuzzy model.

The linguistic value $X_{j,i}^{(R)}$ is defined by the membership function

$$\mu_{X_{j,i}^{(R)}}(x_j) = \begin{cases} \mu_{j,m}(x_j, \omega_{j,m}), & m = R_{i,j}, \quad R_{i,j} > 0 \,, \\ 1, & R_{i,j} = 0 \,, \end{cases} \qquad (90)$$

where $\mu_{j,m}$ is the mth membership function of the jth input, and $\omega_{j,m}$ is the vector of parameters of the membership function $\mu_{j,m}$.

Each row of the matrix R corresponds to one rule. The row index i in (90) corresponds to the number of a particular rule, while the column index j corresponds to the number of an input in the antecedent of that rule. The element $R_{i,j}$ of the matrix R determines the number of the membership function of the jth input in the antecedent of the ith rule. For example, if $R_{2,3} = 5$, then the 5th membership function of the 3rd input is used in the antecedent of the 2nd rule.

We will assume that all the membership functions are of the Gaussian type:

$$\mu_{j,m}(x_j, \omega_{j,m}) = \exp\left(-\frac{(x_j - c_{j,m})^2}{2\sigma_{j,m}^2}\right) \,, \qquad (91)$$

$$\omega_{j,m} = (c_{j,m}, \sigma_{j,m})^{\mathrm{T}} \,,$$

where $c_{j,m}$ and $\sigma_{j,m}$ are the center and width parameters of the mth membership function of the jth input, respectively.

The output of the model is calculated according to [41]:

$$\hat{y} = \frac{\sum_{i=1}^{n_R} w_i(X) f_i(X)}{\sum_{i=1}^{n_R} w_i(X)} = \sum_{i=1}^{n_R} \bar{w}_i(X) f_i(X) , \tag{92}$$

$$\bar{w}_i(X) = \frac{w_i(X)}{w_\Sigma(X)}, \qquad w_\Sigma(X) = \sum_{i=1}^{n_R} w_i(X) , \tag{93}$$

where $w_i(X)$ is the firing strength of the ith rule and $\bar{w}_i(X)$ is the normalized firing strength.

For the current input vector X, the output of the neuro-fuzzy model will be calculated as a combination of the outputs of the local linear models with the weights $\bar{w}_i(X)$ according to (92).

To simplify further transformations, let us introduce the following notation for the membership function of the jth input in the antecedent of the ith rule:

$$\mu_{j,i}^{(R)}(x_j) = \mu_{X_{j,i}^{(R)}}(x_j) , \tag{94}$$

and calculate the firing strength of the ith rule as

$$w_i(X) = \prod_{j=1}^{d} \mu_{j,i}^{(R)}(x_j) . \tag{95}$$

After that, introducing a vector $(n_R(d+1) \times 1)$ of the consequent parameters $\Theta = (\Theta_1^{\mathrm{T}}, \ldots, \Theta_{n_R}^{\mathrm{T}})^{\mathrm{T}}$ and an auxiliary input vector $\Phi = (\bar{w}_i(X)\tilde{X}^{\mathrm{T}}, \ldots, \bar{w}_{n_R}(X)\tilde{X}^{\mathrm{T}})^{\mathrm{T}}$, we can transform (92) into a more compact form:

$$\hat{y} = \Theta^{\mathrm{T}}\Phi . \tag{96}$$

The structure of the neuro-fuzzy network is defined by the matrix R. One membership function can contribute to the calculation of antecedents of several rules, but each rule will have its own linear model in the consequent part, and the values of the consequents are calculated independently for each rule. This is because the consequents (89) of the Takagi–Sugeno fuzzy rules (88) are local linear models, each model corresponding to a specific region in the input space, determined by the combination of the antecedent membership functions of that particular rule.

Each rule must have a unique combination of membership functions with not more than one membership function from each input. According to (88)–(95), the specific structure of the model (the structure of the connections between the first and the second layers which determine the input space partitioning) will be defined by the matrix R. Examples of matrices R and the corresponding rule bases and neuro-fuzzy model structures will be given below.

As shown in [27], the Takagi–Sugeno fuzzy systems can be represented as five-layer fuzzy neural networks. Examples of such neuro-fuzzy networks are

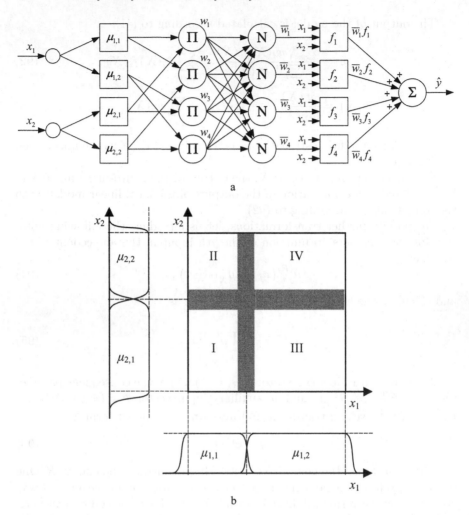

Fig. 9. Neuro-fuzzy network with grid partitioning of the input space: (**a**) Network architecture; (**b**)Membership functions and rule activation hyperboxes

shown in Figs. 9 and 10. The first layer is used for the fuzzification of the inputs. In the second layer, the firing strengths of fuzzy rules are calculated. The third layer is the normalization layer. In the fifth layer, the values of the rule consequents are calculated and multiplied by the normalized firing strength of the respective rules. The fifth layer performs the defuzzification, i.e., the crisp value of the output $\hat{y}$ is obtained according to (92).

In order to demonstrate the input space partitioning and the rule activation hyperboxes (numbered from I to IV), the networks depicted in Figs. 9 and 10 have only two inputs. In the general case, the number of membership

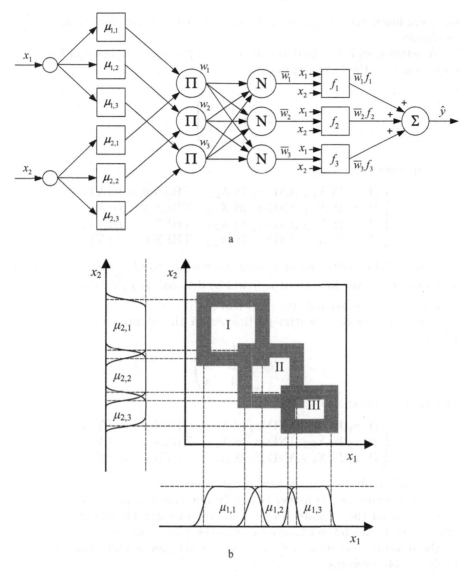

Fig. 10. Neuro-fuzzy network with scatter partitioning of the input space: (**a**) Network architecture; (**b**) Membership functions and rule activation hyperboxes

functions of each input can be arbitrary, and the number of the rules will be determined by the specific structure of the neuro-fuzzy network.

The tunable parameters in the neuro-fuzzy network shown in Figs. 9 and 10 are found only in the first and fourth layers. These are the antecedent parameters (of membership functions) and the consequent parameters (of the local linear models). The nodes with the tunable parameters are shown as squares.

A network with grid partitioning of the input space is shown in Fig. 9. It corresponds to the following matrix:

$$R = \begin{pmatrix} 1 & 1 \\ 1 & 2 \\ 2 & 1 \\ 2 & 2 \end{pmatrix}$$

and four rules:

$$\begin{cases} \text{IF } x_1 \text{ IS } X_{1,1} \text{ AND } x_2 \text{ IS } X_{2,1} & \text{THEN } \hat{y} = f_1(X), \\ \text{IF } x_1 \text{ IS } X_{1,1} \text{ AND } x_2 \text{ IS } X_{2,2} & \text{THEN } \hat{y} = f_2(X), \\ \text{IF } x_1 \text{ IS } X_{1,2} \text{ AND } x_2 \text{ IS } X_{2,1} & \text{THEN } \hat{y} = f_3(X), \\ \text{IF } x_1 \text{ IS } X_{1,2} \text{ AND } x_2 \text{ IS } X_{2,2} & \text{THEN } \hat{y} = f_4(X). \end{cases}$$

In the calculation of the firing strengths according to (95), the first membership function of the first input $\mu_{1,1}$ will be denoted by $\mu_{1,1}^{(R)}$ in the first rule, and by $\mu_{1,2}^{(R)}$ in the second rule [(90) and (94)].

The input space is scatter-partitioned in the network in Fig. 10. The matrix

$$R = \begin{pmatrix} 1 & 1 \\ 2 & 2 \\ 3 & 3 \end{pmatrix}$$

defines three rules:

$$\begin{cases} \text{IF } x_1 \text{ IS } X_{1,1} \text{ AND } x_2 \text{ IS } X_{2,1} & \text{THEN } \hat{y} = f_1(X), \\ \text{IF } x_1 \text{ IS } X_{1,2} \text{ AND } x_2 \text{ IS } X_{2,2} & \text{THEN } \hat{y} = f_2(X), \\ \text{IF } x_1 \text{ IS } X_{1,3} \text{ AND } x_2 \text{ IS } X_{2,3} & \text{THEN } \hat{y} = f_3(X). \end{cases}$$

Each membership function is used in only one rule.

The structure of the model is usually not known a priori. It can be determined from the available data by means of fuzzy clustering algorithms [25, 42], or it can be constructed during the process of learning, as is done in the resource-allocating [22], incremental [43], self-constructing [31], and evolving [44] networks.

Since the antecedent and consequent parameters of the neuro-fuzzy model are unknown in the general case, they must be determined by means of learning methods.

To simplify further transformations, let us introduce a vector ($n_\Omega \times 1$) of the antecedent parameters of the neuro-fuzzy model

$$\Omega = (\Omega_1^T, \dots, \Omega_d^T)^T, \tag{97}$$

where $\Omega_j = (\omega_{j,1}^T, \dots, \omega_{j,n_j^{(\mu)}}^T)^T$ is the vector $(n_j^{(\mu)} n_\omega \times 1)$ of the parameters of membership functions of the jth input, $n_j^{(\mu)}$ is the number of membership functions of the jth input, and n_ω is the number of parameter per membership function [$n_\omega = 2$ for the Gaussian function (91)].

The total number of the antecedent parameters is

$$n_\Omega = n_\omega \sum_{j=1}^{n_x} n_j^{(\mu)} . \tag{98}$$

The problem of learning in a neuro-fuzzy network consists in choosing the parameter vectors $\Omega(k)$ and $\Theta(k)$ that would minimize the error function at every iteration k

$$E(k) = \frac{1}{2} e^2(k) , \tag{99}$$

where $e(k) = y(k) - \hat{y}(k)$ is the modeling error, $y(k)$ is the actual output of the modeled system (target value, or the training sample) at time instant k, and $\hat{y}(k)$ is the output of the neuro-fuzzy model according to (96):

$$\hat{y}(k) = \Theta^{\mathrm{T}}(k-1)\Phi(k) , \tag{100}$$

where

$$\Phi(k) = (\bar{w}_1(X(k))\tilde{X}^{\mathrm{T}}(k), \ldots, \bar{w}_{n_R}(X(k))\tilde{X}^{\mathrm{T}}(k))^{\mathrm{T}} . \tag{101}$$

The tuning of the antecedent parameters is a nonlinear optimization problem, since the system output depends nonlinearly on the components of the vector Ω, as can be easily seen in (91)–(95). The learning in neuro-fuzzy networks is usually implemented via the error back-propagation [32], based on the gradient descent optimization method. But this method converges quite slowly. Much higher rate of convergence can be achieved with the second-order methods [45], such as the Hartley [46] or Marquardt algorithm [47]. The second-order procedures are usually used for batch learning, when the network parameters are updated once per number of training samples, comprising one training data set (usually several hundred samples or more). One cycle of batch learning is called an "epoch." The network parameters are updated at the end of every epoch.

Here we use the ideas of the second-order learning to derive adaptive real-time learning procedures. It is assumed that the network parameters are updated every time a training sample is presented. One iteration corresponds to one presentation of a training sample, followed by the immediate parameter update.

We can write the generalized Hartley–Marquardt algorithm for tuning of the antecedent parameters in the form

$$\Omega(k+1) = \Omega(k) + \lambda(J(k)J^{\mathrm{T}}(k) + \eta(k)L)^{-1}J(k)e(k) , \tag{102}$$

$$J(k) = \nabla_\Omega \hat{y}(k) = (\nabla_{\Omega_1}^{\mathrm{T}} \hat{y}(k), \ldots, \nabla_{\Omega_d}^{\mathrm{T}} \hat{y}(k))^{\mathrm{T}} , \tag{103}$$

where $J(k)$ is the gradient vector ($n_\Omega \times 1$) of the system output with respect to the parameters of all membership functions at time instant k, $\eta(k)$ is a scalar regularizing parameter, L is a diagonal positively defined regularizing matrix $(n_\Omega \times n_\Omega)$, and λ is a scalar gain.

The vector $J(k)$ consists of the gradients of the system output with respect to the parameters of the membership functions of all inputs. In turn, the gradient $\nabla_{\Omega_j} \hat{y}$ consists of the gradients of the system output with respect to the parameters of the membership functions of the jth input (the discrete time k is left out for clarity):

$$\nabla_{\Omega_J} \hat{y} = \left(\nabla^{\mathrm{T}}_{\omega_{j,1}} \hat{y}, \ldots, \nabla^{\mathrm{T}}_{\omega_{j,n_j^{(\mu)}}} \hat{y} \right)^{\mathrm{T}}$$

$$= \left(\frac{\partial \hat{y}}{\partial v_{j,1}}, \frac{\partial \hat{y}}{\partial \sigma_{j,1}}, \ldots, \frac{\partial \hat{y}}{\partial v_{j,n_j^{(\mu)}}}, \frac{\partial \hat{y}}{\partial \sigma_{j,n_j^{(\mu)}}} \right)^{\mathrm{T}}, \quad j = 1, \ldots, d. \quad (104)$$

The partial derivatives, constituting the gradients $\nabla_{\Omega_j} \hat{y}$, can be calculated using the chain rule, as is usually done with the back-propagation method [32] (the arguments of the functions are left out for simplicity):

$$\frac{\partial \hat{y}}{\partial c_{j,m}} = \left(\sum_{\substack{i=1 \\ \forall i (R_{i,j}=m)}}^{n_R} \left(\frac{\partial \hat{y}}{\partial \bar{w}_i} \cdot \frac{\partial \bar{w}_i}{\partial w_i} \cdot \frac{\partial w_i}{\partial \mu_{j,i}^{(R)}} \right) \right) \frac{\partial \mu_{j,m}}{\partial c_{j,m}}, \quad (105)$$

$$\frac{\partial \hat{y}}{\partial \sigma_{j,m}} = \left(\sum_{\substack{i=1 \\ \forall i (R_{i,j}=m)}}^{n_R} \left(\frac{\partial \hat{y}}{\partial \bar{w}_i} \cdot \frac{\partial \bar{w}_i}{\partial w_i} \cdot \frac{\partial w_i}{\partial \mu_{j,i}^{(R)}} \right) \right) \frac{\partial \mu_{j,m}}{\partial \sigma_{j,m}}, \quad (106)$$

where $j = 1, \ldots, d, m = 1, \ldots, n_j^{(\mu)}$,

$$\frac{\partial \hat{y}}{\partial \bar{w}_i} = f_i, \quad (107)$$

$$\frac{\partial \bar{w}_i}{\partial w_i} = \frac{\partial (w_i / w_\Sigma)}{\partial w_i} = \frac{w_\Sigma - w_i}{w_\Sigma^2}, \quad (108)$$

$$\frac{\partial w_i}{\partial \mu_{j,i}^{(R)}} = \prod_{\substack{n=1 \\ \forall n (n \neq j)}}^{d} \mu_{n,i}^{(R)}, \quad (109)$$

$$\frac{\partial \mu_{j,m}}{\partial c_{j,m}} = \frac{\mu_{j,m}}{\sigma_{j,m}^2} (x_j - c_{j,m}), \quad \frac{\partial \mu_{j,m}}{\partial \sigma_{j,m}} = \frac{\mu_{j,m}}{\sigma_{j,m}^3} (x_j - c_{j,m})^2. \quad (110)$$

Introducing the following notation

$$\delta_{j,m} = \sum_{\substack{i=1 \\ \forall i (R_{i,j}=m)}}^{n_R} \left(\frac{\partial \hat{y}}{\partial \bar{w}_i} \cdot \frac{\partial \bar{w}_i}{\partial w_i} \cdot \frac{\partial w_i}{\partial \mu_{j,i}^{(R)}} \right)$$

$$= \sum_{\substack{i=1 \\ \forall i (R_{i,j}=m)}}^{n_R} \left(f_i \frac{w_\Sigma - w_i}{w_\Sigma^2} \prod_{\substack{n=1 \\ \forall n (n \neq j)}}^{d} \mu_{n,i}^{(R)} \right), \quad (111)$$

and substituting (110) and (111) into (105) and (106), we obtain

$$\frac{\partial \hat{y}}{\partial c_{j,m}} = \delta_{j,m} \frac{\mu_{j,m}}{\sigma_{j,m}^2}(x_j - c_{j,m}), \quad \frac{\partial \hat{y}}{\partial \sigma_{j,m}} = \delta_{j,m}\frac{\mu_{j,m}}{\sigma_{j,m}^3}(x_j - c_{j,m})^2. \quad (112)$$

To reduce the computational complexity of (102), we can use the matrix inversion lemma:

$$(J\, J^{\mathrm{T}} + \eta\, L)^{-1} = \eta^{-1}L^{-1} - \frac{\eta^{-1}L^{-1}\, J\, J^{\mathrm{T}}\eta^{-1}L^{-1}}{1 + J^{\mathrm{T}}\eta^{-1}L^{-1}\, J}. \quad (113)$$

Having done the following evident transformations

$$\left(\eta^{-1}L^{-1} - \frac{\eta^{-1}L^{-1}\, J\, J^{\mathrm{T}}\, \eta^{-1}L^{-1}}{1 + J^{\mathrm{T}}\eta^{-1}L^{-1}\, J}\right) J$$

$$= \left(\eta^{-1}L^{-1} - \frac{\eta^{-2}\, L^{-1}J\, J^{\mathrm{T}}L^{-1}}{1 + \eta^{-1}J^{\mathrm{T}}L^{-1}\, J}\right) J$$

$$= \left(\frac{\eta^{-1}L^{-1} + \eta^{-2}\, J^{\mathrm{T}}L^{-1}J\, L^{-1} - \eta^{-2}\, L^{-1}J\, J^{\mathrm{T}}L^{-1}}{1 + \eta^{-1}J^{\mathrm{T}}L^{-1}\, J}\right) J$$

$$= \left(\frac{L^{-1} + \eta^{-1}J^{\mathrm{T}}L^{-1}J\, L^{-1} - \eta^{-1}L^{-1}J\, J^{\mathrm{T}}L^{-1}}{\eta + J^{\mathrm{T}}L^{-1}\, J}\right) J$$

$$= \frac{L^{-1}J + \eta^{-1}J^{\mathrm{T}}L^{-1}J\, L^{-1}J - \eta^{-1}L^{-1}J\, J^{\mathrm{T}}L^{-1}J}{\eta + J^{\mathrm{T}}L^{-1}\, J}$$

$$= \frac{L^{-1}J}{\eta + J^{\mathrm{T}}L^{-1}\, J} \quad (114)$$

and substituting the result of (114) into (102), we obtain [48]

$$\Omega(k+1) = \Omega(k) + \lambda\frac{L^{-1}\, J(k)}{\eta(k) + J^{\mathrm{T}}(k)L^{-1}\, J(k)}e(k). \quad (115)$$

When $\eta(k) = 0$ and L is the identity matrix, the (115) can be regarded as a nonlinear modification of the Widrow–Hoff algorithm [49]. Since the regularizing matrix L is usually diagonal, its inversion is trivial and does not increase the computational load significantly. When the matrix L is constant, its inverse can be computed beforehand. The transformation (114) eliminates the inversion of the matrix $(n_\Omega \times n_\Omega)$ in (102), and the resulting equation (10) is strictly equivalent to (102).

To provide smoothing properties to the learning algorithm, let us rewrite the denominator of (115) as follows:

$$\eta(k) = \alpha\left(\eta(k-1) + J^{\mathrm{T}}(k-1)L^{-1}\, J(k-1)\right) = \alpha\, q(k-1), \quad (116)$$
$$q(k) = \eta(k) + J^{\mathrm{T}}(k)L^{-1}\, J(k), \quad (117)$$

where $0 \leq \alpha \leq 1$ is a forgetting factor.

Substituting (116) and (117) into (115), we obtain

$$\begin{cases} \Omega(k+1) = \Omega(k) + \lambda\, q^{-1}(k)\, L^{-1}\, J(k)\, e(k)\,, \\ q(k+1) = \alpha\, q(k) + J^{\mathrm{T}}(k) L^{-1}\, J(k), \quad 0 \le \alpha \le 1. \end{cases} \tag{118}$$

It can be easily seen in (96) that $\Phi = \nabla_\Theta \hat{y}$, so we can write the learning algorithm for the consequent parameters similar to (118):

$$\begin{cases} \Theta(k+1) = \Theta(k) + \sigma\, r^{-1}(k)\, M^{-1}\, \Phi(k) e(k)\,, \\ r(k+1) = \beta\, r(k) + \Phi^{\mathrm{T}}(k)\, M^{-1}\, \Phi(k), \quad 0 \le \beta \le 1\,, \end{cases} \tag{119}$$

where σ and M are the scalar gain and the regularizing matrix, respectively.

To demonstrate the applicability of the proposed algorithms, a neuro-fuzzy network was trained to predict the Mackey–Glass chaotic time series [1]. In the experiment, the network was trained in two modes: batch sequential learning (parameters are updated as every training sample is presented, but the training is performed for many epochs on the same training set) and online learning (parameters are updated as every sample is presented, the training is performed for one "epoch," and the number of iterations equals the size of the training set in the batch mode times the number of epochs in the batch mode).

In the batch mode, 500 values from the generated data for $k = 118, \ldots, 617$ were used as the training data set, and the succeeding 500 values for $k = 618, \ldots, 1117$ were then used as the testing data set. A four input-one output neuro-fuzzy network with 2 Gaussian membership functions per input and 16 rules, forming a grid partition, was created. The network was trained for 100 epochs (500 iterations each) with the procedures (118), and (119). The parameters of the learning algorithm were as follows: $\alpha = \beta = 0.95$, $\lambda = \rho = 1$, L and M were the identity matrices of the corresponding dimensions. Initial values were $q(0) = 10,000$, $r(0) = 1$. The resulting plots are shown in Fig. 11. The two curves, representing the actual (thick line) and predicted (dashed line) values, are almost indistinguishable. The forecast error is shown in Fig. 11 as a thin line below the plots of the actual and predicted values. To estimate the quality of the results obtained, we used the root mean squared errors on the training data set ($\mathrm{RMSE_{TRN}}$) and testing data set ($\mathrm{RMSE_{CHK}}$).

The same network was also trained using the ANFIS learning rules [29]. The results were also compared with those obtained using the Neuro-Fuzzy Function Approximator (NEFPROX) in [50]. All the results are summarized in Table 2.

The ANFIS hybrid learning rule provides the best results among the methods. It employs the recursive least squares estimate (RLSE) for tuning of the consequent parameters, and is more complex computationally.

This complexity increases quadratically as the number of tuning parameters grows, in contrast to the proposed learning procedures (118) and (119),

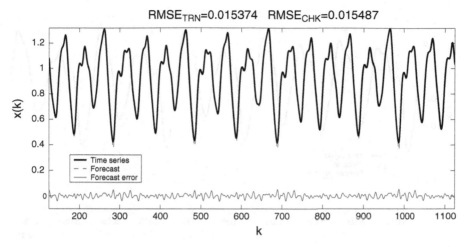

Fig. 11. The Mackey–Glass time series, six-step ahead forecast, and forecast error (batch learning, 100 epochs)

whose computational complexity increases linearly when the regularizing matrices L and M are diagonal. Thus, for systems with large number of tuning parameters, the proposed method should be preferred.

The NEFPROX system [50] also includes time-consuming procedures, while providing less accurate results in comparison to those provided by the proposed method (118), (119). In NEFPROX, which is based on the fuzzy perceptron [51] and is capable of extracting fuzzy rules from data, the emphasis is placed on the interpretability of the rules.

The other ANFIS learning rule, based only on the the back-propagation algorithm, was unable to provide satisfactory approximation after 100 training epochs. At the same time, this learning rule is more computationally expensive than the proposed procedures (118) and (119) with diagonal regularizing matrices, because the length of the step in the parameter space in the ANFIS back-propagation learning rule is calculated as the Euclidean norm, involving the operation of extraction of the square root.

In the online mode of learning, the same neuro-fuzzy network was trained with the procedures (118) and (119) for 50,000 iterations (50,000 training

Table 2. Comparison of different learning methods in the Mackey–Glass time series prediction

Method	# Epochs	# Rules	RMSE$_{TRN}$	RMSE$_{CHK}$
Hybrid ANFIS	100	16	0.0083246	0.0083428
Back-propagation ANFIS	100	16	0.30724	0.30717
NEFPROX	216	129	0.0315	0.0332

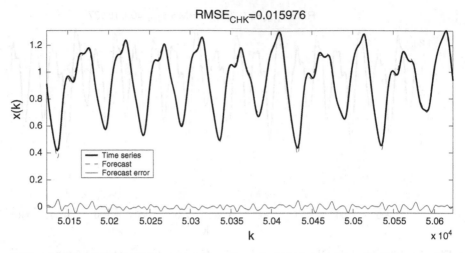

Fig. 12. The Mackey–Glass time series, six-step ahead forecast, and forecast error (online learning, 50,000 iterations)

samples for $t = 118, \ldots, 50{,}117$. The parameters of the learning procedures were the same as in the previous run. After 50,000 iterations the training was stopped, and the succeeding 500 points for $t = 50{,}118, \ldots, 50{,}617$ were used as the testing data set to compute the forecast (Fig. 12).

The ANFIS learning rule relies on the exponentially recursive least squares algorithm for the tuning of the consequent parameters. This algorithm is known to be unstable in the real-time operation, and the forgetting factor must be carefully chosen between 0.95 and 1.0 to avoid instability. Forgetting factors close to 1 may prevent the model from adequate tracking of the rapidly changing parameters. In this experiment, forgetting factors smaller than 0.999 caused numerical instability of the ANFIS learning algorithm. The proposed learning methods retain numerical stability for any values of the forgetting factors between 0 and 1.

6 Modeling and Forecasting of Chaotic Sequences Using Neo-Fuzzy Kolmogorov's Networks

The requirements for the improvement of the approximation properties as well as the acceleration of information processing call for the development of new architectures and learning algorithms that would outperform the conventional ones used in the problems of modeling and forecasting of chaotic signals.

In this section, a new architecture, called Fuzzy Kolmogorov's Network (FKN) containing two layers of neo-fuzzy neurons (NFNs) [52, 53] in both the hidden and output layer parameters, is constructed, so it can be trained with very fast and computationally efficient procedure. This architecture is based on the Kolmogorov's superposition theorem (KST) [54], stating that *any* continuous function of d variables can be *exactly* represented by superposition of continuous functions of one variable and addition:

$$f(X) = \sum_{l=1}^{2d+1} g_l \left[\sum_{j=1}^{d} \Psi_{l,j}(x_j) \right],$$

(120)

where $X \in [x_1^{\min}, x_1^{\max}] \times \cdots \times [x_d^{\min}, x_d^{\max}]$, $g_l(\cdot)$ and $\Psi_{l,j}(\cdot)$ are some continuous univariate functions, and $\Psi_{l,j}(\cdot)$ are independent of $f(\cdot)$. Aside from the exact representation, the KST can be used for the construction of parsimonious universal approximators, and has thus attracted the attention of many researchers in the field of soft computing.

Hecht-Nielsen was the first to propose a neural network implementation of KST [55], but did not consider how such a network can be constructed. Computational aspects of approximate version of KST were studied by Sprecher [56, 57]. Igelnik and Parikh [58] proposed the use of spline functions for the construction of Kolmogorov's approximation. Yam et al. [59] proposed the multiresolution approach to fuzzy control, based on the KST, and proved that the KST representation can be realized by a two-stage rule base. They demonstrated that the exponential growth in number of rules can be avoided via the two-stage fuzzy inference, but did not show how such a rule base could be created from data. Lopez-Gomez and Hirota developed the Fuzzy Functional Link Network (FFLN) [60] based on the fuzzy extension of the Kolmogorov's theorem. The FFLN is trained via fuzzy delta rule, whose convergence can be quite slow.

The KST-based universal approximator proposed here has a simple structure and optimal linear learning procedures with high rate of convergence. The FKN (Fig. 13) comprises two layers of NFNs [52] and is described by the following equations [61–63]:

$$\hat{y} = \hat{f}(X) = \hat{f}(x_1, \dots, x_d) = \sum_{l=1}^{h} f_l^{[2]}(o^{[1,l]}),$$

$$o^{[1,l]} = \sum_{j=1}^{d} f_j^{[1,l]}(x_j), \quad l = 1, \dots, h,$$

(121)

where h is the number of hidden layer neurons, $f_l^{[2]}(o^{[1,l]})$ is the lth nonlinear synapse in the output layer, and $o^{[1,l]}$ is the output of the lth NFN in the hidden layer, and $f_j^{[1,l]}(x_j)$ is the jth nonlinear synapse of the lth NFN in the hidden layer.

The equations for the hidden and output layer synapses are

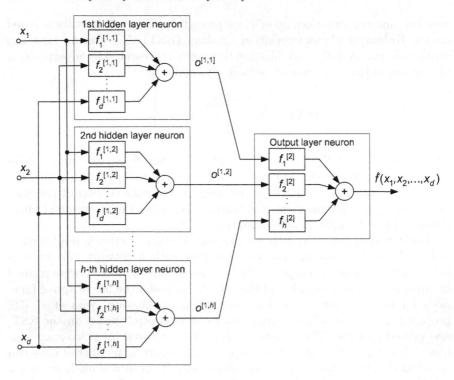

Fig. 13. FKN with d inputs and h neo-fuzzy neurons in the hidden layer

$$f_j^{[1,l]}(x_j) = \sum_{p=1}^{m_1} \mu_{j,p}^{[1]}(x_j) w_{j,p}^{[1,l]},$$

$$f_l^{[2]}(o^{[1,l]}) = \sum_{i=1}^{m_2} \mu_{l,i}^{[2]}(o^{[1,l]}) w_{l,i}^{[2]}, \quad l = 1,\dots,n, \quad j = 1,\dots,d, \quad (122)$$

where m_1 and m_2 are the number of membership functions per input in the hidden and output layers respectively, $\mu_{j,h}^{[1]}(x_j)$ and $\mu_{l,i}^{[2]}(o^{[1,l]})$ are the membership functions, and $w_{j,p}^{[1,l]}$ and $w_{l,i}^{[2]}$ are the tunable synaptic weights.

Nonlinear synapse is a single input–single output fuzzy inference system with crisp consequents, and is thus a universal approximator [34] of univariate functions. It can provide a piecewise-linear approximation of any functions $g_l(\cdot)$ and $\Psi_{l,j}(\cdot)$ in (120). So the FKN, in turn, can approximate any function $f(x_1,\dots,x_d)$ on the input space hyperbox.

The output of the FKN is computed as the result of two-stage fuzzy inference:

$$\hat{y} = \sum_{l=1}^{h} \sum_{i=1}^{m_2} \mu_{l,i}^{[2]} \left[\sum_{j=1}^{d} \sum_{p=1}^{m_1} \mu_{j,p}^{[1]}(x_j) w_{j,p}^{[1,l]} \right] w_{l,i}^{[2]}. \quad (123)$$

The description (123) corresponds to the following two-level fuzzy rule base:

IF x_j IS $X_{j,p}$ THEN $o^{[1,1]} = w_{j,p}^{[1,1]}d$ AND ... AND $o^{[1,h]} = w_{j,p}^{[1,h]}d$,

$$j = 1, \ldots, d, \quad p = 1, \ldots, m_1,\qquad (124)$$

IF $o^{[1,l]}$ IS $O_{l,i}$ THEN $\hat{y} = w_{j,p}^{[2]}h$, $\quad l = 1, \ldots, h, \quad i = 1, \ldots, m_2,$ (125)

where $X_{j,p}$ and $O_{l,i}$ are the antecedent fuzzy sets in the first and second level rules, respectively. Each first level rule contains n consequent terms $w_{j,p}^{[1,1]}d, \ldots, w_{j,p}^{[1,h]}d$, corresponding to h hidden layer neurons.

Total number of rules is

$$N_R^{\mathrm{FKN}} = d \cdot m_1 + h \cdot m_2,\qquad (126)$$

that is, it depends *linearly* on the number of inputs d.

The rule base is complete, as the fuzzy sets $X_{j,p}$ in (124) completely cover the input hyperbox with m_1 membership functions per input variable. Due to the linear dependence (126), this approach is feasible for input spaces with high dimensionality d without the need for clustering techniques for the construction of the rule base.

Straightforward grid-partitioning approach with m_1 membership functions per input produces $(m_1)^d$ fuzzy rules, leading to combinatorial explosion and being practically not feasible for $d > 4$.

The set of rules (124), (125) in an FKN can be interpreted as a grid-partitioned fuzzy rule base

IF x_1 IS $X_{1,1}$ AND ... AND x_d IS $X_{d,1}$ THEN $\hat{y} = \hat{f}(c_{1,1}^{[1]}, \ldots, c_{d,1}^{[1]})$,

$$\vdots$$

IF x_1 IS X_{1,m_1} AND ... AND x_d IS X_{d,m_1} THEN $\hat{y} = \hat{f}(c_{1,m_1}^{[1]}, \ldots, c_{d,m_1}^{[1]})$, (127)

with total of $(m_1)^d$ rules, whose consequent values are equal to the outputs of the FKN computed on all the possible d-tuples of prototypes of the input fuzzy sets $c_{1,1}^{[1]}, \ldots, c_{d,m_1}^{[1]}$.

The weights of the FKN are determined by means of a batch-training algorithm proposed in [62–64]. A training set containing N samples is used. The minimized error function is

$$E(t) = \sum_{k=1}^{N} [y(k) - \hat{y}(t, k)]^2 = [Y - \hat{Y}(t)]^{\mathrm{T}} [Y - \hat{Y}(t)],\qquad (128)$$

where $Y = [y(1), \ldots, y(N)]^{\mathrm{T}}$ is the vector of target values, and $\hat{Y}(t) = [\hat{y}(t, 1), \ldots, \hat{y}(t, N)]^{\mathrm{T}}$ is the vector of network outputs at epoch t.

Yamakawa et al. [52] proposed the use of gradient descent-based learning for the NFN. Although this method can be directly applied to the output layer, it would also require the use of the back-propagation technique [32] for the hidden layer. Besides that, the gradient descent-based learning procedure converges very slowly.

However, since the nonlinear synapses (122) are linear in parameters, we can employ direct linear least squares (LS) optimization instead of derivative-based methods. To formulate the LS problem for the output layer, rewrite (123) as

$$\hat{y} = W^{[2]^T} \varphi^{[2]}(o^{[1]}), \quad W^{[2]} = \left[w_{1,1}^{[2]}, w_{1,2}^{[2]}, \ldots, w_{n,m_2}^{[2]} \right]^T, \qquad (129)$$

$$\varphi^{[2]}(o^{[1]}) = \left[\mu_{1,1}^{[2]}(o^{[1,1]}), \mu_{1,2}^{[2]}(o^{[1,1]}), \ldots, \mu_{n,m_2}^{[2]}(o^{[1,n]}) \right]^T.$$

The LS solution will be

$$W^{[2]} = \left(\Phi^{[2]^T} \Phi^{[2]} \right)^{-1} \Phi^{[2]^T} Y^{[2]}, \quad Y^{[2]} = Y, \qquad (130)$$

where $\Phi^{[2]} = \left[\varphi^{[2]}(o^{[1]}(1)), \ldots, \varphi^{[2]}(o^{[1]}(N)) \right]^T$.

Now we have to determine the hidden layer weights. The use of triangular membership functions enables the linearization of the second layer around $o^{[1,l]}$:

$$f_l^{[2]}(o^{[1,l]}) = a_l^{[2]}(o^{[1,l]})o^{[1,l]} + b_l^{[2]}(o^{[1,l]}), \qquad (131)$$

$$a_l^{[2]}(o^{[1,l]}) = \frac{w_{l,q+1}^{[2]} - w_{l,q}^{[2]}}{c_{l,q+1}^{[2]} - c_{l,q}^{[2]}}, \quad b_l^{[2]}(o^{[1,l]}) = \frac{c_{l,q+1}^{[2]} w_{l,q}^{[2]} - c_{l,p}^{[2]} w_{l,q+1}^{[2]}}{c_{l,q+1}^{[2]} - c_{l,q}^{[2]}}, \qquad (132)$$

where $w_{l,q}^{[2]}$ and $c_{l,q}^{[2]}$ are the weight and center of the pth membership function in the lth synapse of the output layer, respectively. The membership functions in an NFN are chosen such that only two adjacent membership functions q and $q+1$ fire at a time [53].

With respect to (121), (123), and (131), we obtain the following expression for the linearized FKN:

$$\hat{y} = \sum_{l=1}^{n} \sum_{i=1}^{d} \sum_{h=1}^{m_1} a_l^{[2]}(o^{[1,l]}) \mu_{i,h}^{[1]}(x_i) w_{i,h}^{[1,l]} + \sum_{l=1}^{n} b_l^{[2]}(o^{[1,l]}). \qquad (133)$$

Rewrite the previous equation as follows:

$$\hat{y} = W^{[1]^T} \varphi^{[1]}(x) + \theta^{[1]}(x),$$

$$W^{[1]} = \left[w_{1,1}^{[1,1]}, w_{1,2}^{[1,1]}, \ldots, w_{d,m_1}^{[1,1]}, \ldots, w_{d,m_1}^{[1,h]} \right]^T,$$

$$\varphi^{[1]}(x) = \left[\varphi_{1,1}^{[1,1]}(x_1), \varphi_{1,2}^{[1,1]}(x_1), \ldots, \varphi_{d,m_1}^{[1,1]}(x_d), \ldots, \varphi_{d,m_1}^{[1,h]}(x_d) \right]^T,$$

$$\varphi_{j,p}^{[1,l]}(x_j) = a_l^{[2]}(o^{[1,l]}) \mu_{j,p}^{[1,l]}(x_j), \quad \theta^{[1]}(x) = \sum_{l=1}^{h} b_l^{[2]}(o^{[1,l]}). \qquad (134)$$

Introducing vector $\Theta^{[1]} = \left[\theta^{[1]}(x(1)), \ldots, \theta^{[1]}(x(N))\right]^{\mathrm{T}}$ and matrix $\Phi^{[1]} = \left[\varphi^{[1]}(x(1)), \ldots, \varphi^{[1]}(x(N))\right]^{\mathrm{T}}$ and noticing that $\hat{Y} = W^{[1]^{\mathrm{T}}}\Phi^{[1]} + \Theta^{[1]}$, we can formulate the LS problem for the hidden layer weights:

$$Y - (\Phi^{[1]}W^{[1]} + \Theta^{[1]}) = (Y - \Theta^{[1]}) - \Phi^{[1]}W^{[1]} = 0 . \tag{135}$$

The solution of the LS problem is

$$W^{[1]} = \left(\Phi^{[1]^{\mathrm{T}}}\Phi\right)^{-1}\Phi^{[1]^{\mathrm{T}}}Y^{[1]}, \quad Y^{[1]} = Y - \Theta^{[1]} . \tag{136}$$

The solutions (130) and (136) are not unique when matrices $\Phi^{[q]^{\mathrm{T}}}\Phi^{[q]}$ are singular ($q = 1$, 2 is the layer number). To avoid this, instead of (130) and (136) at every epoch t we find

$$W^{[q]}(t) = \left(\Phi^{[q]^{\mathrm{T}}}(t)\Phi^{[q]}(t)\right)^{+}\Phi^{[q]^{\mathrm{T}}}(t)Y^{[q]}(t) , \tag{137}$$

where $(\cdot)^{+}$ is the symbol of matrix pseudo-inverse [64].

The FKN is trained via a two-stage derivative-free optimization procedure without any nonlinear operations. In the forward pass, the output layer weights are calculated. In the backward pass, calculated are the hidden layer weights. The number of tuned parameters in the hidden layer is $S_1 = dm_1h$, in the output layer $S_2 = hm_2$, and total $S = S_1 + S_2 = h(dm_1 + m_2)$. Thus, in the forward pass, a matrix $S_2 \times S_2$ is inverted, and in the backward pass inverted is a matrix $S_1 \times S_1$. For comparison, the nonlinear LS methods, such as the Gauss–Newton and Levenberg–Marquardt procedures, require the inversion of a matrix $S \times S$. Since the number of calculations in matrix inversion is proportional to S^3 and it will always hold that $S^3 > S_1^3 + S_2^3$, the proposed training method is much faster.

Hidden layer weights are initialized deterministically using the formula

$$w_{p,j}^{[1,l]} = \exp\left\{-\frac{j\left[m_1(l-1)+p-1\right]}{d(m_1h-1)}\right\}, \tag{138}$$
$$p = 1, \ldots, m_1, \quad j = 1, \ldots, d, \quad l = 1, \ldots, h,$$

broadly similar to the parameter initialization technique based on rationally independent random numbers, which was proposed in [58] for the Kolmogorov's spline network. The output layer weights of the FKN are initialized with zeros.

We used the FKN in two experiments. In the first one, it was trained to predict the time series generated by the chaotic Mackey–Glass time-delay differential equation (84). From the generated data, 500 values for $t = 118, \ldots, 617$ were used as the training data set, and the next 500 for $t = 618, \ldots, 1117$ as the checking data set. The FKN used for prediction had 4 inputs as in the previous experiments, 9 neurons in the hidden layer with 3 membership functionsper input, and 1 neuron in the output layer with

Table 3. Results of six-step ahead prediction of the Mackey–Glass time series

Network	Parameters	Epochs	RMSE$_{TRN}$	RMSE$_{CHK}$
FKN	153	50	0.0028291	0.004645
MLP	145	50	0.002637	0.003987

5 membership functionsper synapse (153 adjustable parameters altogether). In the six-step ahead prediction, it demonstrated a performance similar to that of a two-hidden layer perceptron with 145 parameters trained with the Levenberg–Marquardt procedure. Both networks were trained for 50 epochs.

Root mean squared error on the training and checking sets (RMSE$_{TRN}$ and RMSE$_{CHK}$) was used to estimate the accuracy of predictions. The results are listed in Table 3. Actual time series, the prediction provided by the FKN, and prediction error are shown in Fig. 14.

It can be readily seen that the results are very close in accuracy, but the computational complexity of the FKN and its learning algorithms is about

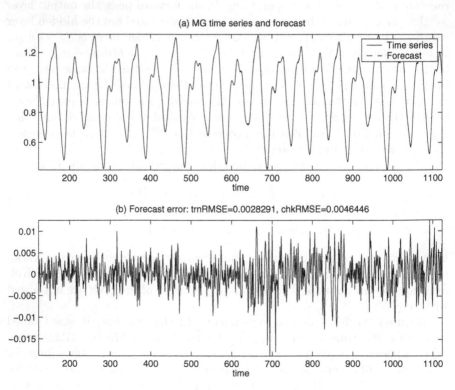

Fig. 14. Mackey–Glass time series prediction

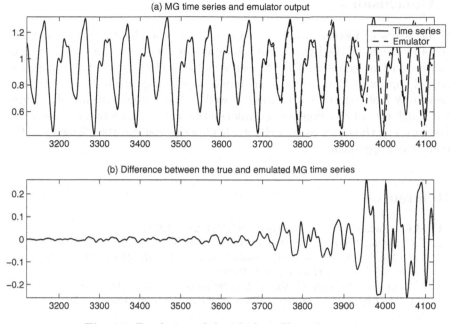

Fig. 15. Emulation of the Mackey–Glass time series

a order of magnitude lower than that of the Multi Layer Perceptron (MLP) with the Levenberg–Marquardt procedure. This circumstance makes the FKN more preferable for the identification and forecasting in real time.

The second experiment consisted in the emulation of the Mackey–Glass time series by the FKN. The values of the time series were fed into the FKN only during the training stage. three thousand values for $t = 118, \ldots, 3117$ were used as the training set. The FKN had 17 inputs corresponding to the delays from 1 to 17, 5 neurons in the hidden layer with 5 membership functions per input, and 1 neuron in the output layer with 7 membership functions per synapse (total 460 adjustable parameters). The FKN was trained to predict the value of the time series one step ahead.

The training procedure converged after 52 epochs with the final value of $\text{RMSE}_{\text{TRN}} = 0.000318$, and the last 17 values of the time series from the training set were fed to the inputs of the FKN. Then the output of the network was connected to its inputs through the delay lines, and subsequent 1,000 values of the FKN output were computed. As can be seen from Fig. 15, the FKN captured the dynamics of the real time series very well. The difference between the real and emulated time series becomes visible only after about 500 time steps. The emulated chaotic oscillations remain stable, and neither fade out nor diverge. In such a way, the FKN can be used for long-term chaotic time series predictions.

7 Conclusions

Some issues of identification and forecasting of the chaotic signals and systems by means of the hybrid computational intelligence techniques were considered in this chapter. The efficiency of the architectures that implement kernel approximation (radial basis function networks, neuro-fuzzy networks, neo-fuzzy Kolmogorov's networks) as well as specialized learning algorithms with high rate of convergence and information processing were demonstrated. Examples of the processing of chaotic time series confirm the efficiency of the considered approach.

References

1. M.C. Mackey, L. Glass: Science **197**, 287–289 (1977)
2. F.C. Moon: *Chaotic Vibrations* (Wiley, New York, 1987)
3. M.C. Mackey, L. Glass: *From Clocks to Chaos. The Rhythms of Life* (Princeton University Press, Princeton, NJ, 1988)
4. P. Berge, Y. Pomeau, C. Vidal: *L'ordre dans le Chaos* (Hermann, Paris, 1988)
5. B.B. Mandelbrot: *Die fraktale Geometrie der Natur* (Birkhaeuser Verlag, Basel, 1991)
6. E. Ott: *Chaos in Dynamical Systems* (Cambridge University Press, New York, 1993)
7. R.M. Crownower: *Introduction to Fractals and Chaos* (Jones and Bartlett Publishers, Boston, 1995)
8. H.I. Abarbanel, T.W. Frison, L.S Tsimring: IEEE Signal Process. Mag. **N3**, 49–65 (1998)
9. D.A. White, D.A. Sofge (eds): *Handbook of Intelligent Control Neural, Fuzzy, and Adaptive Approaches* (Van Nostrand Reinhold, New York, 1992)
10. S. Abe: *Neural Networks and Fuzzy Systems: Theory and Applications* (Kluwer Academic Publishers, Boston, 1997)
11. Da. Ruan (ed): *Intelligent Hybrid Systems: Fuzzy Logic, Neural Networks, and Genetic Algorithms* (Kluwer Academic Publishers, Boston, 1997)
12. Yu. Osana, M. Hagiwara: Successive learning in heteroassociative memories using chaotic neural networks. In: *Proc. IEEE Int. Joint Conf. on Neural Networks "IJCNN'98,"* Anchorage, Alaska, 1998, pp 1107–1112
13. S. Haykin: *Neural Networks. A Comprehensive Foundation* (Prentice Hall, Upper Saddle River, NJ, 1999)
14. I. Rojas, J. Gonzalez, A. Canas, A.F Diaz, F.J. Rojas, M. Rodriguez: Int. J. Neural Syst. **10**(N5), 353–364 (2000)
15. L. Ljung: *System Identification—Theory for the User.* 2nd edn (Prentice Hall, Upper Saddle River, NJ, 1999)
16. R. Isermann: *Digitale Regelsysteme* (Springer, Berlin Heidelberg New York, 1988)
17. T. Kohonen: *Self-Organizing Maps* (Springer, Berlin Heidelberg New York, 1995)
18. E.S. Chang, S. Chen, B. Mulgrew: IEEE Trans. Neural Networks **7**, 190–194 (1996)

19. S.A. Billings, X. Hong: Neural Networks **11**, 479–493 (1998)
20. O. Nelles, S. Ernst, R. Isermann: Automatisierungstechnik **6**, 251–262 (1997)
21. Ye. Bodyanskiy, O. Chaplanov, V. Kolodyazhniy: Adaptive quadratic radial basis function network for time series forecasting. In: *Proc. 10th East-West Fuzzy Colloquim*, Zittau, Germany, 2002, pp 164–172
22. J. Platt: Neural Comput. **3**, 213–225 (1991)
23. Ye. Bodyanskiy, V. Kolodyazhniy, A. Stephan: An adaptive learning algorithm for a neuro-fuzzy network. In: *Computational Intelligence. Theory and Applications*, ed by B. Reusch. Lecture Notes in Computer Science, vol 2206 (Springer, Berlin Heidelberg New York, 2001), pp 68–75
24. J.-S.R. Jang, C.-T.Sun: IEEE Trans. Fuzzy Syst. **4**, 156–159 (1993)
25. S. Chiu: J. Intelligent Fuzzy Syst. **2**, 267–278 (1994)
26. S. Horikawa, T. Furuhashi, Y. Uchikawa: IEEE Trans. Neural Networks **3**, 801–860 (1992)
27. J.-S. R. Jang, C.-T. Sun, E. Mizutani: *Neuro-Fuzzy and Soft Computing—A Computational Approach to Learning and Machine Intelligence* (Prentice Hall, Upper Saddle River, NJ, 1997)
28. C.-T. Lin, C.S.G. Lee: IEEE Trans. on Comput. **40**, 1320–1336 (1991)
29. J.-S. R. Jang: IEEE Trans. Syst. Man, Cybern. **23**, 665–685 (1993)
30. J.-S. R. Jang: Neuro-Fuzzy Modeling: Architectures, Analyses, and Applications, Ph.D. Dissertation, EECS Department, University of California at Berkeley (1992)
31. C.-F. Juang, C.-T. Lin: *IEEE Trans. Fuzzy Syst.* **6**, 12–32 (1998)
32. D.E. Rumelhart, G.R. Hinton, R.J. Williams: Learning internal representation by error propagation. In: *Parallel Distributed Processing*, ed by D.E. Rumelhart, J.L. McClelland (MIT Press, Cambridge, MA, 1986), pp 318–364
33. L.-X. Wang, J.M. Mendel: Back-propagation fuzzy systems as nonlinear dynamic system identifiers. In: *Proc. 1st IEEE Int. Conf. on Fuzzy Systems*, 1992, pp 1409–1416
34. B. Kosko: Fuzzy systems as universal approximators. In: *Proc. 1st IEEE Int. Conf. on Fuzzy Systems*, San Diego, CA, 1992, pp 1153–1162
35. V. Kreinovich, G.C. Mouzouris, H.T. Nguyen: Fuzzy rule based modeling as a universal approximation tool. In: *Fuzzy Systems: Modeling and Control*, ed by H.T. Nguyen, M. Sugeno (Kluwer Academic Publishers, Boston, 1998), pp 135–195
36. L.-X. Wang: Fuzzy systems are universal approximators. In: *Proc. 1st IEEE Int. Conf. Fuzzy Systems*, San Diego, CA, 1992), pp 1163–1170
37. T. Takagi, M. Sugeno: IEEE Trans. Syst., Man, Cybern. **15**, 116–132 (1985)
38. H. Ying: IEEE Trans. Fuzzy Syst. **6**, 582–587 (1998)
39. H. Ying: IEEE Trans. Syst., Man, Cybern. **28**, 515–520 (1998)
40. L.A. Zadeh: Fuzzy Sets, Inf. Control **8**, 338-353 (1965)
41. L.H. Tsoukalas, R.E. Uhrig: *Fuzzy and Neural Approaches in Engineering* (Wiley-Interscience, New York, 1997)
42. J.C. Bezdek: *Pattern Recognition with Fuzzy Objective Function Algorithms* (Plenum, New York, 1981)
43. B. Fritzke: Incremental neuro-fuzzy systems. In: *Proc. SPIE's Optical Science, Engineering and Instrumentation '97: Applications of Fuzzy Logic Technology IV*, San Diego, CA, 1997

44. Q. Song, N. Kasabov: Dynamic evolving neural-fuzzy inference system (DEN-FIS): On-line learning and application for time-series prediction. In: *Proc. 6th Int. Conf. on Soft Computing*, Iizuka, Fukuoka, Japan, 2000, pp 696–701

45. A.J. Shepherd: *Second-Order Methods for Neural Networks: Fast and Reliable Training Methods for Multi-Layer Perceptrons* (Springer-Verlag, London, 1997)

46. H. Hartley: Technometrics **3**, 269–280 (1961)

47. D.W. Marquardt: SIAM J. Appl. Math. **11**, 431–441 (1963)

48. P. Otto, Ye. Bodyanskiy, V. Kolodyazhniy: Integrated Computer-Aided Eng. **10**, 399–409 (2003)

49. B. Widrow, M.E. Hoff: IRE Western Electric Show and Convention Record Part 4, 96–104 (1960)

50. D. Nauck, R. Kruse: A neuro-fuzzy approach to obtain interpretable fuzzy systems for function approximation. In: *Proc. IEEE Int. Conf. on Fuzzy Systems*, 1998, pp 1106–1111

51. D. Nauck: A fuzzy perceptron as a generic model for neuro-fuzzy approaches. In: *Proc. 2nd German GI-Workshop Fuzzy-Systeme'94*, Munich, Germany, 1994

52. T. Yamakawa, E. Uchino, T. Miki, H. Kusanagi: A neo-fuzzy neuron and its applications to system identification and prediction of the system behavior. In: *Proc. 2nd Int. Conf. on Fuzzy Logic and Neural Networks "IIZUKA-92,"* Iizuka, Japan, 1992, pp 477–483

53. T. Miki, T. Yamakawa: Analog implementation of neo-fuzzy neuron and its on-board learning. In: *Computational Intelligence and Applications*, ed by N.E. Mastorakis (WSES Press, Piraeus, 1999), pp 144–149

54. A.N. Kolmogorov: Dokl. Akad. Nauk SSSR **114**, 953–956 (1957)

55. R. Hecht-Nielsen: Kolmogorov's mapping neural network existence theorem. In: *Proc. IEEE Int. Conf. on Neural Networks*, San Diego, CA, 1987, vol 3, pp 11–14

56. D.A. Sprecher: Neural Networks **9**, 765–772 (1996)

57. D.A. Sprecher: Neural Networks **10**, 447–457 (1997)

58. B. Igelnik, N. Parikh: IEEE Trans. Neural Networks **14**, 725–733 (2003)

59. Y. Yam, H.T. Nguyen, V. Kreinovich: Multi-resolution techniques in the rules-based intelligent control systems: A universal approximation result. In: *Proc. 14th IEEE Int. Symp. on Intelligent Control/Intelligent Systems and Semiotics ISIC/ISAS'99*, Cambridge, MA, Sept. 15–17, 1999, pp 213–218

60. A. Lopez-Gomez, S. Yoshida, K. Hirota: Int. J. Fuzzy Syst. **4**, 690–695 (2002)

61. V. Kolodyazhniy, Ye. Bodyanskiy: Universal approximators employing neo-fuzzy neurons. In: *Proc. 8th Fuzzy Days*, Dortmund, Germany, 2004 [CD-ROM]

62. V. Kolodyazhniy, Ye. Bodyanskiy: Fuzzy neural networks with Kolmogorov's structure. In: *Proc. 11th East-West Fuzzy Colloquium*, Zittau, Germany, 2004, pp 139–146

63. V. Kolodyazhniy, Ye. Bodyanskiy: Fuzzy Kolmogorov's network. In: *Proc. 8th Int. Conf. Knowledge-Based Intelligent Information and Engineering Systems International (KES 2004)*, Wellington, New Zealand, 2004, pp 764–771

64. A. Albert: *Regression and the Moore-Penrose Pseudoinverse* (Academic, New York, 1972)

Fuzzy–Chaos Hybrid Controllers
for Nonlinear Dynamic Systems

Keigo Watanabe, Lanka Udawatta, and Kiyotaka Izumi

Abstract. Controlling of chaos is an interesting research topic while employing of deterministic chaos for controlling is more interesting. This chapter focuses on employing and utilizing of inherent chaotic features in a nonlinear dynamical system in a useful manner. When it comes to employing deterministic chaos, there are tremendous advantages such as low-energy consumption, robustness of the controller performance, information security, and simplicity of employing chaos whenever it has chaotic attractive features in the original systems itself. If the original system does not have chaotic properties, deterministic chaos will be introduced to the system. Keeping these objectives, the control algorithm is constructed in order to control nonlinear systems, which exhibit chaotic behavior. We introduce two phases of control: First phase uses open-loop control forming a chaotic attractor or using chaotic inherent features in a system itself. Fuzzy model based controller is employed under state feedback control in the second phase of control. The Henon map and the three-dimensional Lorenz attractor, which have chaotic attractive features in their original systems, are taken into consideration so as to utilize the benefits of chaos. Then, a two-link manipulator is considered to illustrate the design procedure with employing deterministic chaos. Simulation results show the effectiveness of the proposed controller.

1 Introduction

Recent advances in artificial intelligence (AI) backed by various soft computing techniques have been significantly explored and applied to practical situations in machine learning, systems control, intelligent planning and scheduling, uncertain reasoning, data mining, natural landscape understanding and translation, computer vision, virtual reality, and games [1]. Not only these soft computing techniques can be employed to handle this kind of complicated situations, but also we can apply them for achieving better performance to control complex systems, especially by using the concepts in neural networks, fuzzy reasoning, and evolutionary computation. Complicated dynamics in nonlinear systems makes real-time implementation difficult and therefore AI techniques are an attractive alternative in controlling such systems. Gleaning the searching and learning abilities of evolutionary computation [2] to fulfill the ultimate control objective with any fuzzy or neuro controllers is a rich option. These techniques have been applied to many problems in different instances and succeeded.

K. Watanabe et al.: *Fuzzy–Chaos Hybrid Controllers for Nonlinear Dynamic Systems,*
StudFuzz **187**, 481–506 (2006)
www.springerlink.com

Fuzzy reasoning [3–5] and chaos theory [6–8] are two important fields in this research area, especially when dealing with nonlinear systems control. In particular, research over the past few decades shows a rapid development of fuzzy model based control theory which brings up scientist into a new era in controlling nonlinear systems [5, 9]. In fact, the concept can be applied very well to nonlinear dynamical systems [10–12]. On the other hand, chaos theory plays an important role in analyzing dynamical systems. We basically focus on these two important fields and bring a novel concept for controlling nonlinear dynamical systems.

In the field of controlling nonlinear dynamical systems that show chaotic features, the approaches adopted so far can be categorized into two groups in large. One approach is to curb the chaotic dynamics of a system in order to reduce the system to a manageable level [13], so that conventional nonlinear control methodologies can be employed. The other approach is to intentionally activate chaotic dynamics in a normal nonlinear dynamic system to exploit some of the latent characteristics of chaotic attractors [14–16]. It ia stated that in a nonlinear dynamical system, a chaotic attractor can be formed by an appropriate usage of open-loop control [6, 17].

This research simply focuses on employing useful chaos while ensuring the stability condition of the overall system. Here, we introduce two phases of control. First phase uses an open-loop controller. Once the system has entered a specified area, the open-loop control is cut off and the second phase of control scheme is adopted. Moreover, a conceptual control algorithm is presented by introducing a fuzzy model based controller by employing powerful linear matrix inequalities (LMIs) available to date for stability analysis. Here, the second phase of control is carried out by constructing a convex fuzzy attractive domain while eliminating the local minima problem with the help of evolutionary computation [14, 18] or recently developed semidefinite programming algorithms. For this purpose, the state feedback gain scheduling of the control system in the second phase is achieved by solving a set of LMIs via an optimization technique based on evolutionary computation. In addition, the present method has the advantage of solving LMIs either using evolutionary computation as proposed in [19] or recently developed powerful semidefinite programming tools based on convex optimization algorithms available to date in mathematical literature.

The rest of the chapter is organized as follows: Theoretical review on chaos and fuzzy systems is given in Sect. 2, whereas Sect. 3 focuses on the concept of the present controller. Further details on hybrid controller design are given in Sect. 4. Stability and gain scheduling of the closed-loop system are presented in Sect. 5. Three design examples: the Henon map, the Lorenz attractor, and a two-link manipulator with simulation results are presented in Sects. 6, 7, and 8 respectively. Finally, concluding remarks are given in Sect. 9.

2 Review of Chaos and Fuzzy Systems

2.1 Chaotic Systems

Even though there is no generally accepted definition for chaos, it is worth to absorb something from the history to explore the subject. In 1963, Edward Lorenze, a meteorologist and the first experimenter in chaos, described his model of weather prediction phenomena with a set of nonlinear differential equations. He was working on a problem of weather prediction, with a set of 12 equations to model and predict the weather [6]. Unfortunately, system did not forecast the weather conditions as expected. However, his computer program did theoretically predict what the weather might be. From this set of equations, it is concluded that the system does sensitively depend on initial conditions. At that time, there were only two kinds of order previously known: a steady state, in which the variables never change, and periodic behavior, in which the system goes into a loop, repeating itself indefinitely. Lorenz's equations were definitely in order, they always followed a spiral. They never settled down to a single point, but since they never repeated the same thing, they were not periodic either. He called the image he got when he graphed the equations the Lorenz attractor (see Fig. 1). As a result, Lorenz's discoveries were not acknowledged until years later, when they were rediscovered by others. But it gave a break through, leading to a new research field called chaos. For example, Hennon map is one of the available chaotic systems for analyzing chaos (see Fig. 1b).

Recently, development of chaos theory brings up scientist into a new era in analyzing nonlinear systems. It is known that the chaos exhibits a deterministic random behavior. Yet it needs more investigation on such nonlinear systems in designing control algorithms. Among the various methods available

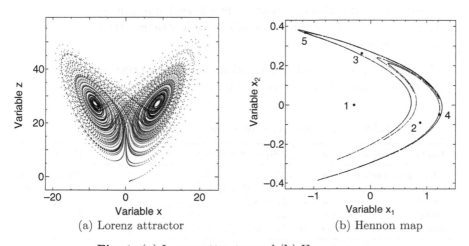

(a) Lorenz attractor   (b) Hennon map

Fig. 1. (a) Lorenz attractor and (b) Hennon map

to analyze such nonlinear systems, there exist fixed point analysis, linearization, Poincare map, homoclinic orbits and Horseshoes, Lyapunov method, spectral analysis, fractal, etc. [7–12, 14, 18, 20]. Still these types of analytical methods are categorized under classical techniques of exploring chaos. In general, chaotic systems have the following properties:

- Spectrum of a chaotic system is not solely of discrete frequencies, but has a continuous, broad-band nature. This noise-like spectrum is characteristic of chaotic systems. The analytical calculation of the Fourier transform is quite complicated for practical systems, but numerical methods are straightforward. When the data are discrete, it is possible to employ an algorithm like FFT.
- Chaotic systems have the property of sensitive dependence on initial condition so called butterfly effect. For a given two different initial conditions, arbitrarily close to one another, the two trajectories generated by each starting point will be diverged for a given time. Therefore, the long-term behavior of the chaotic system can never be predicted, but still it is confined to a specific domain $\Gamma \in \Re^n$ and the trajectory is bounded. This confirms the property: deterministic systems exhibit random behavior.
- Another classical technique for analyzing dynamical systems is from Poincare map [7]. It replaces the flow of an nth-order continuous-time system with an $(n-1)$th-order discrete-time system. It observes the motion stroboscopically, viewing the phase-space. This phenomenon is explored by Henri Poincare and some details on the Poincare map for various attractors are given in Table 1.
- For an attractor, construction must outweigh expansion such that the Lyapunov exponents behave as given in Table 1.
- Next method of classifying attractors is to use the concept of dimension: capacity, information dimension, correlation dimension, and Lyapunov dimension. Classification of attractors by dimension is quite straightforward. Simple attractor has integer dimension, while strange attractors have fractional dimension.

The first consumer product to exploit chaos theory was produced in 1993 by Goldstar Company in the form of a revolutionary washing machine. Later,

Table 1. Different types of attractors and their properties

Attractor	Poincare map	Lyapunov exponent	Spectrum
Equilibrium	Does not exit	$0 > \lambda_1 \geq \cdots \geq \lambda_n$	Does not exit
Periodic	$1 \leq$ points	$\lambda_1 = 0, 0 > \lambda_2 \geq \cdots \geq \lambda_n$	Components
2-periodic	$1 \leq$ closed curves	$\lambda_1 = \lambda_2 = 0, 0 > \lambda_3 \geq \cdots \geq \lambda_n$	Components
k-periodic	$1 \leq (k-1)$-tori	$\lambda_1 = \cdots = \lambda_k = 0, 0 > \lambda_{k+1} \geq \cdots \geq \lambda_n$	Components
Chaotic	Continuum points	$\lambda_1 > 0; \sum_{l=1}^{n} \lambda_i < 0$	Continuous

there are many theoretical and practical advancements in this research area [21, 22].

2.2 Fuzzy Reasoning Based Systems

Fuzzy logic and fuzzy set theory provide a rich and meaningful addition to standard logic, defining a useful rule base, especially for the purpose of intelligent systems and engineering control. The basic idea of "fuzzy logic" was suggested by Prof. L.A. Zadeh [3, 4] and the first implementation of fuzzy logic controller was reported by Mamdani and Assilian [23, 24]. Later, Takagi–Sugeno (TS) fuzzy models became more popular due to its simplicity and easiness of implementing on practical controllers [9]. The main feature of TS-type fuzzy model is that it expresses the local dynamics of each fuzzy rule by linear dynamical model. In fact, research over the past two decades shows a rapid development of fuzzy model based control theory that brings up scientist into a new era in controlling nonlinear systems [25–27].

Fuzzy logic or fuzzy reasoning is a departure from classical two-valued sets and logic that uses "soft" linguistic (large, hot, tall) system variables and a continuous range of truth values in the interval [0, 1], rather than strict binary (1 or 0) decisions and assignments. If X is a collection of objects denoted by x, then a fuzzy set A in X is defined as a set of ordered pairs

$$A = \{(x, \mu_A(x)) \mid x \in X\} \tag{1}$$

where μ_A is called the membership for the fuzzy set A. Function μ_A maps each element of X to a membership grade (or membership value) between 0 and 1 (included).

When the system to be controlled has complex nonlinear dynamics it is not easy to control such a nonlinear system using conventional methods; therefore, fuzzy reasoning based controllers are an attractive alternative. Fuzzy controllers provide a new direction toward the realization of controlling such a class of nonlinear systems. Blending of chaos with fuzzy rules, in order to capture the hidden nonlinearities of the system to be controlled, will be useful in developing such controllers. For example, in [28], the design criterion of a genetic algorithm (GA) based neural fuzzy controller is presented for an antibreak system. In practical situations, design engineer needs to define the desired fuzzy membership functions according to application. Typical examples are Gaussian and triangular membership functions as shown in Fig. 2.

3 Concept of the Controller

Inherent chaotic characteristics can be useful in moving a system to various points in state space confining to a specific domain $\Gamma \in \Re^n$. If the original

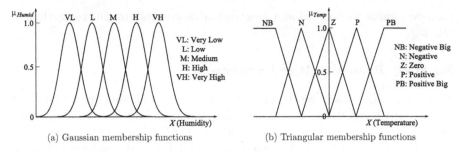

(a) Gaussian membership functions   (b) Triangular membership functions

Fig. 2. Typical membership functions

system does not have chaotic attractive features, a suitable open-loop control input will be introduced to create chaos in order to fulfill the desired control objective. In this concept, this feature is promoted to drive the system states to a predefined convex domain $C \in \Re^{n+1}$ with aid of an appropriate open-loop control excitation on the nonlinear system. The domain C must be a convex region which consists of a well-defined fuzzy rule-base and the construction details are given in the next section. Once it reaches to the predefined fuzzy attractive domain, open-loop input is cut off and a fuzzy model based controller is employed under state feedback control to achieve desired target. This design concept is shown in Fig. 3 for a two-dimensional case. Here it is intended to drive the system states from point P_1 to point P_3. In the first phase of control, chaotic attractive features drive the system from point P_1 to point P_2. Then, from P_2 to P_3 is systematically achieved via a fuzzy model based regulator.

The feedback controller design is based on multiple linearizations around a single equilibrium point, i.e., so called off-equilibrium linearizations. It is known that the off-equilibrium linearization will significantly improve the transient dynamics of the control system for a general control problem [5]. Rather, it is interesting to note that such a technique is useful for constructing

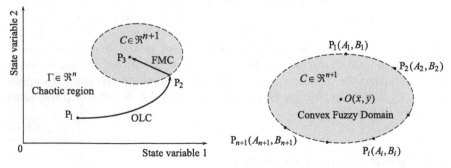

Fig. 3. Fuzzy – chaos hybrid control scheme. The *left* part denotes the concept of this control method and the *right* part shows the detailed convex fuzzy attractive domain, where OLC denotes the open-loop control and FMC is the fuzzy model based control

a globally stable fuzzy attraction domain without trial and error compared to the method proposed in [6, 17].

3.1 Open-Loop Control Using Chaos

Let the trajectory $\phi_t : \Re^n \mapsto \Re^n$ be a solution to a linearized model with initial condition α_0 and $\hat{\lambda}_1, \hat{\lambda}_2, \ldots, \hat{\lambda}_n$ be the eigenvalues of the linearized equation. Taking,

$$m_i(t) = \exp[\hat{\lambda}_i t] \tag{2}$$

The dynamical system trajectory ϕ_t will be chaotic and attractive if the Lyapunov exponents $\lambda_1, \lambda_2, \ldots, \lambda_n$ of the above linearized model are such that

$$\lambda_1 > 0 \text{ (one of the Lyapunov exponents)}, \quad \sum_{i=1}^{n} \lambda_i < 0 \tag{3}$$

where

$$\lambda_i = \lim_{t \to \infty} \frac{1}{t} \ln[m_i(t)] \tag{4}$$

Since the control technique of the first phase is realized through the chaotic attractor $\Gamma \in \Re^n$, robustness of the system is guaranteed even for large disturbances. Chaotic attractors have particular domains such that the system states always drive toward the safe attractive region. Therefore, this advantage cannot be gained by classical control schemes like PI, PID, and others.

3.2 Closed-Loop Control Using Fuzzy Model Based Regulators

Nonlinear dynamic continuous-time systems (CS) can be described by nonlinear differential equations (or difference equations for discrete-time systems (DS)) as

$$\begin{aligned} \dot{x} &= F(x, u) \quad \text{for CS} \\ x(t+1) &= F(x, u) \quad \text{for DS} \end{aligned} \tag{5}$$

where $x \in \Re^n$ is the state vector and $u \in \Re^m$ gives the control input vector of the systems. The equilibrium points $(\bar{x}, \bar{u})$ (or fixed points) of the dynamic system satisfy

$$\begin{aligned} \varepsilon &= \{(x, u) \in \Re^{n+m} \mid F(\bar{x}, \bar{u}) = 0\} \quad \text{for CS} \\ \varepsilon &= \{(x, u) \in \Re^{n+m} \mid \bar{x} = F(\bar{x}, \bar{u})\} \quad \text{for DS} \end{aligned} \tag{6}$$

Here, it is proposed to select a suitable set of off-equilibrium points [23] such that all the subsystems compose a convex region $C \in \Re^{n+1}$ which keeps the equilibrium point $(\bar{x},\ \bar{u}) \in C$ approximately on the center of gravity of the convex region as shown in Fig. 3. More generally, n-dimensional state space system needs at least $n+1$ off-equilibrium points which represent the convex region to ensure the stability of a particular subsystem [19, 29–34]. Neglecting higher order terms, we obtain a linearized model around any arbitrary point $(x_0, u_0) \in C$ as follows:

$$
\begin{aligned}
\dot{x} &= A_0(x - x_0) + B_0(u - u_0) + F(x_0, u_0) && \text{for CS} \\
x(t+1) &= A_0(x - x_0) + B_0(u - u_0) + F(x_0, u_0) && \text{for DS}
\end{aligned}
\tag{7}
$$

where

$$
A_0 = \frac{\partial F}{\partial x}(x_0, u_0)
$$

$$
B_0 = \frac{\partial F}{\partial u}(x_0, u_0)
$$

For example, two-dimensional state space model needs three off-equilibrium points such that the equilibrium point lies on the center of mass of an equilateral triangle keeping its corners on the three off-equilibrium points. The dynamics of the nonlinear system are approximated near an arbitrary point $(x_0, u_0) \in C$. Then, (7) can be rewritten in the form

$$
\begin{aligned}
\dot{x} &= A_0 x + B_0 u + d_0 \\
x(t+1) &= A_0 x + B_0 u + d_0
\end{aligned}
\tag{8}
$$

where $d_0 = F(x_0, u_0) - A_0 x_0 - B_0 u_0$.

Note here that an arbitrary point (x_0, u_0) need not be an equilibrium point $(\bar{x}, \bar{u})$.

Fuzzy models from Takagi–Sugeno consist of a set of IF–THEN rules for the above approximated systems. The ith plant rule of each subsystems for both continuous-time and discrete-time fuzzy systems is given by

$$
\begin{aligned}
&\text{IF } z_1(t) \text{ is } M_{i1} \text{ and } \ldots \text{ and } z_p(t) \text{ is } M_{ip} \\
&\text{THEN } \left\{ \begin{aligned}
\dot{x}(t) &= A_i x(t) + B_i u(t) + d_i \\
x(t+1) &= A_i x(t) + B_i u(t) + d_i \\
y(t) &= C_i x(t), \quad i = 1, \ldots, r
\end{aligned} \right.
\end{aligned}
\tag{9}
$$

where r is the number of fuzzy rules and M_{ij} $(i = 1, \ldots, r$ and $j = 1, \ldots, p)$ are the fuzzy sets. The state vector is $x(t) \in \Re^n$, the input vector is $u(t) \in \Re^m$, and the output vector is given by $y(t) \in \Re^q$. A_i, B_i, and C_i are the system parameter matrices and d_i is the offset term of the ith fuzzy model. For a given state, $z_1(t), \ldots, z_p(t)$ are the premise variables (or antecedent inputs).

Subjecting to the parallel distributed compensation, we can design the following fuzzy regulators:

Regulator rule i:

IF $z_1(t)$ is M_{i1} and ... and $z_p(t)$ is M_{ip}

THEN $\boldsymbol{u}(t) = -K_i[\boldsymbol{x}(t) - \boldsymbol{x}_r] + \boldsymbol{u}_r,\quad i = 1,\dots,r$ \hfill (10)

for the fuzzy models (9), where $\boldsymbol{x}_r$ is a state reference trajectory, $\boldsymbol{u}_r$ is the corresponding input trajectory, and K_i is the local feedback gain matrix. Thus, the fuzzy regulator rules have linear state-feedback laws in the consequent parts and the overall fuzzy regulator can be reduced to

$$\boldsymbol{u}(t) = -\sum_{i=1}^{r} h_i(\boldsymbol{z}(t)) K_i[\boldsymbol{x}(t) - \boldsymbol{x}_r] + \boldsymbol{u}_r \tag{11}$$

where

$$\boldsymbol{z}(t) = [z_1(t),\dots,z_p(t)] \tag{12}$$

$$w_i(\boldsymbol{z}(t)) = \prod_{j=1}^{p} M_{ij}(z_j(t)) \tag{13}$$

$$h_i(\boldsymbol{z}(t)) = \frac{w_i(\boldsymbol{z}(t))}{\sum_{l=1}^{r} w_l(\boldsymbol{z}(t))} \tag{14}$$

for all t, in which $M_{ij}(z_j(t))$ denotes the confidence (or grade) of membership of $z_j(t)$ in M_{ij}.

4 Fuzzy–Chaos Hybrid Controller

In order to control the original nonlinear system with a chaotic input in the open-loop system and a fuzzy controller in the closed-loop system, a fuzzy–chaos hybrid control scheme is proposed here. Such a control scheme can be considered in two cases, depending on the choice of equilibrium points as the reference. In this case, the fuzzy–chaos hybrid control can be implemented by

$$\text{IF } \sum_{i=1}^{r} w_i(\boldsymbol{z}(t)) \equiv 0 \quad \text{THEN } \boldsymbol{u}(t) = \hat{\boldsymbol{u}}(t)$$

$$\text{ELSE} \tag{15}$$

$$\boldsymbol{u}(t) = -\sum_{i=1}^{r} h_i(\boldsymbol{z}(t)) K_i[\boldsymbol{x}(t) - \bar{\boldsymbol{x}}] + \bar{\boldsymbol{u}}$$

where $\hat{\boldsymbol{u}}(t)$ is an open-loop input to make the original nonlinear system chaotic and $(\bar{\boldsymbol{x}}, \bar{\boldsymbol{u}})$ is the prespecified equilibrium point which would be stabilized. Figure 4 shows the hybrid controller. If the stabilized equilibrium point

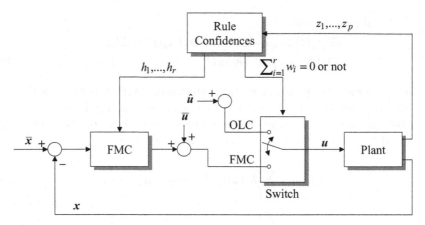

Fig. 4. Fuzzy–chaos hybrid control scheme

is arbitrary among all the equilibrium points, the above fuzzy–chaos hybrid control can be modified as follows:

$$\text{IF } \sum_{i=1}^{r} w_i(\boldsymbol{z}(t)) \equiv 0 \quad \text{THEN } \boldsymbol{u}(t) = \hat{\boldsymbol{u}}(t)$$

$$\text{ELSE} \tag{16}$$

$$i_{\max} = \max\{h_1(\boldsymbol{z}(t)), \ldots, h_r(\boldsymbol{z}(t))\}$$

$$\boldsymbol{u}(t) = -K_{i_{\max}}[\boldsymbol{x} - \bar{\boldsymbol{x}}_{\max}] + \bar{\boldsymbol{u}}_{\max}$$

where $i_{\max}$ denotes the rule number that has largest rule confidence, $K_{i_{\max}}$ is the corresponding feedback gain matrix, and $(\bar{\boldsymbol{x}}_{\max}, \bar{\boldsymbol{u}}_{\max})$ is an equilibrium point existing in the fuzzy attractive domain constructed by using the $i_{\max}$th rule. Schematically, further explanation on the largest confidence when it comes to the controller design is given in Fig. 5.

5 Stability of the Closed-Loop Controller

The stability of a nonlinear control system is systematically checked by the well-known Lyapunov stability theorems. Here, the stability of the closed-loop system of (9) is explained ensuring the stability. Off-equilibrium models $P_1, P_2, \ldots, P_{n+1}$ associated with the convex domain are shown in Fig. 3.

Tanaka et al. [27] or Tanaka and Wang [35] have verified that the equilibrium of the fuzzy system described by the TS fuzzy model (9) is asymptotically stable in the large if there exists a common positive definite matrix P and semidefinite matrix Q such that for a decay rate problem in CS

$$G_{ii}^{\mathrm{T}} P + P G_{ii} + (s - 1)Q + 2\alpha P < 0 \tag{17}$$

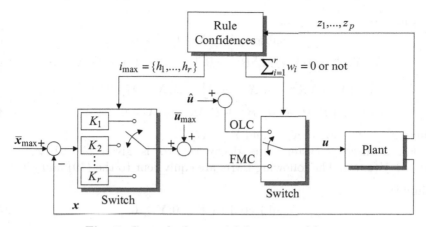

Fig. 5. Control scheme with largest confidence

$$\left(\frac{G_{ij} + G_{ji}}{2}\right)^{\mathrm{T}} P + P\left(\frac{G_{ij} + G_{ji}}{2}\right) - Q + 2\alpha P \leq 0 \quad (i < j) \qquad (18)$$

where feedback gains are denoted by $K_i (i = 1, \ldots, r)$ such that $G_{ij} = A_i - B_i K_j$, if $u(t) = -K_i x(t)$ and $1 < s < r$. Note that as far as the speed response is concerned, it is important to set the largest Lyapunov exponent $\alpha > 0$ and the condition given by $\dot{V}(x(t)) \leq -2\alpha V(x(t))$ for all trajectories is equivalent to (17) and (18).

For a discrete fuzzy control system, the system will be asymptotically stable if there exist a common positive definite matrix P and a common positive semidefinite matrix Q such that

$$G_{ii}^{\mathrm{T}} P G_{ii} - \alpha^2 P - (s - 1)Q < 0 \qquad (19)$$

$$\left(\frac{G_{ij} + G_{ji}}{2}\right)^{\mathrm{T}} P\left(\frac{G_{ij} + G_{ji}}{2}\right) - \alpha^2 P - Q \leq 0 \quad (i < j) \qquad (20)$$

where $\alpha < 1$.

Notice that if r is large, it might be difficult to find the common positive definite matrix P and a common positive semidefinite matrix Q. In such situations, this relaxed stability condition with s improves performance when it determines the common P and Q in order to find the gains [27]. Engineering design often involves several objectives. Therefore, determination of feedback gains of the closed-loop system can be carried out by either an evolutionary computational based algorithm as proposed in [18, 19], or recently developed interior-point methods using LMI tool up to date [33]. Moreover, LMIs are presented to fulfill a systematic design and the system designer has the flexibility of including and optimizing other constraints [5, 36] when using any evolutionary computation approach [2, 37].

For a continuous dynamical system, (17) and (18) can be rearranged to formulate the following optimization problem:

Maximize α

$$\text{subjected to } X > 0, Y \geq 0 \tag{21}$$

$$-XA_i^T - A_iX + M_i^T B_i^T + B_iM_i - (s-1)Y - 2\alpha X \geq 0 \tag{22}$$

$$2Y - XA_j^T - A_iX - XA_j^T - A_jX + M_j^T B_j^T$$
$$+B_iM_j + M_i^T B_j^T + B_jM_i - 4\alpha X \geq 0 \quad (i < j) \tag{23}$$

where $X = P^{-1}$, $M_i = K_iX$, and $Y = XQX$.

For a discrete system, the decay rate α is given by $\Delta V(\boldsymbol{x}(t)) \leq (\alpha^2 - 1)V(\boldsymbol{x}(t))$. The following LMIs are equivalent to the (19) and (20):

Minimize

$$\beta \text{ subjected to } X > 0, Y \geq 0 \tag{24}$$

$$\begin{bmatrix} \beta X - (s-1)Y & XA_i^T - M_i^T B_i^T \\ A_iX - B_iM_i & X \end{bmatrix} > 0 \tag{25}$$

$$\begin{bmatrix} \beta X + Y & \\ 1/2\{A_iX + A_jX - B_iM_j - B_jM_i\} & \\ & \\ 1/2\{A_iX + A_jX - B_iM_j - B_jM_i\}^T & \\ X & \end{bmatrix} \geq 0 \tag{26}$$

where $X = P^{-1}$, $M_i = K_iX$, and $Y = XQX$ and β is selected as $\alpha^2 = \beta$ $(\alpha < 1)$.

6 Design Example 1: Henon Map

In this example, the chaotic system, Henon map, is presented to illustrate the proposed design procedure. The nonlinear dynamic equations of the Henon map are given by

$$\begin{aligned} x_1(t+1) &= -1.4x_1^2 + x_2 + 1 \\ x_2(t+1) &= 0.3x_1 \end{aligned} \tag{27}$$

For simplicity, the uncontrolled Henon map, starting from $(-0.3, 0.0)$ is given in Fig. 1. Here, the points 1, 2, 3, 4, and 5 denote the first consequent five points respectively.

The open-loop control input to the system (27) u is selected as in (28) in implementing the desired control algorithm:

$$\begin{aligned} x_1(t+1) &= -1.4x_1^2 + x_2 + 1 + u \\ x_2(t+1) &= 0.3x_1 \end{aligned} \tag{28}$$

Fixed points of the system of difference (27) are satisfied as the (6), resulting two fixed points (0.6314, 0.1894) and $(-1.1314, -0.3394)$. Therefore, we can design two convex regions that correspond to two fixed points.

6.1 Construction of Two Fuzzy Attractive Domains

Here we select a triple point subregion $(A_i x(t) + B_i u(t)$ for $i = 1, 2, 3)$ such that it surrounds the fixed point $(x_a = 0.6314,\ x_b = 0.1894)$ as shown in Fig. 6. The equilibrium point lies on the center of mass of an equilateral triangle having the coordinates (x_{a1}, x_{b1}), (x_a, x_{b2}), and (x_{a3}, x_{b1}) in order to determine the common P and common Q. In order to construct the fuzzy attractive domain that corresponds to a particular fixed point, it is necessary to determine the maximum size of the convex region. By varying the length $x_a - x_{a1}$ of the triangle as shown in Fig. 6, it is possible to change the size of the fuzzy attractive domain with respect to a particular fixed point. By increasing the width $x_a - x_{a1}$, it is possible to increase the size of the convex domain of the first fixed point (stability details are given in Table 2).

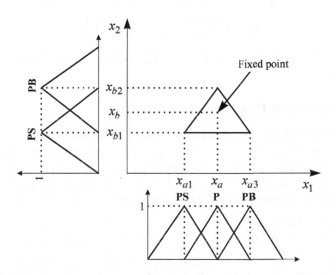

Fig. 6. Tripple point subsystem $(i = 1, 2, 3)$ and its membership functions

Table 2. Design parameters for various $x_a - x_{a1}$

$x_a - x_{a1}$	Common P and Q	Stability
0.05	Exist	Stable
0.1	Exist	Stable
0.2	Exist	Stable
0.3	Exist	Stable
0.4	Does not exist	Not guaranteed
> 0.4	Does not exist	Not guaranteed

For example, taking $(x_a - x_{a1}) = 0.2$, three linearized models corresponded to the first fixed point are given as

$$A_1 = \begin{bmatrix} -1.4878 & 1 \\ 0.3 & 0 \end{bmatrix}, \quad B_1 = \begin{bmatrix} 1 \\ 0 \end{bmatrix}; \quad A_2 = \begin{bmatrix} -1.7678 & 1 \\ 0.3 & 0 \end{bmatrix}, \quad B_2 = \begin{bmatrix} 1 \\ 0 \end{bmatrix}$$

$$A_3 = \begin{bmatrix} -2.0478 & 1 \\ 0.3 & 0 \end{bmatrix}, \quad B_3 = \begin{bmatrix} 1 \\ 0 \end{bmatrix}$$

The same procedure is repeated to select the next subdomain $(A_i x(t) + B_i u(t)$ for $i = 4, 5, 6)$ around the second fixed point and the three linearized models are given as

$$A_4 = \begin{bmatrix} 3.4478 & 1 \\ 0.3 & 0 \end{bmatrix}, \quad B_4 = \begin{bmatrix} 1 \\ 0 \end{bmatrix}; \quad A_5 = \begin{bmatrix} 3.1678 & 1 \\ 0.3 & 0 \end{bmatrix}, \quad B_5 = \begin{bmatrix} 1 \\ 0 \end{bmatrix}$$

$$A_6 = \begin{bmatrix} 2.8878 & 1 \\ 0.3 & 0 \end{bmatrix}, \quad B_6 = \begin{bmatrix} 1 \\ 0 \end{bmatrix}$$

The points $P_1(A_1, B_1), P_2(A_2, B_2), \ldots, P_6(A_6, B_6)$ on the Henon map are shown in Fig. 7.

6.2 Calculation of Feedback Gains

Gain scheduling of the above problem can be formulated as an optimization problem with the LMIs of (25) and (26) and it is solved by using an optimization technique based on evolutionary computation [18, 19]. Here we obtain the common P and common Q for the first fixed point guaranteeing the stability. We obtained the P_1 and Q_1 as follows at $\beta = 0.86$ $(s = 3)$:

$$P_1 = \begin{bmatrix} 699.633 & 71.9250 \\ 71.9250 & 812.871 \end{bmatrix}, \quad Q_1 = \begin{bmatrix} 27.3824 & 0.9216 \\ 15.8108 & 38.907 \end{bmatrix}$$

Similarly, P_2 and Q_2 matrices associated with the second fixed point can be obtained as follows at $\beta = 0.96$ $(s = 3)$:

$$P_2 = \begin{bmatrix} 633.277 & 29.3221 \\ 29.3221 & 656.3649 \end{bmatrix}, \quad Q_2 = \begin{bmatrix} 55.4408 & 30.5980 \\ 30.5980 & 76.9469 \end{bmatrix}$$

The gains K_1, K_2, and K_3 are obtained as below providing the global stability of the fuzzy control system for the first subsystem:

$$K_1 = [-0.8292 \ 1.0092], \quad K_2 = [-2.1818 \ 0.5875],$$
$$K_3 = [-2.0672 \ 1.4699]$$

Similarly, the gains K_4, K_5, and K_6 are obtained for the second subsystem as follows:

$$K_4 = [3.4272 \ 0.7360], \quad K_5 = [3.8298 \ 1.5356],$$
$$K_6 = [2.6047 \ 0.8615]$$

6.3 Simulation Results

Based on the methodology, as proposed in Sect. 2, the Henon map is controlled by a fuzzy–chaos hybrid controller. Since the system has been already a chaotic, it can be used for the first phase of control with no input ($\hat{u} = 0$). Once the system states reach to one of the above two subdomains, fuzzy controller will drive the system toward the fixed point. For example, by using the above gains K_i ($i = 1, 2, 3$), the fuzzy controller is constructed from the following IF–THEN rule base:

$$
\begin{aligned}
&R_1: \text{ IF } x_1 \text{ is PS AND } x_2 \text{ is PS} \quad \text{THEN } K = K_1 \\
&R_2: \text{ IF } x_1 \text{ is P AND } x_2 \text{ is PB} \quad \text{THEN } K = K_2 \\
&R_3: \text{ IF } x_1 \text{ is PB AND } x_2 \text{ is PS} \quad \text{THEN } K = K_3
\end{aligned}
\tag{29}
$$

where PS, P, and PB represent the words positive, small, positive, and positive big, respectively. In order to verify the above design procedure in Sects. 2 and 3, the proposed fuzzy–chaos hybrid controller is applied to the chaotic system (27). Here we allow the system to drive its states toward one of the above two fixed points chaotically. The rule base (8) was employed here to construct the controller as follows:

$$
\text{IF } \sum_{i=1}^{6} w_i(z(t)) \equiv 0
$$
$$
u - \hat{u} = 0
$$
$$
\text{ELSE}
$$
$$
u = -K_{i_{\max}}[x - \bar{x}_{i_{\max}}] \quad (i = 1, \dots, 6)
\tag{30}
$$

where $K_{i_{\max}}$ is the gain which corresponds to $w_{i_{\max}}(k)$ ($i = 1, \dots, 6$).

Figure 7 shows the resulting trajectory of the chaotic system controlled by the proposed controller starting from $(-0.3, 0)$.

7 Design Example 2: Lorenz Attractor

The system of equations of Lorenz attractor is given by the following differential equations:

$$
\begin{aligned}
\frac{dx}{dt} &= -\delta(x - y) \\
\frac{dy}{dt} &= -xz + rx - y \\
\frac{dz}{dt} &= xy - bz
\end{aligned}
\tag{31}
$$

Here, the terms δ, r, and b have the values $\delta = 10, r = 28$, and $b = 8/3$, respectively.

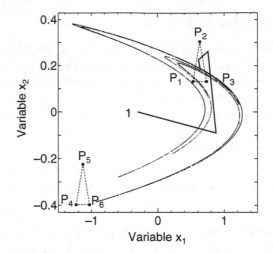

Fig. 7. Trajectory of the controlled Henon map starting from point $1(-0.3, 0)$

Phase trajectory of the uncontrolled Lorenz attractor starting from $(1, 1, 1)$ is shown in Fig. 1. System of differential equations in (31) has three fixed points numerically $(0, 0, 0)$, $(8.4853, 8.4853, 27)$, and $(-8.4853, -8.4853, 27)$.

The control input u_1 is selected to control the system shown in (31) and it is given as in (32) in implementing the desired control system:

$$
\begin{aligned}
\dot{x}_1 &= -\delta(x_1 - x_2) + u_1 \\
\dot{x}_2 &= -x_1 x_3 + r x_1 - x_2 \\
\dot{x}_3 &= x_1 x_2 - b x_3
\end{aligned}
\tag{32}
$$

where, x_1, x_2, and x_3 represent the states of the variables x, y, and z, respectively, i.e., the system state vector is defined by $\boldsymbol{x}^{\mathrm{T}} = [x_1 \; x_2 \; x_3]$.

7.1 Construction of the Fuzzy Attractive Domain

The fixed point $(8.4853, 8.4853, 27)$ is selected as the reference point in order to illustrate the design procedure. Therefore, the convex region which corresponds to this fixed point is constructed using four subsystems $(A_i \boldsymbol{x}(t) + B_i \boldsymbol{u}(t)$ for $i = 1, 2, 3, 4)$ such that it surrounds the fixed point $(8.4853, 8.4853, 27)$ as explained in [29, 32]. The equilibrium point is kept the center of mass of a tetrahedron having the coordinates of corners P_1, P_2, P_3, and P_4 as shown in Table 3 in order to determine the common P and common Q.

Based on the methodology proposed in Sect. 2, the Lorenz attractor is controlled by a fuzzy–chaos hybrid controller. Since the system has been already a chaotic, it can be used for the first phase of control with no input $(\hat{u} = 0)$ and a simplified version of the control algorithm is presented in the

Table 3. Coordinates of the fuzzy attractive domain

P_i	x_1	x_2	x_3
P_1	−1.5091	2.7139	22.3235
P_2	18.4858	2.7184	23.2978
P_3	8.4839	20.0168	22.3233
P_4	8.4882	8.4795	38.9072

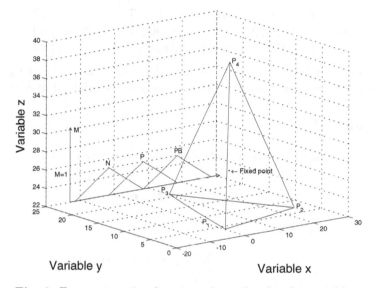

Fig. 8. Fuzzy attractive domain and membership for variable x

next subsection. Once the system states reach to the fuzzy attractive domain, fuzzy controller will drive the system toward the fixed point. Constructed fuzzy attractive domain using the concept explained in [29] is shown in Fig. 8. For example, triangular fuzzy memberships for the variable x_1 (i.e., x) is also given in Fig. 8. Here, the terms PB, P, and N represent the words positive big, positive, and negative, respectively. The rule base for variable x_1 can be simply implemented as follows:

$$
\begin{aligned}
R_1 &: \text{IF } x_1 \text{ is N AND } x_2 \text{ is } \ldots \text{ AND } x_3 \text{ is } \ldots \text{ THEN } K = K_1 \\
R_2 &: \text{IF } x_1 \text{ is PB AND } x_2 \text{ is } \ldots \text{ AND } x_3 \text{ is } \ldots \text{ THEN } K = K_2 \\
R_3 &: \text{IF } x_1 \text{ is P AND } x_2 \text{ is } \ldots \text{ AND } x_3 \text{ is } \ldots \text{ THEN } K = K_3 \\
R_4 &: \text{IF } x_1 \text{ is P AND } x_2 \text{ is } \ldots \text{ AND } x_3 \text{ is } \ldots \text{ THEN } K = K_4
\end{aligned} \tag{33}
$$

Four linearized models corresponding to the selected fixed point are as follows:

$$
A_1 = \begin{bmatrix} -10 & 10 & 0 \\ 5.6765 & -1 & 1.5091 \\ 2.7139 & -1.5091 & -2.6667 \end{bmatrix} ; \qquad B_1 = \begin{bmatrix} 1 \\ 0 \\ 0 \end{bmatrix}
$$

$$A_2 = \begin{bmatrix} -10 & 10 & 0 \\ 4.7022 & -1 & -18.4858 \\ 2.7184 & 18.4858 & -2.6667 \end{bmatrix} ; \quad B_2 = \begin{bmatrix} 1 \\ 0 \\ 0 \end{bmatrix}$$

$$A_3 = \begin{bmatrix} -10 & 10 & 0 \\ 5.6767 & -1 & -8.4839 \\ 20.0168 & 8.4839 & -2.6667 \end{bmatrix} ; \quad B_3 = \begin{bmatrix} 1 \\ 0 \\ 0 \end{bmatrix}$$

$$A_4 = \begin{bmatrix} -10 & 10 & 0 \\ -10.9072 & -1 & -8.4882 \\ 8.4795 & 8.4882 & -2.6667 \end{bmatrix} ; \quad B_4 = \begin{bmatrix} 1 \\ 0 \\ 0 \end{bmatrix}$$

These four linearized models are used to determine the gains of the fuzzy attractive domain in the second phase of control as explained in Sect. 2.

7.2 Controller

Simplified version of the total controller can be expressed as follows:

$$\text{IF } \sum_{i=1}^{4} w_i(\boldsymbol{z}(t)) \equiv 0 \text{ THEN } \boldsymbol{u}(t) = u_1 = 0$$

$$\text{ELSE} \tag{34}$$

$$\boldsymbol{u}(t) = u_1 = -\sum_{i=1}^{4} h_i(\boldsymbol{z}(t)) K_i [\boldsymbol{x}(t) - \bar{\boldsymbol{x}}]$$

7.3 Simulation Results

The optimization algorithm presented in (21)–(23) is applied to the system $(A_i \boldsymbol{x}(t) + B_i \boldsymbol{u}(t)$ for $i = 1, 2, 3, 4)$ in order to determine the desired gains of the feedback controller. Using an optimization technique based on the evolutionary computation as explained in [19] solves the LMIs. The following common P and Q are obtained:

$$P = \begin{bmatrix} 9992.43 & 197.338 & 297.654 \\ 197.338 & 7393.03 & 8.59565 \\ 297.654 & 8.59565 & 7361.84 \end{bmatrix}$$

$$Q = \begin{bmatrix} 6454.36 & 648.724 & 1872.85 \\ 648.724 & 283.821 & 496.334 \\ 1872.85 & 496.334 & 3520.69 \end{bmatrix}$$

Following gains were obtained at $\alpha = 1.0$:

$$K_1 = \begin{bmatrix} -3.8845 & 13.6663 & 0.7498 \end{bmatrix}$$
$$K_2 = \begin{bmatrix} 4.2118 & 13.4469 & -1.3512 \end{bmatrix}$$
$$K_3 = \begin{bmatrix} 8.0471 & 13.4239 & 7.8422 \end{bmatrix}$$
$$K_4 = \begin{bmatrix} 7.8300 & 3.1779 & -1.8896 \end{bmatrix}$$

Note that the triangular fuzzy membership functions are selected similar to that in [32] keeping the desired width of each triangle.

The controlled three-dimensional view of the Lorenz attractor starting from $(1, 1, 1)$ is presented in Fig. 9 under the proposed control scheme.

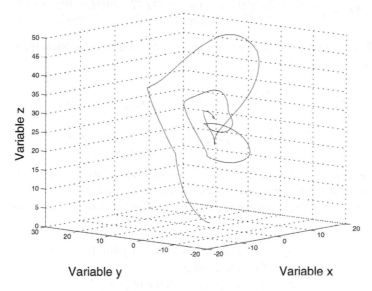

Fig. 9. Three-dimensional view of the controlled Lorenz attractor

8 Design Example 3: Two-link Robot Arm

8.1 Manipulator Modeling

Two-link manipulator in Fig. 10 is taken into consideration in order to employ deterministic chaos in the first phase of control. Joint torque vector $[\tau_1 \ \tau_2]^T$ of the manipulator is given in the dynamic (35) as follows:

$$\begin{bmatrix} \tau_1 \\ \tau_2 \end{bmatrix} = M \begin{bmatrix} \ddot{\theta}_1 \\ \ddot{\theta}_2 \end{bmatrix} + (\dot{\theta}_1 \dot{\theta}_2) B + C \begin{bmatrix} \dot{\theta}_2^2 \\ \dot{\theta}_1^2 \end{bmatrix} + F \begin{bmatrix} \dot{\theta}_1 \\ \dot{\theta}_2 \end{bmatrix} + G \qquad (35)$$

where inertia matrix is given by

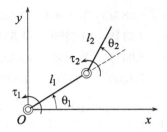

Fig. 10. Two-link manipulator

$$M = \begin{bmatrix} m_2l_2^2 + 2m_2l_1l_2c_2 + (m_1 + m_2)l_1^2 & m_2l_2^2 + m_2l_1l_2c_2 \\ m_2l_2^2 + m_2l_1l_2c_2 & m_2l_2^2 \end{bmatrix}$$

B, C, F, and G represent the Coriolis, centrifugal, viscous friction, and gravity terms, respectively:

$$B = \begin{bmatrix} -m_2l_1l_2s_2 \\ 0 \end{bmatrix}; \qquad C = \begin{bmatrix} 0 & -m_2l_1l_2s_2 \\ m_2l_1l_2s_2 & 0 \end{bmatrix}$$

$$F = \begin{bmatrix} f_1 & 0 \\ 0 & f_2 \end{bmatrix}; \qquad G = \begin{bmatrix} m_2l_2gc_{12} + (m_1 + m_2)l_1gc_1 \\ m_2l_2gc_{12} \end{bmatrix}$$

Here, the terms c_i, s_i, and c_{ij} are given by $\cos\theta_i$, $\sin\theta_i$, and $\cos(\theta_i + \theta_j)$, respectively [38, 39]. The above dynamical system can be implemented using a set of nonlinear differential equations by introducing state variables as follows:

$$\begin{aligned} x_1 &= \theta_1 &\quad \text{angle of joint 1;} \\ x_2 &= \dot{\theta}_1 &\quad \text{angular velocity of joint 1;} \\ x_3 &= \theta_1 &\quad \text{angle of joint 2;} \\ x_4 &= \dot{\theta}_2 &\quad \text{angular velocity of joint 2.} \end{aligned}$$

Defining the system state vector as $x^T = [x_1 \ x_2 \ x_3 \ x_4]$, it is found that one of the equilibrium points is at $[-\pi/2, 0, 0, 0]$. In the next section, we construct a convex domain around $[-\pi/2, 0, 0, 0]$ so as to illustrate the design procedure explained in Sect. 2. Therefore, it requires minimum number of five off-equilibrium points ($n + 1 = 5$) to construct the fuzzy attractive domain.

8.2 Design of Open-Loop Controller

As proposed in Sect 2, if the original system does not have chaotic properties, deterministic chaos will be introduced into the system. The open-loop controller can be constructed by applying a periodic torque ($\tau_1 = A + B\cos\omega t$) to the joint 1 and the position of the attractive domain will be changed according to the DC component (A), amplitude (B), and the frequency (ω). The manipulator parameters are

Table 4. Different A, B, and ω values for constructing different attractors

Torque τ_1	A	B	ω
T_1	-18	4	10
T_2	-15	4	10
T_3	-10	4	10
T_4	0	4	10
T_5	10	4	10

$$m_1 = 3 \text{ kg}, \qquad m_2 = 4 \text{ kg}$$
$$l_1 = 0.3 \text{ m}, \qquad l_2 = 0.3 \text{ m}$$
$$f_1 = 0.8 \text{ kg s}^{-1}, \qquad f_2 = 0.8 \text{ kg s}^{-1}$$
$$g = 9.81 \text{ m s}^{-2}$$

Table 4 gives different values for A, B and ω for constructing corresponding attractors in Fig. 11. In this design, we set the torque parameters of joint 1 as $A = 0$, $B = 4$, and $\omega = 10$ in order to drive the system state vector $x^T = [\theta_1, \dot{\theta}_1, \theta_2, \dot{\theta}_2]$ toward the equilibrium point $[-\pi/2, 0, 0, 0]$ forming a chaotic attractor around $[-\pi/2, 0, 0, 0]$.

8.3 Design of Closed-Loop Controller

In this case, we construct the convex domain around the equilibrium point $[-\pi/2, 0, 0, 0]$ as proposed in [19]. Gain scheduling of the closed-loop controller can be carried out using the LMIs. Fuzzy attractive domain is constructed with five off-equilibrium points P_i, $(i = 1, 2, \ldots, 5)$ and the points are

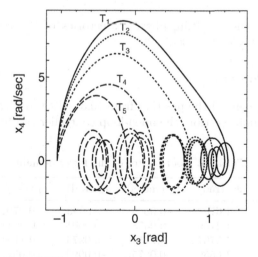

Fig. 11. Construction of different attractors corresponding to torques from $\tau_1 = T_1$ to $\tau_1 = T_5$

given in Table 5. Rest of the design procedure of the closed-loop controller is similar to the two examples illustrated above. After the optimization process based on the evolutionary computation for determining the common P and common Q, the following common P and Q were obtained at $\alpha = 9.99$:

$$P = \begin{bmatrix} 67.2854 & 9996.0412 & 3497.1003 & 409.0776 \\ 9996.0412 & 3023.2453 & 2058.9513 & 471.4612 \\ 3497.1003 & 2058.9513 & 732.5273 & 9999.7477 \\ 409.0776 & 471.4612 & 9999.7477 & 2636.7672 \end{bmatrix}$$

$$Q = \begin{bmatrix} 3858.9581 & 103.2063 & 5996.9195 & 2204.3101 \\ 103.2063 & 5025.5096 & 5936.5655 & 920.3096 \\ 5996.9195 & 5936.5655 & 1887.9925 & 1042.4371 \\ 2204.3101 & 920.3096 & 1042.4371 & 2255.1776 \end{bmatrix}$$

Following gains were obtained:

$$K_1 = \begin{bmatrix} -0.9903 & -3.2081 & 0.4705 & 0.5111 \\ 2.6385 & 0.1614 & -1.2647 & 1.4867 \end{bmatrix}$$

$$K_2 = \begin{bmatrix} -1.0152 & -2.6399 & -1.2049 & -1.1647 \\ -1.5773 & -2.1621 & 0.4410 & 1.0741 \end{bmatrix}$$

$$K_3 = \begin{bmatrix} -4.8227 & -3.1958 & -1.1226 & -1.0493 \\ -0.5121 & -0.7764 & 1.7699 & 1.2797 \end{bmatrix}$$

$$K_4 = \begin{bmatrix} -0.0710 & -2.1731 & -2.7916 & -1.2194 \\ 2.3181 & -0.9752 & 4.2792 & 0.9927 \end{bmatrix}$$

$$K_5 = \begin{bmatrix} -0.8204 & -2.5124 & -0.9432 & -1.6543 \\ -3.3369 & -2.2102 & 0.5640 & -0.0133 \end{bmatrix}$$

Figure 12 shows the resulting phase trajectories of the controlled system starting from $[-2\pi/5,\ 0,\ -\pi/3,\ 0]$.

8.4 Position Control

It is supposed to control the position from $(0.3, -1.8)$ to $(0.2, -0.5)$. As proposed in the previous section, the open-loop controller can be constructed by

Table 5. Coordinates of the fuzzy domain

P_i	x_1 (rad)	x_2 (rad/s)	x_3 (rad)	x_4 (rad/s)
P_1	-1.6708	-0.0578	0.0993	-0.0701
P_2	-1.4709	-0.0578	-0.0001	0.0944
P_3	-1.5707	0.1155	-0.0974	0.1154
P_4	-1.6559	-0.0875	-0.0999	-0.0577
P_5	-1.4747	0.0917	0.1000	-0.0577

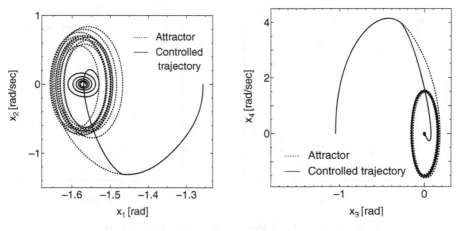

Fig. 12. Controlled phase trajectories

applying a periodic torque ($\tau_1 = 15 + 4\cos 10t$) to the joint 1 keeping the second joint as underactuated condition ($\tau_2 = 0$). The manipulator parameters are unchanged. For position control, fuzzy attractive domain points are given in Table 6. The common P, common Q, and feedback gains of the fuzzy controller are as follows:

$$P = \begin{bmatrix} 1264.4601 & 9992.9647 & 3773.5421 & 4226.5646 \\ 9992.9647 & 5412.4919 & 4033.8524 & 818.5849 \\ 3773.5421 & 4033.8525 & 713.5202 & 9999.8147 \\ 4226.5646 & 818.5849 & 9999.8147 & 3280.0809 \end{bmatrix}$$

$$Q = \begin{bmatrix} 1302.1423 & 1349.6125 & 2505.7693 & 4506.9873 \\ 1349.6125 & 1834.4586 & 5682.6848 & 2506.8759 \\ 2505.7693 & 5682.6848 & 5321.6209 & 6244.7572 \\ 4506.9873 & 2506.8759 & 6244.7573 & 2630.8178 \end{bmatrix}$$

Here, $\alpha = 8.5$

Table 6. Coordinates of the fuzzy domain for position control

P_i	x_1 (rad)	x_2 (rad/s)	x_3 (rad)	x_4 (rad/s)
P_1	−0.8299	−0.0577	5.3539	0.0911
P_2	−0.6299	−0.0576	5.4399	0.0911
P_3	−0.7301	0.1155	5.5376	0.1154
P_4	−0.8018	0.0843	5.3401	−0.0577
P_5	−0.6325	−0.0965	5.5401	−0.0577

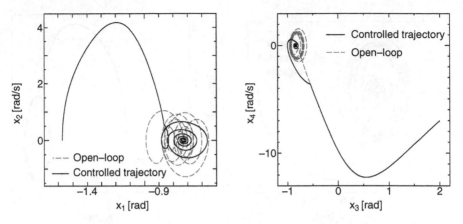

Fig. 13. Controlled phase trajectories: position control

$$K_1 = \begin{bmatrix} -1.2569 & -4.6736 & -3.6240 & -4.0328 \\ -3.3365 & -2.5719 & 2.3216 & 5.05332 \end{bmatrix}$$

$$K_2 = \begin{bmatrix} -5.6195 & -6.5955 & -1.7853 & -0.8748 \\ -1.4306 & -1.7681 & 2.6193 & 1.4209 \end{bmatrix}$$

$$K_3 = \begin{bmatrix} -3.3699 & -3.8742 & -1.2319 & -1.0718 \\ -7.7198 & -5.4325 & 2.8976 & 6.4647 \end{bmatrix}$$

$$K_4 = \begin{bmatrix} 1.6034 & -4.2976 & 1.0405 & -2.8926 \\ -1.3812 & 0.2503 & 0.7967 & 2.8737 \end{bmatrix}$$

$$K_5 = \begin{bmatrix} -1.4642 & -6.1780 & -3.9840 & -2.6404 \\ 0.2588 & -1.6819 & -2.1839 & -0.1882 \end{bmatrix}$$

Controlled phase trajectories of the two-link manipulator for position control are given in Fig 13.

9 Conclusions

A methodology for controlling chaotic systems using fuzzy model based regulators has been presented to employ and utilize inherent chaotic features of nonlinear dynamical systems. This approach differs from the others in regarding well-defined stability criteria for the closed-loop stability with the aid of a fuzzy model based regulator. Without deforming chaotic features in a nonlinear system, a systematic design procedure has been presented. Designer can have the choice of selecting one of the optimization algorithms mentioned in this chapter when constructing the fuzzy attractive domain, in which a global solution is obtained so as to achieve the desired stability condition of the closed-loop system. Simulation results given by three different examples show the effectiveness of the present controller and the ability of employing and utilizing deterministic chaos. According to the results, the tracking performance of the proposed fuzzy–chaos hybrid controller confirms the applicability of the proposed algorithm.

References

1. S. Russell, P. Norvig: *Artificial Intelligence: A Modern Approach* (Prentice Hall, New-Delhi, India, 2003)
2. K. Watanabe, M.M.A. Hashem: *Evolutionary Computations: New Algorithms and Their Applications to Evolutionary Robots* (Springer, Berlin Heidelberg New York, 2004)
3. L.A. Zadeh: J. Inf. Control **8**, 338–353 (1965)
4. L.A. Zadeh: IEEE Comput **21**(4), 83–93 (1988)
5. R. Palm, D. Driankov: *Model Based Fuzzy Control* (Springer, Berlin Heidelberg New York, 1997)
6. T.L. Vincent, W.J. Grantham: *Nonlinear and Optimal Control Systems* (Wiley, New York, 1997)
7. T. Kapitaniak: *Chaos for Engineers—Theory, Applications and Control* (Springer, Berlin Heidelberg New York, 1998)
8. A. Mees, C. Sparrow: Proc. IEEE **75**(8), 1058–1007 (1987)
9. T. Takagi, M. Sugeno: IEEE Trans. Syst. Man Cyber. **15**(1), 116–132 (1985)
10. T. Taniguchi, K. Tanaka: J. Advanced Comput. Intell. **3**(2), 68–74 (1999)
11. S.Y. Yi, M.J. Chung: IEEE Trans. Fuzzy Syst. **6**(2), 216–225 (1998)
12. H. Ying: IEEE Trans. Fuzzy Syst. **6**(2), 226–234 (1998)
13. Y. Joo, L. Shieh, G. Chen: IEEE Trans. Fuzzy Syst. **7**(4), 394–408 (1999)
14. L. Udawatta, K. Watanabe, K. Kiguchi, K. Izumi: IEEE Trans. Fuzzy Syst. **10**(3), 401–411 (2002)
15. Y. Nakamura, T. Suzuki, M. Koinuma: IEEE Trans. Robotics Control **13**(6), 853–862 (1997)
16. T. Suzuki, Y. Nakamura: Control of manipulators with free-joints via averaging method. In: *Proc. of IEEE Int. Conf. on Robotics and Automation (ICRA'97)*, 1997, pp 2998–3005
17. T.L. Vincent: IEEE Control Syst. Magazine **17**(6), 65–76 (1997)
18. L. Udawatta, K. Watanabe, K. Kiguchi, K. Izumi: J. Appl Soft Comput **4**, 25–34 (2004)
19. K. Watanabe, L. Udawatta, K. Izumi: A solution of fuzzy regulator problems via evolutionary computations. In: *Proc. of SICE '99*, Morioka, Japan, 1999, pp 835–836 (in Japanese)
20. T.S. Parkera, L.O. Chua: Proc. IEEE **75**(8), 982–1007 (1987)
21. C.T. Lin, C.P. Jou: IEEE Trans. Neural Networks **10**(4), 846–859 (1999)
22. L. Udawatta, K. Watanabe, K. Kiguchi, K. Izumi: Fuzzy–chaos hybrid controller for 3-dimensional attractors. In: *Proc. of IIZUKA 2000 (6th International Conference on Soft Computing)*, Iizuka, Fukuoka, Japan, 2000, pp 459–464
23. E.H. Mamdani: Fuzzy sets, twenty years of fuzzy control: Experiences gained and lessons learnt. In: *Proc. of IEEE International Conf. on Fuzzy Systems*, 1993, pp 339–344

506   K. Watanabe et al.

24. H. Ying: *Inte. J. Fuzzy Syst.* **3**, 192–197 (2000)
25. D. Driankov, R. Palm: *Advances in Fuzzy Control* (Springer, Berlin Heidelberg New York, 1998)
26. X. Ma, Z. Sun Y. He: IEEE Trans. Fuzzy Syst. **6**(1), 41–51 (1998)
27. K. Tanaka, T. Ikeda, H.O. Wang: IEEE Trans. Fuzzy Syst. **6**(2), 250–265 (1998)
28. Y. Lee, S.H. Zak: IEEE Trans. Evolutionary Comput. **6**(2), 198–211 (2002)
29. L. Udawatta, K. Watanabe, K. Kiguchi, K. Izumi: Construction of multi-dimensional fuzzy attractive domains for fuzzy model-based regulators. In: *Proc. of 1st Annual Conference of SOFT Kyushu Chapter*, 1999, pp 35–40
30. L. Udawatta, K. Watanabe, K. Kiguchi, K. Izumi: Control of robot manipulators via chaotic attractors and fuzzy model-based regulators, In: *Proc. of 5th International Symposium on Artificial Life and Robotics*, vol 2, Oita, Japan, 2000, pp 673–676
31. L. Udawatta, K. Watanabe, K. Kiguchi, K. Izumi: Control of underactuated manipulators via chaos. In: *Proc. of Robotics and Mechatronics*, Kumamoto, Japan, 2000, 1A1-28-027, pp 1–2
32. K. Watanabe, L. Udawatta, K. Kiguchi, K. Izumi: Control of chaotic systems using fuzzy model-based regulators. In: *Lecture Notes in Artificial Intelligence* (Springer, Berlin Heidelberg New York, 1999) pp 248–256
33. S. Boyd, L.E. Ghaoui, E. Feron, V. Balakrishnan: *Linear Matrix Inequalities in Systems and Control Theory* (SIAM, Philadelphia, PA) 1994
34. H.S.M. Coxeter: *Regular Polytopes* (Dover Publishers, New York, 1973)
35. K. Tanaka, H.O. Wang: *Fuzzy Control Systems Design and Analysis: A Linear Matrix Inequality Approach*, (Wiley, Chichester, UK, 2001)
36. S. Limanond, J. Si: IEEE Trans. Neural Networks, **9**(6), 1422–1429 (1998)
37. T. Nanayakkara, K. Watanabe, K. Izumi: Evolving in dynamic environment through adaptive chaotic mutation. In: *Proc. of 4th International Symposium on Artificial Life and Robotics*, vol 2, Oita, Japan, 1997, pp 520–523
38. J.J. Craig: *Introduction to Robotics: Mechanics and Control* (Addison-Wesley, Reading, MA, 1989) pp 170–220
39. K.S. Fu, R.C. Gonazalez, C.S.G. Lee: *Robotics, Control, Sensing, Vision, and Intelligence* (McGraw-Hill, New York, 1987)

Fuzzy Model Based Chaotic Cryptosystems

Chian-Song Chiu and Kuang-Yow Lian

Abstract. In this chapter, we address a fuzzy model based chaotic cryptosystem. For the crytosystem, the plaintext (message) is encrypted using the superincreasing sequence formed by chaotic signals at the drive system side. The resulting ciphertext is embedded to the output or state of the drive system and is sent to the response system end. The plaintext is retrieved via the synthesis approach for signal synchronization. We show that the chaotic synchronization problem can be solved using linear matrix inequalities. The advantages of this crytosystem are the systematic methodology of fuzzy model based design suitable for well-known Lure type discrete-time chaotic systems; flexibility in selection of chaotic signals for secure key generator; flexibility of masking the ciphertext using either the state or output; multiuser capabilities; and a time-varying superincreasing sequence. In light of the above advantages, the chaotic communication structure has a higher-level of security compared to traditional masking methods. In addition, numerical simulations and DSP-based experiments are carried out to verify the validity of theoretical results.

1 Introduction

Many fuzzy model based design, in recent years, are carried out using Takagi–Sugeno (TS) fuzzy models [1, 2]. The main concept is that the fuzzy IF–THEN rules have consequent parts which represent local linear models. Then the overall output of the fuzzy system represents the input to output relationship of any general nonlinear system in an interesting region. Using analogous discussions to that of linear control, many objectives are achievable in a unified manner. The stability of such fuzzy model based systems relies on finding a common symmetric positive matrix and results in a linear matrix inequalities (LMIs) problem [3]. Powerful numerical toolboxes are then used to solve these problems. Since this control approach proposes the advantage of straighforwardness for investigating nonlinear systems with ripe analysis tools of linear control, mass research has focused on this topic.

On the other hand, chaotic dynamics are deterministic, but extremely sensitive to initial conditions. Even infinitesimal changes in initial condition will lead to an exponential divergence of orbits. The pioneering work of Carroll and Pecora [4] has led to many works regarding synchronization of two chaotic systems [5, 6], where two chaotic systems with suitable coupling

C.-S. Chiu and K.-Y. Lian: *Fuzzy Model Based Chaotic Cryptosystems*,
StudFuzz **187**, 507–525 (2006)
www.springerlink.com

produce identical oscillations. Many theories [5–9] have been proposed to achieve the synchronized manner from master–slave configuration. This master–slave configuration consists of the original chaotic system as a *drive system* to provide a *driving signal* to synchronize another system called the *response system*. In addition, chaotic signals are typically broadband, noise-like, and difficult to predict; therefore, they can be used in various context for masking information-bearing waveforms. They can also be used as modulating waveforms in spread spectrum systems. The idea of chaotic masking [10, 11] is to directly add the message in a noise-like chaotic signal at the transmitter's end, while chaotic modulation [12–15] is by injecting the message into a chaotic system as a spread-spectrum transmission. Later, at the receiver, a coherent detector with some signal processing is employed to recover the message. But the signal masking or parameter modulation approach to chaotic communications only provides a lower level of security as stated in [16]. Therefore recent works have considered the use of basic cryptosystem theory to add to the security of chaotic communications [17–19].

In this chapter, a higher level security methodology, namely, the fuzzy model based chaotic cryptosystem is proposed. First, Lure type discrete-time chaotic systems are exactly represented by TS fuzzy models. Then using a chaotic signal, which can be flexibly chosen to be (a) an output of the TS fuzzy chaotic drive system or (b) any state in which the synchronization error approaches zero, a superincreasing sequence is generated. The cryptosystem methodology is briefly given as follows. The plaintext (message) is encrypted using the superincreasing sequence at the drive system side which results in the ciphertext. In light of the previous works [20–22], the ciphertext may be added to the output or state of the drive system. Then this ciphertext embedded scalar signal is sent to the response system end. Following the design of a response system, chaotic synchronization between the drive and response system is achieved by solving LMIs. Since synchronization is ensured, which means internal states of the drive and response system are same, we regenerate the same superincreasing sequence and recover the ciphertext at the response system end. Finally, using the regenerated superincreasing sequence, the ciphertext is decrypted into the plaintext. This chaotic cryptosystem approach has several advantages: (i) the TS fuzzy model representation is general for most well-known Lure type discrete-time chaotic systems; (ii) the synchronization problem using LMIs are systematic and straightforward, which may be solved by powerful software toolboxes; (iii) the superincreasing sequence generated by chaotic signals is time varying; (iv) there exists a flexible choice for whether the ciphertext is embedded in the state or output of the drive system; and (v) multiuser capabilities. In light of the above advantages, the proposed chaotic cryptosystem has a systematic design, whereas a higher level of security is provided compared to traditional signal masking methods.

The rest of the chapter is organized as follows: In Sect. 2, the chaotic cryptosystem structure and algorithm are introduced. In Sect. 3, the TS fuzzy model synchronization method is given. Section 4 illustrates the chaotic cryptosystem using discrete-time chaotic system as an example, while two numerical simulations are performed. In Sect. 5, DSP-based experiments are also carried out to verify the theoretical results. Finally, some conclusions are given in Sect. 6.

2 Chaotic Cryptosystem Structure

In traditional chaotic communications, the synchronization of drive-response system was exploited. This direct signal masking based approach often proposes setbacks in security, as stated in [16]. Therefore applying cryptosystem theory, the message encoded using a chaotic signal provides a higher level of security and multiuser capabilities. Now, we introduce some basic cryptosystem terminology.

The composite message vector N to be transmitted is the *plaintext*, which is encoded by using the *superincreasing sequence* S_i. The encoded plaintext results in the *ciphertext* $E(\cdot)$. The process of recovering the plaintext from the ciphertext is the decryption function $\widehat{E}(\cdot)$. Encryption and decryption process use the keys K and $\widehat{K}$, respectively. The definition of a superincreasing sequence is as follows:

Definition 1. *A real sequence* $\{S_i\}_{i=1}^{\ell}$ *is called a superincreasing sequence if the following is satisfied* $S_j > \sum_{i=1}^{j-1} S_i$, $\ell \geq j > 1$ *and all* $S_i > 0$.

Note that in traditional superincreasing problems [23, 24], the sequence is a set of positive integers. The superincreasing sequence used here is modified where a set of positive real numbers is considered.

The proposed secure communication framework consists of three main components:

(a) *Chaotic model component:* This consists of a TS fuzzy model representing discrete-time chaotic systems.
(b) *Encrypting component:* This consists of a superincreasing sequence $\{S_i\}$, $i = 1,\ldots,\ell$, where ℓ is the number of messages, and a plaintext $N = [n_1 \ n_2 \ldots n_\ell]$ with the message $n_i \in \{0, 1\}$. The superincreasing sequence along with plaintext combines into a ciphertext as follows:

$$E(N(t), K(t), K(t-1),\ldots, K(t-\ell+1)) = S(t) N(t)^{\mathrm{T}} \equiv E(t),$$

where $E(\cdot)$ is an encryption function that makes use of the superincreasing sequence $S(t) = [S_1(t) \ S_2(t) \ \ldots \ S_\ell(t)]$ formed by the key signal $K(t-i)$, $i = 0,\ldots,\ell - 1$. The sequence of the key, i.e., $K(t)$, $K(t-1)$, $\ldots$, $K(t-\ell+1)$, may be (i) output of the TS fuzzy chaotic drive system or (ii) any state in which the synchronization error approaches zero.

(c) *Decrypting component:* This consists of a TS fuzzy model discrete-time chaotic response system generating the same key as the encryption component. Achieving $\widehat{E}(\cdot) \to E(\cdot)$ and $\widehat{K}(\cdot) \to K(\cdot)$, the plaintext is obtained from decrypting the ciphertext $\widehat{E}(t)$ as follows:

$$\widehat{N}(t) = D(\widehat{K}(t), \widehat{E}(N(t), \widehat{K}(t), \widehat{K}(t-1), \ldots, \widehat{K}(t-\ell+1))),$$

where $D(\cdot)$ is an decryption function that makes use of the recovered key $\widehat{K}(t-i)$, $i = 0, \ldots, \ell-1$.

The structure above is depicted in a block diagram in Fig. 1. We now explain in details the algorithm for the chaotic cryptosystem.

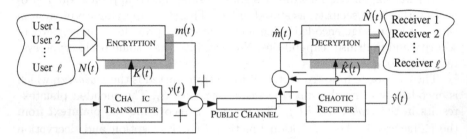

Fig. 1. Block diagram of chaotic encryption methodology

Drive system

1. Generate a superincreasing sequence $\{S_i(t)\}$, where $S_1(t) = |K(t)| + \tau$, $S_j(t) = \sum_{i=1}^{j-1} S_i(t) + |K(t-j+1)| + \tau, \forall j = 2, \ldots, \ell$ and $\tau > 0$.
2. Form the encryption function $E(t) = S(t)N(t)^{\mathrm{T}}$.
3. Modify encryption function $E(t)$ into $\xi(t) = (E - H(t)/2)/(\gamma H(t)/2)$, where $H = \sum_{i=1}^{\ell} S_i$ and γ is a scalar such that $\xi(t) \in (-0.01\ 0.01)$ is sufficiently small as to not destroy the chaotic characteristics of the masking signal.
4. Add $\xi(t)$ to the masking signal and send a scalar coupling signal to response system.

Response system

1. Design a TS fuzzy response system to recover the signal $\hat{\xi}(t)$ from synchronization.
2. Generate a superincreasing sequence $\{\widehat{S}_i(t)\}$, $i = 1, \ldots, \ell$, from the key signal obtained from synchronization, i.e., $\widehat{S}_1(t) = |\widehat{K}(t)| + \tau$, $\widehat{S}_j(t) = \sum_{i=1}^{j-1} \widehat{S}_i(t) + |\widehat{K}(t-j+1)| + \tau, \forall j = 2, \ldots, \ell$ and $\tau > 0$.

3. Demodify $\hat{\xi}(t)$ to obtain $\widehat{E}(t) = \hat{\xi}(t)(\gamma \widehat{H}(t)/2) + \widehat{H}(t)/2$.
4. Decrypt the message using the superincreasing sequence from Step 3 of the response system design and the following algorithm:

$$\widehat{V} = \widehat{E}$$
$$\text{For } i = \ell \text{ down to } 1$$
$$\text{Begin}$$
$$\text{If } \widehat{V} - \widehat{S}_i > -\varepsilon$$
$$\hat{n}_i = 1$$
$$\widehat{V} = \widehat{V} - \widehat{S}_i$$
$$\text{Else } \hat{n}_i = 0$$
$$\text{END}$$

where $0 < \varepsilon < \tau$.

Notice that if $n_\ell(t) = 1$ (or $n_\ell(t) = 0$), then $E(t) \geq S_\ell(t)$ or $(E(t) \leq S_\ell(t) - \tau)$. After the transient time, the response system is synchronized to the drive system. This means that $\widehat{E}(t) = E(t)$ and therefore $\widehat{S}_\ell(t) = S_\ell(t)$. Finally $\hat{n}_\ell(t) = 1$ (or $\hat{n}_\ell(t) = 0$) from the above algorithm and the other plaintext $\hat{n}_{\ell-1}(t) \ldots \hat{n}_1(t)$ are recovered by the iteration loop.

Remark 1. When the cryptosystem simultaneously serves ℓ users, the plaintext N consists of multiusers' messages, i.e., one bit of each user is transmitted in one time. If considering only one user service, ℓ bits of the user's message are simultaneously transmitted. This means that the advantages of the cryptosystem are (i) multiuser capabilities, (ii) high data transmission rate for one user, (iii) high-frequency efficiency, and (iv) a higher level of security compared to traditional chaotic communications.

Remark 2. In comparison, traditional cryptosystems often take constant parameters as the secure key, such as chaotic parameters. In contrast, the proposed approach uses time-varying internal states of chaotic systems as the secure key such that the cryptosystem has more flexibility and higher security. Moreover, since the selection of $K(\cdot)$ is not public, the secure key for decryption $\widehat{K}(\cdot)$ cannot be easily obtained from a pure chaotic synchronization. In other words, the cryptosystem still works even though attackers have gotten the ciphertext $\xi(t)$ and internal states.

3 Takagi–Sugeno Fuzzy Modeling for Chaotic Systems

Since the chaotic cryptosystem in the previous section illustrates the importance of obtaining same keys $\widehat{K}(\cdot) = K(\cdot)$ and the same ciphertexts

$\widehat{\xi}(\cdot) = \xi(\cdot)$, we now give detailed explanation of the synchronization methodology. Consider a general discrete-time nonlinear dynamic equation as follows:

$$x(t+1) = f(x(t)), \tag{1}$$

where $x \in R^n$ is the state vector and $f(\cdot)$ is a nonlinear function appropriate dimension. Then the fuzzy model is composed of the following rules:

Plant Rule i: IF $z_1(t)$ is F_{1i} and $\cdots$ and $z_g(t)$ is F_{gi}
THEN $x(t+1) = A_i x(t) + b_i(t)$, $i = 1, 2, \ldots, r$, \tag{2}

where $z_1(t) \sim z_g(t)$ are the premise variables which consist of the states of the system; F_{ji} $(j = 1, 2, \ldots, g)$ are the fuzzy sets; r is the number of fuzzy rules; A_i is a system matrix with appropriate dimension; and $b_i(t) \in R^n$ denotes the bias term.

Now focus on how to construct a TS fuzzy model (2) which exactly represents the nonlinear system (1). First, consider a scalar nonlinear system $x(t+1) = a(x)$ where $a(x)$ is only dependent on a state variable x, and it is assumed that x varies among a universe of discourse Ω. Assume that the nonlinear term can be represented in the form $\phi(x)x$, i.e., we assume $\phi(x) = a(x)/x$ exists. Notice that the bias term $b_i(t)$ in (2) arises if we cannot extract x from the nonlinear term. Then the nonlinear system is represented by

Rule i: IF z is F_i THEN $x(t+1) = d_i x$,

which has the following inferred output:

$$x(t+1) = \sum_{i=1}^{r} \mu_i(z) \, d_i x.$$

The membership function F_i is assumed to be normalized and is denoted by $\mu_i(z)$, i.e., satisfying $1 \geq \mu_i(z) \geq 0$ and $\sum_{i=1}^{r} \mu_i(z) = 1$. To represent $\phi(x) = \sum_{i=1}^{r} \mu_i(z) \, d_i$ exactly, we must suitably assign $\mu_i(z)$ and d_i. As an example, for the rule base containing only two rules, i.e., $r = 2$, we can choose

$$\mu_1 = \frac{-d_2}{d_1 - d_2} + \frac{1}{d_1 - d_2}\phi(x), \qquad \mu_2 = 1 - \mu_1,$$

where d_1 and d_2 are determined such that $\mu_i(z) \in [0\ 1]$ for all $x \in \Omega$.

Second, consider a system $x(t+1) = f(x)$, where the vector $f(x)$ with a nonlinear term $\phi(z) x_k$ appear at some lth entry. Accordingly, the system may be expressed as

Rule i: IF z is F_i THEN $x(t+1) = \bar{A}_i x + A_0 x$, where $A_0 x$ denotes the linear part and the normalized membership function for F_i is denoted by $\mu_i(z)$. Then $\mu_i(z)$ and d_i can be determined according to the above case. The fuzzy system is inferred as follows:

$$x(t+1) = \sum_{i=1}^{r} \mu_i(z)\, \bar{A}_i x + A_0 x \, ,$$

where $\sum_{i=1}^{r} \mu_i(z)\, \bar{A}_i x$ is to represent $\phi(x)\, x_k$. Notice that all entries of $\bar{A}_i$ are zeroes except for the (l, k)-entry is d_i. Then we have

$$x(t+1) = \sum_{i=1}^{r} \mu_i(z)\, A_i x = f(x)$$

by defining $A_i = \bar{A}_i + A_0$.

Third, the basic modeling technique is extended to deal with a vector function including two nonlinear terms, i.e., $f(x) = f_1(x) + f_2(x) + A_0 x$. The nonlinear terms are, respectively, assumed with the following forms: $f_1(x) = \sum_{i_1=1}^{q_1} \mu_{1i_1}(z_1)A_{1i_1}x$ and $f_2(x) = \sum_{i_2=1}^{q_2} \mu_{2i_2}(z_2)A_{2i_2}x$. Then, the following extension is obtained:

Rule k: IF z_1 is $F_{1i_1(k)}$ and z_2 is $F_{2i_2(k)}$

$$\text{THEN} \quad x(t+1) = (A_{1i_1(k)} + A_{2i_2(k)} + A_0)x \tag{3}$$

where the indices $i_1(k)$ and $i_2(k)$ are defined as follows:

$$i_1(k) = \left[\frac{k-1}{q_2}\right] + 1 \qquad i_2(k) = k - \left[\frac{k-1}{q_2}\right] q_2 \, .$$

where $[\cdot]$ denotes the floor function. Let $\mu_{1i_1(k)}$ and $\mu_{2i_2(k)}$ denote the normalized membership functions of F_{1i_1} and F_{2i_2}, respectively. A formula for the multiindex is

$$\sum_{i_1=1}^{q_1} \sum_{i_2=1}^{q_2} \mu_{i_1}\mu_{i_2} = \sum_{k=1}^{q_1 q_2} \mu_{i_1(k)}\mu_{i_2(k)} \, . \tag{4}$$

For the fuzzy system (3), the inferred output is

$$x(t+1) = \sum_{k=1}^{q_1 q_2} \mu_{1i_1(k)}(z_1)\mu_{2i_2(k)}(z_2)(A_{1i_1(k)} + A_{2i_2(k)} + A_0)x \, . \tag{5}$$

According to (4), the right-hand side of (5) leads to

$$\sum_{i_1=1}^{q_1} \sum_{i_2=1}^{q_2} \mu_{1i_1}(z_1) \mu_{2i_2}(z_2) (A_{1i_1} + A_{2i_2} + A_0) x$$

$$= \sum_{i_1=1}^{q_1} \sum_{i_2=1}^{q_2} \mu_{1i_1}(z_1) \mu_{2i_2}(z_2) A_{1i_1} x$$

$$+ \sum_{i_1=1}^{q_1} \sum_{i_2=1}^{q_2} \mu_{1i_1}(z_1) \mu_{2i_2}(z_2) A_{2i_2} x + A_0 x$$

$$= \sum_{i_1=1}^{q_1} \mu_{1i_1}(z_1) A_{1i_1} x + \sum_{i_2=1}^{q_2} \mu_{2i_2}(z_2) A_{2i_2} x + A_0 x$$

$$= f_1(x) + f_2(x) + A_0 x \, ,$$

which is equal to $f(x)$. Note that μ_{1i_1} and μ_{2i_2} are regarded as normalized membership functions by assuming $\sum_{i_1=1}^{q_1} \mu_{1i_1} = \sum_{i_2=1}^{q_2} \mu_{2i_2} = 1$ and $1 \geq \mu_{km}(z) \geq 0$ for each k and m.

We do not extend the above fuzzy modeling technique to a more general nonlinear system, since most well-known continuous-time and discrete-time chaotic systems have at most two nonlinear terms. Actually, the well-known Lure type discrete-time chaotic systems can be exactly represented by TS fuzzy models with only one premise variable. To further simplify the design of the response system, an intuitive choice is to let the premise variable as the output signal. In light of the above, the general form of TS fuzzy models for chaotic systems is written as follows:

$$\text{Plant Rule } i\colon \text{IF } y(t) \text{ is } F_i \quad \text{THEN } x(t+1) \\ = A_i x(t) + b_i(t), \quad i = 1, 2, \ldots, r \, , \tag{6}$$

where the scalar variable $y(t)$ will be taken as the output signal. The overall inferred output can be written as

$$x(t+1) = \sum_{i=1}^{r} \mu_i(y(t)) \{A_i x(t) + b_i(t)\} \, ,$$

$$y(t) = C x(t) \, ,$$

where $\mu_i(y(t)) = \frac{\omega_i(y(t))}{\sum_{i=1}^{r} \omega_i(y(t))}$ with $\omega_i(y(t)) = F_i(y(t)) \geq 0$.

TS fuzzy models for several well-known Lure type discrete-time chaotic systems are given in Table 1 and the details can be found in [20].

4 Chaotic Cryptosystem Using Discrete-time Systems

In this section, the ciphertext is masked by either state or output of a chaotic system, and the modulation process is carried out by injecting the masking signal into the fuzzy chaotic transmitter. Then the masked signal is sent to the fuzzy chaotic receiver, where the ciphertext is extracted according to the masking methods. This idea is explained in the following:

Table 1. Takagi–Sugens fuzzy model of lure type discrete-time chaotic systems

Chaotic Systems	Dynamical Equations	Fuzzy Sets	System Matrices	Bias Terms
Hénon map	$x_1(t+1) = 1.4 - x_1^2(t) + 0.3x_2(t)$ $x_2(t+1) = x_1(t)$	$F_1(x_1) = \frac{1}{2}(1 + x_1/d)$ $F_2(x_1) = \frac{1}{2}(1 - x_1/d)$ $d = 2$	$A_1 = \begin{bmatrix} -d & 0.3 \\ 1 & 0 \end{bmatrix}; A_2 = \begin{bmatrix} d & 0.3 \\ 1 & 0 \end{bmatrix}$	$b_1 = b_2$ $= \begin{bmatrix} 1.4 \\ 0 \end{bmatrix}$
Lozi map	$x_1(t+1) = 3 - 1.8|x_1(t)| + x_2(t)$ $x_2(t+1) = 0.25x_1(t)$	$F_1(x_1) = |x_1|/d$ $F_2(x_1) = 1 - |x_1|/d$ $d = 3.5$	$A_1 = A_2 = \begin{bmatrix} 0 & 1 \\ 0.25 & 0 \end{bmatrix}$	$b_1 = \begin{bmatrix} 3-1.8d \\ 0 \end{bmatrix}$ $b_2 = \begin{bmatrix} 3 \\ 0 \end{bmatrix}$
Cubic map	$x_1(t+1) = 2.77x_1(t) - x_1^3(t)$ $\quad -0.2x_2(t)$ $x_2(t+1) = x_1(t)$	$F_1(x_1) = \frac{1}{2}(1 + x_1^2/d)$ $F_2(x_1) = \frac{1}{2}(1 - x_1^2/d)$ $d = 5$	$A_1 = \begin{bmatrix} 2.77-d & -0.2 \\ 1 & 0 \end{bmatrix}$ $A_2 = \begin{bmatrix} 2.77+d & -0.2 \\ 1 & 0 \end{bmatrix}$	$b_1 = b_2$ $= 0$
Three-dimensional system	$x_1(t+1) = 1 + x_2(t) + x_1^2(t)$ $x_2(t+1) = 0.33x_1(t) + x_3(t)$ $x_3(t+1) = -0.33x_1(t)$	$F_1(x_1) = \frac{1}{2}(1 + x_1/d)$ $F_2(x_1) = \frac{1}{2}(1 - x_1/d)$ $d = 5$	$A_1 = \begin{bmatrix} d & 1 & 0 \\ 0.33 & 0 & 1 \\ -0.33 & 0 & 0 \end{bmatrix}$ $A_2 = \begin{bmatrix} -d & 1 & 0 \\ 0.33 & 0 & 1 \\ -0.33 & 0 & 0 \end{bmatrix}$	$b_1 = b_2$ $= \begin{bmatrix} 1 \\ 0 \\ 0 \end{bmatrix}$

4.1 Ciphertext Masked by State

First, consider the ciphertext be masked by chaotic states. Although users' messages can be encrypted by the algorithm in Sect. 2, higher security is required in some applications. To this end, let the fuzzy chaotic transmitter with ciphertext ξ embedded to be

$$\begin{aligned}
\text{Transmitter Rule } i: \text{ IF } \bar{y}(t) \text{ is } F_i \\
\text{THEN } x(t+1) = A_i \bar{x}(t) + b_i(t) \\
\bar{x}(t) = x(t) + M_s \xi(t) \\
\bar{y}(t) = C\bar{x}(t) ,
\end{aligned}$$

where $\bar{x}(t) \in R^n$ represents the composed states that mask the ciphertext ξ; $M_s = [m_1 \dots m_n]^T \in R^n$ with $m_j \in [0,1]$ for $j = 1, 2, ..., n$, denotes a public state masking key which specifies the state used to mask the ciphertext; and $\bar{y}(t)$ is the coupling signal which is transmitted to a receiver through a public channel. The inferred output of the transmitter is

$$\begin{aligned}
x(t+1) &= \sum_{i=1}^{r} \mu_i(\bar{y}(t))[A_i x(t) + b_i(t) + A_i M_s \xi(t)] , \\
\bar{y}(t) &= Cx(t) + CM_s \xi(t) .
\end{aligned} \tag{7}$$

The modulation form (7) can be regarded as an extension of modulated chaotic communications. To recover the message, the fuzzy receiver is designed as

Receiver Rule i : IF $\overline{y}(t)$ is F_i
$$\text{THEN } \widehat{x}(t+1) = A_i x(t) + b_i(t) + L_i(\overline{y}(t) - \widehat{y}(t))$$
$$\widehat{y}(t) = C\widehat{x}(t) ,$$

where $\widehat{x}(t)$ denotes the estimate of state $x(t)$; $\widehat{y}(t)$ denotes the estimate of output $\overline{y}(t)$; and $L_i \in R^n$ is a design vector. The fuzzy inferred receiver can be expressed in the form

$$\widehat{x}(t+1) = \sum_{i=1}^{r} \mu_i(\overline{y}(t))[A_i \widehat{x}(t) + b_i(t) + L_i(\overline{y}(t) - \widehat{y}(t))] ,$$
$$\widehat{y}(t) = C\widehat{x}(t) . \tag{8}$$

Define the error signal $\widetilde{x}(t) = x(t) - \widehat{x}(t)$ and $\widetilde{y}(t) = \overline{y}(t) - \widehat{y}(t)$. From (7) and (8), the error system is obtained as

$$\widetilde{x}(t+1) = \sum_{i=1}^{r} \mu_i(\overline{y}(t)) \left[(A_i - L_i C)\widetilde{x}(t) + (A_i - L_i C)M_s \xi(t) \right] , \tag{9}$$
$$\widetilde{y}(t) = C\widetilde{x}(t) + CM_s \xi(t) . \tag{10}$$

For simplicity, an exact linearization technique is used here to yield $(A_1 - L_1 C) = (A_i - L_i C) = \overline{A}$, for $i > 1$. Then the state masking key M_s will be properly chosen such that the ciphertext can be extracted from $\widetilde{y}$ as (10).

Theorem 1. *Consider the chaotic transmitter (7) and receiver (8). If there exist a common positive definite matrix P, $\alpha > 0$, and design gain L_i such that matrix $\overline{A}$ is nilpotent as well as the following LMI problem*

$$\begin{array}{l} \underset{P, W_i}{minimize} \; \alpha \\ subject \; \alpha > 0, \; P > 0 \\ \begin{bmatrix} \alpha I & (PA_1 - W_1 C - PA_i - W_i C)^T \\ PA_1 - W_1 C - PA_i - W_i C & P \end{bmatrix} > 0 \quad (11) \\ for \; all \; 1 < i \leq r \end{array}$$

with $W_i = PL_i$, is feasible. Then the state masking key M_s is properly chosen from (for $j = 0, 1, \ldots, n - 1$)

$$C\overline{A}^j M_s = \begin{cases} u_m, & if \;\; j = l , \\ 0, & otherwise , \end{cases} \tag{12}$$

where $l \in \{1, 2, \ldots, n\}$ is the number of the steps up to the initial recovery of the message and $u_m \neq 0$ is an unmasking constant. The ciphertext is recovered according to $\xi(t - l) = \frac{1}{u_m}\widetilde{y}(t)$ as well as all states of chaotic transmitter and receiver are synchronized after n steps.

Proof. For the chaotic communication system satisfying LMIs (11), the error system (9) is reduced to

$$\tilde{x}(t+1) = \overline{A}\tilde{x}(t) + \overline{A}M_s\xi(t) , \tag{13}$$

where $\overline{A} = A_1 - L_1 C = A_i - L_i C$ and the gains L_i is obtained as $L_i = P^{-1}W_i$ from the solutions P and W_i. This leads to the solution of the error system (13) to be

$$\tilde{x}(t) = \overline{A}^t\tilde{x}(0) + \sum_{j=1}^{t} \overline{A}^j M_s\xi(t-j) ,$$

$$\tilde{y}(t) = C\overline{A}^t\tilde{x}(0) + CM_s\xi(t) + \sum_{j=1}^{t} C\overline{A}^j M_s\xi(t-j) . \tag{14}$$

Due to this matrix $\overline{A}$ is nilpotent, i.e., $\overline{A}^n = 0$, then after n steps $(t \geq n)$ the output error becomes

$$\tilde{y}(t) = \sum_{j=0}^{n-1} C\overline{A}^j M_s\xi(t-j) .$$

According to (12), the recovered ciphertext $\xi(t-l) = \frac{1}{u_m}\tilde{y}(t)$. Moreover, since synchronization is achieved, $\widehat{K}(t) \to K(t)$ after the transient time.

Remark 3. A stronger result is obtained if the LMI problem (11) is feasible and $\overline{A}$ is stable. In this case, the state masking key M_s can be properly chosen to satisfy $\overline{A}M_s = 0$ such that the error solution (14) becomes

$$\tilde{x}(t) = \overline{A}^t\tilde{x}(0) ,$$

$$\tilde{y}(t) = C\overline{A}^t\tilde{x}(0) + CM_s\xi(t) .$$

Then, the ciphertext is extracted from $\xi(t) = \tilde{y}(t)/(CM_s)$.

When the conditions in Theorem 1 or Remark 1 are satisfied, the chaotic transmitter and receiver are synchronized. In other words, the internal states of the chaotic system $x_1(t), \ldots, x_n(t)$ can be taken as the secure key $K(t)$, while the decryption secure key $\widehat{K}(t)$ is available in the receiver end. Therefore, the cryptosystem components mentioned in Sect. 2 are complete. The following is an application example.

Example 1. Here we use the discrete-time Henon map to illustrate the above chaotic cryptosystem design. The Henon map is as follows:

$$x_1(t+1) = -x_1^2(t) + 0.3x_2(t) + 1.4 ,$$
$$x_2(t+1) = x_1(t) ,$$
$$y(t) = x_1(t) . \tag{15}$$

Take $x_1(t)$ as the premise variable of the fuzzy rules, Henon map in the fuzzy representation (6) consists of $x(t) = [x_1(t) \quad x_2(t)]^T$, $C = [1 \quad 0]$, the fuzzy sets $F_1(y(t)) = \frac{1}{2}(1 + \frac{y(t)}{d})$ and $F_2(y(t)) = \frac{1}{2}(1 - \frac{y(t)}{d})$, and

$$A_1 = \begin{bmatrix} -d & 0.3 \\ 1 & 0 \end{bmatrix}, \quad A_2 = \begin{bmatrix} d & 0.3 \\ 1 & 0 \end{bmatrix}, \quad b_1 = b_2 = \begin{bmatrix} 1.4 \\ 0 \end{bmatrix},$$

where $d = 2$ and $x_1(t) \in [-d \ d]$. According to Theorem 1, the design vectors are obtained as $L_1 = [-d \quad 1]^T$ and $L_2 = [d \quad 1]^T$. In turn, the closed-loop error dynamics (13) is written as

$$\tilde{x}(t+1) = \begin{bmatrix} 0 & 0.3 \\ 0 & 0 \end{bmatrix} \tilde{x}(t) + \begin{bmatrix} 0 & 0.3 \\ 0 & 0 \end{bmatrix} \begin{bmatrix} m_1 \\ m_2 \end{bmatrix} \xi(t).$$

Thus, the ciphertext will be masked by the state $x_1(t)$, i.e., we select the state masking key as $M_s = [1 \quad 0]^T$. Moreover, assume that the number of multiusers are set to 8 and the plaintexts are randomly binary $[0 \ 1]$. Through the transmission, the encryption and decryption algorithm in Sect. 2 are applied, whereas the state $x_1(t)$ be taken as the output which generates the secure key $K(t)$. Then, applying the chaotic transmitter (7) and receiver (8), the cryptosystem is completed. In Fig. 2, we illustrate the chaotic coupling signal; scaled ciphertext $m(\cdot)$, $\hat{m}(\cdot)$ (dotted line); encrypting function $E(\cdot)$, $\hat{E}(\cdot)$ (dotted line); and chaotic phase portrait, respectively. In Figs. 3a–b, $u1 \sim u4$ are the first four user plaintext transmitted (total of eight users) and error between the recovered plaintext and original plaintext of users $u1 \sim u4$.

4.2 Ciphertext Masked by Output

Let the ciphertext $\xi(t)$ to be directly added into the output of the chaotic system. The chaotic transmitter is expressed in a TS fuzzy representation as

> Transmitter Rule i: IF $\bar{y}(t)$ is F_i
> THEN $x(t+1) = A_i x(t) + b_i(t) + L_i M_o \xi(t)$
> $\bar{y}(t) = C x(t) + M_o \xi(t)$, $i = 1, 2, \ldots, r$,

where the gains L_i, $i = 1, 2, \ldots, r$, will be determined later, and M_o is a public output masking key which masks the ciphertext by a constant. The fuzzy inferred result for the chaotic transmitter is obtained as

$$x(t+1) = \sum_{i=1}^{r} \mu_i(\bar{y}(t)) \left\{ \overline{A}_i x(t) + b_i(t) + L_i \bar{y}(t) \right\},$$

$$\bar{y}(t) = C x(t) + M_o \xi(t), \tag{16}$$

where $\overline{A}_i = A_i - L_i C$. To recover the ciphertext, the chaotic fuzzy receiver is designed as

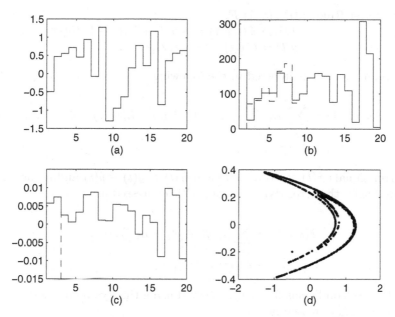

Fig. 2. (a) Chaotic coupling signal; (b) encryption function $E(\cdot)$, $\hat{E}(\cdot)$ (*dotted line*); (c) scaled ciphertext $m(t)$, $\hat{m}(t)$; and (d) phase portrait

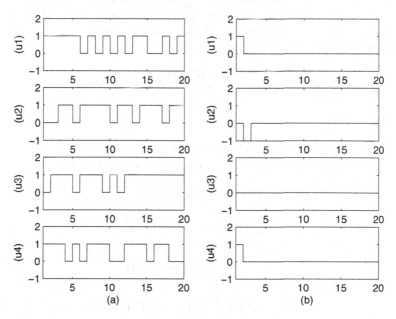

Fig. 3. (a) Plaintext transmitted my users 1 to 4 (total eight users) and (b) error between original message and recovered message of user 1 to 4

Receiver Rule i: IF $\overline{y}(t)$ is F_i
$$\text{THEN } \hat{x}(t+1) = A_i\hat{x}(t) + b_i(t) + L_i\left(\overline{y}(t) - \hat{y}(t)\right)$$
$$\hat{y}(t) = C\hat{x}(t), \quad i = 1, 2, \ldots, r.$$

The overall receiver is inferred in the following:

$$\hat{x}(t+1) = \sum_{i=1}^{r} \mu_i(\overline{y}(t)) \left\{A_i x(t) + b_i(t) + L_i\left(\overline{y}(t) - \hat{y}(t)\right)\right\},$$
$$\hat{y}(t) = C\hat{x}(t). \tag{17}$$

Let error signals $\tilde{x}(t) \equiv x(t) - \hat{x}(t)$ and $\tilde{y}(t) \equiv \overline{y}(t) - \hat{y}(t)$ and according to (16) and (17), the error dynamics of $\tilde{x}(t)$ is expressed as

$$\tilde{x}(t+1) = \sum_{i=1}^{r} \mu_i(\overline{y}(t))(A_i - L_iC)\tilde{x}(t), \tag{18}$$
$$\tilde{y}(t) = C\tilde{x}(t) + M_o\xi(t). \tag{19}$$

The stability conditions for (19) is derived using the Lyapunov method. The main result is addressed here.

Theorem 2. *Consider the chaotic transmitter (16) and receiver (17). The ciphertext can be recovered from $\xi(t) = \frac{1}{M_o}\tilde{y}(t)$ and all states of chaotic transmitter and receiver are synchronized in an asymptotic manner if there exist a common positive definite matrix P and gains L_i, for $i = 1, 2, \ldots, r$, such that the following LMIs are satisfied:*

$$\begin{bmatrix} P & (PA_i - W_iC)^{\mathrm{T}} \\ PA_i - W_iC & P \end{bmatrix} > 0, \quad \text{for all } i, \tag{20}$$

where $W_i \equiv PL_i$.

Proof. Given a Lyapunov function candidate as $V(\tilde{x}(t)) = \tilde{x}^{\mathrm{T}}(t)P\tilde{x}(t) > 0$. Taking difference of $V(t)$ along the error dynamics (18) yields

$$\triangle V(\tilde{x}(t)) = V(\tilde{x}(t+1)) - V(\tilde{x}(t))$$
$$= \sum_{i=1}^{r} \mu_i^2(\overline{y}(t))\tilde{x}^{\mathrm{T}}(t)[\overline{A}_i^{\mathrm{T}}P\overline{A}_i - P]\tilde{x}(t)$$
$$+ \sum_{i<j}^{r} \mu_i(\overline{y}(t))\mu_j(\overline{y}(t))$$
$$\cdot\tilde{x}^{\mathrm{T}}(t)[\overline{A}_i^{\mathrm{T}}P\overline{A}_j + \overline{A}_j^{\mathrm{T}}P\overline{A}_i - 2P]\tilde{x}(t), \tag{21}$$

where $P > 0$ and $\overline{A}_i = A_i - L_iC$. Notice that if $\overline{A}_i^{\mathrm{T}}P\overline{A}_i - P < 0$, then $\overline{A}_i^{\mathrm{T}}P\overline{A}_j + \overline{A}_j^{\mathrm{T}}P\overline{A}_i - 2P < 0$. This means if there are P and L_i such that the conditions (20) are held, then $\overline{A}_i^{\mathrm{T}} P \overline{A}_i - P < 0$ by Schur complement. Let $-Q$ denote the maximum negative definite matrix of $\overline{A}_i^{\mathrm{T}} P \overline{A}_i - P$ for all i. Then $\Delta V(\tilde{x}(t)) \leq -\tilde{x}^{\mathrm{T}}(t) Q \tilde{x}(t) < 0$. Thus, the synchronization error $\tilde{x}(t)$ converges to zero as $t \to \infty$. According to the (19), $\tilde{y}(t)$ converges to $M_o \xi(t)$ as $t \to \infty$.

Since the convergent rate of the synchronization error $\tilde{x}(t)$ affects the transmission performance, the decay rate design for chaotic cryptosystems can be performed by solving LMIs problems as follows:

Chaotic Cryptosystem with decay rate:

$$\underset{P, W_i}{\text{minimize}} \beta$$

$$\text{subject to } P > 0, 0 < \beta < 1$$

$$\begin{bmatrix} \beta P & (PA_i - W_i C)^{\mathrm{T}} \\ PA_i - W_i C & P \end{bmatrix} > 0, \quad \text{for all } i \,,$$

where $W_i \equiv PL_i$. The (21) becomes $\Delta V(\tilde{x}(t)) \leq -(1 - \beta) V(\tilde{x}(t))$, with parameter β tuning the decay rate.

Example 2. Consider the above chaotic cryptosystem using the discrete-time Henon map (15). Assume the cryptosystem serves eight users whose the plaintexts are randomly binary $[0 \ 1]$. Let the state $x_1(t)$ be the output which generates the secure key $K(t)$. The cyphertext, which is obtained from the encryption algorithm in Sect. 2, is added at the output of the drive system $\overline{y}(t)$. According to the fuzzy representation of Henon map in Example 1 and Theorem 2, the chaotic transmitter (16) and receiver (17) are designed with gain vectors $L_1 = [-2.1384 \ 5.6608]^{\mathrm{T}}$ and $L_2 = [2.1384 \ 5.6608]^{\mathrm{T}}$. Note that the parameter of the decay rate is solved as $\beta = 0.1$. For simplicity, we set the output masking key as $M_o = 1$. Figure 4 shows the chaotic coupling signal; scaled ciphertext $m(\cdot)$, $\hat{m}(\cdot)$ (dotted line); encrypting function $E(\cdot)$, $\widehat{E}(\cdot)$ (dotted line); and chaotic phase portrait, respectively. In Figs. 5a–b, $u1 \sim u4$ are the first four user plaintext transmitted (total of eight users) and error between the recovered plaintext and original plaintext of users $u1 \sim u4$.

5 DSP-Based Experiments

To verify the theoretical results, we carry out DSP-based experiments on the chaotic cryptosystem. Here, the hardware used is the DSpace DS1102 single board system which is based on Texas Instruments TMS320C31 DSP. Note that a minimum bit rate of the DSP is necessary to avoid quantization errors. In other words, we must have at least a bit rate of 12 to ensure that the signal generated by the DSP maintains chaotic characteristics. This is an important feature in order to achieve synchronization and generate the superincreasing

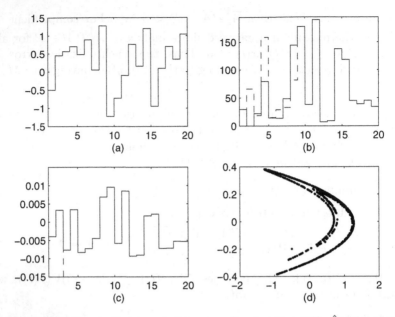

Fig. 4. (a) Chaotic coupling signal; (b) encryption function $E(\cdot)$, $\hat{E}(\cdot)$ (*dotted line*); (c) scaled ciphertext $m(t)$, $\hat{m}(t)$; and (d) phase portrait

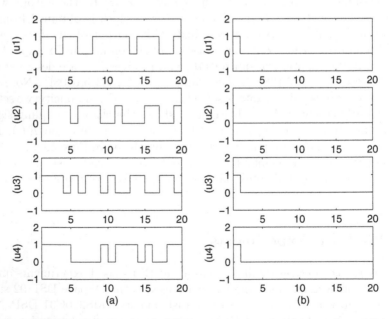

Fig. 5. (a) Plaintext transmitted my users 1 to 4 (total eight users); and (b) error between original message and recovered message of user 1 to 4

sequence. To increase the performance, the step size of t may be set to a small value. But for simplicity, the step size of t is set as 1 s. The coupling signal is externally connected from the output of the drive system through D/A channel to the A/D channel leading to the response system. The chaotic system structure and parameters are same as the numerical simulations. First, let the ciphertext masked by state as Examples 1, then we obtain the results as shown in Fig. 6, where we illustrate the scaled ciphertext $(\xi(t) \times 10)$, chaotic coupling signal, the synchronization of $x_1(t)$ and $\hat{x}_1(t)$, and the message of user 2 along with the error of recovered message. Next, let the ciphertext masked by output as Examples 2, we obtain the results as shown in Fig. 7. From the oscilloscope images, after the transient responses, the plaintext is recovered exactly. Note here that the recovering of ciphertext and decoding into plaintext is implemented in a real-time sense. Therefore comparing the numerical and experimental results, we are able to conclude the consistency of the theoretical derivation.

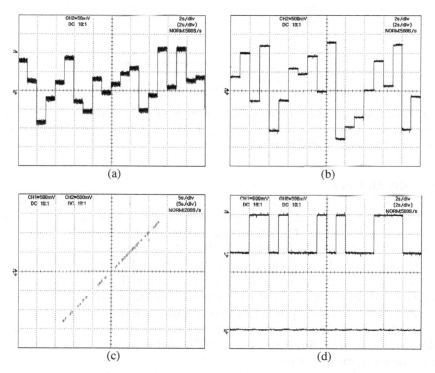

Fig. 6. Oscilloscope images of (**a**) scaled ciphertext $(\xi(t) \times 10)$; (**b**) chaotic coupling signal; (**c**) synchronization of $x_1(t)$ and $\hat{x}_1(t)$; and (**d**) the message of user 2 along with the error of recovered message

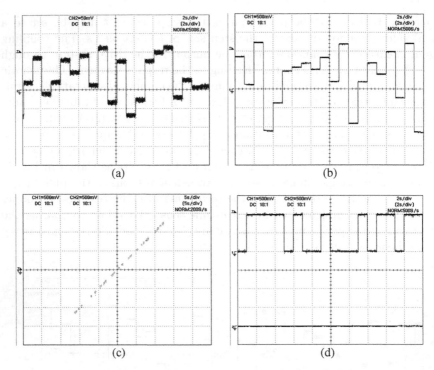

Fig. 7. Oscilloscope images of (**a**) scaled ciphertext ($\xi(t) \times 10$); (**b**) chaotic coupling signal; (**c**) synchronization of $x_1(t)$ and $\hat{x}_1(t)$; and (**d**) the message of user 2 along with the error of recovered message

6 Conclusions

In this study, we have proposed a systematic but flexible design for a chaotic cryptosystem based on TS fuzzy models. Extending the properties of synchronization, chaotic communications have been proposed in the sense of signal masking and encryption. In addition, the ciphertext may be embedded in the state or output of the drive system which again enhances the flexibility of design. Compared to traditional signal masking based chaotic communications, this cryptosystem provides message transmitting for multiusers, whereas a basic encrypting algorithm makes the channels more secure. Moreover, numerical simulations and DSP-based experiments are consistent with the theoretical results.

Acknowledgment

This work was supported by the National Science Council, R.O.C., under Grant NSC-89-2213-E033-006.

References

1. T. Takagi, M. Sugeno: IEEE Trans. Syst. Man Cybern. **15**, 116–132 (1985)
2. H.O. Wang, K. Tanaka, M.F. Griffin: IEEE Trans. Fuzzy Syst. **4**, 14–23 (1996)
3. S. Boyd, L.E. Ghaoui, E. Feron, V. Balakrishnan: *Linear Matrix Inequalities in System and Control Theory* (SIAM, Philadelphia, PA, 1994)
4. T.L. Carroll, L.M. Pecora: IEEE Trans. Circuits Syst. **38**, 453–456 (1991)
5. C. Chen, X. Dong: *From Chaos to Order Methodologies, Perspectives and Applications* (World Scientific, Singapore, 1998)
6. M. Lakshmanan, K. Murali: *Chaos in Nonlinear Oscillators: Controlling and Synchronization* (World Scientific, Singapore, 1996)
7. O. Morg¨ul, E. Solak: Phys. Rev. E **54**, 4803–4811 (1996)
8. H.J.C. Huijberts, H. Nijmeijer, A. Yu. Pogromsky: An observer point of view on synchronization of discrete-time systems. In: *Proc ISCAS*, 2000, pp 491–494
9. G. Grassi, S. Mascolo: IEEE Trans. Circuits Syst. II **46**, 478–483 (1999)
10. K.M. Cuomo, A.V. Oppenheim, S.H. Strogatz: IEEE Trans. Circuits Syst. II **40**, 626–633 (1993)
11. L.J. Kocarev, K.D. Halle, K. Eckert, L.O. Chua, U. Parlitz: *Int. J. Bifurcation Chaos* **2**, 709–713 (1992)
12. T.L. Liao, N.S. Huang: IEEE Trans. Circuits Syst. I **46**, 1144–1150 (1999)
13. C.W. Wu, L.O. Chua: Int. J. Bifurcation Chaos **3**, 1619–1627 (1993)
14. K.S. Halle, C.W. Wu, M. Itoh, L.O. Chua: Int. J. Bifurcation Chaos **3**, 469–477 (1993)
15. K.Y. Lian, T.S. Chiang, P. Liu: Int. J. Bifurcation Chaos **10**, 2193–2206 (2000)
16. K. Short: Int. J. Bifurcation Chaos **4**, 959–977 (1994)
17. M. Brucoli, D. Cafagna, L. Carnimeo, G. Grassi: Int. J. Bifurcation Chaos **9**, 2027–2037 (1999)
18. G. Grassi, S. Mascolo: IEEE Trans. Circuits Syst. II **46**, 478–483 (1999)
19. T. Yang, C.W. Wu, L.O. Chua: IEEE Trans. Circuits Syst. I **44**, 469–472 (1997)
20. K.Y. Lian, T.S. Chiang, P. Liu, C.S. Chiu: IEEE Trans. Fuzzy Syst. **9**, 539–553 (2001)
21. K.Y. Lian, P. Liu, C.S. Chiu, T.S. Chiang: Int. J. Bifurcation Chaos **13**, 215–225 (2003)
22. K.Y. Lian, P. Liu, C.S. Chiu, T.S. Chiang: Int. J. Bifurcation Chaos **12**, 835–846 (2002)
23. M.R. Garey, D.S. Johnson: *Computers and Intractability: A Guide to the Theory of NP-Completeness* (WH Freeman, San Francisco, CA, 1976)
24. D.E.R. Denning: *Cryptography and Data Security* (Addison Wesley, New York, 1982)

Evolution of Complexity

Pavel Ošmera

Abstract. The strength of physical science lies in its ability to explain phenomena as well as make prediction based on observable, repeatable phenomena according to known laws. Science is particularly weak in examining unique, nonrepeatable events. We try to piece together the knowledge of evolution with the help of biology, informatics, and physics to describe a complex evolutionary structure with unpredictable behavior. Evolution is a procedure where matter, energy, and information come together. Our research can be regarded as a natural extension of Darwin's evolutionary view of the last century. We would like to find plausible uniformitarian mechanisms for evolution of complex systems. Workers with specialized training in overlapping disciplines can bring new insights to an area of study, enabling them to make original contributions. This chapter describes evolution of complexity as a basic principle of evolutionary computation.

1 Introduction

Optimization is an important aspect of many scientific and engineering problems. Recent optimization techniques model principles of natural evolution. Evolutionary algorithms apply selection and mutation operators to a population of states to guide the population to an optimal solution of the objective function.

Classic optimization methods often lead to unacceptably poor performance when applied to real-world circumstances. A more robust optimization technique is required. Applying the logical aspects of the evolutionary process to optimization offers several distinct advantages. There exists a large body of knowledge about the process of natural evolution that can be used to guide simulations. This process is well suited for solving problems with unusual constraints where heuristic solutions are not available or generally lead to unsatisfactory results. Often revolution has an interdisciplinary character. Its central discoveries often come from people straying outside the normal bounds of their specialties.

Naturalistic explanations of life's origin are speculative [1]. But does this mean that such inquiries are impotent or without value? The same criticism can be made of any attempt to reconstruct unique events of the past. We cannot complete our knowledge without answering some of the fundamental questions about nature. How does life begin? What is turbulence? Above

P. Ošmera: *Evolution of Complexity*, StudFuzz **187**, 527–578 (2006)
www.springerlink.com   © Springer-Verlag Berlin Heidelberg 2006

all, in a universe ruled by entropy, it had been drawing inexorably toward greater and greater disorder, how does order arise? Although the various speculative origin scenarios may be tested against data collected in laboratory experiments, these models cannot be tested against the actual events in question, i.e., the origin. Such scenarios, then, must ever remain speculation, not knowledge. There is no way to know whether the results from these experiments tell anything about the way life itself originated. In a strict sense, these speculative reconstructions are not falsifiable; they may only be judged plausible or implausible. In the familiar Popper's sense of what science is, a theory is deemed scientific if it can be checked or tested by experiment against observable, repeatable phenomena. Behavior of complex nonlinear systems with unpredictable behavior can be explained by a relatively simple and transparent system—a magnetic pendulum (see Fig. 1). The idea is to set the pendulum swinging and guess which attractor will win. Even with just three magnets placed in a triangle, the pendulum's motion cannot be predicted. The unexpected behavior can be extended to physiological and psychiatric medicine, economic forecasting, and perhaps the evolution of society. A physicist could not truly understand turbulence or complexity unless he understood pendulums. The chaos began to unite the study of different systems. A simulation brings its own problem: the tiny imprecision build into each calculation rapidly takes over, because this is a system with sensitive dependence on initial conditions. But people have to know about disorder if they are going to deal with it. Classical scientists want to discover regularities. It is not easy to find the grail of science, the Grand Unified Theory or the "theory of everything." On the other hand, there is a trend in science toward reductionism, the analysis of system only in terms of their constituent

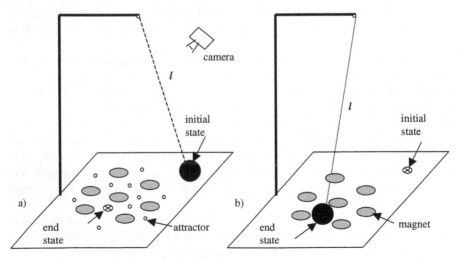

Fig. 1. Magnetic pendulum

parts: quarks, chromosomes, or neurons. Some scientists believe that they are looking for the whole.

In the pendulum, energy is transformed between potential energy and kinetic energy. Assuming that we have removed all magnets from the platform, the pendulum will eventually stop swinging and come to rest in a perfectly vertical plane. When we pull the pendulum up off that vertical axis, we create potential energy. When we release the pendulum from our grip (from an initial state), this potential energy converts into kinetic energy. The first law of thermodynamics is the conservation of energy that states that energy may neither be created nor be destroyed—it is simply transformed from one state to another. In the magnetic pendulum, the forces of gravity and magnetism act on the pendulum to convert back and forth between potential and kinetic energy in a wild form. The second law of thermodynamics states that the entropy (disorder) of a closed system must always increase. There is friction on the pendulum when it pushes through the air (air resistance). This friction prevents anything from having perpetual motion. The potential and kinetic energies of the pendulum will transform into heat until the pendulum stops swinging. By placing varying numbers of magnets in varying positions on the base plate, we can create different movements and patterns in the swing of the pendulum. By flipping the magnets so that the north or south poles face up further creates different movements and patterns in the swing of the pendulum. By using the combination of varying the positions of the magnets on the plate, varying the number of magnets on the plate, and varying the upfacing poles, we can create virtually an endless pattern of crazy and unpredictable swings of the pendulum.

Magnetic fields are most easily understood in terms of magnetic field lines. These field lines define the direction and strength of the magnetic field at any location in 3D nonlinear space. These magnetic lines have both direction and strength—the closer we are to a magnetic source, the stronger the field lines. The magnetic field lines always begin on the north poles of a magnet and end on the south poles. The magnetic field of a magnetic dipole is approximately proportional to the inverse cube of the distance from the dipole. Therefore, if we double the distance from the magnet, then the magnetic field strength will be reduced by a factor of 8. This magnetic system is very complex. If we know the initial state we cannot predict the final state. Even with just three magnets on the base plate, we cannot predict the motion. On the other hand, if we know the end state we cannot derive the history to the initial state. The same problem is with life's origin. There was nobody with a camera to record life's history. We may wish that a crack team of scientific observers had been present to record and detail the origin of life when it occurred. But since there were no observers, and since we cannot go back to investigate the primitive earth, we must do what we can to gain after-the-fact evidence of what may have occurred. Mimicking the early earth is tricky business. Our conclusion about life's origin can be only speculative—a hypothesis that cannot be confirmed because life is a nonrepeatable event. We should not expect any meaningful results within laboratory time. Millions of years of simulation might be required for any detectable progress. This method would obscure the complex chemical interactions sought for observation by allowing literally thousands of different reactions to go on simultaneously.

The measure of the amount of information is called *entropy*. The term "entropy" was deliberately chosen for the name of the measure of average information, because of its similarity to entropy in thermodynamics. In thermodynamics, entropy is a measure of the degree of disorder of a system, and disorder is clearly related to information. We can consider the interpretation of entropy as the amount of information, or equivalently the "degree of surprise," which is obtained when a particular event has occurred. We expect that the information will depend on probability of event. If the event is certain to occur, and there is no surprise when the event is found to occur, then no information is received. We see that the amount of information is proportional to the logarithm of the probability. This arises essentially because, for independent events, probabilities are multiplicative, while information is additive. Information has both qualitative and quantitative aspects. We implicitly assume that the probabilities being considered remained unchanged with the passage of time. In many applications this assumption is justified but this stationary approach cannot be accepted to living systems (organisms).

The unit of information is known as the bit. The amount of information conveyed in an event depends on the probability of the event. For example, information is more higher if the speaker says "I am very pleased to tell you that you have just won the lottery" than in the statement "Your name is Mr. Brown" (if you are Mr. Brown). Quantity of information is proportional to the amount of recording medium required to store it. Redundancy is the presence of more symbols in a message than is strictly necessary. Redundancy is an important concept in information theory.

An autonomous agent is a self-reproducing system able to perform at least one thermodynamic work cycle. Communities of agents will coevolve to an edge of chaos between overrigid and overfluid behavior. An evolving biosphere is all about the coming into existence in the universe of complex, diversifying, ever-changing initial and boundary conditions that constitute coevolving autonomous agents. An autonomous agent is necessarily a nonequilibrium device, which stores energy. Work is the constrained release of energy. Molecular autonomous agents are parallel processing molecular dynamical systems. Species evolve in a chaotic regime, each species changing the adaptive peaks on its landscape that retreat—due to adaptive moves of other species—faster than each species can attain the peaks on its own landscape. There are, in short, dimly understood laws that allow the coevolutionary construction of accumulating complexity (see Fig. 2).

Shannon's information theory was developed initially to quantify telephonic traffic and had been greatly extended since then. We know the human

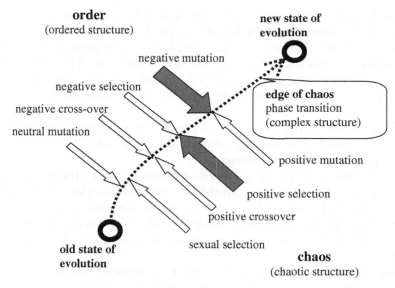

Fig. 2. Evolution on the edge of chaos

genome harbors some 30,000 structural genes, each encoding into the linear sequence of amino acids, thereby constructing a protein. Humans have about 260 different cell types: liver, nerve, muscle, etc. Life cannot be longtime predicted and leads in unexpected directions [2].

An autonomous agent is a physical system, such as a bacterium, that can act on its own behalf in an environment. All free-living cells and organisms are clearly autonomous agents. Autonomous agents must be displaced from thermodynamic equilibrium. Work cycles cannot occur at equilibrium. In complex chemical reaction systems, self-reproducing molecular systems form with high probability. Life is an emergent collective behavior of complex chemical networks [1, 3–5]. Enzymes catalyze, or speed up, chemical reactions. As the molecular diversity of a reaction system increases, a critical threshold is reached at which collectively autocatalytic, self-reproducing chemical reaction networks emerge spontaneously.

Darwin's theory starts with life already here [6]. Darwin assumed gradualism. Most variations would be minor. The no-free-lunch theorem says that, averaged over all possible fitness landscape, no search procedure outperforms any other search procedure. In the absence of any knowledge, or constraint, on the fitness landscape, on average, any search procedure is as good as any other. But life uses mutation, recombination, and selection. These search procedures seem to be working quite well. Our universe is vastly nonrepeating (nonergodic).

Another example of complex system is a sand pile. If we drop the sand slowly on the table the sand will gradually pile up. We will obtain many small avalanches and progressively fewer large avalanches. These huge avalanches are exactly the signature of the famous "butterfly effect" seen in the weather, where a small initial change can have large-scale consequences. The spreading avalanches constitute sensitivity to initial conditions. The same body of theory predicts that most species go extinct soon after their formation, while some live a long time. The predicted species lifetime distribution is a power law [4].

The sequence of DNA bases can be arbitrary and can encode "information." The symmetry allows the arbitrariness of bases to be consistent with the template replication mechanism.

Natural life (n-life) is chemical and is written in the language of chemistry. Two chemicals can be separated by a potential barrier. The transition state occurs at the top of the energy barrier between the two energy wells. Chemical reactions always tend, spontaneously, to proceed toward chemical equilibrium. Human emotions are conscious manifestations of the brain's chemistry.

Some processes occur spontaneously, some conceivably do not. Heat spontaneously diffuses from the hot to the cold object, cooling the former and warming the latter. The classical thermodynamic concepts of temperature, pressure, and entropy were reduced to statistical features of idealized sets of gas particles: temperature becoming the average kinetic energy of the particles, pressure the momentum transferred to the walls of the vessel, and entropy a measure of the number of microstates per macrostate. The compact description of the equilibrium is about as compact as you can get. A few macroscopic variables (temperature, pressure, and volume) suffice.

One modern example of a compact description of something is a computer program. Then the concept of a compact description becomes the concept of shortness of the program. In order to maximize compression, we must get all redundancy out of both the input symbol string and the symbol string representing the program. But as the length of the most compact description increases, bit by actual bit, its information content increases, bit by bit. This is a quite different approach to compare with Shannon's information theory. Thus, for each bit in reduction of the entropy of the system by our measurement, the information content of the most description increases, on average, exactly as rapidly. The sum of the entropy of the system plus the observer's knowledge about that system is a constant for an equilibrium system. We have to record the information about the system somewhere, say, in the memory. To erase a memory-stored bit has a minimal energy cost that exactly balances the work we could get from the system by using the stored information about the system. The complexity of the record is related to the reduction in entropy of the measured system. So evolution is a procedure where matter, energy, information, and indeed, work, come together (see Fig. 3). We suspect that the triad of matter, energy, and information is

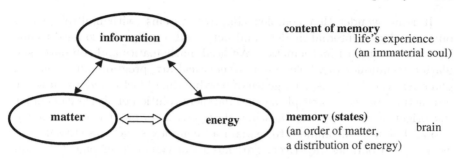

Fig. 3. Information as a feature of matter and energy

insufficient. But we lack a coherent concept of organization, its emergence, and self-constructing propagation and self-elaboration.

2 Self-Organization and Adaptation of Complex Systems

In dynamical systems, transition can be found: order, complexity, and chaos. Analogously, water can exist in solid, transitional, and fluid phases [5, 6]. In nonlinear systems, a chaos theory tells us that the slightest uncertainty in our knowledge of initial conditions will often grow inexorably, and our predictions are nonsense. Complex adaptive systems share certain crucial properties (nonlinearity, complex mixture of positive and negative feedback, nonlinear dynamics, emergence, collective behavior, spontaneous organization, etc.). In the natural world, such systems include brains, immune systems, ecology, cells, developing embryos, and ant colonies. In the human world, they include cultural and social systems. Each of these systems is a network of a number of "agents" acting in parallel. In a brain, the agents are nerve cells; in ecology, the agents are species; in a cell, the agents are organelles such as the nucleus and the mitochondria; in an embryo, the agents are cells; and so on. Each agent finds itself in the environment produced by its interactions with the other agents in the system. It is constantly acting and reacting to what the other agents are doing. There are emergent properties, the interaction of a lot of parts, the kinds of things that the group of agents can do collectively, something that the individual cannot. There is no master agent, for example, a master neuron in the brain. Complex adaptive systems have a lot of levels of organization (hierarchical structures), with agents at any level serving as building blocks for agents at a higher level. The immune system is governed by local interaction between cells and antibodies; there is no central controller in distributed control. Similar behavior can be found in the development of the Internet. We can use biological laws to describe the development of the Internet.

It is no wonder that complex adaptive systems (with multiple agents, building blocks, internal models, and perpetual novelty) are so hard to analyze using standard mathematics. We need mathematics and computer simulation techniques (a whole new mathematical art: programming) that emphasize internal models, emergence of new building blocks, and a rich web of interactions between multiple agents. Parallel evolution of agents can explain the origin of the eye. The Darwin's theory with its gradual process is not wrong, but it is incomplete to explain all complex structures. Evolution is working in a parallel way. Every part has a Darwinian development, but due to positive feedback, the complex structure can be changed by an avalanche jump. This can explain why paleontologists have not found the missing links of species. The development of the mobile phone proceeded in the same way. In the beginning, it was a parallel independent development of the phone and the radio. Relatively late they were joined together by a massive avalanche jump to create the mobile phone.

We now have a good understanding of chaos and fractals showing how simple systems with simple parts can generate very complex behaviors. The edge of chaos is a special region onto itself, the place where you can find systems with life-like, complex behavior (see Figs. 2 and 4). Living systems are actually very close to this edge-of-chaos phase transition, where things are much looser and more fluid. A natural selection is not an antagonist of self-organization. It is a force that constantly pushes emergent, self-organizing systems toward the edge of chaos from a chaos area. A mutation and a crossover are opposite forces pushing the systems from an order to chaos areas. Evolution always seemed to lead to the edge of chaos. The complex evolutionary structure has been described in [2, 7–9]. A random genetic crossover or mutation may give a species the ability to run much faster than before. The agent starts changing, then it induces changes in one of its

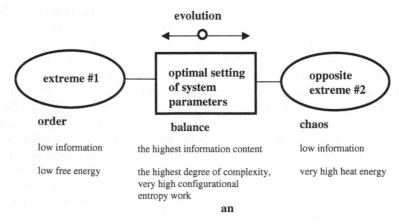

Fig. 4. Evolutionary setting of system parameters

neighbors, and finally you get an avalanche of changes until everything again stops changing. Systems get to the edge of chaos through adaptation: each individual (agent) tries to adapt to all the others. Coevolution can also get them there; the whole system coevolves to the edge of chaos. In ecosystems or ecosystem models, three regimes can be found: ordered regime, chaotic regime, and edge of chaos like a phase transition. When the system is at the phase transition, then, of course, order and chaos are in balance. There is an evolutionary metadynamics, a process that would tune the internal organization of each agent so that they all reside at the edge of chaos. The maximum fitness occurs right at the phase transition.

The original DNA would have been coding for a living creature; but if errors then accumulate in it, it will soon be coding for a nonliving creature. Errors (mistakes) can be divided into three classes: advantageous mistakes, which make things better; neutral mistakes, which make no difference; and harmful mistakes, which make things worse. The first two kinds of mistake underlie all evolutionary changes in the world. If selection removes more errors than mutation introduces, the quality of life improves. Life has many methods of dealing with error. Errors can be prevented from happening to begin with; they can be corrected; they can be disguised; they can be purged by natural selection.

3 Basic Principle of Evolutionary Computation

Scientific discussion of evolution dates back to 200 years [1]. Jean Baptiste de Lamarck wrote extensively about evolution. Lamarck was the first person to support the idea of evolution with logical arguments and was also the first person to put forth an hypothesis concerning the mechanisms of evolutionary change. He suggested that living organisms have the ability to change gradually over many generations by the inheritance of structures that have become larger and more highly developed as a result of continued use or, conversely, have diminished in size as a result of disuse. Only a part of evolutionary changes has been related with the mechanisms proposed by Lamarck (see Fig. 5).

Darwin suggested that slight variation among individuals significantly affects the change that a given individual will survive and reproduce. He called this differential reproductive success of varying individuals *natural selection*. Darwin recognized that the reproductive rates of organisms are so high that they would result in enormous population increases if all the offspring survived. Therefore, Darwin reasoned, mortality must increase as population density increases and competition for space, food, shelter, and other environmental necessities becomes severe, and predation and disease become more prevalent. Individuals affect future generation not only through their own offspring but also by helping the survival of relatives who contain the same genes as a result of descent from a common ancestor.

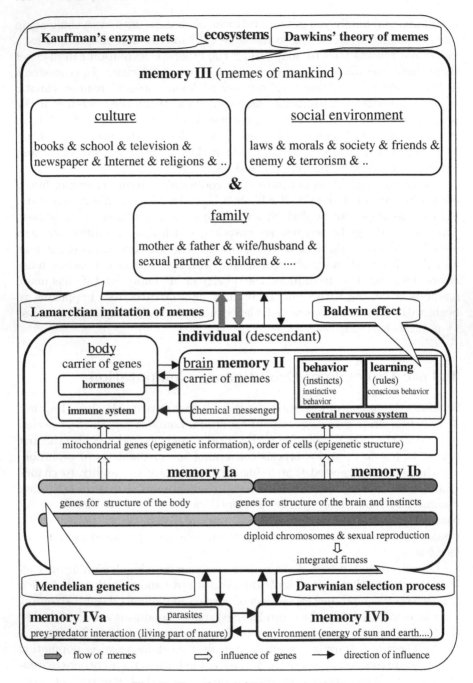

Fig. 5. Complex evolutionary structure

Gregor Mendel (1822–1884) accurately observed patterns of inheritance and proposed a mechanism to account for some of the patterns. Genes determine individual traits. Various kinds of offspring appear in proportion that can be predicted from Mendel's laws. We often use the term Mendelian genetics (see memory I in Fig. 5) to refer to the most basic patterns of inheritance in sexually reproducing organisms with more than one chromosome. Mendel gave his classic paper *Experiments in Plant Hybrids* for *Natural Science Society* in Brno in 1865. Mendel observed that the spherical seed trait was dominant, being expressed over the dented seed trait, which he called recessive. The physical appearance of a character is its phenotype, which Mendel correctly supposed to be the result of the genotype, or genetic constitution, of organism showing the phenotype. Genetics after Mendel: alleles and their interaction.

In diploid organisms, chromosomes come in pairs (memory Ia and memory Ib in Fig. 5). One of each chromosome pair derives from each parent; it does not matter whether, for example, the dominant allele was contributed by the mother or the father. Sex is determined by differences in the chromosomes; but such determination operates in a bewildering variety of ways. Normal human females carry two X chromosomes, normal males carry one X and one Y chromosome. Persons who have some other number of sex chromosomes may develop abnormally.

Fifty years after Darwin, the American psychologist James Baldwin [3] said that nature selection was not merely a law of biology but applied to all the sciences of life and mind, and he coined the term "social heredity" to describe the way individuals learn from society by imitation and instruction. Baldwin explained how intelligent behavior, imitation, and learning can affect selection pressure on the genes (Baldwin effect). There is no Lamarckian "inheritance of acquired characteristics" in the sense of passing the result of learning onto the next generation through genes (memory II in Fig. 5). The Baldwin effect creates new kinds of creatures that are capable of adapting to change far more quickly than their predecessors.

In some ways it is obvious that ideas and cultures evolve. Changes are gradual and build on what went on before. Inventions do not spring out of nowhere but depend on previous inventions, and so on. If I invent a brilliant new recipe for a new meal, I can pass it on to you and you can pass it on to your granny and she can pass it to her best friend. Also, this is not inheritance in the biological sense and genes are not affected. There is difference between "copy the product" and "copy the instructions." Let us suppose that you play a beautiful piece of music for your friends and one of them wants to learn it too. You could either play it many times until your friend can copy it accurately (copy the product) or simply hand him the written music in a book (copy the instructions). In the second case, the individual playing styles of each pianist will not affect because copies of the written music are passed on. In the first case the process appears Lamarckian but in the second case it does not.

The term "meme" first appeared in 1976, in Richard Dawkin's best-selling book *The Selfish Gene* [3]. The term "selfish" here means that the genes act only for themselves; their only interest is their own replication; all they want is to be passed on to the next generation. Many memes are passed from parent to child. Parents teach their children many of the rules of their own society. Children get their first language from their parents and usually their religion too. Where memes are transmitted horizontally they can travel quite independently of the genes. Modern industrial life is a world of horizontal transmission. Our main sources of information are sources that did not exist in our long evolutionary past: schools, radio, television, newspapers, books and magazines, Internet, and lots and lots of friends and acquaintances widely spread around the city, the country, and even the world. Dawkins also introduced the important distinction between "replicators" and "vehicles." Memes are stored in human brains (or books or inventions) and passed on by imitation. Memes are copied from one person to another and spread by imitation. Dawkins discussed their propagation by jumping from brain to brain via a process which in the broad sense can be called *imitation*. Everything that is passed from a person to another person in this way is a meme. This includes all the words in your vocabulary, the stories you know, the skills and habits you have picked up from others, and the games you like to play. It includes the songs you sing and the rules you obey. Memes and genes are not the same. Genes are instructions for making proteins, stored in the cells of the body and passed on in reproduction. Their competition drives the evolution of the biological world. Memes are instructions for carrying out behavior, stored in brain—memory II (or other memories)—and passed on by imitation. Their competition drives the evolution of the mind. Both genes and memes are replicators and must obey the general principles of evolutionary theory. Genes are instructions encoded in molecules of DNA; memes are instructions embedded in human brains, or in artefacts as books, pictures, Internet, WWW pages, etc. People are different. Their ability to imitate creates a second replicator that acts in its own interests and can produce behavior that is memetically adaptive but biologically maladaptive. People will pick up their memes from films, radio, books, Internet, and television long before they have even produced any children. There are many indirect measures of horizontal transmission such as literacy rates, or the availability of telephones, mobile phones, radios, hi-fi systems, fax machines, compact disks, audiotapes and videotapes, and computers.

If this religious behavior helped people acquire more mates, then any genes that inclined them to be more religious in the first place would also flourish. In this way genes for religious behavior would increase because of religious memes. Religions also dictate sexual practices, promote certain kinds of cooperative behavior, and regulate aggression and violence. Religions build theories about the world and then prevent them from being tested. Religions provide nice, appealing and comforting ideas, and cloak them in a mask of "truth, beauty, and goodness." The theories can then thrive in spite of being untrue, ugly, or cruel. Some religions positively encourage murder and war against people of other faiths [3].

In the biological world, sexual species work by copying the instructions. The genes are the instructions that are copied, the phenotype is the result and is not copied. Dawkins [3] described organisms as vehicles for the genes, built to carry them around and protect them. Pictures, books, tools, and buildings are meme vehicles (memory III in Fig. 5). In our own bodies, thousands of genes cooperate to produce muscles and nerves, liver and brain, and to result in a machine that effectively carries all genes around inside it.

Many people seem to hate the idea that human sexuality can be explained in terms of genetic advantage. All sexual behavior is culturally determined. The modern sexual behavior is meme-driven. We enjoy sex because animals that enjoyed sex in the past passed on their genes. But evolution has also given us intelligence, which has enabled us to work out the function of sex and manipulate things so as to get pleasure of sex without the cost of child care. On this view, birth control and sex for fun, and many other aspects of modern sexual life, are mistakes which the genes have not eliminated. We know that women prefer to mate with high-status men, and that these men leave more offsprings, either by having more wives or by fathering children by women who are not their wives.

Meme evolution is faster then gene evolution. Sexual memes have influence in our lives in ways that have little or nothing to do with genes. Catholicism's taboo against birth control has been extremely effective in filling the world with millions of Catholics who bring up their children to believe that condoms and pills are evil, and that God wants them to have as many children as possible. Birth rates are highest in the developing countries and lowest in the technologically advanced ones. All the great religions of the world began as small-scale cults, usually with a charismatic leader, and over the year a few of them spread to take in billions of people all across the planet. The Catholic God is watching at all times and will punish people who disobey His commandments with most terrible punishments—burning forever in hell, for example. These threats cannot easily be tested because God and hell are invisible, and the fear is inculcated from early childhood.

For most of the past two or three million years memes have evolved slowly. Our sexual desires still follow the dictates of genetic evolution while memetic evolution changes the rules. The consequences of memetic evolution is that humans can be more altruistic than their genes alone would dictate. Altruism is defined as behavior that benefits another creature at the expense of the one carrying it out. In other words, altruism means doing something that costs time, effort, or resources, for the sake of someone else. This might mean providing food for another animal, giving a warning signal to protect others while putting yourself at risk, or fighting an enemy to save another animal from harm. Many animals live social and cooperative lives, parents lavish devotion on their offspring, and many mammals spend hours every day grooming their friends and neighbors.

Humans are uniquely cooperative and spend a great deal of their time doing things that benefit others as well as themselves. They have moral sensibilities and strong sense of right and wrong. They are altruists. Their behavior is controlled by rules. Among many versions of the rules are "Be good to those who imitate you," "Be good to children," "Be good to your cultural ancestors" or more generally, "Be good to your close cultural relatives," and so on. These rules can be stored by "fuzzy rules" (left part of memory II in Fig. 6). People who come close to death and survive are often changed by the experience, becoming more caring of others and less concerned with themselves. The altruism trick permeates religious teachings. Many believers are truly good people. In the name of their faith they help their neighbors, give money to the poor, and try to live honest and moral lives. As Dawkins points out [3], good Catholics have faith; they do not need proof. Indeed, it is a measure of how spiritual and religious you are that you have faith enough to believe in completely impossible things without asking questions. The religious memes were just behaviors, ideas, and stories that were copied from one person to another in long history of human attempts to understand the world. They were successful because they happened to come together into mutually supportive gangs that included all the right tricks to keep them safely stored in millions of brains, books, and buildings, and repeatedly passed on to more. That is why they are still with us, and why millions of people's behavior is routinely controlled by ideas that are false or completely untestable. Religious memes have therefore played an important role in the development of human societies. Language provides a good example of cultural evolution. The evolutionary history of various languages can be accurately traced. Family trees of languages have been constructed that are comparable with the genetic family trees based on differences in DNA.

The environment itself is constantly changing because of all these developments, and so the process is never static. Artificial neural networks (right part of memory II in Fig. 6) have demonstrated that many of the features of human memory can be simulated in computers. Effective memes will be those that cause high-fidelity, long-lasting memory. Memes may be successful at spreading largely because they are memorable rather than because they are important or useful. Apart from that, effective transmission of memes depends critically on human preferences, attention, emotions, and desires (desire for sex, for food, for avoiding danger, and for excitement and power).

In 1989 the World Wide Web (WWW) was invented. The Internet had already been expanding for many years. Memes can now be stored on the hard disk of a computer. For robots to become like humans to have human-like artificial intelligence and artificial consciousness—they would need to

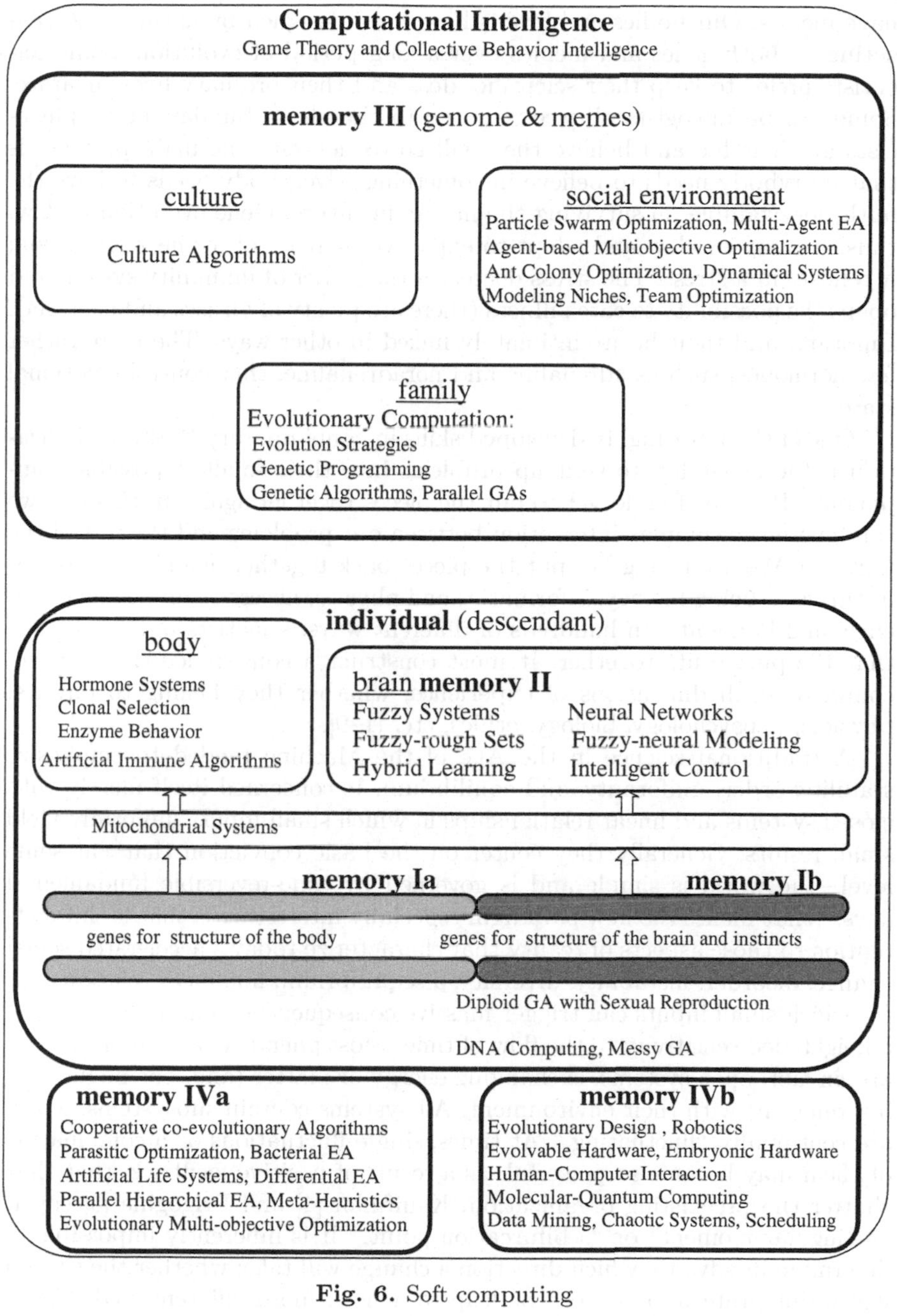

Computational Intelligence
Game Theory and Collective Behavior Intelligence

memory III (genome & memes)

culture

Culture Algorithms

social environment
Particle Swarm Optimization, Multi-Agent EA
Agent-based Multiobjective Optimalization
Ant Colony Optimization, Dynamical Systems
Modeling Niches, Team Optimization

family
Evolutionary Computation:
Evolution Strategies
Genetic Programming
Genetic Algorithms, Parallel GAs

individual (descendant)

body
Hormone Systems
Clonal Selection
Enzyme Behavior
Artificial Immune Algorithms

brain **memory II**
Fuzzy Systems   Neural Networks
Fuzzy-rough Sets   Fuzzy-neural Modeling
Hybrid Learning   Intelligent Control

Mitochondrial Systems

memory Ia   **memory Ib**

genes for structure of the body   genes for structure of the brain and instincts

Diploid GA with Sexual Reproduction

DNA Computing, Messy GA

memory IVa
Cooperative co-evolutionary Algorithms
Parasitic Optimization, Bacterial EA
Artificial Life Systems, Differential EA
Parallel Hierarchical EA, Meta-Heuristics
Evolutionary Multi-objective Optimization

memory IVb
Evolutionary Design , Robotics
Evolvable Hardware, Embryonic Hardware
Human-Computer Interaction
Molecular-Quantum Computing
Data Mining, Chaotic Systems, Scheduling

Fig. 6. Soft computing

have memes. Our bodies and brains have been designed by natural selection acting on both genes and memes over a long period of evolution. Some scientists prefer to keep their scientific ideas and their ordinary lives separate. Some can be biologists all week and go to church on Sunday, or be physicists all their life and believe they will go to heaven. The main problem is that everybody needs to believe in something. Everybody needs to have the faith, for example, in surviving the nature in future. Gene evolution is slow; thus, it cannot follow and adapt to explosive increase of meme-information; it reflexes in a stress. The stress decreases the power of immunity system and opens the possibility to catch illness (there are plenty of viruses and bacteria). Emotions and thought are intimately linked in other ways. There are rather few hormones, such as adrenaline and noradrenaline, that control emotional states.

One of the most highly developed skills in contemporary Western civilization is the capability to split up problems into their smallest possible components. But we often forget to put the pieces together again. In this way we can ignore the complex interaction between our problems and the rest of the universe. We are trying "to put the pieces back together again"—the pieces in this case being biology, informatics and physics, necessity and change, science, and humanity. In hundreds of different ways, scientists have expressed that the puzzle fits together. It must construct a consistence that can accommodate all dimensions of experience, whether they belong to physics, physiology, psychology, biology, ethics, etc. [1–9].

A traditional science in the Age of the Machine tended to emphasize stability, order, uniformity, and equilibrium. It concerned itself mostly with closed systems and linear relationships in which small inputs uniformly yield small results. Generally they center on the basic conviction that—at some level—the world is simple and is governed by time-reversible fundamental laws. What makes the new paradigm especially interesting is that it shifts attention to those aspects of reality that characterize today's accelerated social change: disorder, instability, diversity, disequilibrium, nonlinear relationships (in which small inputs can trigger massive consequences), and temporality—a heightened sensitivity to the flow of time. Most phenomena of interest to us are, in fact, open systems, exchanging energy or matter (and, one might add, information) with their environment. All systems contain subsystems, which are continually "fluctuating." At times, single fluctuations or a combination of them may become so powerful, as a result of positive feedback, that they shatter the preexisting organization. Kauffman [4] and Prigogine [5] call it "a singular moment" or "a bifurcation point." It is inherently impossible to determine in advance which direction a change will take: whether the system will disintegrate into "chaos" or leap to a new, more differentiated, higher level of "order" or organization which is called "dissipative structure."

A living system appears very complex from the thermodynamic point of view. Certain reactions are close to equilibrium, while others are not. Here we have a remarkable case of the convergence of two sciences: biology and physics. It seems reasonable to assume that some of the first stages moving toward life were associated with the formation of mechanisms capable of absorbing and transforming chemical energy. However, it is necessary to emphasize that such an evolutionary structure implies a drastic simplification of the situation defined simply in terms of competition between self-replicating processes in the environment where only a limited amount of needed resources exists (for example, a prey–predator interaction).

Today, even physics tells us that irreversible processes play a constructive and indispensable role. It is generally known that stability and simplicity are exceptions. Today we are going from deterministic, reversible processes to stochastic and irreversible ones. Certain events go only one way—not because they cannot go the other way, but because it is extremely unlikely that they go backward. Initial conditions corresponding to a single point in unstable systems correspond to infinite information and are therefore impossible to find or observe. No equilibrium brings "order out of chaos." However, as it was already mentioned, the concept of order (or disorder) is more complex than it was thought to be. Open systems evolve to higher and higher forms of complexity (see Fig. 4). Molecular biology shows that not everything in a cell is alive in the same way. Some processes reach equilibrium, others are dominated by regulatory enzymes far from equilibrium.

Evolution always seemed to lead to the complexity. The complex evolutionary structure in Fig. 5 that was described in [7, 8] can be transformed to the structure of the computational intelligence (see Fig. 6).

Life is clearly a chemical phenomenon, and only molecules can spontaneously undergo a complex chemical reaction with one another. The first source of chemistry's power is a simple variety. The second source of power is reactivity: the structure A can manipulate the structure B to form something new—a structure C. A growth of complexity really does have something to do with far-from-equilibrium systems building themselves up, cascading to higher and higher levels of organization. Once they have accumulated a sufficient diversity of objects at the higher level, they go through a kind of autocatalytic phase transition and get an enormous proliferation of things at that level. Life is a natural expression of complex matter. It is a very deep property of chemistry and catalysis and is far from equilibrium [4].

4 Biologically Inspired Computing

Among all of natural computing approaches, biologically inspired systems are the oldest and may be the most popular ones. It is the field of investigation that draws upon metaphors or theoretical models of biological systems in order to design computational tools or systems for solving complex problems. Adaptation, self-organization, prediction, communication, and optimization have to be performed by biological organisms, in parallel and in different scales and structural levels, in order to survive and maintain life. Nature was first in the solution of very complicated problems. Our complex problems, in areas as engineering, computing, and economics, are usually characterized by the absence of a complete mathematical model of relevant phenomena. There is the existence of a large number of variables to be adjusted and conditions to be simultaneously satisfied, the presence of high degrees of nonlinearity, and formulation of multipurpose tasks with a combinatorial explosion of candidate solutions.

An evolution of information in memory I (see Fig. 5) can be simulated by genetic algorithms (GAs) with diploid chromosomes and limited lifetime or sexual reproduction [10]. Diploid chromosomes increase the efficiency and robustness of GAs, and they can better track optimal parameters in a changing environment. The sex is distinguished by a two-bit value stored in the chromosome [9]. It can be demonstrated that standard GAs with haploid chromosomes are unable to correctly locate optimal solutions for time-dependent objective functions. The adaptation of GAs depends on the speed of landscape changes through time. Results of different GAs with sexual reproduction given in [9] are very promising. For simple problems GAs with limited lifetime can be used. They prefer a shorter lifetime and many new generations. There is an optimal lifetime where almost all individuals in the population have high fitness. This strategy is used in nature by viruses and bacteria. The sexual reproduction is typical for complicated creatures. This phenomenon can be explained by a need for a higher age to start the reproduction (the learning time).

In [3] two different meme transformation processes were described. The first transformation process is called the copy, where a given meme is simply copied into itself with allowed mutation. The second transformation process is the migration. This strategy consists in looking for a fitness meme in neighboring compartments. If the fitness of such a meme is greater by a threshold than the fitness of the current meme, then the fittest meme is copied to the current compartment with allowed mutation. Neural networks (part of memory II in Fig. 6) are adapted by a prescribed number of epochs. The basic capability of neural networks is to learn patterns from examples. When such a learned neural network gets an input similar to the learned one, it is able to give a similar output, which means classifying unknown input patterns.

A learning process should represent a quality of learning ability by a size of neighborhood search around a chromosome. The concept of chromosome (memory I) is supported with memes (memory III). Both of them (memes and chromosomes) determine an integrated fitness. A message (idea) is broadcasted throughout a subpopulation of the so-called recipients. These recipients incorporate the broadcasted message into their memes. The Dawkins' theory of memes [3] was considered as an extension of Baldwin effect. An application of Dawkins' theory of memes represents a plausible possibility of how to explain and interpret many features of social behavior of animals. Each population chromosome is automatically accompanied by a meme (memory I and III). The memes are carriers of information about best solutions that have already been achieved.

Employing of memes is very important in the evolution process as a further source of information about the best solution that should be found. At the beginning of evolution the memes are empty (there is no information). During the evolution process they can receive an information by so-called memetic interaction. In the reproduction process, created offspring obtain memes from one of two parental chromosomes (generally offspring are educated by mothers). Many different strategies can be used in the process of evaluation of meme and chromosome. It seems that memes might be very important for permanent adapting of the best evolutionary solution when the goal of evolution (or environment where populations live) is slowly changing. In [9] was observed that GA only with learning has a great inertia; it is unable to well adapt to a new evolution goal. The alternative approach based on expert system (fuzzy rules—memory II in Fig. 6) is not applied widely because of similar difficulties of creation and development of knowledge bases. A transition from one application to another requires replacement of rules set (heuristics) and algorithm of fitness function calculation. Furthermore, an exactness of received solution often is not a satisfactory one.

5 Order vs. Complexity in the Question of Information

It is widely held that in physical sciences the laws of thermodynamics have had a unifying effect similar to that of the theory of evolution in the biological sciences [1]. What is intriguing is that the predictions of one seem to contradict the predictions of the other. The second law of thermodynamics suggests a progression from order to disorder, from complexity to simplicity, in the physical universe. Yet biological evolution involves a hierarchical progression to increasingly complex forms of living systems, seemingly in contradiction to the second law of thermodynamics. Thermodynamics is an exact science, which deals with energy. Our world seethes with transformations of matter and energy. The concept of entropy S gives us a more quantitative way to describe the tendency for energy flow in a particular direction. The entropy change for a system is defined mathematically as the flow of energy divided by the temperature:

$$\Delta S \geq \frac{\Delta Q}{T} \tag{1}$$

where ΔS is the change in entropy, ΔQ is the heat flow into or out of the system, and T is the absolute temperature in Kelvin. From the first law of thermodynamics we derive that the change of energy of system ΔE is equal to the work done on (or by) the system ΔW and the heat flow into (or out of) the system ΔQ: $\Delta E = \Delta Q + \Delta W$. There is another way to view entropy. The entropy of a system is a measure of probability of given arrangement of mass and energy within it. The energy of universe is constant; the entropy of the universe tends toward a maximum. But the earth is not an isolated system, since it is open to energy flow from the sun. It seems safe to conclude that systems near equilibrium (whether isolated or closed) can never produce the degree of complexity intrinsic in living systems. The possibilities that a configurational entropy works are more promising, however, if one considers a system subjected to energy flow which may maintain it far from equilibrium, and its associated disorder. The formation and maintenance of complex systems at energy levels well removed from equilibrium requires continuous and configurational work to be done on the system [1]. This continuous work requires energy and/or mass flow through the system, apart from which the system will return to an equilibrium condition (lowest Gibbs free energy, see Fig. 4), with the decomposition of complex molecules into simple ones. Mass, such as water and carbon dioxide, also flows through plants, providing necessary raw materials, but not energy. In collecting and storing useful energy, plants serve the entire biological world. While the maintenance of living systems is easily rationalized in terms of thermodynamics, the origin of such living systems is quite another matter. Thus, it is thermodynamically possible to develop complex living forms, assuming that the energy flow through the system can somehow be effective in organizing the simple chemicals into the complex arrangement associated with life. In existing living systems, the coupling of the energy flow to the organizing "configurational work" occurs through the metabolic function of DNA, enzymes, etc.

Only recently has it been appreciated that the distinguishing feature of living systems is complexity rather than order (see Fig. 4). For example crystals are very orderly, spatially periodic arrangements of atoms (or molecules) but they carry very little information. By definition then, a periodic structure has order. An aperiodic chemical structure (in an organism) can have complexity if the temperature of the structure is approximately between 0 and 100°C. Informational molecules have a low degree of order but a high degree of specified complexity. It should be noted that aperiodic polypeptides or polynucleotides do not necessarily represent meaningful information or biologically useful function. A random (chaotic) arrangement of DNA bases is aperiodic but contains little, if any, useful information since it is devoid of meaning. Thus, informational macromolecules may be described as being aperiodic and in a specified sequence. The information content in a given sequence of unit, be they digits in a number, letters in a sentence, or amino acids in a polypeptide or protein, depends on the minimum number of instructions needed to specify or describe the structure. Many instructions are needed to specify a complex, information-bearing structure as DNA. Only a few instructions are needed to specify an ordered structure as a crystal. In this case we have a description of the initial sequence or unit arrangement, which is then repeated.

Now consider the components of the Gibbs free energy [Eq. (2)], where the change in enthalpy $\Delta H = \Delta E + P\Delta V$ is principally the result of changes in the total bonding energy ΔE, with the $P\Delta V$ term assumed to be negligible (P is pressure and ΔV is volume change) [1].

$$
\begin{array}{ccccc}
\Delta G & = & \Delta H & - & T\Delta S & - & T\Delta S_c \\
\text{Gibbs} & & \text{Chemical} & & \text{Thermal} & & \text{Configurational} \\
\text{free energy} & & \text{work} & & \text{entropy work} & & \text{entropy work}
\end{array} \tag{2}
$$

We will refer to this enthalpy component ΔH as the chemical work. The change in the entropy ΔS that accompanies the polymerization reaction may be divided into two distinct components which correspond to the changes in the thermal energy distribution ΔS_t and the mass-structure distribution ΔS_c. The change in configurational entropy "coding" goes from a random arrangement to a specified sequence. Configurational entropy S_c is concerned only with the arrangement of mass in the system. We shall be especially interested in the sequencing of amino acids in polypeptides (or proteins) or of nucleotides in polynucleotides (e.g., DNA). The entropy S of a system is given by

$$
S = k \ln \Omega_t \Omega_c = S_t + S_c \tag{3}
$$

where k is Boltzmann's constant and Ω_t and Ω_c refer to the number of ways energy and mass, respectively, may be arranged in a system. If we want to convert a random polymer into an informational molecule, we can determine the increase in information I by finding the difference between the configuration entropy with states Ω_{cr} for initial random polymer and informational molecule with coding states Ω_{ci}:

$$
I = R_{max} - R = k \ln \Omega_{cr} - k \ln \Omega_{ci} \tag{4}
$$

where R_{max} is the greatest possible value of the randomness R under given conditions. Information I is a measure of order. Order can be defined as absence of randomness. We are trying to find the general principles underlying the emergence of order from chaos. In thermodynamics, entropy is a measure of disorder of a system, and disorder is clearly related to information. It is not easy to define a measure of complexity. Complexity is the place where matter, energy, and information come together. Maybe the measure of complexity has something to do with minimum of matter, energy, and information that are necessary to create a complex thing. In this case, information can be described in the minimal length of algorithm that is necessary in a creative procedure of work to be done. One modern example of a compact description of something is a computer program. In order to maximize compression, we must get all redundancy out. We can use the automatic generation of programs with grammatical evolution [11]. In the beginning, presumably, the universe was simple, homogeneous, featureless, and almost isotropic. Now it is vastly complex. The fact that a biosphere builds up astounding complexity and diversity suggests that our current knowledge is missing something fundamental. Because the biosphere is a part of the universe, the universe becomes complex too. The heart of the mystery concerns a proper understanding of terms as order or organization, disorder, chaos, information, and entropy.

Arranging a pile of bricks into the configuration of a house requires work. One would hardly expect to accomplish this work with energy of dynamite or hurricane, however. Not only must energy flow through the system, it must be coupled with procedure of work to be done. We must have instructions (e.g., information in a "a cookery book") of how to build the house. Regularity or order cannot serve to store the large amount of information required by complex and living systems. A highly irregular, but specified, structure is required rather than an ordered structure. Coupling the energy flow through the system to do the chemical and thermal entropy work is much easier than doing the configurational entropy work. The uniform failure in literally thousands of experimental attempts to synthesize protein or DNA demonstrates the difficulty in achieving a high degree of information content, or specified complexity from undirected flow of energy through a system without evolution of information content. Thermal entropy, however, seems to be physically independent from the information content of complex systems which we have analyzed and called configurational entropy.

However, when we examine a biological cell or an organism, the situation is quite different: not only are these systems open, but they exist only because they are open. The feed of the flux of matter and energy is coming to them from the outside world. The free energy E can create P different organisms (species), every species with N_i copies that are created by DNA information I_i, where Q_{in_i} is the metabolic heat inside of an organism released by its activities and Q_{out} is the metabolic heat that is lost. T_i is the temperature, m_i is the mass, V_i is the volume, W is a scrap (waste), and c_1, c_2 are constants [7, 8]:

$$E = E_{\text{structure}} + E_{\text{heat}} = \sum_{i=1}^{p}(c_1 N_i V_i o_i + Q_{in\_i}) + W + Q_{out} \qquad (5)$$

where $Q_{in\_i} = c_2 N_i m_i T_i$ and o_i is the measure of energy order (defined by the energy density $= E_{\text{structure}\_i}/V_i$). On the earth there appeared a complex biosphere with the food chains that must satisfy (5). An increase of energy ΔE is covered by sun or earth activity. In a far-from-equilibrium condition, various types of self-organization processes may occur. Evolution (with fitness F_e) has a tendency to maximize the accumulated energy in living systems with a complex evolution structure (see Fig. 5):

$$F_e = \max_{\text{through\_evolution}}\left\{c_1 \cdot \sum_{i=1}^{p} N_i V_i o_i\right\} \qquad (6)$$

In open systems evolution increases the measure of complexity. There is an increase of an accumulated energy, information, and a complex matter.

A living system appears very complex from the thermodynamic point of view. Certain reactions are close to equilibrium, while others are not. Open systems evolve to higher and higher forms of complexity. Molecular biology shows that not everything in a cell is alive in the same way. Some processes reach equilibrium and others are dominated by regulatory enzymes far from equilibrium [5].

6 Overview of Evolutionary Algorithms

Evolutionary computation is generally considered as a consortium of genetic algorithms (GA), genetic programming (GP), and evolutionary strategies (ES). There are new GPs: grammatical evolution [11], grammatical swarm [12], and chemical genetic programming [13].

One model of social learning that has attracted interest in recent years is drawn from a swarm metaphor. Two popular variants of swarm models exist: those inspired by studies of social insects such as ant colonies and those inspired by studies of the flocking behavior of birds and fish. The essence of these systems is that they exhibit flexibility, robustness, and self-organization [12]. Although the systems can exhibit remarkable coordination of activities between individuals, this coordination does not stem from a "center of control" or a "directed" intelligence, rather it is self-organizing and emergent. Social "swarm" researchers have emphasized the role of social learning processes in these models. In the context of particle swarm optimization (PSO) [12, 14], a swarm can be defined as a population of interacting elements that is able to optimize some global objective through collaborative search of a space. These elements (particles) move (fly) in an n-dimensional search space, in an attempt to uncover ever-better solutions to the problem of interest. In essence, social behavior helps individuals to adapt to their environment, as it ensures that they obtain access to more information than that captured by their own senses. It is interesting to note that this approach of PSO is completely devoid of any crossover operator characteristic of GAs. Each particle has a memory of the best location in the search space that it has found so far (pbest). Each particle knows the best location found to date by all the particles in the population (or in an alternative version of the algorithm, a neighborhood around each particle) (gbest). At each step of the algorithm, particles are displaced from their current position by applying a velocity vector to them. The velocity size and direction are influenced by the velocity in the previous iteration of algorithm (simulates "momentum") and the location of a particle relative to its pbest and gbest [12]. Therefore, at each step, the size and direction of each particle's move are functions of its own history (experience), and the social influence of its peer group.

Differential evolution (DE) [15] is a rather unknown approach to numerical optimization, which is very simple to implement and requires little or no parameter tuning. After generating and evaluating an initial population the solutions are refined by a DE operator as follows. Choose for each individual genome j three individuals k, l, and m randomly from the population. Then calculate the difference of the chromosomes in k and l, scale it by multiplication with a parameter f, and create an offspring by adding the results to the chromosome of m. The only additional twist in this process is that not the entire chromosome of offspring is created in this way, but that genes are partly inherited from individual j.

There are other soft computing methods [9]:

Multiagent evolutionary algorithms
Agent-based multiobjective optimization
Ant colony optimization, team optimization, culture algorithms [16]
Fuzzy logic, neural networks, fuzzy-rough sets, fuzzy-neural modeling
Hybrid learning, intelligent control, cooperative coevolutionary algorithms
Parasitic optimization, bacterial evolutionary algorithms (BEA)
Artificial immune algorithms [17], artificial life systems
Parallel hierarchical evolutionary algorithms [18]
Meta-heuristics, evolutionary multiobjective optimization [14, 19]
Evolvable control [9], embryonic hardware
Human–computer interaction, molecular-quantum computing
Data mining, chaotic systems, scheduling, etc.

7 Parallel Grammatical Evolution with Sexual Selection

Grammatical evolution (GE) [11] can be considered to be a form of grammar-based GP. In particular, Koza's GP has enjoyed considerable popularity and widespread use. Unlike a Koza-style approach, there is no distinction made at this stage between what he describes as function (operator in this case) and terminals (variables). Koza originally employed Lisp as his target language. This distinction is more of an implementation detail than a design issue. Grammatical evolution can be used to generate programs in any language, using Backus-Naur Form (BNF). BNF grammars consist of terminals, which are items that can appear in the language, i.e., $+$, $-$, sin, log, etc., and nonterminal, which can be expanded into one or more terminals and nonterminals. A nonterminal symbol is any symbol that can be rewritten to another string, and conversely a terminal symbol is one that cannot be rewritten. The major strength of GE with respect to GP is its ability to generate multiline functions in any language. Rather than representing the programs as parse trees, as in GP, a linear genome representation is used. A genotype–phenotype mapping is employed such that each individual's variable length byte strings contain the information to select production rules from a BNF grammar. The grammar allows the generation of programs in an arbitrary language that are guaranteed to be syntactically correct. The users can tailor the grammar to produce solutions that are purely syntactically constrained, or they may incorporate domain knowledge by biasing the grammar to produce very specific forms of sentences.

The GE system in [11] codes a set of pseudorandom numbers, which are used to decide which choice to take when a nonterminal has one or more outcomes. Because the GE mapping technique employs a BNF definition, the system is language independent, and, theoretically, can generate arbitrarily complex functions. There is quite an unusual approach in GEs, as it is possible for certain genes to be used two or more times if the wrapping operator is used. BNF is a notation that represents a language in the form of production rules. It is possible to generate programs using the grammatical swarm optimization (GSO) technique [12] with a performance similar to the GE, given the relative simplicity of GSO, the small population sizes involved, and the complete absence of a crossover operator synonymous with program evolution in GP or GE. Grammatical evolution was one of the first approaches to distinguish between the genotype and phenotype. GE evolves a sequence of rule numbers that are translated, using a predetermined grammar set into a phenotypic tree.

There is a new method of GP [13], named chemical genetic programming (CGP), which enables evolutionary optimization of the mapping from genotypic strings to phenotypic trees. In biological cells information is derived from DNA to give each cell its functionality. This process is called translation. A series of metabolic reactions, catalyzed by several enzymes, translate genetic information into proteins. Each amino acid is a biochemical building block, so together the amino acids form the fundamental set of functional units in a cell. In the CGP a cell is evolved, and includes a DNA string that codes genetic information and smaller molecules for the mapping from DNA code to computational functionality. Genetic modification of a cell's DNA allows the DNA code and genotype-to-phenotype translation to coevolve. Building an optimal translation table enhances evolution within a population while maintaining the necessary diversity to explore the entire search space. The collection of grammatical rewriting rules, or translations, is stored in each cell's translation table. Using the breeding techniques of conventional GP, GE or CGP is an effective alternative. In CGP however, the translation relation is dynamic, and the system is allowed to optimize both the combination of symbols in the genotypic string as well as the use of the translation set to create a cell's phenotypic tree. CGP can create good programs with a higher probability. Artificial cells in CGP are represented with a numerical string (DNA), transcription units (tRNAs), a rule set (amino acids), and a translation table (aminoacyl-tRNAs), enabling coevolution between the codes on DNA and their functionality. Cells are evolved using mutation, crossover, and molecular exchange operations. Unlike conventional GP, CGP operators, such as * and ^, are treated as terminal symbols. Each amino acid in the initial set is syntactically valid, which ensures that all amino acids created by translating elements from this set are also valid. The CGP is able to ignore the entire class of solutions that do not form executable structures. Through evolution, beneficial amino acids will be propagated throughout the population, while destructive amino acids will be eliminated. The diversity of amino acids will gradually decrease, while the diversity of translations will increase. The complexity of translation will grow in time, as productive combinations of amino acids can be merged into a single translation. The CGP is a suitable method for solving symbolic regression problems. CGPs use feedback to determine the optimal content of the translation table, and CGP has an adaptive cell in which the final functional tree is created by the translation of symbol sequences in DNA using a set of evolved amino acid rewriting rules.

The genotype-to-phenotype mapping is a critical point in designing an evolutionary system. This mapping provides the building blocks that a system is allowed to work with progressing toward its objective.

Our approach uses a parallel GE (PGE) [20, 21]. A population is divided into several subpopulations that are arranged in the hierarchical structure. Every subpopulation has two separate parts: a male group and a female group. Every group uses quite a different type of selection. In the first group a classical type of GA selection is used. In the second group only different individuals can be included. It is similar to a harem arrangement. This strategy increases an inner adaptation of PGE. The following text explains why we used this approach. It is a biologically inspired computing. Analogy would lead us one step further, namely, to the belief that the combination of GE with sexual selection will occur. On the principle of the sexual reproduction we can create a PGE with a hierarchical structure.

Gene numbers are related to complexity because, generally speaking, a more complex life form uses more instructions than a simpler life form. Sexual life form could evolve to be more complex than clonal life forms [24]. Complex life is a problem to be explained. It is something of a puzzle. The double helix is not the only form of redundancy in the system of Mendelian inheritance that we and other complex life forms use. We carry two sets of double-helical molecules, one inherited from our father and the other from our mother. This condition is called *diploidy*. Ploidy refers to the number of DNA sets that an individual inherits. Maternal errors could be corrected using paternal codes, and paternal errors using maternal codes. Why do we have two DNA sets? It is not logically inevitable. Sex does require a doubling and halving of the gene numbers through the cycle of life and reproduction. Sex is not the same as reproduction. It is any process that mixes genes from different individuals. The offspring would be fine as long as it did not inherit errors in the same bit of DNA from both its mother and its father. Many proteins involved in reproduction are evolving unusually rapidly. The champion is lysin, a protein that determines whether a sperm can enter an egg. It is evolving as much as 25 times faster than gamma interferon, a protein important in the immune systems of mammals.

The code in an RNA virus is very economically written. It has no junk and contains overlapping genes in which the same bit of code is used in more than one gene. Modern RNA viruses are parasites and depend on more complex creatures. The RNA viruses may only be able to survive by running off a huge number of offsprings. It is known to make over 1,000 offspring copies from each parent [24].

Scientists are unable to perform the reverse translation, from body to DNA code, except in one or two easy cases. It seems that biological mechanisms have been unable to do it either. Cloning is simpler than sex, in which genes of two parents are combined and shuffled, and some of them then disposed of. DNA is routinely copied, without sex, in the cells of our bodies. Sex is indeed expensive. It halves the rate of reproduction. Ironically, successful clones are the cause of their own demise. As the clone increases in frequency, it becomes more vulnerable to disease. A disease can spread easily between members of the same clone, and as the clone becomes more common, the disease has more changes to evolve to infiltrate the target.

Bacteria are mainly clonal, but they also use mock-sexual processes from time to time. They are swapping components of DNA at random. Bacterial sex is not connected with reproduction. There were two bacteria before the sex act, and two after it. Maybe there is a compensating advantage that can increase the quality of the offspring. Sex is a puzzle that has not yet been fully explained. There are two theories at present. One is that sex helps us in our evolutionary battle against parasites. The other plausible theory is that sex helps to remove bad genes. The trick of sex is to combine errors from different individuals and create some error-free offsprings. The evolution helps organisms in the struggle of complex life against copying error. The overriding copy is known as dominant, the overridden copy as recessive. When a recessive gene is rare, however, it can persist unseen because most of the individuals who harbor it will have only one copy. Because family members are genetically more similar to one another than they are to strangers, sex in the family raises the odds of uniting two copies of a harmful recessive gene. Problems with incest are due to recessive genes [25].

In the animal branch of the tree of life, it is indeed the most complex forms that rely on sex. Sex has become a reproductive process as the complexity of life has advanced from the single-celled to the many-celled stage. If sex is a powerful force to purge error, life can evolve to be more complex by existing biological processes. Is the complexity of life on earth limited? Can new gene technologies increase the complexity of life? Why are slime molds so oversexed? Slime molds have 13 sexes, which are determined by three genes. But there is some power that strongly constrains the number of sexes. Maybe two sexes are optimal.

The formalities of Mendelian inheritance were needed when sex became associated with reproduction. Paternal care was an another step up in the evolution of complexity, because life with parental care is more complex than life without it. Evolution does not obey human notions of morality, nor is human morality a reflection of some natural law. Parents can pass nongenetic information (an experience as meme) to the offspring. There are human cultural rules that preempt the possibility of social conflict (see memory III in Fig. 5). In humans, males choose among females as well as females among males. Mate choice in favor with error-free members of the opposite sex also increases the efficiency of the purge. Many organisms compete for territories, and only the organisms that successfully occupy a territory may manage to breed. Infection diseases or meteoric agencies, for example, can make or break our future. A limited lifetime is evolutionary tuned to provide an acceptable error rate with a sufficient adaptability to the changing environment. Darwinian natural selection was synthesized in a theory capable of explaining how the environment controls gene combinations. Natural selection can act only by the preservation and accumulation of infinitesimally small and inherited modifications, each profitable to the preserved being. A further modification can possibly be produced through natural selection by the slow and gradual accumulation of numerous, slight, yet profitable, variations. Instincts vary slightly in a state of nature too. They are of the highest importance to each animal (see memory II in Fig. 5).

Full advantage of sex is seen at the biochemical level. Individuals who are able to scramble and recombine their genes with a mate can maintain a perpetual genetic flux. Two separate mating strategies are underway: monogamy and the current strongly polygamous arrangement. A population can only evolve from promiscuity to monogamy if incorruptible couples consistently have more surviving children then libertines do. Humans, taken as species, cannot be described as exclusively monogamous. Divorce rates and extramarital affairs attest to that. The males of many species are larger and more irascible than their mates. To help them in their battle against rivals, warrior males throughout the animal kingdom have often become heavy weights, equipped with weapons enabling them to stab, ram, kick, or wrestle. For those, which compete for harems, the reward for being a successful male is proportionately high and so the conflict become that more serious. Competition between the lusty males is therefore intense, and success will favor only the heaviest and most belligerent of them. The association between sex and symmetry has been discovered in several other kinds of animals. Parasites and diseases are causing deviations in the host's development. Such abnormalities may not be corrected and will be carried into adulthood as, for example, a slightly deformed body or a badly configured tail.

The conflicts of interests between males and females have been resolved with a whole spectrum of parental relationships. Sex creates more variability in each generation, and organisms practising sexual reproduction in a changing and unpredictable environment would have great advantages in the battle to be able to survive. The most vigorous individuals, or those who have most successfully struggled with their conditions of life, will generally leave most progene. But success will often depend on having special weapons or means of defense, or the charms of the males, and the slightest advantage will lead to victory. There is little doubt that the sexual process has been responsible for creating this fantastic treasury of living things, including us humans with our powerful brains. If adult and ageing individuals faithfully passed on copies of their DNA to their offspring, after a few generations the accumulated errors would make life untenable. Sex is most prevalent in species, which live in the complex and very variable environments where microbes and parasites abound. Species are defined as groups that cannot interbreed. From time to time organisms evolve to give up sex, reproducing asexually instead. When this happens, any genetic differences between a parent and child are, by definition, due to mutation only. But, in fact, most mutations are neutral. They have no effect. They change the DNA sequence of a gene, but they do not affect the information.

For reasons that remain mysterious, the loss of sex is almost followed by swift extinction. An organism does not necessarily have to reproduce to spread his genes. Instead, he can devote himself to helping his relations spread theirs. Such a support often explains apparently altruistic behavior, such as that of the ants, bees, and wasps which slave away for the good of the colony and never reproduce themselves. With animals having separated sexes there will be in most cases a struggle between the males for possession of the females.

8 Origin of Complexity

Matter has an innate tendency to self-organizing and generating complexity. This tendency has been at work since the birth of the universe, when a pinpoint of featureless matter budded from "nothing" at all [20]. Irreversibility and nonlinearity characterize phenomena in every field of complexity. Nonlinearity causes small changes on one level of organization to produce large effects (anomalies) at the same or higher levels. The smallest of events can lead to the most massive consequences. For example, the slightest shift in the position of the attractor leads to a different trajectory (see Fig. 1). We can see an emergent property, which manifests as the result of positive and negative feedback. But global features of the system cannot be understood only by analyzing the parts separately. Deterministic chaos arises from the infinitely complex fractal structure. A fractal's form is the same no matter what length scale we use. By using the techniques of parallelism and massive parallelism in computer simulations we come a little closer to explaining the basic principles of complex systems. Our attention is directed to the most efficient algorithms of turbulence simulation, which can help us understand a behavior of very complex fractal objects as a whirl. Chaotic systems are exquisitely sensitive to initial conditions, and their future behavior can only be reliably predicted over a short time period. Moreover, the more chaotic system, the less compressible its algorithmic representation.

Turbulence is regarded as one of the "grand challenge" problems in contemporary high-performance computing [22, 23]. Despite this astonishing progress during the 50 years since the visionary work of von Neumann, simulating turbulent fluid flow in realistic way is still largely beyond the capability of today's computers. In essence, the common underlying theme linking complexity of nature with computation depends on the emergence of a complex organized behavior from many simpler cooperative and conflicting interactions between the microscopic components, irrespective of whether they are spinning electrons, atoms, etc.

Earthquakes, avalanches, and financial crashes do have a common fingerprint: the distribution of events follows a single power law [4, 26]. This power law means that the physics of small avalanches is the same as that of large ones. Self-organization is a natural consequence of time evolution of vast aggregates of simple agents. By making these agents interact in a more complex way we could create an even greater variety of behavior, such as spiral structures reminiscent of galaxies, hurricanes, tornadoes, and particles of matter. Nonliving things, for instance crystals, are capable of self-reproduction during growth. Evolution on the edge of chaos (see Fig. 2) can be extended for nonliving systems. The negative forces are caused by negative fluctuation and positive forces are caused by positive fluctuation and by selection as an influence of boundary conditions.

Fractals seem to be very powerful in describing natural objects on all scales. Fractal dimension and fractal measure are crucial parameters for such description [26, 27]. Many natural objects have self-similarity or partial self-similarity of the whole object and its part. Different physical quantities describing properties of fractal objects in E-dimensional Euclidean space with a fractal dimension D were defined in [26]. Fractal dimension D depends on the interrelation between the number of repetition and reduction of individual object. There is relationship between the dimensionality and fractal properties of the matter, which contains the constant of golden mean $\phi = (\sqrt{5} - 1)/2 = 0.6180339887$. Constant ϕ is a special case of fractal dimension D defined by the condition $D(D - E + 2) = 1$ for $E = 3$ [26]. Links between inverse coupling constants of various interactions (gravitational, electromagnetic, weak, and strong) in the three-dimensional Euclidean space are discussed in [27]. Different properties of particles (and interactions between them) correspond to the specific values of a fractal dimension. The values $D = 0$, $E - 2$, $E - 1$, E play the most important role in such analysis [27].

9 Parallel Evolutionary Optimization of Controllers with a System's Identification

A very strong trend to digital computer control is evident in most areas of application. Large process applications are found in power plants, paper mills, refineries, chemical plants, steel mills, etc. Aircraft control, manufacturing, machine tools and robots, environmental control, and transportation are but a few of an expanding list of applications. Digital control offers important advantages in flexibility of modifying controller characteristics or of adapting the controller in the case when plant dynamics change with operating conditions. In multivariable systems, with more than one input and output, modern techniques for optimizing system performance or reducing interactions between feedback loops can be implemented.

Many approaches to the control of a plant or a process rely on the identification of model of the plant or process to be controlled. As the need for control is extended to systems of increasing complexity which are also often highly nonlinear, the ability to produce a plant model which is adequate over all operating conditions becomes a more and more demanding task. A different approach to the problem is provided by PID controllers or rule-based fuzzy control in which the plant model is replaced by a certain number of control rules, given by experts or deduced by observation. Controlling of the plant then turns into a process of plant state observation and invoking of appropriate rule from a stored rule base. At the heart of this process is an inference mechanism, by which the correct control action is inferred from the observed state. A key step in the application of fuzzy methods to process control is the inference mechanism. Essentially, this means that given, the plant-state as a fuzzy set, it is necessary to infer the correct control action from a rule base.

An advantage of fuzzy systems is that they are based on linguistic rules, and therefore it is often easy to understand the underlying functionality of the systems. The process for designing fuzzy systems is usually heuristic. Also, approximate position and shape of the membership functions may be determined by common sense, but exact (optimal) position and shape are unknown. An automatically designed fuzzy system can sort out important features from insignificant ones and also discover new relationships.

Automatic design of fuzzy controllers should optimize membership function and the rule base at the same time. A parallel genetic algorithm (PGA) is an extremely powerful optimization technique that could be used to optimize the parameters of controllers. To increase the efficiency of a GA the influence of migration in a multilevel distributed GA (MDGA) can be added. MDGAs use the power of the computers better than one-level distributed GAs [18].

PGAs are used to optimize both the structure and the associated parameters of controllers. Intelligence should emerge from mutual interaction among competence modules.

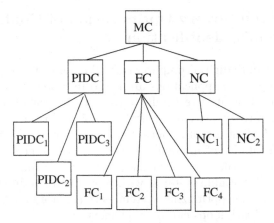

Fig. 7. Two-level structure of the MDGA

We must consider the mechanisms linking GA to the solved problem. There are two such mechanisms: encoding solutions to the problems on chromosomes, and evaluating function that returns a measurement of the worth of chromosome to the context of the problem. The evaluation function is the link between the GA and the problem to be solved. Most of the computer time in GAs is spent on evaluating objective functions. The initial population for traditional GA is usually chosen randomly. We used a real number encoding scheme instead of a traditional binary one [18]. It is very useful to restrict the range of each gene (parameter). This action makes the GA much more effective, because it does not waste its time looking through a statistic's range or searching on a scale that is insignificant with respect to the statistic. To increase the efficiency of a GA, the influence of migration in hybrid and distributed GAs was tested [9, 18].

To increase the efficiency of GAs, a fourth operator was introduced—migration. Several small separate populations that are on an individual PC in a computer network create new individuals, from which the best are included in one common population on a master PC. The best strings are passed to the master population when subpopulation reaches some local minimum. This seems better than copying the best strings to neighboring subpopulations at regular intervals. In the second-level structure, we can create a cluster of independent subpopulations [clusters of PID-controllers (PIDCs), fuzzy controllers (FCs), and neural controllers (NCs)] running on the first level (see Fig. 7), from which information is sent to a higher level (to master controller MC). With this structure, we must solve the following questions: what is sent, when it is sent, where it is sent, and whether the direction of data flow is only from a lower level to higher level or in both directions (independent or dependent development of subpopulations).

GAs are highly parallel procedures, therefore allowing for fairly simple implementations on any kind of machine architecture. However, as we observed earlier, there is a disadvantage of using small distribution populations, as usually done on distributed-memory machines, whenever centralized organization can deliver the best search results for the problem at hand. The way of resolving this search quality performance trade-off is by distributing the population through the processors and performing periodic exchanges of individuals between subpopulations. Another way of approaching this trade-off: a "semidistributed" population is a distribution that, by organizing processors in clusters, achieves some centralization without generating much contention or excessive communication. The initial population of the MDGA is randomly generated. Members of the population are sorted according to the cost function values, and the population is split into the two parts (not the same rise). The group containing the best solutions is called "elite." The elite group stays unchanged during iteration, and the members of the other group are regenerated. The algorithm guarantees that every new created member is originated by using parents with better value of the cost function than the substituted member. This modification is described in [24]. Genetic algorithms in hierarchical structures run independently, but each GA (except the "top master") occasionally sends its best solution to the upper level of master GA. The GA periodically checks its master's best solution. When the GA has a better solution than its upper-level master, it sends this solution to the upper level. The period of checking is given in a number of iterations, and its value depends on the level the GA is working in. The master GA (e.g., every GA except the bottom level) checks periodically contributions of the lower levels and adds them to its elite.

One possible representation of a hierarchical structure in the chromosome is shown in Fig. 8.

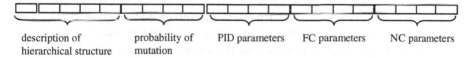

| description of
hierarchical structure | probability of
mutation | PID parameters | FC parameters | NC parameters |

Fig. 8. Chromosome representation of hierarchical structure

9.1 An Artificial Immune System

The immune system is a complex, self-organizing and highly distributed system that has no centralized control and which uses learning and memory when solving particular tasks. An artificial immune system (AIS) fully exploits self-organizing properties of the vertebrate immune system. The biological immune system is an efficient natural protection system whose primary function is to generate multiple antibodies from the antibody gene libraries and to keep it alive even if the unknown foreign pathogen infects the body. The AIS has several desired features for optimization purposes, like robustness, flexibility, learning ability, and memory. The AIS is self-organizing, since it determines survival of newly created clones, and it determines its own size. This is referred to as meta-dynamics of the system. These characteristics are often useful for description of hierarchical systems.

Two distinct algorithms have emerged as successful implementation of AIS: the immune network model described by Jerne and the negative selection algorithm developed by Forrest. The immune system implements two types of selection [28]: negative selection and clonal selection. These two processes have to be dealt with separately by researchers. Negative selection, which operates on lymphocytes maturing in thymus (called T-cells), ensures that these lymphocytes do not respond to self-proteins. This is achieved by killing any T-cell that binds to a self-protein while it is maturing. The second selection process, called clonal selection, operates on lymphocytes that have matured in the bone marrow (called B-cells). Any B-cell that binds to a pathogen is stimulated to copy itself. The copying process is subject to high probability of errors (hypermutation). The combination of copying with mutation and selection amounts to an evolutionary algorithm that gives rise to B-cells that are increasingly specific to the invading pathogen. If the same or similar pathogens invade in the future, the immune system will respond much more quickly because it maintains a memory of successful responses from previous infections. A constrain-handling approach based on emulation of an immune system (particularly, using the negative selection model) can be incorporated into parallel GA [17].

9.2 Design of Evolvable Controllers

In recent years, there has been growing interest in using intelligent approaches such as fuzzy, neural network, evolutionary methods, and their combined technologies for the Proportional-Integral-Derivative (PID) controller [9, 28, 29]. The PID controller has been widely used owing to its simplicity and robustness in chemical processes, power plants, and electrical systems. Its popularity is also due to its easy implementation in hardware and software. However, with only PID parameters, it is very difficult to control a plant with complex dynamics, such as large dead time, inverse response, and highly nonlinear characteristics. That is, since the PID controller is usually poorly tuned, a higher degree of experience and technology is required for tuning in the actual plant. Also, a number of approaches have been proposed to implement mixed control structures that combine a PID controller with fuzzy logic or neural networks. We try to adapt the hierarchical structure of controller modules to dynamic environment [9].

The use of classical PID controllers is preferred whenever nonlinear techniques are not strictly required. A GA is used to optimize both the structure and the associated parameters of controllers. Intelligence should emerge from mutual interaction among competence modules.

We tested several versions of GAs:

Standard GA is one population with the size equal to 40 individuals. Individuals are sorted by their fitness. The higher the fitness, the higher is the probability of selecting the individual to be a parent (a roulette wheel). The second parent is selected in the same manner. Then, the crossover, mutation, and correction operators are applied. The reproduction is repeated until the worse half of population is replaced.

GA with two subpopulations. The sexual approach where the male and the female are distinguished [3]. Every subpopulation has the size = 20 individuals. Individuals are sorted by their fitness. The first parent is selected from the male subpopulation while the second parent is selected from the female subpopulation. The selection probability of the first parent is performed with a uniform distribution function, while the selection probability of the second parent is different, using a modified roulette wheel approach that more often preferred a better individual. Crossover, mutation, and correction operators are then applied [9]. The reproduction is repeated until the worse half of every subpopulation is replaced.

Parallel GA can have a two-level structure. The first level of hierarchical structure is created by several populations with different GAs [18, 30]. The best or random individuals from the first level are sent to the second level. At this level, the standard GA with the elitism runs. This two-level structure allows us to find a better solution than that found by GAs in the first level; the best solution from the first level can never be lost but only overtaken in the second level.

We used the following modifications of GAs:

Limited lifetime. Any individual is not removed (replaced) until the minimum lifetime is reached [21]. The minimum lifetime is randomly generated when the individual is born. Individuals can survive for several generations even if they are not good. They have an opportunity to improve their fitness during the lifetime. This approach can slow down the evolution process, but it also prevents potentially good individuals to be lost.

HC operator uses a hill-climbing approach. In randomly selected individuals and in randomly selected genes, several small modifications are carried out; the best modification with the best fitness is retained for further use.

Adaptive version of PGA with AIS is given in [17]. We have used a constraint-handling approach based on the emulation of the immune system (particularly, using the negative selection model) that was incorporated into a parallel GA.

Our cost function has three parts [9]:

- the sum of the square control error (LMS),
- the overshoot penalty,
- the oscillation penalty of a control action.

The computer simulation with an evolutionary optimization helps identify the nonlinear systems and simultaneously can change parameters of controllers. Parallel GAs with the diploid chromosomes and an AIS can increase the efficiency and robustness of systems, and thus they can track better optimal parameters in a changing environment. From the experimental session it can be concluded that modified standard GAs with two subpopulations can design controllers much better than classical versions of GAs.

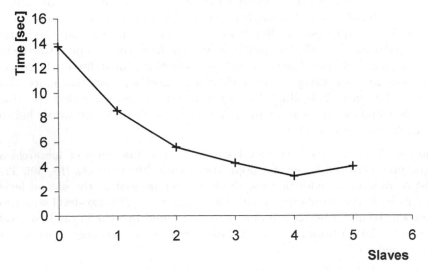

Fig. 9. The dependence of computation-time on the number of slaves

The HC modification of standard GA can improve the speed of the convergence, but the lifetime modification can support the higher variety of population. In the beginning, the HC modifications have better results in most cases, but the lifetime modifications have slower convergence; in the end, they can find better solutions than the HC modifications. The dependence of computation-time and generations on the slaves in parallel GA is in Fig. 9 and Fig. 10.

It can be demonstrated that standard GAs with haploid chromosomes are unable to correctly locate optimal solutions for time-dependent objective functions. An interesting fact was found, namely that sexual reproduction (SR) and the immune system (IS) use two types of selections working in parallel (IS: clonal and negative selection; SR: female and male selection [9, 21]). Every selection mechanism solves one task (for example, convergence or adaptation). The system as a whole has then both desired features.

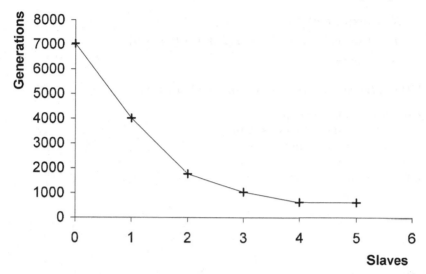

Fig. 10. The dependence of the number of generations on the number of slaves

10 Parallel Grammatical Evolution with Backward Processing

Our approach uses a PGE [11, 12]. A population is divided into several sub-populations that are arranged in the hierarchical structure. Every subpopulation has two separate parts: a male group and a female group. Every group uses quite a different type of selection. In the first group a classical type of GA selection is used. In the second group only different individuals can be included. It is a biologically inspired computing similar to a harem arrangement. This strategy increases an inner adaptation of PGE. The following text explains why we used this approach. Analogy would lead us one step further, namely, to the belief that the combination of GE with a sexual reproduction. On the principle of the sexual reproduction we can create a PGE with a hierarchical structure.

10.1 Backward Processing of the GE

The PGE is based on the GE [11], where BNF grammars consist of terminals and nonterminals. Terminals are items which can appear in the language. Nonterminals can be expanded into one or more terminals and nonterminals. Grammar is represented by the tuple $\{N,T,P,S\}$, where N is the set of nonterminals, T the set of terminals, P a set of production rules which map the elements of N to T, and S is a start symbol which is a member of N. For example, below is the BNF used for our problem:

$$N = \{\text{expr, fnc}\}$$
$$T = \{\,\sin, \cos, +, -, /, *, X, 1, 2, 3, 4, 5, 6, 7, 8, 9\}$$
$$S = <\text{expr}>$$

and P can be represented as four production rules:

1. <expr> := <fnc><expr>
     <fnc><expr><expr>
     <fnc><num><expr>
     <var>
2. <fnc> := sin
    cos
    +
    *
    −
    $U-$
3. <var> := X
4. <num> := 0, 1, 2, 3, 4, 5, 6, 7, 8, 9

The symbol $U-$ denotes a unary minus operation. The production rules and the number of choices associated with each are given in Table 1.

There are notable differences when compared with [11]. We do not use two elements <pre_op> and <op>, but only one element <fnc> for all functions with n arguments. There are no rules for parentheses; they are substituted by a tree representation of the function. The element <num> and the rule <fnc><num><expr> were added to cover generating numbers. The rule <fnc><num><expr> is derived from the rule <fnc><expr><expr>. Using this approach we can generate the expressions more easily. For example when one argument is a number, then $+(4, x)$ can be produced, which is equivalent to $(4 + x)$ in an infix notation. The same result can be received if one of <expr> in the rule <fnc><expr><expr> is substituted with <var> and then with a number, but it would need more genes.

There are not any rules with parentheses because all information is included in the tree representation of an individual. Parentheses are automatically added during the creation of the text output.

Table 1. The number of available choices for each production rule

Rule no.	Choices
1	4
2	6
3	1
4	10

10.2 Reduction of the Search Space

If the GE is not restricted anyhow, the search space can have infinite number of solutions. For example, the function $\cos(2x)$ can be expressed as $\cos(x+x)$, $\cos(x+x+1-1)$, $\cos(x+x+x-x)$, $\cos(x+x+0+0+0+\ldots)$, etc. It is desired to limit the number of elements in the expression and the number of repetitions of the same terminals and nonterminals.

In our system every application of an element from the set will decrease its number of possible uses (see Table 2). In this case, when this counter is 0, the element is removed from the set. It means that the number of production rules is decreased and the gene value will have quite a different influence on the result.

Table 2. Number of available choices for each production rule

Rule	All choices	Availabe choices
A	2	0
B	5	3
C	6	3
D	3	3

Consider two genes with values 20 and 81; without reducing the space the results would be

$$20 \bmod 4 = 0 \Rightarrow A$$
$$81 \bmod 4 = 1 \Rightarrow B$$

When rule A is temporarily removed from the set:

$$20 \bmod 3 = 2 \Rightarrow D$$
$$81 \bmod 3 = 0 \Rightarrow B$$

When a rule is removed from the set the result of modulo operation codes a different rule. Result 0 codes rule A before reduction and rule B after reduction. Therefore it is better to have a sorted list of rules (see Table 3). Using the sorted list without reduction the results would be

$$20 \bmod 4 = 0 \Rightarrow C$$
$$81 \bmod 4 = 1 \Rightarrow B$$

When rule A is removed from the sorted list:

$$20 \bmod 3 = 2 \Rightarrow D$$
$$81 \bmod 3 = 0 \Rightarrow C$$

Table 3. Sorted table of available choices for each production rule

Rule	All choices	Availabe choices
C	6	3
B	5	3
D	3	3
A	2	0

The same result of modulo operation 0 codes the same rule C. The probability that rule C is removed before rule A is low because the number of all choices for rule C is higher. When the number of available choices for rule C is zero then the result of modulo operation has still a different mapping from genotype to phenotype. The use of a list sorted by available choices simplifies analysis of genotype or phenotype.

10.3 Grammatical Evolution

The chromosome is represented by a set of integers filled with random values in the initial population. Gene values are used during chromosome translation to decide which terminal or nonterminal to pick from the set. When selecting a production rule there are four possibilities; we use gene_value mod 4 to select a rule. However, the list of variables has only one member (variable X) and gene_value mod 1 always returns 0. A gene is always read; no matter if a decision is to be made, this approach makes some genes in the chromosome somehow redundant. Values of such genes can be random, but genes must be present.

Figure 11 shows the genotype–phenotype translation scheme. Body of the individual is shown as a linear structure, but in fact it is stored as a one-way tree (child objects have no links to parent objects). In the diagram we use abbreviated notations for nonterminal symbols: f, <fnc>; e, <expr>; n, <num>; v, <var>.

The column description in Fig. 11 is as follows:

A. Objects of the individual's body (resulting trigonometric function),
B. Genes used to translate the chromosome into the phenotype,
C. Modulo operation, divisor is the number of possible choices determined by the gene context,
D. Result of the modulo operation,
E. State of the individual's body after processing a gene on the corresponding line,
F. Blocks in the chromosome and corresponding production rules,
G. Block marks added to the chromosome.

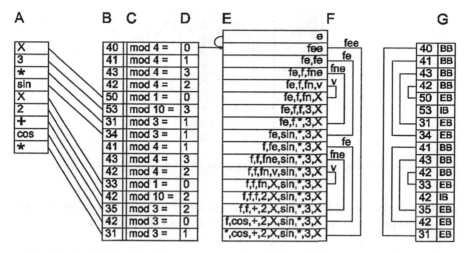

Fig. 11. Relations between genotype (column B) and phenotype (column A)

Since operation modulo takes two operands, the resulting number is influenced by gene value and by gene context (Fig. 11C = Fig. 11 column C). Gene context is the number of choices determined by the currently used list (rules, functions, variables). Therefore, genes with same values might give different results of modulo operation depending on what object they code. On the other hand, one terminal symbol can be coded by many different gene values as long as the result of modulo operation is the same $(31 \bmod 3) = (34 \bmod 3) = 1$. In the example given (Fig. 11A), the variables set has only one member X. Therefore, the modulo divider is always 1 and the result is always 0; a gene which codes a variable is redundant in that context (Fig. 11D). If the system runs out of genes during phenotype–genotype translation then the chromosome is wrapped and genes at the beginning are reused.

10.4 Processing the Grammar

The processing of the production rules is done backwards—from the end of the rule to the beginning (Fig. 12). For example, production rule <fnc><expr1> <expr2> is processed as <expr2><expr1><fnc>. We use <expr1> and <expr2> at this point to denote which expression will be the first argument of <fnc>.

The main difference between <fnc> and <expr> nonterminals is in the number of real objects they produce in the individual's body. Nonterminal <fnc> always generates one and only one terminal; on the contrary, <expr> generates an unknown number of nonterminal and terminal symbols. If the phenotype is represented as a tree structure then a product of the <fnc> nonterminal is the parent object for handling all objects gener-

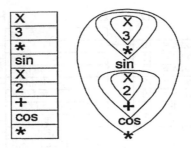

Fig. 12. Proposed backward notation of a function tree structure

ated by <expr> nonterminals contained in the same rule. Therefore the rule
<fnc><expr1><expr2> can be represented as a tree (Fig. 13).

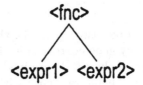

Fig. 13. Production rule shown as a tree

To select a production rule (selection of a tree structure) only one gene
is needed. To process the selected rule a number of n genes is needed and
finally to select a specific nonterminal symbol again one gene is needed. If
the processing is done backwards the first processed terminals are leafs of
the tree and the last processed terminal in a rule is the root of a subtree.
The very last terminal is the root of the whole tree. Note that in a forward
processing (<fnc><expr1><expr2>) the first processed gene codes the rule,
the second gene codes the root of the subtree, and the last are leafs.

When using the forward processing and coding of the rules described
in [11] it is not possible to easily recover the tree structure from genotype. This is caused by <expr> nonterminals using an unknown number
of successive genes and the last processed terminal being just a leaf of the
tree. The proposed backward processing is shown in Fig. 11E.

10.5 Phenotype-to-Genotype Projection

Using the proposed backward processing system the translation to a phenotype subtree has a certain scheme. It begins with a production rule (selecting
the type of the subtree) and ends with the root of the subtree (in our case
with a function) (Fig. 11F). In the genotype this means that one gene used to

select a production rule is followed by n genes with different contexts, which are followed by one gene used to translate <fnc>. Therefore, a gene coding a production rule forms a pair with a gene-coding terminal symbol for <fnc> (root of the rule). Those genes can be marked when processing the individual. This is an example of a simple marking system:

BB: Begin block (a gene coding a production rule)
IB: Inside block
EB: End block (a gene coding a root of a subtree)

The EB and BB marks are pair marks and in the chromosome they define a block (Fig. 11G). Such blocks can be nested but they do not overlap (the same way as parentheses). The IB mark is not a pair mark, but it is always contained in a block (IB marks are presently generated by <num> nonterminals). Given a BB gene a corresponding EB gene can be found using a simple LIFO method.

A block of chromosome enclosed in a BB–EB gene pair then codes a subtree of the phenotype. Such a block is fully autonomous and can be exchanged with any other block or it can serve as a completely new individual.

Only BB genes code the tree of an individual's body, while EB and IB genes code the terminal symbols in the resulting phenotype. The BB genes code the structure of the individual; changing their values can cause change of the applied production rule. Therefore, change (e.g., by mutation) in the value of a structural gene may trigger change of context of many or all following genes.

This simple marking system introduces a phenotype feedback to phenotype; however, it does not affect the universality of the algorithm. It is not dependent on the used terminal or nonterminal symbols; it only requires the result to be a tree structure. Using this system it is possible to introduce a progressive crossover and mutation.

10.6 Crossover

When using GE the resulting phenotype coded by one gene depends on the value of the gene and on its context. If a chromosome is crossed at random point, it is very probable that the context of the genes in the second part will change. This way crossover causes destruction of the phenotype, because the newly added parts code a different phenotype than in the original individual.

This behavior can be eliminated by using a block marking system. Crossover is then performed as an exchange of blocks. The crossover is made always in an even number of genes, where the odd gene must be a BB gene and the even gene must be a EB gene. The starting BB gene is presently chosen randomly; the first gene is excluded because it encapsulates (together with the last used gene) the whole individual.

The operation takes two parent chromosomes and the result is always two child chromosomes. It is also possible to combine the same individuals, while the resulting child chromosomes can be entirely different.

Given the parents,

1. $\cos(x+2) + \sin(x*3)$
2. $\cos(x+2) + \sin(x*3)$

the operation can produce children

3. $\cos(\sin(x*3) + 2) + \sin(x*3)$
4. $\cos(x+2) + x$

This crossover method works similar to direct combining of phenotype trees; however, this method works purely on the chromosome. Therefore, phenotype and genotype are still separated. The result is a chromosome, which will generate an individual with a structure combined from its parents. This way we receive the encoding of an individual without backward analysis of his phenotype. To perform a crossover the phenotype has to be evaluated (to mark the genes), but it is neither used nor known in the crossover operation (also it does not have to exist).

10.7 Mutation

Mutation can be divided into mutation of structural (BB) genes and mutation of other genes. Mutation of one structural gene can affect other genes by changing their context. Therefore, the structural mutation amount should be very low. On the other hand, the amount of mutation of other genes can be set very high and it can speed up searching an approximate solution.

Given an individual

$$\sin(2+x) + \cos(3*x)$$

and using only mutation of nonstructural genes, it is possible to get

$$\cos(5-x) * \sin(1*x)$$

Therefore, the structure does not change, but we can get a lot of new combinations of terminal symbols. The divided mutation allows using the benefits of high mutation while eliminating the risk of damaging the structure of an individual.

10.8 Population Model

The system uses three populations forming a simple tree structure (Fig. 14). There is a master population and two slave populations which simulate different genders. The links among the populations lead only one way—from bottom to top.

Fig. 14. The population model

10.8.1 Female Population

When a new individual is to be inserted in a population a check is performed whether it should be inserted. If a same or similar individual already exists in the population then the new individual is not inserted. In a female population every genotype and phenotype occurs only once. The population maintains a very high diversity; therefore the mutation operation is not applied to this population. Removing the individuals is based on two criteria. The first criterion is the age of an individual—length of stay in the population. The second criterion is the fitness of an individual; using the second criterion a maximum population size is maintained. Parents are chosen using the tournament system selection.

10.8.2 Male Population

New individuals are not checked so duplicate phenotypes and genotypes can occur; also the mutation is enabled for this population. Mutation rate can be safely set very high (30%) provided that the structural mutation is set very low (less then 2%). For a couple of best individuals the mutations are nondestructive. If a protected individual is to be mutated, a clone is created and added to the population. If the system stagnates in a local solution the mutation rate is raised using a linear function depending on the number of cycles for which the solution was not improved. Parents are chosen using a logarithmic function depending on the position of an individual in a population sorted by fitness. A histogram of selections (Fig. 15) shows that the best individuals ($X = 0$) are chosen three times; that is, each of them has three children. For every selected male parent a new selection of female parent is made.

10.8.3 Master Population

The master population is superior to the male and female populations. Periodically the subpopulations send over their best solutions. Moreover, the master population performs another evolution on its own. Parents are selected using the tournament system. The master population uses the same system of mutations as the male population, but for removing individuals from the population only the fitness criterion is used. Therefore, the master population also serves as an archive of best solutions of the whole system (Fig. 16).

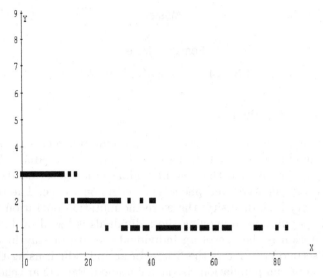

Fig. 15. A histogram of selections of parents depending on a position in sorted population

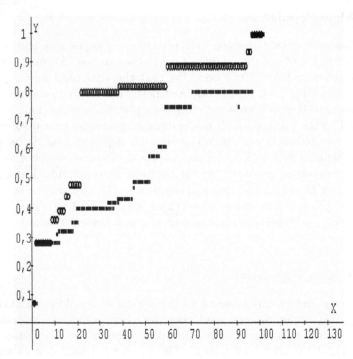

Fig. 16. Example of a maximum fitness history of the three-population system; circles mark the master population, squares mark the two subpopulations, and full circles indicate the same fitness for both the master and F and M subpopulations

10.9 Fitness Function

Around the searched function there is defined an equidistant area of a given size. Fitness of an individual's phenotype is computed as the number of points inside this area divided by the number of all checked points (a value in $\langle 0, 1 \rangle$). This fitness function forms a strong selection pressure; therefore the system finds an approximate solution very quickly.

10.10 Results

Given a sample of 100 points in the interval $[0, 2\pi]$ and using the block marking system described in Sect. 10.5, PGE has successfully found the searched function $\sin(2 * x) * \cos(2 + x)$ on the majority of runs. The graph of Fig. 17 shows maximum fitness in the system for ten runs and an average (in bold). The system also found a function with fitness better then 0.8 in less then 40 generations using three populations with size of 100 individuals each (Fig. 18). On the other hand, the same system with phenotype-to-genotype projection disabled (Fig. 19). The majority of runs did not find the searched function within 120 generations.

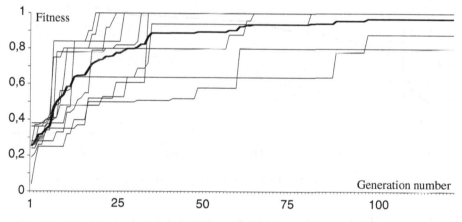

Fig. 17. Convergence of the PGE using backward processing (average in bold)

We have simplified the generation of numbers by adding a new production rule, thus allowing the generation of functions containing integer constants. The described parallel system together with phenotype-to-genotype projection improved the speed of the system. The progressive crossover and mutation eliminates destroying partial results and allows us to generate more complicated functions (e.g., $\sin(2 * x) * \cos(2 + x)$).

We have described a parallel system, parallel grammatical evolution (PGE), that can map an integer genotype onto a phenotype with the backward coding. PGE has proved successful for creating trigonometric identities.

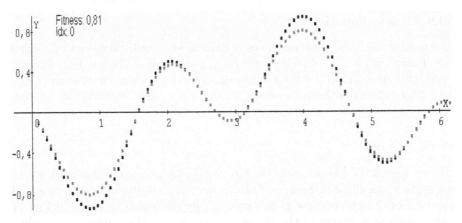

Fig. 18. A generated function with fitness $0.81 - \sin(\sin(-9+x)) * \cos(-11+2*x)$ (light) and searched function $\sin(2*x) * \cos(2+x)$ (bold)

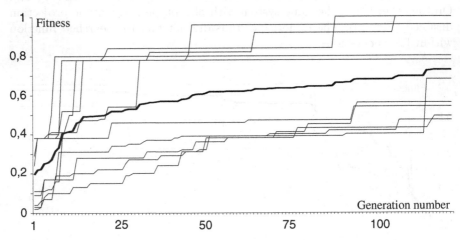

Fig. 19. Convergence of the PGE using forward processing (average in bold)

Parallel GEs with the sexual reproduction can increase the efficiency and robustness of systems, and thus they can track better optimal parameters in a changing environment. From the experimental session it can be concluded that modified standard GEs with two subpopulations can design PGE much better than classical versions of GEs.

The PGE algorithm was tested with the group of 6 computers in the computer network (see Fig. 21). Five computers calculated in the structure of five subsystems MR1, MR2, MR3, MR4, and MR5 and one master MR. The male subpopulation M of MR in the higher level follows the convergence of the subsystem. In Fig. 20 is presented 10 runs of the PGE-program. The shortest time of computation is only 10 generation. All calculation were

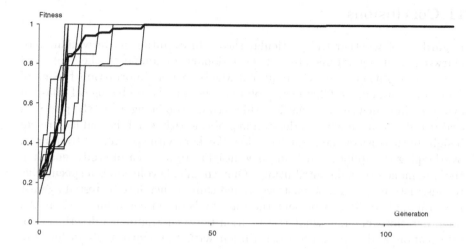

Fig. 20. Convergence of the PGE with 5 PC using backward processing (average in bold)

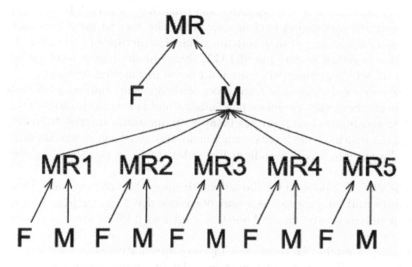

Fig. 21. The parallel structure of PGE with 6 computers

finished before 40 generation. This is better to compare with backward processing on one computer (see Fig. 17). The forward processing on one computer was the slowest (see Fig. 18).

The parallel grammatical evolution can be used for the automatic generation of programs. We are far from supposing that all difficulties are removed but first results with PGEs are very promising.

11 Conclusions

Regardless of whether the particular theory of evolution is Darwinian, neo-Darwinian, or something else, an evolutionary preamble to the biological phase of evolution is clearly required. Maybe a new information theory of evolution can help us. Chemical evolution, then, is the prebiological phase of evolution in which the very earliest things came into being. Over the past several decades, our growing understanding of the early earth has added crucial insight to theories of chemical evolution. Workers with specialized training in overlapping disciplines can bring new insights to an area of study, enabling them to make original contributions. Only chemical evolution is a speculative reconstruction of a unique past event, and cannot therefore be tested against recurring nature. It can present the ways life's origin could have arisen. An open question remains: what is a relation between Shannon's description of information and a procedure description with the shortest algorithm and without redundancy?

The computer simulation helps identify the conditions under which the evolution of the living world is running forever. Parallel GAs can increase the efficiency and robustness of systems, and thus they can track better optimal parameters in a changing environment. It is not easy to say which individual modifications in parallel and hierarchical structure are the best. If we join them together by the parallel GA, then—in the higher level—it is not important which method will contribute more to the final solution.

The increased awareness from other scientific communities, such as biology and mathematics, promises new insights and new opportunities. There is much to accomplish and there are many open questions. Interest from diverse disciplines continues to increase and simulated evolution is becoming more generally accepted as a paradigm for optimization in practical engineering problems.

The PGE can be used for the automatic generation of programs. This can help us to find information as a part of complexity. I am far from supposing that all difficulties are removed but first results with PGEs are very promising [21].

Finally, I would like to present my very speculative origin scenario. It is possible that all very complex systems exist in anomalous states, as turbulence with a whirl's structure. These anomalous states have a hierarchical structure. Maybe 3D matter is a first anomalous stage after a collision of the 12D "superstring" spaces. At the second level is the origin of living systems. At the third level is a brain with a consciousness. There is no greater anomaly in nature than matter that can live and can have a consciousness.

If a fractal description with the fractal dimension is plausible for us, we can imagine that all objects of the universe are fractal whirls with a different fractal dimension. If we see whirl structures in a macro-world (as spiral galaxies) and a real world (as tornado, whirlpools, and hurricanes), it can be probable that particles in a micro-world have the same fractal-whirl structure [22, 23]. To create a fractal whirl we need a minimum value of energy—a quantum of energy. There can be a relation between Planck's constant and the fractal dimension of the whirl (a special kind of turbulence). The value of frequency of a whirl's vibrations increases the accumulated energy of a whirl in coincidence with physical law for photon's energy. Photons can be whirls with a very small mass. Perhaps our universe is not a superstring space but a superwhirl space. Whirl structures can explain magnetism, gravity, etc [22, 23]. Whirls can attract each other using their different polarities. Whirls with their rotation have inertia, which explains what the mass of matter can be. We can see, for example, the fractal-whirl structure on Jupiter's weather. Perhaps it will be a plausible speculation but research is needed to test it.

References

1. Ch.B. Thaxton, W.L. Bradley, R.L. Olsen: *The Mystery of Life's Origin: Reassessing Current Theories* (Philosophical Library, New York, 1984)
2. P. Ošmera: Evolution of system with unpredictable behavior. In: *Proc. MENDEL 2004*, Brno, Czech Republic, 2004, pp 1–6. P. Ošmera: Genetic algorithms and their applications, The Habilit Work (in Czech language), 2002
3. R. Dawkins: *The Selfish Gene* (Oxford University Press, Oxford, 1976)
4. S.A. Kauffman: *Investigations* (Oxford University Press, New York, 2000)
5. I. Prigogine, I. Stengers: *Order out of Chaos* (Flamingo, London, 1985)
6. M.M. Waldrop: *Complexity—The Emerging Science at Edge of Order and Chaos* (Viking, New York, 1993)
7. P. Ošmera: Complex adaptive systems. In: *Proc. MENDEL 2001*, Brno, Czech Republic, 2001, pp 137–143
8. P. Ošmera: Complex evolutionary structures. In: *Proc. MENDEL 2002*, Brno, Czech Republic, 2002, pp 109–116
9. P. Ošmera: Evolvable Controllers using Parallel Evolutionary Algorithms. In: *Proc. MENDEL 2003*, Brno, Czech Republic, 2003, pp 126–132
10. J. Sanchez-Velazco, J.A. Bullinaria: Sexual selection with competitive/cooperative operators for genetic algorithms. In: *Proc. NCI 2003*, Cancun, Mexico, 2003, pp 308–316
11. M. O'Neill, C. Ryan: *Grammatical Evolution: Evolutionary Automatic Programming in an Arbitrary Language* (Kluwer Academic Publishers, Boston, 2003)
12. M. O'Neill, A. Brabazon, C. Adley: The automatic generation of programs for classification problems with grammatical swarm. In: *Proc. CEC 2004*, Portland, OR, 2004, pp 104–110
13. W. Piaseczny, H. Suzuki, H. Sawai: Chemical genetic programming—Evolution of amino acid rewriting rules used for genotype-phenotype translation. In: *Proc. CEC 2004*, Portland, OR, 2004, pp 1639–1646
14. X. Hu, Y. Shi, R. Eberhart: Recent advances in particle swarm. In: *Proc. CEC 2004*, Portland, OR, 2004, pp 90–95
15. S. Patarlini, T. Krink: High performance clustering with differential evolution. In: *Proc. CEC 2004*, Portland, OR, 2004, pp 2004–2005

16. B. Peng, R.G. Reynolds: Cultural algorithms: Knowledge learning in dynamic environments. In: *Proc. CEC 2004*, Portland, OR, 2004, pp 1751–1758
17. C.A. Coello, N.C. Cortés: A parallel implementation of an artificial immune system to handle constrain in genetic algorithms. In: *WCCI 2002*, Hawaii, pp 819–824
18. P. Ošmera, I. Šimoník, J. Roupec: Multilevel distributed genetic algorithms. In: *Proc. Int. Conf. IEE/IEEE on Genetic Algorithms*, Sheffield, UK, 1995, pp 505–510
19. K.I. Smith, R.M. Everson, J.E. Fielsend: Dominance measures for multi-objective simulated annealing. In: *Proc. CEC 2004*, Portland, OR, 2004, pp 23–30
20. P. Ošmera, J. Roupec: Limited lifetime genetic algorithms in comparison with sexual reproduction based gas. In: *Proc. MENDEL 2000*, Brno, Czech Republic, 2000, pp 118–126
21. P. Ošmera, O. Popelka, T. Panaček: Parallel Grammatical Evolution with Backward Processing, Proceedings of MENDEL'2005, Brno, Czech Republic, (2005) pp 27–28.
22. P. Ošmera: Evolution of the Universe Structures, Proceedings of MENDEL'-2005, Brno, Czech Republic, 2005, pp 1–6 32.
23. P. Ošmera:The Vortex-fractel Theory of the Gravitation, Proceedings of MENDEL'2005, Brno, Czech Republic, (2005), pp 7–14
24. M. Ridley: *Mendel's Demon* (Phoenix, London, 2001)
25. O. Judson: *Dr. Tatiana's Sex Advice to All Creation: The Definitive Guide to the Evolutionary Biology of Sex* (Vintage, London, 2003)
26. P. Coveney, R. Highfield: *Frontiers of Complexity—The Research for Order in a Chaotic World* (Faber and Faber, London, 1996)
27. O. Zmeskal, M. Nezadal, M. Buchnicek: Chaos, Solitons Fract. **17**, 113–119 (2003)
28. H.D. Kim, P.W. Hong, J.I. Park: Auto-tuning of reference model based PID controller using immune algorithm. In: *Proc. CEC 2002*, Honolulu, Hawaii, 2002, pp 509–516
29. F. Cupertino, D. Naso, L. Salvatore, B. Turchiano: Design of cascaded controllers for DC drivers using evolutionary algorithms. In: *Proc. CEC 2002*, Honolulu, Hawaii, 2002, pp 309–316
30. P. Ošmera, R. Matousek: Automatic optimal design of fuzzy controllers. In: *Proc. GALESIA'97*, Glasgow, UK, 1997, pp 439–443
31. O. Zmeskal, M. Nezadal, M. Buchnicek: Coupling constants in fractal and cantorian physics, Solitons Fract. (in press)

Problem Solving via Fuzziness-Based Coding of Continuous Constraints Yielding Synergetic and Chaos-Dependent Origination Structures

Osamu Katai, Tadashi Horiuchi, and Toshihiro Hiraoka

Abstract. Based on the comparison of artificial systems with natural systems to elucidate the differences of their characteristics, we will propose a framework of a double-layered architecture of a problem solving system for constraint satisfaction problems, where the upper layer has characteristics corresponding to the artificial systems and the lower layer has characteristics corresponding to the natural systems. These two layers are derived by "fuzzy coding" (coding by fuzziness) in order to "decompose" continuous constraints for problem reduction. Thereafter, two different approaches to problem solving by the layered system architecture are proposed. One way of problem solving is to make use of the synergetic and tacit of the known layered structure. The other way is to focus on the chaotic phenomena through the interaction between the two layers. Moreover, considerations are made on the cause and meaning of these chaos phenomena in order to suggest some directions to make good use of it.

1 Introduction

Recently, it has become more and more difficult for the artificial systems based on traditional system methodologies to cope with real world problems which are very complex and large-scaled. Hence, it is important to realize symbiotic ways of constructing systems by integrating the "self-organizing and diversity-generating" characteristics in redundant and complex natural systems and the "rationality & logicality" oriented characteristics in artificial systems.

In this research, we first focus on the essential differences of these characteristics between artificial systems and natural systems; while "centralized, sequential and logic-based" processes play the main role in the artificial systems; "distributed, concurrent and self-organizing" processes play the main role in the natural systems.

Then we focus on the constraint satisfaction problems involving continuous variables and introduce fuzziness as coding schema for these variables to reduce the complexity of constraints, which results in a layered problem solving system architecture that consist of the following two kinds of layers, either of which corresponds to artificial systems or natural systems, respectively "differentiated layer" and "homogeneous layer". Between these two

O. Katai et al.: *Problem Solving via Fuzziness-Based Coding of Continuous Constraints Yielding Synergetic and Chaos-Dependent Origination Structures*, StudFuzz **187**, 579–601 (2006)
www.springerlink.com   © Springer-Verlag Berlin Heidelberg 2006

layers, we introduce an emergence mechanism based on the notion of "tacit knowing" by Polanyi [1] and the principle of "synergetics" by Haken [2].

That is, not only the "bottom up" interaction from a homogeneous layer to a differentiated layer but also the "top down" interaction from a differentiated layer to a homogeneous layer are set between these layers.

From the viewpoint of emphasizing the process of problem solving rather than its result, two different approaches are proposed, that is, "distributed concurrent approach" and "centralized sequential approach". Especially, in the latter approach (centralized sequential approach), while it is shown by Brouwer's fixed point theorem in topology in mathematics that the constraint propagation dynamics without coding via fuzziness illustrates simple and stable (periodic) behavior, it is clarified that the "hybrid" systems consisted of the differentiated layer ("symbolic layer"), and the homogeneous layer ("continuous layer") illustrates quite complex behavior caused by the "chaotic" structure ("fuzzy symbolic dynamics") involved in the two layers.

Finally, we discuss construction of more flexible constraint satisfaction problem solving system architectures which are capable of adapting in a self-organizing manner to the changes of the problem structures and the environments by integrating the above two approaches "symbiotically" so as to utilize the meaning of the chaotic phenomena in the hybrid system.

More concretely, we note the system characteristics that various solution sequences satisfying the constraints locally are produced along with the temporal progress of the fuzzy symbolic dynamics (centralized sequential approach), and then the possibility of making use of the above characteristics to the distributed concurrent approach is examined on the basis of the potentiality to search for a wide variety of solutions in the centralized sequential approach.

2 Artificial Systems and Natural Systems

In this research, we first focus on the essential differences of the characteristics between artificial systems and natural systems in order to consider symbiotic ways of constructing systems by integrating characteristics of natural systems and those of artificial systems; while "centralized, sequential and logic-based" processes play the main role in artificial systems, "distributed, concurrent and self-organizing" processes play the main role in natural systems.

In general, in traditional artificial systems usually developed in engineering field, we pursue goal-oriented function formation with a special emphasis on rationality and logicality where high efficiency and cost performance are given priority over flexibility and redundancy.

From the viewpoints of problem solving systems, these systems can be considered to consist of functionally differentiated elements which aim to solve the subproblems.

On the other hand, in natural systems such as biological systems essentially involving rich redundancy, adaptive and self-organizing processes play an important role. And goals are regarded not to be set in advance but to be formed through pattern formation processes based on the emergence principles such as synergetics [2].

Hence, flexibility and adaptability to environmental changes are given priority over efficiency and cost performance. From the viewpoints of problem solving systems, it can be considered that these systems are composed of homogeneous elements which have self-organizing ability related to rich redundancy.

3 Layered Problem Solving System Architecture Based on Fuzzy Coding of Continuous Constraints

Based on the above discussions about the differences of the characteristics between artificial systems and natural systems, we will propose a framework of the hybrid problem solving system by integrating the characteristics of natural systems and those of artificial systems in a symbiotic way without losing both merits.

In this research, we focus on the constraint-oriented problem solving as a framework of problem solving, and we propose a system architecture for solving constraint satisfaction problems involving not only discrete variables but also continuous variables.

From constraint-oriented perspectives on problem solving, a fuzzy set can be interpreted as a set of intervals (constraint intervals), and we call it "constraint-interval fuzzy set" [3]. As shown in Fig. 1a, it is given as an ordered collection of "crisp" intervals on the universe of discourse each of which represents a constraint called "constraint interval." The grade axis (in the traditional fuzzy set theory) is now regarded to be an "ordinal" scale axis. Such a constraint-interval fuzzy set can be represented on two-dimensional Cartesian coordinates by its lower and upper bounds as shown in Fig. 1b which we call MinMax graph.

By introducing this notion of fuzzy set, a continuous variable can be coded by fuzzy sets (fuzzy labels). This coding by fuzziness has more flexibility than the usual crisp coding which divides the domain of variable (universe of discourse) into distinct (exclusive) intervals, because in this "fuzzy coding" there remains a room for the choice of fuzzy labels and also for the selection of constraint interval form the selected fuzzy label.

The types of using fuzziness for coding continuous variables can be listed as follows:

1. fuzziness on ambiguous concepts (words) in natural language expressions;
2. fuzziness on operational variables in fuzzy control rules acquired by experts through their experience;

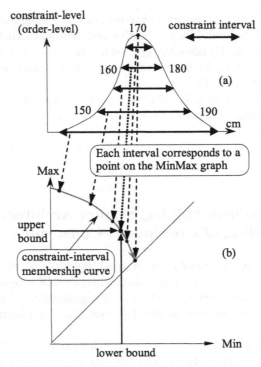

constraint-level
(order-level)

constraint interval

Each interval corresponds to a
point on the MinMax graph

constraint-interval
membership curve

Fig. 1. A constraint-interred fuzzy set and its MinMax graph representation

3. fuzziness originated from constraints imposed on the problem in question;
   (a) fuzziness originated from "crisp constraints";
   (b) fuzziness originated from "fuzzy constraints".

Here, a "fuzzy constraint" means a constraint which is given as a "multi-level" constraint ranging from a loose level (low-constraint level) to a strict level (high-constraint level), and thus there remains a room to select an appropriate constraint level suitable for solving problems. In this case, we think that fuzziness in each variable is derived by decomposing a constraint ("joint constraint") on the combination of several variables (whenever it is crisp or fuzzy) into componental (marginal) constraints through "projection" of the joint constraint onto componental variables.

Namely, if we have a crisp constraint relation C on a pair of variables, say x and y, we can approximate the constraint C by introducing appropriate constraint interval fuzzy sets imposed on x and y, respectively. This decomposition is then translated to coding of C into the label correspondence relation between fuzzy sets and the link correspondence relation between constraint intervals as shown in Fig. 2.

In the case where the constraint C itself is fuzzy, i.e., when it is given as a multi-level constraint, we can encode it into the label correspondence relation

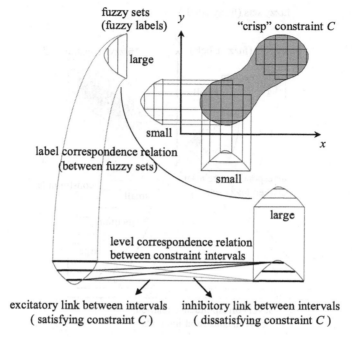

fuzzy sets
(fuzzy labels)

"crisp" constraint C

large

small

label correspondence relation
(between fuzzy sets)

small

large

level correspondence relation
between constraint intervals

excitatory link between intervals   inhibitory link between intervals
(satisfying constraint C)   (dissatisfying constraint C)

Fig. 2. Decomposition by coding of "crisp" constraint

between fuzzy sets and the link correspondence relation between constraint intervals by a similar approximation method as shown in Fig. 3.

This "fuzzy coding" which decomposes the constraint into two kinds of correspondence relations (i.e. those between fuzzy sets and those between constraint intervals) enables us to solve constraint satisfaction problems involving continuous variables in the double-layered system where symbolic (discrete) variables are treated on the upper layer and continuous variables are treated on the lower layer.

That is, by introducing such a fuzzy coding between the upper layer and the lower layer, we propose a double-layered system architecture as shown in Fig. 4 for solving constraint satisfaction problems involving continuous variables.

The upper layer is called "differentiated layer", and this corresponds to artificial systems and consists of heterogeneous elements, where "symbolic, goal-oriented, highly efficient" computation corresponding to the information processing on the left side of human brain ("L-mode computation") is performed.

On the other hand, the lower layer is called "homogeneous layer", and this corresponds to natural systems and consists of a large number of homogeneous elements, where "continuous, bottom up, adaptable" computation

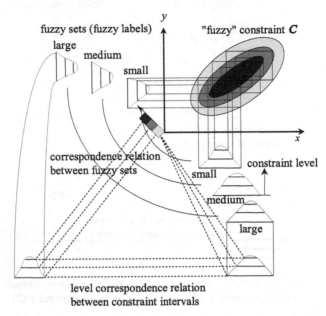

fuzzy sets (fuzzy labels)

Fig. 3. Decomposition by coding of a "fuzzy" constraint

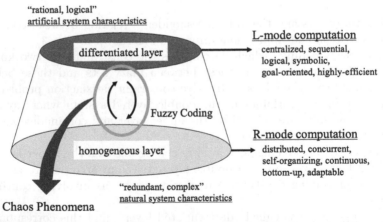

Fig. 4. A double-layered problem solving system architecture consisting of two layers of different functionalities

corresponding to the information processing on the right side of human brain ("R-mode computation") is executed.

Between these two layers, we introduce an emergence mechanism based on the notion of "tacit knowing" by Polanyi [1] and the principle of "synergetics" by Haken [2]. That is, not only the "bottom up" interaction from the homogeneous layer to the differentiated layer but also the "top down" inter-action from the differentiated layer to the homogeneous layer are set between these layers.

4 Two Approaches in the Proposed System Architecture

From the viewpoint of emphasizing the process of problem solving rather than its result, two different approaches are proposed as follows.

Distributed concurrent approach [4]: In this approach, the constraint sat-isfaction processes are executed in a multiple and concurrent manner by introducing the principle of "synergetics" in the homogeneous layer. More details of this approach and an application to design problems will be explained in Sect. 4.1.

Centralized sequential approach [5]: In this approach, the sequential con-straint propagations are performed repeatedly as a method of constraint satisfaction in the homogeneous layer. We will discuss this approach in detail including the simulation results in Sect. 4.2.

Especially, in the latter approach (centralized sequential approach), while it is shown by Brouwer's fixed point theorem in topology in mathematics that the constraint propagation dynamics without coding via fuzziness illustrates simple and stable (periodic) behavior, it is clarified that the "hybrid" systems consist of the differentiated layer ("symbolic layer") and the homogeneous layer ("continuous layer") illustrates quite complex behavior caused by the "chaotic" structure ("fuzzy symbolic dynamics") involved in the two layers.

Both of the above approaches are regarded as process-oriented problem solving approaches. In the former approach of "distributed concurrent ap-proach", problem solving will be considered to be an order formation process based on the principle such as synergetics, and the global patterns obtained through the process correspond to the solutions (convergent solutions).

In the latter approach of "centralized sequential approach", we will con-sider problem solving by constraint propagation process which changes the focus variable (constraint) from one to another, and the derived solutions through the process can be considered as a new notion of process-oriented solutions including traditional solutions as convergence solutions.

In such solutions, global consistencies are not always satisfied but local consistencies which we focus on are satisfied at a moment, and the focus point is moved to another variable successively through constraint propaga-tion process.

Hence, the solution itself will change to have a dynamic meaning from a static one. In other words, this means not to relax the problem side but to relax the notion of solutions at a meta level, and this kind of relaxation seems to happen frequently in our real world.

4.1 Double-Layered Architecture
for Autonomous Decentralized Problem Solving

4.1.1 Double-Layered Architecture Derived
from Decomposition of Constraints

When the constraint regions become more complex, we have to approximate them with several combinations of fuzzy sets. In this case, the problem of constraint satisfaction is reduced to the following two parts (steps): first, to search for the pair of fuzzy sets (fuzzy labels) for each (joint) constraint, and second, to search for the constraint interval (constraint level) in each fuzzy set.

The above way of treating crisp and fuzzy constraints by the combination of qualitative (discrete) labels and quantitative (continuous) interval constraints results in the double-layered architecture for problem solving shown in Fig. 5.

The upper layer treats the qualitative part of constraint satisfaction problems, and the lower layer treats the quantitative part. That is, on the upper layer, the structural (hard and global) constraints are treated, and on the lower layer, the elastic (soft and local) constraints are treated.

On the upper layer, the global consistency of the solution should be maintained. Thus, we will use a logic-based approach for this constraint

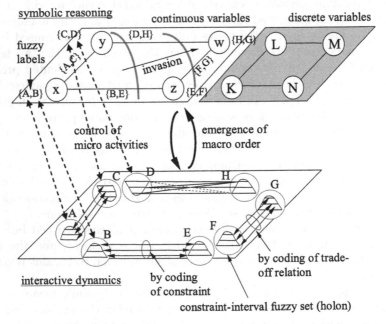

Fig. 5. A double-layered problem solving system architecture of an autonomous decentralized system

satisfaction. More precisely, we will use an invasion process to search the fuzzy sets (fuzzy labels) by regarding the problem as a "consistent labeling problem (CLP)" [6, 7].

On the lower layer, the processing of constraints is done in a decentralized and autonomous manner. We regard each constraint-interval fuzzy set as a "holon," a term which Koestler used to stand for a processing unit (subsystem) which can self-organize the interrelationship among other units (holons [8]). Each constraint interval (in each holon) makes local judgement to change its activation level by referring to those of other neighboring holons. Each constraint interval in each holon receives an external activation input which is dependent on the level of preference on the interval constraint.

However, it is merely a trigger to holon activation, and the final configuration of the activation levels of holons is ruled by a complex interrelationship among all the holons. Namely, each holon is dependent on all other holons, and vice versa.

4.1.2 Introduction of the Synergetic Approach

It should be noted that the "macro order" is not necessarily attained merely through local (micro) processing of constraints. Concerning this macro order formation, Haken's famous slaving principle, the leading principle in Synergetics [8], is extremely suggestive, for it says that the macro order is derived from the collective behavior of primitive components of the system, and that it rules the behavior of each primitive component (primitive behavior).

Due to this "feedback loop" between the macro order and the primitive (micro) behavior, it may be the case that only the most dominant macro order will survive among the existing macro orders.

Here, the following should be noted: it is not the case that once the macro order is organized by the collective micro behavior of holons, the macro order then governs the micro (primitive) behavior; rather the organization of the macro order and the governing of micro behavior proceed "simultaneously" (concurrently) by interacting with each other through the above feedback loop.

According to the general observations on holonic systems (decentralized autonomous systems) given above, we arrive at the architecture for the mutual interaction between the upper (structural) and the lower (elastic) layers as follows.

1. Each holon on the lower layer interacts only with its neighboring holons, and if its activation level exceeds a certain threshold value, the degree of activation is transmitted to the upper layer through linkage relations between the two layers.
2. On the upper layer, the invasion procedure starts at the unit which has labels with high-activation levels.

3. The labels with low-activation levels are eliminated from the upper layer before the synthesis procedure starts. This operation is sometimes called "filtering" in constraint satisfaction problem solving.
4. After the application of each synthesis procedure, we temporally terminate the activities of the inconsistent labels corresponding to the redundant units that will never be referred to in the future synthesis procedure.
5. After each synthesis procedure (calculation of the labels on the new front by eliminating inconsistent label pairs), we decrease the threshold values attached to the links between the two layers that are associated with the labels appearing on the current front.
6. If a label pair is eliminated from the upper layer, the corresponding link weight (of activation transmission linkage) is decreased.

4.1.3 Application to Structural Design Problems

Let us consider the case of structural design problems. In many cases, structural design problems include both continuous variables and "discrete variables". For example, size and amount of displacement of parts, stress on parts, etc. are usually continuous variables, whereas materials, manufacturing methods, joint styles among parts, etc. are regarded to be discrete variables relating to "structural selection." In this section, we will consider an application of the proposed system to such a structural design problem including both continuous variables and discrete variables.

As an illustrative example, we consider a simple beam design problem shown in Fig. 6.

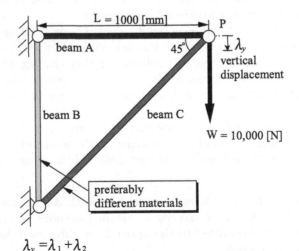

$$\lambda_y = \lambda_1 + \lambda_2$$

λ_1 : displacement component caused by beam B
λ_2 : displacement component caused by beam C

Fig. 6. A beam design problem

The design parameters are the cross-sectional area A_1 of beams A and C, the cross-sectional area A_2 of beam B, and two discrete variables representing the material of beam B and that of beams A and C. The maximum weight W to be supported at point P and the length L of beam A are given. As the objective functions, we consider the horizontal displacement l_x, the vertical displacement l_y, the stress on each beam, the total weight, and also the total cost.

Suppose that we select the material of beams from among three kinds of metals listed in Table 1. Here we consider the situation where the material of beam B and that of beams A and C should preferably be different.

Table 1. Candidate materials and their properties of the beams

Material	Soft iron	Cast iron	Aluminum
Young's modulus E (Gpa)	210	100	72
Cost function C ($\$\,m^{-3}$)	18	14	12
Specific gravity ρ ($g\,cm^{-3}$)	7.86	7.18	2.69

We have seven constraint equations on these parameters listed in Table 2.

We approximated the constraint region by three groups of rectangles, each of which consists of three-leveled interval constraints as shown in Fig. 7.

Even if the variables which appear in constraint expressions correspond to the same physical quantity, these variables are dealt as being different. Fig. 8 shows the constraint network on the upper layer and the advancement of invasion process.

Here we will explain the treatment of discrete variables in the proposed system. Once we set the candidate value of a discrete variable (e.g. E_2: the

Table 2. Constraint equations on the beam design problem

Displacement: $\lambda_x = \frac{WL}{E_2 A_2}$; $\lambda_y = WL\left\{\frac{1}{E_1 A_1} + \frac{1}{E_2 A_2}\left(1 + \frac{\sqrt{2}}{4}\right)\right\}$

Stress: $\sigma_{AC} = \frac{W}{A_2}$; $\sigma_{BC} = \frac{\sqrt{2}W}{A_2}$; $\sigma_{AB} = \frac{W}{A_1}$

Total weight : $M = \rho_1 A_1 L + \rho_2 A_2 L(1 + \sqrt{2})$

Total cost: $C = \alpha_1 A_1 L + \alpha_2 A_2 L(1 + \sqrt{2})$

Given parameters: W: maximum weight to be supported

L: length of beam A

Design parameters:

A_1: cross-sectional area of beams A and C $\quad$ A_2: cross-sectional area of beam B

E_1: Young's modulus of beams A and C $\quad$ E_2: Young's modulus of beam B

ρ_1: specific gravity of beams A and C $\quad$ ρ_2: specific gravity of beam B

α_1: cost function of beams A and C $\quad$ α_2: cost function of beam B

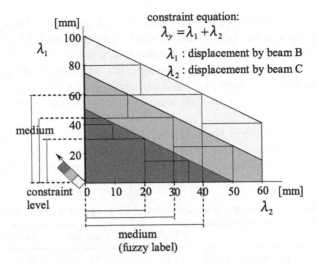

Fig. 7. Decomposition by coding of a fuzzy constraint

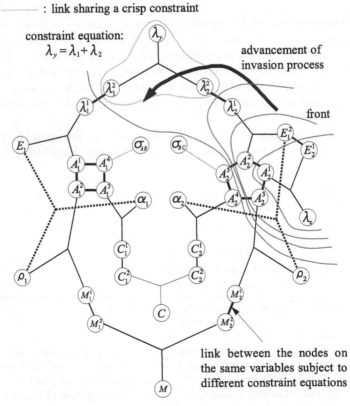

Fig. 8. A constraint network on the upper layer and an advancement of the invasion process

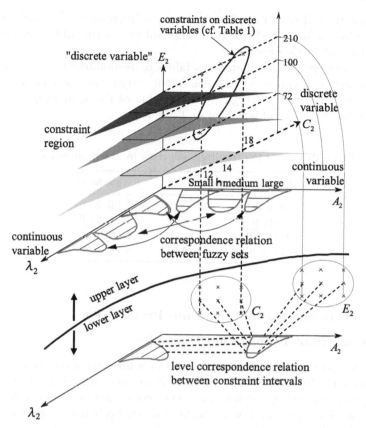

Fig. 9. The upper and the lower layers obtained by coding a constraint on the combination of continuous variables A_2 and l_2 and discrete variable E_2 and also by coding a constraint on the combination of discrete variables E_2 and C_2

young modulus of beam B), the constraint region consisted of the continuous variables (e.g. λ, A_2) is determined, as shown in Fig. 9.

This has the same structure as that of the "fuzzy constraint" where the multi-leveled constraint is considered according to the constraint level (the choice of a decision maker), as shown in Fig. 3. Therefore, the relation between continuous variables can be treated in the same way as in Fig. 3. Each candidate value of the discrete variables can be regarded as a specialized fuzzy set where the "constraint interval" is given as a "point"; the lower and the upper bound of the constraint interval and the relating continuous variables can be treated in the same way as in Fig. 3. Finally, the relation between discrete variables (e.g. E_2, λ_2) can be treated as the constraint on the upper layer as shown in Fig. 9.

By executing the system implemented on transputers, we finally obtained 11 different solutions on the upper layer. For each of the solutions, we have relevant activated constraint intervals on the lower layer. For each variable, we can select an activated constraint interval as a permissible solution of the constraint satisfaction problem.

In the case you have different variables corresponding to the same physical quantity, you can select the consistent interval as the common part of the selected intervals on these variables. One of the final precise solutions obtained by this method is shown in Table 3.

Table 3. One of the resultant solutions by the proposed system

Design parameters	Tolerance
A_1 (cm^2)	21.0–24.0
A_2 (cm^2)	22.0–28.0
Material of beam B	Soft iron
Material of beams A&C	Cast iron

4.2 Problem Solving by Constraint Propagation

4.2.1 Constraint Propagation Dynamics

Constraint propagation is one of the methods for solving constraint satisfaction problems. In this method, an input value on a variable is at first locally propagated to another variable via a joint constraint on these variables and then such a local propagation is executed successively until a globally consistent solution is obtained.

Considering the constraint region $C(x, y)$ as a constraint relation, we suppose that an interval (constraint interval) instead of a value is locally propagated one after another. Since the output interval (propagated interval) is computed from the input interval I and the function f derived from the constraint region $C(x, y)$, the process of interval propagation can be regarded as a dynamical process. Hence, we call it "constraint propagation dynamics" (Fig. 9).

First we take note of "inverse monotone relation" of the constraint propagation, i.e.,

$$I' \subseteq I \Rightarrow f(I) \subseteq f(I')$$

Here "$\subseteq$" means the inclusion relation between intervals. By one operation of constraint propagation, the inverse monotone relation is realized, and by two operations of constraint propagation, the monotonous relation is realized.

Constraint intervals in constraint-interval fuzzy set are represented on two-dimensional Cartesian coordinates by its lower and upper bounds. On this MinMax graph, we consider the area (set of intervals) where the constraint propagation can be executed endlessly and we call it "eternally recurrent area".

By the above relation, we can now show that the eternally recurrent area is a closed area surrounded by right up curves and horizontal or vertical line segments. Hence, using Brouwer's fixed point theorem [9] given below, we can now prove the existence of at least one fixed point I satisfying $f(I) = I$.

Brouwer's fixed point theorem: If X is a simplex $|\sigma_n|$, an arbitrary continuous mapping $f : |\sigma_n| \rightarrow |\sigma_n|$ has at least one fixed point.

We mainly mentioned the case when the constraint network consisted of a single loop of variables as illustrated in Fig. 10.

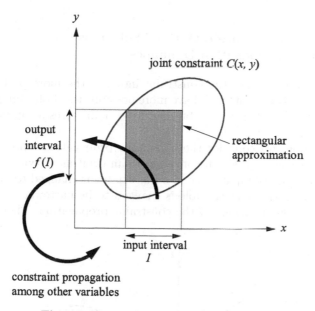

Fig. 10. Constraint propagation dynamics

But in general, constraint networks consist of multiple loops of variables sharing some common variables.

In such a case as shown in Fig. 11, when two different intervals x_l, x_m come to a common variable x through different loops, we decide to take an average value (interval) of x_l and x_m as a compromise value for the input interval to x.

It is expected that such constraint propagation systems do not in general show complex but stable behavior that converges to periodic solutions, because those systems are basically composed of loops each of which has a stable property as explained above.

And we confirmed it by carrying out the computer simulation of constraint propagations for some cases when the constraint network consists of two loops as shown in Fig. 11.

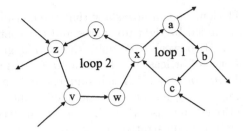

Fig. 11. A constraint network consisting of two loops

4.2.2 Derivation of Pareto Optimal Solutions by Constraint Propagation Dynamics

We think that the wider the constraint interval, the more preferable it is, since wider constraint intervals have more possibilities of obtaining solutions. Hence, it can be said that the obtained fixed point gives a solution which is Pareto optimal.

For example, we consider the constraint propagation among four crisp constraints in one loop, which converges to an orbit as shown in Fig. 12.

In this case, if the input interval on a variable is changed to be wider, the input interval to the next variable is changed to be narrower because of the "inverse monotone relation" of the constraint propagation. Therefore, if the

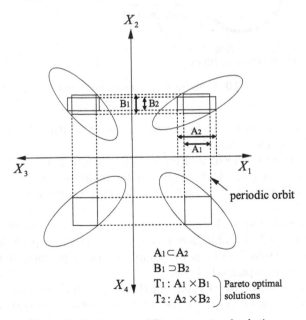

Fig. 12. Derivation of Pareto optimal relations

propagation converges to an orbit, the resultant constraint intervals mean a set of Pareto optimal solutions.

If the constraint region has a simple shape, the fixed point is usually given as a stable equilibrium point of the dynamical system. Thus, in such a case, Pareto optimal solutions can be obtained simply by a series of successive constraint propagations.

4.2.3 Fuzzy Symbolic Dynamics

When fuzziness (constraint-interval fuzzy set) is introduced to the constraint propagation dynamics for coding the variables, we have a more complex constraint propagation with "coding" (selection of a fuzzy label) and "filtering" (selection of a constraint level) by fuzziness.

More precisely, we assume that several fuzzy sets (fuzzy labels) are set up on a variable, say x, as fuzziness inherent to the variable. When an input interval is given to the variable x, one of the fuzzy sets and also one of the constraint intervals belonging to the fuzzy set are selected by employing certain rules for these selections. This constraint interval selected in this way is then used as an input interval to the next constraint propagation illustrated in Sect. 4.1. That is, the output interval is obtained by the approximation of the constraint region using the input interval, and such a constraint propagation will be continued by regarding the output interval as an input interval to next variable (Fig. 13).

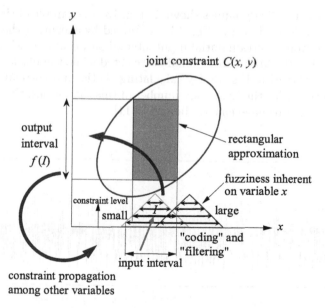

Fig. 13. Fuzzy symbolic dynamics

In such a constraint propagation dynamics, we can analyze the structure of this dynamical system by pursuing the labels (fuzzy sets) selected at each stage of constraint propagation. This characterization of the dynamics is similar to the notion of "symbolic dynamics" [10] since fuzzy label can be regarded as a symbol for coding the value of variables. Therefore, we call such a constraint propagation dynamics "fuzzy symbolic dynamics".

As mentioned in the previous section, the constraint propagation dynamics without "coding" and "filtering" by fuzziness shows rather simple behavior, since it has a stable property to converge to a stable equilibrium point. On the other hand, a fuzzy symbolic dynamics with the mechanism of "coding" (selection of a fuzzy label) and "filtering" (selection of a constraint level) by fuzziness are expected to show quite complex behavior due to a kind of "fluctuations" caused by the coding and the filtering.

4.2.4 Occurrence of Chaotic Behavior

In order to examine the fundamental nature and structure involved in the fuzzy symbolic dynamics, we carried out the computer simulation of constraint propagations for the cases of some simple settings such as (1) constraint network consisting of a single loop of variables as shown in Fig. 10 and (2) constraint network consisting of multiple loops of variables as shown in Fig. 11.

Analysis of Constraint Level

In the fuzzy symbolic dynamics shown in Fig. 13, an example of time series of the constraint level shown in Fig. 14 is obtained by executing the constraint propagation from a certain initial input interval on variable x, where the grey lines show that the fuzzy label "small" is selected while the black lines stands for the label "large" in Fig. 13. By calculating (1) the autocorrelation, (2) the power spectrum, (3) the Lyapunov number of this time series, this dynamical system is judged to have chaotic behavior [11].

Constraint level

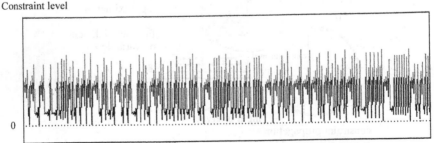

0

Number of constraint propagation   1024

Fig. 14. Time series data of the selected constraint level

Constraint level

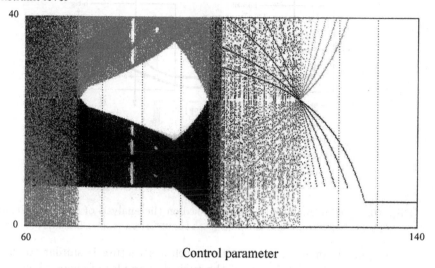

Fig. 15. Bifurcation diagram of fuzzy symbolic dynamics (loop1)

Moreover, in order to know the nature of this dynamics, we calculated the bifurcation diagram in order to elucidate the way how behavior of the dynamics is changed as we change the parameter values of fuzzy sets which correspond to the control parameters of this dynamics. The results is shown in Fig. 15, where the grey points and the black points, respectively, reflect the distinction of "small" and "larege" fuzzy labels.

In this bifurcation diagram, we can readily see the appearance of various and quite different patterns of behavior dependent on the values of parameters such as (1) the periodic solution area, (2) the transition area, (3) the chaotic area, and so on.

Analysis of Label Sequence

In order to know the cause of these chaos phenomena, we examine their qualitative behavior by observing the selected label sequence (symbol sequence) when the fuzzy symbolic dynamics shows chaotic behavior.

More concretely, we derived a state transition diagram as shown in Fig. 16 by defining qualitative states such as 'a-small', 'b-big', etc. where 'a' or 'b' means fuzzy label 'small' or 'large' in Fig. 13, respectively.

In this diagram, there are two main loops (such as unstable attractors) in which the state transits for a while in one loop and sometimes the state switches from one loop to another due to slight differences of input intervals.

Hence, it is considered that chaos phenomena are produced through the repetition of the state transition between these two loops due to a fluctuation by selection of fuzzy labels.

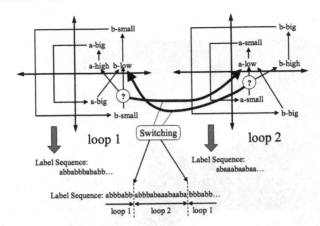

Fig. 16. A state transition diagram based on the analysis of a label sequence

It is quite interesting to note that such a structure is similar to that of the Lorenz model which is one of the typical examples of chaos phenomena. And in both time series data, it is observed that two unstable fixed points exist and that the value of the data switches to another fixed point nearby after diverging from a fixed point.

In other words, the chaotic behavior in fuzzy symbolic dynamics is considered to be caused by the activity in the upper layer (namely, the selection of fuzzy labels) involving a fluctuation that is added to the stable behavior in the constraint propagation dynamics in the lower layer.

The effect of loop structure on the constraint propagation can also be examined by referring to the bifurcation diagram. We analyzed the case of loop structure in Fig. 11. For the case of a single loop, we obtained the diagram in Fig. 17. The double loop case yielded the bifurcation diagram in Fig. 18.

5 Discussion and Conclusion

In this paper, based on the differences of the characteristics between artificial systems and natural systems, we have proposed a framework of double-layered architecture of a problem solving system for constraint satisfaction, where the upper layer called "differentiated layer" has characteristics corresponding to artificial systems and the lower layer called "homogeneous layer" has characteristics corresponding to natural systems.

This dual view of the world (problem) seems to be deeply related to the interesting research on "consciousness-only doctrine" of "Vijnapti-matrata school" in Buddhism done by Toshihiko Izutsu [12]. This doctrine says that everything is created and ceases to hold just in our mind such as "conscious-

constraint level

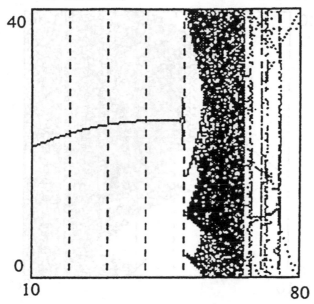

Fig. 17. A bifurcation diagram in a constraint network composed of a single loop of variables (loop1)

ness only". He interpreted the distinction between the two forms of "Vijnapti" (information or intimation) such as

1. Vagvijnapti (verbal intimation or expression)
2. Kayavijnapti (corporal intimation or gesticulation)

as

1. entities on the differentiated layer
2. entities on the homogneous layer.

Moreover, two different approaches called "distributed concurrent approach" and "centralized sequential approach" were proposed from a process-oriented point of view, and the former approach was applied to a structural design problem and the behavior of the latter approach was analyzed through computer simulations.

In the former approach by "distributed concurrent processing", problem solving is regarded to be an order formation process and the global patterns obtained through the process correspond to the solutions (convergence solutions). Different global patterns, however, may be formed due to the difference between initial conditions or fluctuations.

On the other hand, the latter approach by "centralized sequential approach" through a constraint propagation process is expected to explore a

constraint level

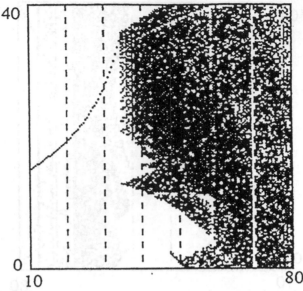

Fig. 18. A bifuraction diagram in a constraint network composed of multiple loops of variables (loop1 and loop2)

variety of solutions successively where global consistencies are not always satisfied but local consistencies which we focus on are always satisfied. In such solutions, the solution concept itself will have a dynamic meaning compared with the traditional one, and hence it may reach a new notion of process-oriented solutions including traditional solutions.

Finally, we discuss how to construct more flexible constraint satisfaction problem solving system architectures which are capable of adapting in a self-organizing manner to the changes of problem structures and the environments by integrating the above two approaches "symbiotically" so as to utilize the meaning of the chaotic phenomena in the hybrid system.

It seems natural to choose a better approach from the above two approaches properly depending on the situations. For instance, we can propose a hybrid method where the former approach can derive more and more solutions (global patterns) by making use of the ability of the latter approach to continue exploring a variety of solutions successively.

More concretely, we note the system characteristics that various solution sequences satisfying the constraints locally are produced along with the temporal progress of the fuzzy symbolic dynamics (centralized sequential approach), and then the possibility of making use of the above characteristics to the distributed concurrent approach is examined on the basis of the potentiality to search for a wide variety of solutions in the centralized sequential approach.

When taking note of the above comments on "consciousness-only doctrine" in Buddhism, the notion of "Engi" ("Pratitya-samutpada", dependent origination), one of the most basic notions in Buddhism will be focused on in this context. Dependent origination or conditioned co-arising emphasizes the interdependence of all things. It says that no beings or phenomena exist on their own but they exist or occur due to their interrelationships with other beings or phenomena [13].

References

1. M. Polanyi: *The Tacit Dimension*, (Routledge & Kegan Paul, London, 1966)
2. H. Haken: *Synergetics—An Introduction, Nonequilibrium Phase Transitions and Self-Organization in Physics, Chemistry and Biology* (Springer, Berlin Heidelberg New York, 1978)
3. O. Katai, M. Ida, T. Sawaragi, S. Iwai: Dynamic and context-dependent treatment of fuzziness from constraint-oriented perspectives. In: *Proc. IFSA'91, Vol on Artificial Intelligence* (1991)
4. O. Katai, T. Horiuchi, S. Matsubara, T. Sawaragi, S. Iwai: Decentralized constraint-oriented problem solving based on fuzzy coding of complex constraints involving continuous variables. In: *Proc. IFSA'95*, 1995, vol 1, pp 133–136
5. O. Katai, T. Horiuchi, T. Sawaragi, S. Iwai, T. Hiraoka: Chaotic structure of fuzzy symbolic dynamics and their relation to constraint-oriented problem solving. In: *Proc. IFES/FUZZ-IEEE'95*, 1995 vol 4, pp 1955–1962
6. R. Seidel: A new method for solving constraint satisfaction problems. In: *Proc. 7th IJCAI*, 1981, pp 338–342
7. E. Tsang: *Foundations of Constraint Satisfaction* (Academic, New York, 1993)
8. A. Koestler: *JANUS—A Summing Up by Arthur Koestler* (Hutchinson, London, 1978)
9. D. R. Smart: Fixed point theorems *Cambridge Tracts in Mathematics*, vol 66 (Cambridge University Press, Cambridge, 1977)
10. H. Bai-lin: *Elementary Symbolic Dynamics* (World Scientific, Singapore, 1989)
11. R. Devaney: *An Introduction to Chaotic Dynamical Systems* (Addison-Welsley, Reading, MA, 1989)
12. S. Jahal Al-Din Ashtiyani, et al. (eds): *Consciousness and Reality: Studies in Memory of Toshihiko Izutsu* (Brill Academic Publishers, Leiden, 1999)
13. E. C. Baptist: *Pticca Samuppada: or, the Buddhist Law of Dependent Origination* (Buddhist Cultural Foundation, 1978) New Delhi, India

Some Applications of Fuzzy Dynamic Models with Chaotic Properties

Alexander Sokolov

Abstract. In this paper an approach to recording and restoring of information is proposed. For this purpose the method for recording and information storing of nonlinear dynamic system trajectories is used. Such systems are based on Takagi–Sugeno recurrent models. Principles of recording, which allow us to reconstruct an arbitrary piece of the bit series, are proposed. Another application belongs to modeling of weak-formalized systems such as economic dynamic processes. We propose to consider the models of multiproject systems as a net of Mamdani rule-based recurrent models. The described approaches are implemented in a Matlab software environment. We investigate the chaotic properties of Mamdani recurrent rule bases as well. The result of investigation of the proposed algorithms are given and analyzed.

1 Introduction

There are many interesting investigations into using fuzzy models, which were proposed in past decades. Among them there are such researches as system analysis with variable structure models, applications of fuzzy recurrent models as chaos generators, fuzzy cluster analyzers, and fuzzy controllers. There are good reasons for using fuzzy control: first, there is no mathematical model available for the process, and, secondly it can satisfy nonlinear control, which can be developed empirically, without complicated mathematics [7]. Chaos control is based on the sensitivity to initial conditions and to the disturbance which is inherent to chaos as a means to stabilize unstable periodic orbits within a chaotic attractor. There are many methods of chaos control [8–10]. Fuzzy controllers are often used for controlling conventional control methods as an extra layer of control in order to improve the effectiveness of the control.

The past few years have been witnessing a sharp growth of interest toward processing, memorizing and storing information in alive systems. Unlike the addressed memory which is now used in computers for writing and reading-out of information the memory of humans and animals is associative, i.e., both 'writing' and 'reading-out' of information are based not only on the number of a memory but also on the content aspect of information [11].

A good number of concepts of realization of the association principle are known to some extent. One of the most popular among them is based on model of dynamic systems and the memorized or recognized objects are

A. Sokolov: *Some Applications of Fuzzy Dynamic Models with Chaotic Properties*,
StudFuzz **187**, 603–625 (2006)
www.springerlink.com

related to basic attractors. The attraction basin of each of the attractors defines the limits of recognition of one image of another. Using a fuzzy recurrent model can describe such iterated mappings.

In this paper we propose two different applications but combined into one part over their closeness to the problem of chaos identification.

2 Reconstruction of Chaotic Orbits with Takagi–Sugeno Recurrent Rule Bases

Let us consider the recurrent Takagi–Sugeno (TS) fuzzy rule base of 0th order [1]

$$R_1: \text{If } x_k = L_1 \text{ then } x_{k+1} = A_1 \,,$$
$$R_2: \text{If } x_k = L_2 \text{ then } x_{k+1} = A_2 \tag{1}$$

$$\ldots$$

$$R_N: \text{If } x_k = L_N \text{ then } x_{k+1} = A_N \,,$$

where $x \in I = [0, 1]$ is a scalar state variable, L_i are linguistic variables (terms), and $A_i \in [0, 1]$ are constants. The transition function $f(x) : I \to I$ can be written in the form

$$f : x_k \to x_{k+1} \,. \tag{2}$$

In Chap. 13 it was proved that three rules such as (1) are necessary and sufficient conditions for producing a chaos mapping (if normality conditions hold for membership functions). Then mapping (2) is chaotic in the sense of Li–Yorke [2]. When we have triangular membership functions mapping (2) is isomorphic to the well-known tent mapping. In such a case we can rewrite mapping (2) as the slopping tent mapping $f(x) : I \to I$, $I = [0, 1]$

$$x_{k+1} = \begin{cases} f_1(x_k) = \dfrac{1}{\lambda} x_k, & \text{if } 0 \le x_k \le \lambda \,, \\ f_2(x_k) = \dfrac{1}{\lambda - 1} x_k + \dfrac{1}{1 - \lambda}, & \text{if } \lambda \le x_k \le 1 \,, \end{cases} \tag{3}$$

where $x_k \in [0, 1]$, $\lambda \in (0, 1)$ (Fig. 1).

Let us consider the following bit sequence:

$$C = \{c_i\}_{i=1}^N \tag{4}$$

with length $N, c_i \in \{0, 1\}$.

It is necessary to restore the source sequence (4) as a bit sequence

$$\tilde{C} = \{\tilde{c}_i\}_{i=M}^N \tag{5}$$

according to the rule

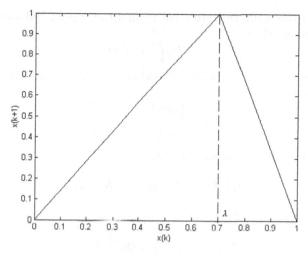

Fig. 1. Slopping mapping

$$\tilde{c}_i = \begin{cases} 1, & \text{if } x_i \geq \lambda \,, \\ 0, & \text{if } x_i \geq \lambda \,, \end{cases} \tag{6}$$

for $i = \overline{M, N}$, $M \leq N$.

Namely, it is necessary to find such a value of x_M that gives the same values for restoring sequence as these in the source one. Besides, it is important to find such a value of λ that gives the maximum member of the restored sequence. The best case is when $M = 1$. The novelty of the proposed approach consists in substitution of chaotic bit series by the initial value x_M which can reconstruct the original orbit with mapping (3). Underlying methodology is based on the so-called backward interval mapping. We propose the following solution of this problem. First of all we propose to use the so-called backward interval mapping.

Let us consider

$$g(x) = f^{-1}(x) \,. \tag{7}$$

Mapping (7) is a contracting mapping if $f(x)$ is chaotic.

For the $(k-1)$th step we can write

$$x_{k-1} = g(x_k) \,. \tag{8}$$

Definition (Wigging and Devaney [3, 4]) *Mapping $f(x) : I \to I$ is chaotic if*

1. it is topologically transitive, i.e., if there exists a $k > 0$ such that $f^k(U) \cap V \neq \emptyset$, where $f^k(U) = \{f^k(x) | x \in U\}$, for any pair of open sets, $U, V \subseteq I$.
2. it is sensitive to the initial conditions, i.e., if there exists a $\delta > 0$ such that $x \in I$ and any neighborhood N of x there exists a $y \in N$ and $n > 0$ such that $|f^n(x) - f^n(y)| > \delta$.
3. The periodic points of f are dense in I.

According to this definition we have to find the topological transitivity for inverse mapping (8) as

$$g^{-n}(U) \bigcap V \neq \varnothing. \tag{9}$$

We use the interval mapping instead of point mapping, namely, we consider the mapping

$$g(I) = \begin{cases} g_1(I) = I_1 \subset I \\ g_2(I) = I_2 \subset I \end{cases}. \tag{10}$$

Because $g(x)$ is contracting mapping,

$$I_1 \bigcup I_2 = I \text{ and } I_1 \bigcap I_2 = \varnothing. \tag{11}$$

The second step of the backward mapping is

$$g^2(I) = \begin{cases} g_1(g(I)) = \begin{cases} g_1(g_1(I)) = I_{11} \subset I_1, \\ g_2(g_1(I)) = I_{21} \subset I_2, \end{cases} \\ g_1(g(I)) = \begin{cases} g_1(g_2(I)) = I_{12} \subset I_1, \\ g_2(g_2(I)) = I_{22} \subset I_2, \end{cases} \end{cases} \tag{12}$$

Lemma. *Let $g(x) = (g_1(x), g_2(x))$ is given, where $g_i, i = 1, 2$ are monotonous and continuous mapping on $g(x) : I \to I$ and $g(x)$ is constructed in the form (10) and it is a contracting mapping ((11) is satisfied), then we have*

$$g^{k+1}(I) = \begin{cases} g_1\left(g^K(I)\right) = I_{1\{K\}} \\ g_2\left(g^K(I)\right) = I_{2\{K\}}, \end{cases} \tag{13}$$

where $\{K\} = \{11\ldots1, 11\ldots2, \ldots, 22\ldots2\}$ is a set of indexes of length K which were used for marking subset of I on the Kth step, and the following conditions are fulfilled:

$$g_1^{K+1}(I) \subset g_1^K(I), K = 0, 1, \ldots,$$
$$g_2^{K+1}(I) \subset g_2^K(I), K = 0, 1, \ldots,$$
$$g_2^K(I) \neq \varnothing, K = 0, 1, \ldots,$$

then

$$I_{1\{K\}} \bigcup I_{2\{K\}} = I,$$

$$\left(\bigcap_{\{K\}} I_{1\{K\}}\right) \bigcap \left(\bigcap_{\{K\}} I_{2\{K\}}\right) = \varnothing,$$

and $g(x) = (g_1(x), g_2(x))$ is a contracting mapping for the set I and all of its subsets.

If $g(x) = (g_1(x), g_2(x))$ satisfies the lemma conditions then $f(x) = g^{-1}(x)$ is chaotic in the sense of the definition of Wigging and Devaney.

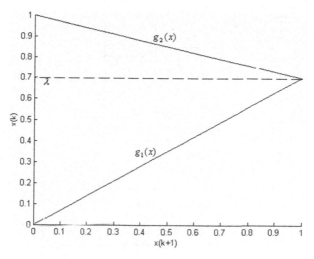

Fig. 2. Backward slopping mapping

For slopping tent mapping (3) we have the following backward mapping [5] (Fig. 2):

$$x_k = \begin{cases} g_1(x_{k+1}) = \lambda x_{k+1} \,, \\ g_2(x_{k+1}) = (\lambda - 1)x_{k+1} + 1 \,. \end{cases} \tag{14}$$

Let us consider the action of this mapping when the argument of function g is interval (Fig. 3). For the initial interval I we have

$$g(I) = \begin{cases} g_1(I) = I_1 = [0, \lambda] \subset I \\ g_2(I) = I_2 = [\lambda, 1] \subset I \,. \end{cases} \tag{15}$$

The following iterative procedure for the backward interval mapping takes place. Let us choose the initial interval according to the rule

$$I_N = \begin{cases} [\lambda, 1], & \text{if } c_N = 1 \,, \\ [0, \lambda], & \text{if } c_N = 0 \,. \end{cases} \tag{16}$$

Then define the possible transitions with the backward interval mapping

$$\tilde{I}_{N-1} = g(I_N) = \begin{cases} g_1(I_N) \subset [0, \lambda] \,, \\ g_2(I_N) \subset [\lambda, 1] \,. \end{cases} \tag{17}$$

The following interval is more precise with respect to the value C_{N-1}:

$$I_{N-1} = \begin{cases} g_2(I_N), & \text{if } C_{N-1} = 1 \,, \\ g_1(I_N), & \text{if } C_{N-1} = 0 \,. \end{cases} \tag{18}$$

Procedure (17), (18) is repeated until we reach the limit of accuracy

$$\text{diam}(I_{M-1}) = \varepsilon \,. \tag{19}$$

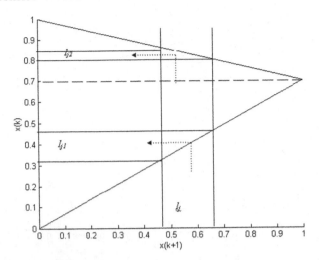

Fig. 3. Backward interval slopping mapping

Then any value $x_M \in I_M$ restores the source sequence (4) with forward mapping (3) from the number of sequence M. Parameter ε is the machine accuracy. In general, it is $\varepsilon = 10^{-15}$ (in MatLab, Delphi programming environment, and so on). So we cannot restore a sequence arbitrary with backward methods (only for instant when the interval becomes ε-length). To increase the quantity of restoring numbers we have to find the optimal value of λ.

According to (14) the total coefficient of contraction of iterative procedure for the backward interval mapping is determined as

$$K = \lambda^n (\lambda - 1)^m , \tag{20}$$

where n is the number of 'zeros' in the source sequence and m is the number of 'ones' (Fig. 4).

The following *statement* is true:

$$\arg \min_{\lambda \in (0,1)} \lambda^n (\lambda - 1)^m = n . \tag{21}$$

Proof. We will find the extremum of function $K(\lambda)$.

$$K'(\lambda) = n\lambda^{n-1}(\lambda - 1)^m + \lambda^n m(\lambda - 1)^{m-1}$$
$$= \lambda^n \left(\frac{n}{\lambda}(\lambda - 1)^m + m(\lambda - 1)^{m-1} \right) .$$

We have $\lambda \neq 0, \lambda \neq 1$ and the expression in brackets is equal to zero in the case $\lambda = n$.

Indeed

$$K'(n) = n^n \left((n - 1)^m + m(n - 1)^{m-1} \right) .$$

With $n + m = 1$ we have

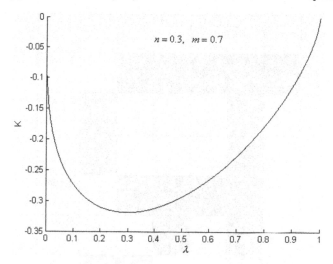

Fig. 4. Coefficient of contraction in the iterative mapping procedure for a special case sequence

$$K'(n) = n^n((n-1)^m + (1-n)(n-1)^{m-1})$$
$$= n^n((n-1)^m - (n-1)^m) = 0 .$$

Statement (21) is proved.

Thus, it is possible to decrease the speed of contraction of intervals for the iterative backward mapping procedure.

Hence, we have a {0,1} sequence and a "machine" which is able to reproduce this sequence starting from certain real x_0. Thus we have to store only x_0 (and the "machine", i.e. the tent mapping). If we use the slopping tent map we need an extra store λ and number of elements in series as well. If we use a simple tent mapping it is necessary to store only x_0 because the maximum elements that can be restored by this method is 48 (in the case of the simple tent). The initial sequence {0,1} is restored according to the algorithm; if chaos of (0,1) sequence is greater then λ we write "one" else "zero". In the simple tent map lambda is equal to 0.5. Parameter λ is the power of slopping. It is used for increasing the number of restoring elements. Labmda is calculated by counting percent of "zeros" and "ones" in the sequence.

Let us consider an application of the described approach to the coding and restoration of the bitmap image (Fig. 5).

Let dark pixels of bitmap be encoded by "ones" and light pixels by "zeros". This bitmap was transformed into a one-dimensional sequence with the length $N = $ Col*Row by reading all lines from top to bottom.

Thus we have the following example of bit sequence (4):

$$C = \{11111001000101000101111001010100110011\} . \tag{22}$$

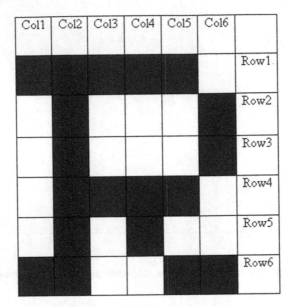

Fig. 5. Graphic presentation of the letter "R"

Let us define the numbers of "zeros" and "ones" in (15) to construct an optimum slopping tent mapping:

$$n = 17/36 = 0.472222222222222 \ ,$$
$$m = 19/36 = 0.527777777777778 \ .$$

Thus in (3) we will have

$$\lambda = 0.472222222222222 \ .$$

According to rule (16) the initial interval I_{36} is defined as

$$I_{36} = [\lambda, 1] = [0.472222222222222, 1] \ .$$

In accordance with iterative rules (17) and (18) the next interval is

$$I_{35} = g_2(I_{36}) = [0.472222222222222, 0.750771604938272] \ .$$

The next step of the iterative procedure gives

$$I_{34} = g(I_{35}) = g_1(I_{35}) = [0.222993827160494, 0.354531035665295].$$

The final result of the implementation of our iterative procedure is

$$I_1 = [0.672604526057511, 0.672604526072895] \ .$$

As we can see

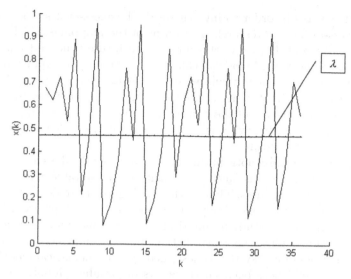

Fig. 6. Graph of chaotic mapping

$$\text{diam}(I_1) = 1.53835832961136e - 011 > \varepsilon = 1e - 15 \;,$$

that is why in our example $M = 1$.

Now if we choose any $x_1 \in I_1$ we can restore a bit sequence (22) using the chaotic mapping (3) and rule (6). For example if $x_1 = 0.67260452606$ then the slopping tent mapping (3) gives the chaotic sequence shown in Fig. 6. The value of λ is now known, and according to rule (6) we can restore all members of the target sequence (22).

The proposed approach can be used for coding and data compression as well. We can increase the number of encoded elements for the bit sequence with a few "ones" or "zeros" in it, because reduction of the total coefficient of contraction takes place in such a case. Obviously that critical factor of the algorithm is the accuracy of software and hardware. It depends on the sensitivity of the chaotic mapping to the initial conditions.

3 Business–Cycles Modeling in Multiproject Systems Based on Recurrent Mamdani Models

One of the most widespread models of the study of the dynamic processes in an economic system and system project management are different equations or a set of such equations. The logistic equation, baker equation and tent mapping are widely used for such a description [6]. By means of such structures one can research business-cycles — a time sequences, defining different features of the models. Many dynamic economic models are simulated by such structures in order to investigate the business-cycles (time series) . The business-cycles are researched in terms of the mappings which link, for instance, profit and its rate in the current and future time under investigation business. The several models, which reflect these intercoupling, are often presented as fuzzy recurrent rule models with linguist variables. A special interest is to study the stability of the time sequences, generated by means of such models. The subject of the recurrent models analysis is the determination of the time series behavior, which can be *converging, divergent, periodic*, or carry the *chaotic* nature.

The presence of the chaotic nature in a dynamic model does not allow us to use it in a long-term forecast and makes such a model suitable only for a short-period study of the business. That is why it is important to determine chaotic characteristics in the economic dynamic models. A given problem becomes particularly urgent for studying multi-project systems, under restrictions of facility conditions, when the contradictions in purpose project take place.

The dynamics of any system is described by a mathematical model, which reflects the dependences between three sets of variable—input, output and state ones. The main characteristic of any dynamic system is in that its behavior depends upon not only the variable acting on it at a given time, but also on the variable acted on it in the past. From well-known knowledge models of 70th—logical, rule-based, frames, neuron and semantic sets—the most convenient ones are rule-based models which are widely used for weak-formalized processes description. These models can naturally describe the individual experience, intuition and behavior. The linguistic variables are very often used for a rough description of objects and processes when their exact description cannot be obtained. At the same time it is necessary to take into account that many weak formalized categories, described by linguistic variables, are not less informative, than an exact description.

A fuzzy recurrent mapping is described by the set of rules $R = \{r_1, r_2 \ldots, r_N\}$, which links the state variables$(x_1, \ldots, x_N)$ of dynamic systems in the current step τ and in the future instant t:

$$
\begin{pmatrix} x_1 \\ x_2 \\ \ldots \\ x_N \end{pmatrix}_t = \begin{pmatrix} r_1 \\ r_2 \\ \ldots \\ r_N \end{pmatrix} \circ \begin{pmatrix} x_1 \\ x_2 \\ \ldots \\ x_N \end{pmatrix}_\tau .
\tag{23}
$$

Rule sets (23) can be written as

$$
r_i : \begin{cases} \text{If } x_1 = A_1^1 \text{ and } x_2 = A_2^1 \cdots [\text{and } x_K = A_K^1] \cdots x_N = A_N^1|_\tau & \text{then } x_i = B^1|_t \\ \text{If } x_1 = A_1^2 \text{ and } x_2 = A_2^2 \cdots [\text{and } x_K = A_K^2] \cdots x_N = A_N^2|_\tau & \text{then } x_i = B^2|_t \\ \qquad \cdots\cdots\cdots \\ \text{If } x_1 = A_1^{K_i} \text{ and } x_2 = A_2^{K_i} \cdots [\text{and } x_K = A_K^{K_i}] \cdots x_N = A_N^{K_i}|_\tau & \text{then } x_i = B^{K_i}|_t \end{cases}
\tag{24}
$$

where K_i is the number of rules in the set r_i; the elements in square brackets are optional; A_i, B are the values of linguistic variables from the suitable term-sets. It is easy to see that the number of rules is in the range $0 \leq K_i \leq \prod_{i=1}^{N} \mathrm{card}(S(x_i))$, where the card $(S(x_i))$ is a power of the term-set of linguistic variable x_i.

Remark. The mapping R usually describes the one time-delay mapping.

It is obvious that analysis of models (23),(24) in a general form is difficult since we deal with a non-linear mapping. The linguistic description of a dynamic model has usually the following form of rules:

$$\text{If } X_k = (x_1 = \mathrm{nb}, x_2 = \mathrm{pm}, \dots, x_n = \mathrm{ze}) \text{ and } U_k = (u_1 = \mathrm{pm},$$
$$u_2 = \mathrm{nb}, \dots, u_m = \mathrm{nm}) ,$$
$$\text{then } X_{k+1} = (x_1 = \mathrm{pb}, \ x_2 = \mathrm{ps}, \dots, x_n = \mathrm{pb}) ,$$

which maps the relations with current and future states depending on the control variables.

Let us rewrite such a mapping as

$$X_{k+1} = X_k \, {}^\circ U_k , \tag{25}$$

where $X_k = (x_1, x_2, \dots, x_n)_k$ is a generalized state vector of the system, $U_k = (u_1, u_2, \dots, u_m)_k$ is a generalized control vector, whose values belong to linguistic variables from suitable term-sets $S = \{\mathrm{nb}, \mathrm{nm}, \dots, \mathrm{ze}, \dots, \mathrm{pm}, \mathrm{pb}\}$, and nb is negative big, nm is negative middle, ze is zero, pm is positive middle, pb is positive big; they all are fuzzy sets with defined membership functions.

Mapping (25) can be represented as a net of transition (Fig. 7) of generalized linguistic states (nodes of the graph) under the action of the generalized linguistic control (ribs of graph).

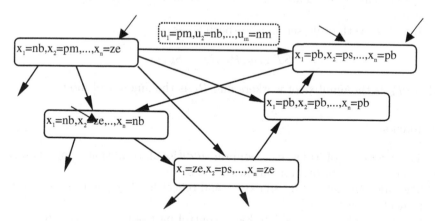

Fig. 7. Fuzzy net of transition

Table 1. Table of linguistic rules $X_{k+1} = X_k \circ U_k$

$U_k \backslash X_k$	nb	nm	ze	pm	pb
nb	nb	Nb	nb	nm	ze
nm	nb	Nb	nm	ze	pm
ze	nb	nb	ze	pb	pb
pm	nm	ze	pm	pb	pb
pb	ze	pm	pb	pb	pb

If N is the dimension of the state vector X, P is the dimension of the control vector U, M is the power of the term-set S, then the maximum number of nodes is M^N, and the number of ribs is $M^N(M^N - 1)/2M^P$.

It is obvious that a direct analysis of such systems is difficult. One of the modern approachs uses the simulation models. In practice instead of general mapping (25) we use its particular forms when vectors X, Y, U are scalar linguistic variables. In this case we can represent the model by the table of linguistic rules. The example of linguistic rules is shown in Table 1.

Let us use these rules for describing of multi-project business system dynamic. Let the project be described as a system of change of one stage to another

$$ST : S_k \to S_{k+1} . \tag{26}$$

Mapping (26) is a set of rules as follows.

If Stage$_k$ is Investment and Human_Resource = 0.9 and Active Money = 0.4 Then Stage$_{k+1}$ = Working_up

This rule defines the conditions under which the transaction from one stage to another can occur. The human_resources and active_money are described in percents.

In this case the rule set can be specified as

$$ST : S_k, P_k, R_k \to S_{k+1} , \tag{27}$$

where P_k is the number of workers and R_k is the amount of money.

Remarks

1. The premises of rules can contain additional conditions, for example, favourable conditions for business, and so on.
2. The rules are defined by an expert and reflect his/her point of view on the project evolution.
3. Some variables are considered as a control part, another parts are considered as external disturbances.

4. Some of stages are fictitious ones. They are rejections of project. Any stage can cross to these stages.

The property of some stages is deliverance of resources P and R after their using. So we can use another rules such as

$$\text{resourse}: S_k \rightarrow P_k \tag{28}$$
$$\text{profit}: S_k \rightarrow R_k . \tag{29}$$

Besides each stage needs in resources:

$$\text{personal}: S_k \rightarrow R_k$$
$$\text{resourse}: S_k \rightarrow R_k .$$

Of course, there are general restrictions on the resources amount –total workers and total money. We can present the dependences between variables as the net shown in Fig 7.

Smulation is a very convenient instrument for investigation of the proposed models. We use for this Fuzzy Logic Toolbox of MatLab/Simulink. First of all let us describe the main parameters of the model.

The Variable Stage

The number of the stage is 7 (stage 8 is the rejection of project).
The membership functions are triangular—$mf_1, \dots, mf_8$.

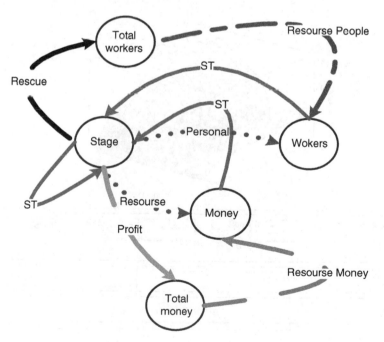

Fig. 8. Net of project description

The Variable Money

The number of values is 3 (not enough, so-so, and enough). The membership functions are triangular—$mf_1, \ldots, mf_3$.

The Variable People

The number of values is 3 (not enough, so-so, enough). The membership functions are triangular—$mf_1, \ldots, mf_3$.

The type of model is Mamdany (because we have linguistic variables in consequents).

The full rule set consists of $3 \times 3 \times 7 = 63$ rules. This mapping is shown in Fig. 9.

The same approach is used for the rest of mappings.

The Matlab scheme for one project modeling is shown in Fig. 10. Red blocks are used for description of the project. These are rule bases and main restrictions. Yellow blocks are the dynamic model for calculating total workers and orange blocks are the dynamic model for calculating money. Sum blocks show that the resource inflow and outflow is defined with the total amount of

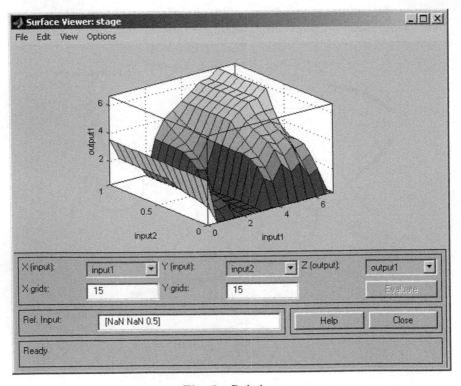

Fig. 9. Rule base

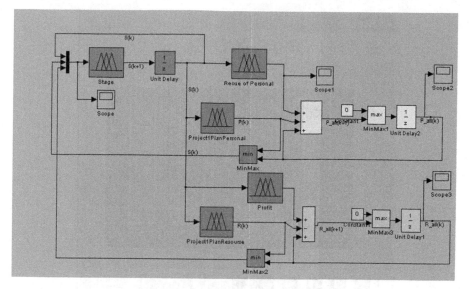

Fig. 10. Project representation

the resource in the previous step plus the amount of resource after finishing the current stage of the project according to rules (28), (29).

The result of simulation is given in Figs. 11, 12.

As can be seen from Fig. 11, after stage 2 we have the rejection of the project due to incorrect resources allocation or a lack of total resources in the system. In Fig. 12 we have the case of correct, fulfilling of all stages.

Simulation of two and more complex projects is the most interesting case. Figs. 13 and 14 show the appropriate models for a two-project system and a three-project system, respectively.

The main feature of these models is common resource supplying. The purpose of investigation is to find the correct distribution of resources under projects. Because we deal with Mamdani models in the project investigation let us add general rules of the chaotic behavior identification in Mamdani recurrent rule bases.

Let us consider the following Mamdani recurrent model:

$$
\begin{aligned}
R_1 &: \text{If } x_k \text{ is } L_1 \text{ then } x_{k+1} \text{ is } L_2 , \\
R_2 &: \text{If } x_k \text{ is } L_2 \text{ then } x_{k+1} \text{ is } L_3 , \\
R_3 &: \text{If } x_k \text{ is } L_3 \text{ then } x_{k+1} \text{ is } L_1 .
\end{aligned}
\tag{30}
$$

According to [1] we have the chaotic behavior in such a model. In this investigation we are planning to find more precise conditions when this model produces the chaos.

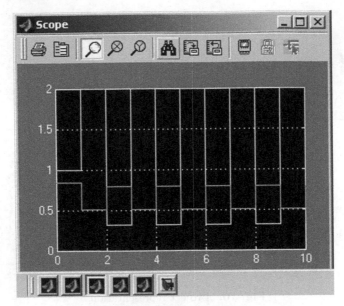

Fig. 11. Stage dynamic

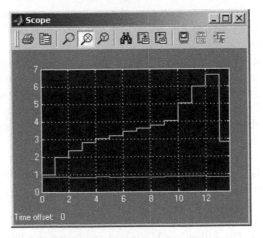

Fig. 12. Executable project

Let a_1, a_2, a_3 be the core positions of the linguistic variables L_1, L_2, L_3, respectively.

For the sake of simplicity let us consider the membership functions of linguistic variables as shown in Fig. 15, and the conditions described in paper [1] are fulfilled.

Each rule in (30) can be considered as a set of fuzzy inference mappings:

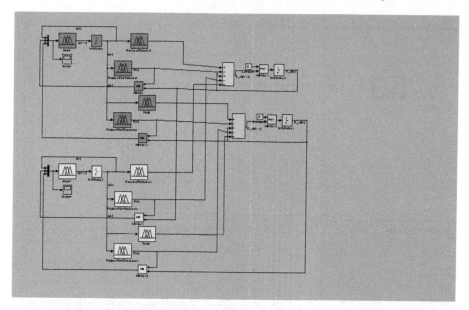

Fig. 13. Two-project system

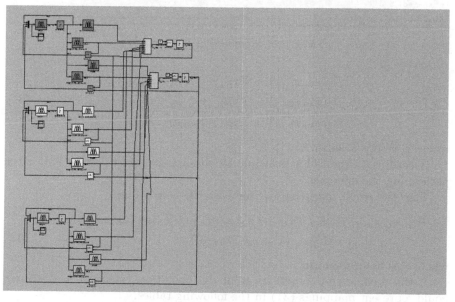

Fig. 14. Three-project system

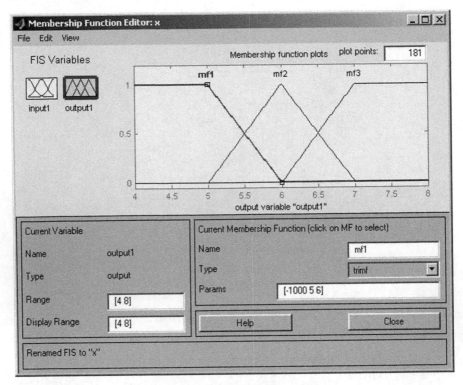

Fig. 15. Membership functions

$$R_1(x_k, x_{k+1}) = T\left(\mu_{L_1}(x_k), \mu_{L_2}(x_{k+1})\right),$$
$$R_2(x_k, x_{k+1}) = T\left(\mu_{L_2}(x_k), \mu_{L_3}(x_{k+1})\right), \tag{31}$$
$$R_3(x_k, x_{k+1}) = T\left(\mu_{L_3}(x_k), \mu_{L_1}(x_{k+1})\right).$$

where T is any T-norm.

According to the FITA principle we could aggregate rule base (31) after performing the inference.

Thus the result mapping for rule base RB is as follows:

$$\mathrm{RB}(x_k, x_{k+1}) = S\left(T\left(\mu_{L_1}(x_k), \mu_{L_2}(x_{k+1})\right), T\left(\mu_{L_2}(x_k), \mu_{L_3}(x_{k+1})\right),\right.$$
$$\left. T\left(\mu_{L_3}(x_k), \mu_{L_1}(x_{k+1})\right)\right), \tag{32}$$

where S is any S-conorm.

If the membership functions fulfil the conditions mentioned above we could represent mappings (31) in the following tables:

Here each cell has the value of the appropriate mapping. The result mapping (32) can be represented as follows:

The output of model (30) has a crisp value which is obtained after defuzzification process. We show below some examples of the mapping $\mathrm{RB}(x_k, x_{k+1})$ for some fixed input values x_k^* as a function of x_{k+1}.

Table 2.

	$x_k \in [0, a_1]$	$x_k \in [a_1, a_2]$	$x_k \in [a_2, a_3]$	$x_k \in [a_3, 1]$
	$R_1(x_k, x_{k+1}) = T\left(\mu_{L_1}(x_k), \mu_{L_2}(x_{k+1})\right)$			
$x_{k+1} \in [a_3, 1]$	0	0	0	0
$x_{k+1} \in [a_2, a_2]$	$\dfrac{a_3 - x_{k+1}}{a_3 - a_2}$	$T\left(\dfrac{a_2 - x_k}{a_2 - a_1}, \dfrac{a_3 - x_{k+1}}{a_3 - a_2}\right)$	0	0
$x_{k+1} \in [a_1, a_2]$	$\dfrac{x_{k+1} - a_1}{a_2 - a_1}$	$T\left(\dfrac{a_2 - x_k}{a_2 - a_1}, \dfrac{x_{k+1} - a_1}{a_2 - a_1}\right)$	0	0
$x_{k+1} \in [0, a_1]$	0	0	0	0

Table 3.

	$x_k \in [0, a_1]$	$x_k \in [a_1, a_2]$	$x_k \in [a_2, a_3]$	$x_k \in [a_3, 1]$
	$R_2(x_k, x_{k+1}) = T\left(\mu_{L_2}(x_k), \mu_{L_3}(x_{k+1})\right)$			
$x_{k+1} \in [a_3, 1]$	0	$\dfrac{x_k - a_1}{a_2 - a_1}$	$\dfrac{a_3 - x_k}{a_3 - a_2}$	0
$x_{k+1} \in [a_2, a_3]$	0	$T\left(\dfrac{x_k - a_1}{a_2 - a_1}, \dfrac{x_{k+1} - a_2}{a_3 - a_2}\right)$	$T\left(\dfrac{a_3 - x_k}{a_3 - a_2}, \dfrac{x_{k+1} - a_2}{a_3 - a_2}\right)$	0
$x_{k+1} \in [a_1, a_2]$	0	0	0	0
$x_{k+1} \in [0, a_1]$	0	0	0	0

Table 4.

	$x_k \in [0, a_1]$	$x_k \in [a_1, a_2]$	$x_k \in [a_2, a_3]$	$x_k \in [a_3, 1]$
	$R_3(x_k, x_{k+1}) = T\left(\mu_{L_3}(x_k), \mu_{L_1}(x_{k+1})\right)$			
$x_{k+1} \in [a_3, 1]$	0	0	0	0
$x_{k+1} \in [a_2, a_2]$	0	0	0	0
$x_{k+1} \in [a_1, a_2]$	0	0	$T\left(\dfrac{x_k - a_2}{a_3 - a_2}, \dfrac{a_2 - x_{k+1}}{a_2 - a_1}\right)$	$\dfrac{a_2 - x_{k+1}}{a_2 - a_1}$
$x_{k+1} \in [0, a_1]$	0	0	$\dfrac{x_k - a_2}{a_3 - a_2}$	1

Analyzing Table 5 and Fig. 16 we can conclude that the maximum value after defuzzification will be obtained when $x_k = a_2$. It is easy to see that we have in this case membership function for the linguistic variable L_3.

Table 5.

$RB(x_k, x_{k+1}) = S(T(\mu_{L_1}(x_k), \mu_{L_2}(x_{k+1})), T(\mu_{L_2}(x_k), \mu_{L_3}(x_{k+1}))$ $T(\mu_{L_3}(x_k), \mu_{L_1}(x_{k+1})))$			
$x_{k+1} \in [a_3, 1]$	0	$\dfrac{x_k - a_1}{a_2 - a_1}$	$\dfrac{a_3 - x_k}{a_3 - a_2}$	0
$x_{k+1} \in [a_2, a_2]$	$\dfrac{a_3 - x_{k+1}}{a_3 - a_2}$	$S\left(\begin{array}{c} T\left(\dfrac{a_2 - x_k}{a_2 - a_1}, \dfrac{a_3 - x_{k+1}}{a_3 - a_2}\right) \\ T\left(\dfrac{x_k - a_1}{a_2 - a_1}, \dfrac{x_{k+1} - a_2}{a_3 - a_2}\right) \end{array}\right)$	$T\left(\dfrac{a_3 - x_k}{a_3 - a_2}, \dfrac{x_{k+1} - a_2}{a_3 - a_2},\right)$	0
$x_{k+1} \in [a_1, a_2]$	$\dfrac{x_{k+1} - a_1}{a_2 - a_1}$	$T\left(\dfrac{a_2 - x_k}{a_2 - a_1}, \dfrac{x_{k+1} - a_1}{a_2 - a_1}\right)$	$T\left(\dfrac{x_k - a_2}{a_3 - a_2}, \dfrac{a_2 - x_{k+1}}{a_2 - a_1},\right)$	$\dfrac{a_2 - x_{k+1}}{a_2 - a_1}$
$x_{k+1} \in [0, a_1]$	0	0	$\dfrac{x_k - a_2}{a_3 - a_2}$	$S(0, 1)$
	$x_k \in [0, a_1]$	$x_k \in [a_1, a_2]$	$x_k \in [a_2, a_3]$	$x_k \in [a_3, 1]$

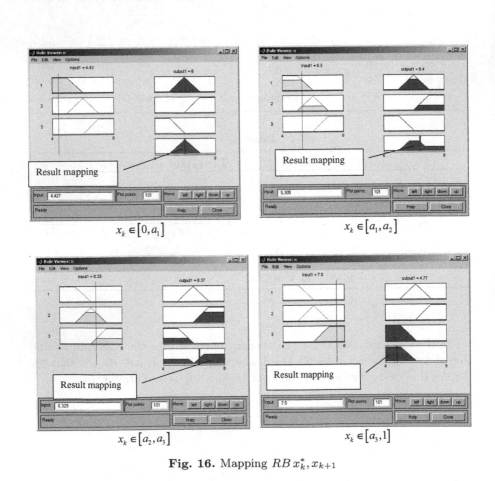

$x_k \in [0, a_1]$

$x_k \in [a_1, a_2]$

$x_k \in [a_2, a_3]$

$x_k \in [a_3, 1]$

Fig. 16. Mapping $RB\, x_k^*, x_{k+1}$

Thus we can write

$$
\begin{aligned}
\mathrm{RB}(a_1, x_{k+1}) &= \mu_{L_2}(x_{k+1}) \,, \\
\mathrm{RB}(a_2, x_{k+1}) &= \mu_{L_3}(x_{k+1}) \,, \\
\mathrm{RB}(a_3, x_{k+1}) &= \mu_{L_1}(x_{k+1}) \,.
\end{aligned}
\tag{33}
$$

Crisp output can be obtained as a procedure of defuzzification applied to expressions (33).

Now after defuzzification we can comment on the transition function of rule base (30). We understand that according to T- and S-norm properties— boundary, monotonicity, commutativity and associativity—we can represent the transition function as a convex function with the maximum value at the point $x_k = a_2$. This value is $\mathrm{d}F(\mu_{L_3}(x_{k+1}))$, where $\mathrm{d}F$ is the defuzzification method. According to proved statements for TS models we can claim that the Mamdani recurrent model is chaotic if conditions of Theorem 4 (Chap. 3) hold for coefficients

$$
\begin{aligned}
A_1 &= \mathrm{d}F(\mu_{L_2}(x_{k+1})) \,, \\
A_2 &= \mathrm{d}F(\mu_{L_3}(x_{k+1})) \,, \\
A_3 &= \mathrm{d}F(\mu_{L_1}(x_{k+1})) \,.
\end{aligned}
\tag{34}
$$

If $\mu_{L_j}(x)$ is isosceles triangles then all methods of defuzzifications give the core positions of appropriate membership functions. In this case we have $A_1 = a_2$, $A_2 = a_3$, $A_3 = a_1$ and we do not need to use Theorem 6 (Chap. 3) to recognize if this mapping is chaotic. We can use the appropriate theorem in these conditions, from [1]. If $\mu_{L_j}(x)$ is an arbitrary triangle (this happens when location of core positions on axe X is arbitrary) the defuzzification procedure gives a point that differs from appropriate core position. In this case we need to use Theorem 4 (Chap. 3) as well in order to recognize if this mapping is chaotic.

Remarks

1. In the paper [5] we dealt with not only triangle membership function, but also with trapezoidal one. So we cannot demand that the defuzzification procedure gives the core position for all membership functions.
2. In the proposed approach we do not restrict the type of membership functions. We demand only finiteness. Using Theorem 4 (Chap. 3) we need to change the definitions of the mapping $f : I \rightarrow I$ in the case of TS model to mapping $\mathrm{d}F(\mathrm{RB}'(x_k, x_{k+1}))$. The parts of the mapping f_1, f_2 in the new mapping are not straight lines but monotonically increasing and decreasing curves of $\mathrm{d}F(\mathrm{RB}(x_k, x_{k+1}))$.

So we can formulate the following theorem

Theorem. *A rule base* (30) *with the mapping* $f : I \to I$ *is chaotic in the sense of Li and Yorke in the interval* $I_{\mathrm{ch}} = [f_2(A_2), A_2] \subseteq I$ *if the following conditions are satisfied:*

(a) $A_1 \in \lfloor 0, f_2^{-1}(a_2))$,

(b) $A_2 \in (a_2, a_3]$, *and* $f_2(A_2) < (a_2)$,

(c) $A_3 \in \begin{cases} [a_1, a_2) \ if \ A_1 \geq a_1 \\ [Z, a_2) \ if \ A < a_1, \ where \ f_1(Z) = Z. \end{cases}$

The generalization of this theorem can be made for an arbitrary number of linguistic variables. Thus, if we have a set of linguistic variables $\varLambda = \{L_1, L_2, \ldots, L_N\}$ and a rule base with the rules

$$\begin{aligned}
R_1 : \quad & \text{if } x_k \text{ is } L_i \text{ then } x_{k+1} \text{ is } L_j \, , \\
R_2 : \quad & \text{if } x_k \text{ is } L_j \text{ then } x_{k+1} \text{ is } L_m \, , \\
R_3 : \quad & \text{if } x_k \text{ is } L_m \text{ then } x_{k+1} \text{ is } L_i \, ,
\end{aligned} \tag{35}$$

we can recognize the chaotic behavior using the representation such as (34) and applying the proposed theorem.

4 Summary

This paper investigates recurrent fuzzy rule bases in chaos-based applications. We have proposed the algorithm of coding and decoding of bit sequences using a sloping tent map. This approach allows us to solve the problem of compression of information that may be a future work. We propose to use the Mamdani models for simulation of economic dynamic processes. The interesting evolution of such an investigation lies in the construction of unified approach for identification of chaos as the Takagi–Sugeno and Mamdani dynamic systems for design of optimal control systems.

References

1. R. Kempf, J. Adamy: Fuzzy Sets Syst. **140**, 259–284 (2003)
2. T.Y. Li, J.A. Yorke: Am. Math. Mont. **82**, 985–992 (1975)
3. P.E. Kloeden: Cycles and chaos in higher dimensional difference equations. In: *Proc. 9th Int. Conf. Nonlinear Oscillations*, Kiev, Naukova Dumka, 1984, vol 2, pp 184–187
4. P.E. Kloeden: Fuzzy Sets Syst. **42**, 37–42 (1991)
5. A. Sokolov, M. Wagenknecht: "Investigation of chaotic behavior of fuzzy Takagi–Sugeno models with regard to simulation and control of technological processes". Scientific Report, Univ. of Zittau/Goerlitz, IPM (2003)
6. M. Yamaguti: Comput. Math. Appl. **28**, 263–267 (1994)
7. A. Harb, I.A. Smadi: J. Vib. Control **10**(7), 797–993 (2004)
8. G. Chen, X. Dong: Int. J. Bifurc. Chaos **3**, 1363–1409 (1993)

9. J.F. Linder, W.L. Ditto: Appl. Mech. Rev. **48**, 795–808 (1995)
10. M.J. Ogorzalek: IEEE Trans. Circuits Syst. **40**, 700–706 (1993)
11. K. Kohonen: *Content-Addressable Memories* (Springer, Berlin Heidelberg New York, 1980)